Laws of Exponents

I $a^m \cdot a^n = a^{m+n}$

II $\dfrac{a^m}{a^n} = a^{m-n} = \dfrac{1}{a^{n-m}} \quad (a \neq 0)$

III $(a^m)^n = a^{mn}$

IV $(ab)^m = a^m b^m$

V $\left(\dfrac{a}{b}\right)^m = \dfrac{a^m}{b^m} \quad (b \neq 0)$

Roots

$a^{1/n} = \sqrt[n]{a}$ (if n is even, $a > 0$, and $\sqrt[n]{a} > 0$)

$\sqrt[n]{a^n} = |a|$ for n even

$\sqrt[n]{a^n} = a$ for n odd

$a^{m/n} = \sqrt[n]{a^m} = (\sqrt[n]{a})^m \quad (a > 0)$

Laws for Radicals

I $\sqrt[n]{ab} = \sqrt[n]{a}\,\sqrt[n]{b} \quad (a > 0, b > 0)$

II $\sqrt[n]{\dfrac{a}{b}} = \dfrac{\sqrt[n]{a}}{\sqrt[n]{b}} \quad (a > 0, b > 0)$

III $\sqrt[cn]{a^{cm}} = \sqrt[n]{a^m} \quad (a > 0)$

Complex Numbers

For $b > 0$, $(\sqrt{-b})^2 = -b$

$i = \sqrt{-1}$; $i^2 = -1$

$\sqrt{-b} = \sqrt{-1}\,\sqrt{b} = i\sqrt{b}$

Complex numbers: $a + bi$

$(b = 0)$ $(b \neq 0)$

Real numbers: $a + bi = a$

Imaginary numbers: $a + bi \quad (a = 0)$

Pure imaginary numbers: $a + bi = bi$

Nonlinear Equations

$ab = 0$ if a equals zero, b equals zero, or a and b equal zero

$x^2 = c$ is equivalent to $x = \sqrt{c}$ or $x = -\sqrt{c}$

$ax^2 + bx + c = 0$ is equivalent to

$$x = \frac{-b \pm \sqrt{b^2 - 4ac}}{2a}$$

For n a natural number, the solution set of

$$[P(x)]^n = [Q(x)]^n$$

contains all the solutions of

$$P(x) = Q(x).$$

Linear Equations in Two Variables

Distance:

$$d = \sqrt{(x_2 - x_1)^2 + (y_2 - y_1)^2}$$

Slope:

$$m = \frac{y_2 - y_1}{x_2 - x_1} \quad (x_1 \neq x_2)$$

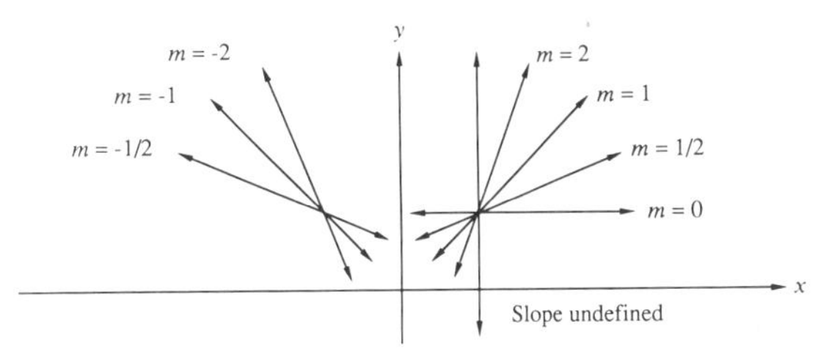

Two line segments with slopes m_1 and m_2 are:

parallel if $m_1 = m_2$

perpendicular if $m_1 m_2 = -1$

Forms of linear equations:

$ax + by + c = 0$	Standard form
$y - y_1 = m(x - x_1)$	Point-slope form
$y = mx + b$	Slope-intercept form
$\dfrac{x}{a} + \dfrac{y}{b} = 1$	Intercept form

(continued on back endpapers)

Algebra for College Students

Students

A *Student Solutions Manual and Study Guide* prepared by Bernard Feldman for this textbook is available from your student bookstore to help you master the course material. It contains detailed solutions, including graphs, for the even-numbered exercises in each section. A Test Problems section will help you determine areas in which you need more practice.

The chapter opener illustrations are: chapter 1: Sumerian measurements and statistics list from the reign of Bur-Sin, circa 2400 B.C., courtesy the Bettmann Archive; chapter 2: an Aztec Calendar stone, courtesy the Bettmann Archive; chapter 3: A Tibetan "wheel of life," including a magic square in its center: chapter 4, an imaginary portrait of Pythagoras from *Calandri's Arithmetic*, 1491; chapter 5: the first attempt to portray Pythagoras as a musician, from F. Gafurius' *Theorica Musice*, 1492; chapter 6: the spirit of Arithmetica between the ancient counter reckoning and the modern algorithm, from the *Margarita Phylosophica*, 1504; chapter 7: *The Gauger*, from Kobel's *Vysirbuch*, 1515; chapter 8: mathematics applied to military science, from *Il Primo Libro delli Qvesiti, et Inventioni diverse*, 1546; chapter 9: practical uses of mathematics in 1560; chapter 10: the explanation of the Quadrant, from *De Quadrante Goemetrica*, 1594; chapter 11: Descartes' explanation of the rainbow, 1656; chapter 12: a Dutch-influenced Japanese explanation of the phases of the moon, 1696; chapter 13: measures of length from the 18th century.

Preface

Algebra for College Students was written to accommodate students with varied backgrounds. The large number of topics provides options for programs that include one or two three-unit courses, or a single five-unit course.

Although Elementary Algebra is a prerequisite, the early chapters provide an intensive review of basic algebraic skills. Special attention is also given to problem-solving, with emphasis on constructing appropriate mathematical models.

If students are prepared in the prerequisite algebra, some of the early chapters can be omitted entirely, or a review can be limited to the chapter summaries and chapter reviews.

Review material

In addition to chapter reviews, four cumulative reviews are included after chapters 5, 7, 10 and 13. The cumulative reviews require students to use a variety of mathematical skills introduced in the previous chapters. Each exercise is referenced to the section in which similar exercises are first considered.

Graded exercises

Exercise sets for each section, chapter reviews, and cumulative reviews are separated into A and B sections. The A sections include basic exercises referenced to specific examples in the text. The B sections include exercises that in general are not referenced to specific examples and thus provide students with a greater challenge.

Reference material

A chapter summary is included at the end of each chapter. In addition, important mathematical properties are exhibited in the front and back endpapers for easy reference. A list of mathematical symbols, which are referenced to appropriate sections in the text, follows this preface.

Answers and partial solutions

The answer section includes answers to all odd-numbered exercises in each section and all the answers to the chapter and cumulative reviews. In addition, the answers for word problems include appropriate mathematical models.

Optional use of calculators

Section 11.3 on exponential functions includes exercises that require the use of a scientific calculator. These exercises are starred and can easily be omitted.

APPENDICES A MORE ABOUT FUNCTIONS 518

B TABLES 535

C FORMULAS FROM GEOMETRY 541

CHAPTER 4 FACTORING POLYNOMIALS 99

CHAPTER 5 RATIONAL EXPRESSIONS 127

CHAPTER 6 EXPONENTS, ROOTS, AND RADICALS 189

Contents

Mathematics Publisher: Kevin Howat
Designer: Wendy Calmenson
Editorial Assistants: Ruth Singer and Sally Uchizono
Print Buyer: Karen Hunt
Managing Designer: Andrew H. Ogus
Proofreader: Karen Stough
Compositor: Graphic Typesetting Service
Assistant Editor: Barbara Holland
Development Editor: Anne Scanlan-Rohrer
Cover: The Mark McGeogh Design Group

Front cover illustration lower left: Kepler planetary system, courtesy Granger Collection.

Printed in the United States of America
1 2 3 4 5 6 7 8 9 10——92 91 90 89 88

Library of Congress Cataloging-in-Publication Data

Drooyan, Irving.
Algebra for college students / Irving Drooyan, Katherine Franklin.
p. cm.
Includes index.
ISBN 0-534-08154-1
1. Algebra I. Franklin, Katherine. II. Title.
QA152.2.D73 1988 87-12476
512.9--dc19 CIP

ISBN 0-534-08154-1

Algebra for College Students

IRVING DROOYAN
Los Angeles Pierce College *Emeritus*

KATHERINE FRANKLIN
Los Angeles Pierce College

Wadsworth Publishing Company
Belmont, California ■ A Division of Wadsworth, Inc.

Chapter 12 on logarithmic functions includes separate sections on using tables (Section 12.3) and using calculators (Section 12.4). Either section can be used to complete the following sections in the chapter. Whether instructors prefer to use tables or calculators, the emphasis in this chapter continues to be on the properties of logarithms and on solving exponential equations. As much attention is given to powers with base e as is given to powers with base 10. Tables for e^x and $\ln x$, as well as the traditional $\log x$ table, are included for students using tables.

Supplementary materials

Supplements include a *Test Items* booklet by Bernard Feldman of Los Angeles Pierce College and a *Computerized Testing System* for IBM or compatible computers; a *Student Solutions Manual and Study Guide* also by Bernard Feldman, consists of worked-out solutions to even-numbered exercises, along with sample tests and corresponding solutions. An "intelligent" tutorial software system by Sergei Ovchinnikov of San Francisco State University is also available for all IBM-PC and compatible computers. *Expert Algebra Tutor* is an interactive program that uses programs, hints, and help screens to tailor lessons specifically to individual students' learning problems in algebra. The result is individualized tutoring strategies with specific page references to problems, examples, and explanations in the textbook.

Acknowledgements

We benefited greatly from the comments and suggestions of the many instructors who reviewed drafts of the manuscript. They include:

Samuel Floyd Barger, Youngstown State University
Harriette J. Stephens, S.U.N.Y. College of Technology at Canton
Judith Staver, Florida Community College of Jacksonville
Thomas Arbutiski, Community College of Allegheny County
Donald Poulson, Mesa Community College
Marion B. Smith, California State University, Bakersfield
Arthur Lieberman, Cleveland State University
Bob C. Denton, Orange Coast Community College
Clark Corbridge, Anchorage Community College
Suzanne Butschun, Tacoma Community College
Carol Achs, Mesa Community College
David E. Steinfort, Grand Rapids Junior College
Thomas J. Woods, Central Connecticut State University
Sandra J. Olson, Winona State University
Robert G. Pumford, Jamestown Community College
William H. Price, Middle Tennessee State University
Jerome J. Cardell, Brevard Community College
Barbara W. Worley, Des Moines Area Community College
Douglas Robertson, University of Minnesota
Jerry Holland, Everett Community College

Mary Catherine Gardner, Grand Valley State College
E. Glenn Kindle, Front Range Community College
Jan Vandever, South Dakota State University
Judy Cain, Tompkins Cortland Community College
Patricia Gilbert, Diablo Valley College
Frank C. Denney, Chabot College
Harriet Beggs, Canton College of Technology
Joseph Buggan, Community College of Allegheny County
Melvin R. Woodard, Indiana University of Pennsylvania
John Leslie, Community College of Allegheny County
Daniel R. Hostetler, University College University of Cincinnati

We especially want to thank William Wooton and Michael Grady, our co-authors on other textbooks, for their generous permission to use material from those books in the preparation of this edition.

Irving Drooyan, Katherine Franklin

Symbols

The section where the symbol is first used is shown in brackets.

$\{a, b\}$	the set whose elements (or members) are a and b [1.1]
A, B, C, etc.	names of sets [1.1]
$\varnothing$	null set or empty set [1.1]
$\in$	is an element of or is a member of [1.1]
$\notin$	is not an element of [1.1]
$\{x \mid \ldots\}$	set of all x such that . . . [1.1]
N	set of natural numbers [1.1]
W	set of whole numbers [1.1]
J	set of integers [1.1]
Q	set of rational numbers [1.1]
H	set of irrational numbers [1.1]
R	set of real numbers [1.1]
$=$	is equal to or equals [1.2]
$\neq$	is not equal to [1.2]
$<$	is less than [1.2]
$\leq$	is less than or equal to [1.2]
$>$	is greater than [1.2]
$\geq$	is greater than or equal to [1.2]
$\lvert a \rvert$	absolute value of a [1.3]
a^n	nth power of a or a to the nth power [2.1]
$P(x)$, $Q(z)$, etc.	P of x, Q of z, etc.; polynomials [2.1]
$\cap$	the intersection of [3.3]
$(\ )$	open interval [3.3]
$[\ ]$	closed interval [3.3]
$\cup$	the union of [3.4]
a^0	equals 1 [6.2]
a^{-n}	$\frac{1}{a^n}$ $(a \neq 0)$ [6.2]

Symbol	Meaning
$a^{1/n}$	nth root of a [6.4]
$\sqrt[n]{a}$	nth root of a [6.5]
$\approx$	is approximately equal to [6.5]
C	set of complex numbers [6.8]
i	imaginary unit [6.8]
$\sqrt{-b}$	$i\sqrt{b}$ $(b > 0)$ [6.8]
$a + bi$	complex number [6.8]
$\pm$	plus or minus [7.2]
(x, y)	ordered pair of numbers; first component is x and second component is y [8.1]
$\begin{vmatrix} a_1 & b_1 \\ a_2 & b_2 \end{vmatrix}$	second-order determinant [9.3]
$\begin{vmatrix} a_1 & b_1 & c_1 \\ a_2 & b_2 & c_2 \\ a_3 & b_3 & c_3 \end{vmatrix}$	third-order determinant [9.4]
$\begin{bmatrix} a_1 & b_1 \\ a_2 & b_2 \end{bmatrix}$	second-order matrix [9.5]
$\begin{bmatrix} a_1 & b_1 & c_1 \\ a_2 & b_2 & c_2 \\ a_3 & b_3 & c_3 \end{bmatrix}$	third-order matrix [9.5]
f, g, etc.	names of functions [11.1]
$f(x)$, etc.	f of x, or the value of f at x [11.1]
f^{-1}	f inverse or the inverse of f [11.4]
$\log_b x$	logarithm to the base b of x [12.1]
$\text{antilog}_b x$	antilogarithm to the base b of x [12.3]
$\ln x$	logarithm to the base e of x [12.3]
s_n	nth term of a sequence [13.1]
S_n	sum of n terms of a sequence [13.1]
Σ	the sum [13.1]
S_∞	infinite sum [13.1]
$n!$	n factorial or factorial n [13.5]
$P(n, n)$	permutations of n things taken n at a time [13.6]
$P(n, r)$	permutations of n things taken r at a time [13.6]
$C(n, r)$ or $\binom{n}{r}$	combinations of n things taken r at a time [13.7]
$[x]$	the greatest integer less than or equal to x [A.3]

Algebra for College Students

1 Review of the Real Number System

We will start our study of intermediate algebra in this chapter by reviewing properties of numbers and mathematical notation that are usually introduced in elementary algebra courses. You may remember some of this material from your previous work in algebra. However, you should understand all the material before continuing to the following chapters.

1.1

DEFINITIONS

Number relationships

In this book we are going to be concerned primarily with the following numbers:

- The **natural numbers,** also called **counting numbers**, consist of the numbers 1, 2, 3, . . . , where the three dots mean "and so on."
- The **whole numbers** consist of the natural numbers and zero. Included are such numbers as 5, 110, and 0.
- The **integers** consist of the natural numbers, their negatives, and zero. Included are such numbers as -7, -3, 0, 5, and 11.
- The **rational numbers** are integers or quotients a/b, where a and b are integers and b does not equal zero. Included are such elements as $-3/4$, $2/3$, 3, and -6. All rational numbers can be written as terminating or repeating decimals. For example, $-3/4$ is equivalent to the terminating decimal -0.75, and $2/3$ is equivalent to the repeating decimal 0.666. . . .
- The **irrational numbers** are numbers with decimal representations that are nonterminating and nonrepeating. An irrational number cannot be represented in the form a/b, where a and b are integers. Included are such elements as $\sqrt{15}$, $-\sqrt{7}$ and π.
- The **real numbers** consist of all rational and irrational numbers.

Sets and symbols

A set is a collection of objects. In algebra, we work primarily with collections, or sets, of numbers. Each item in a set is called an **element** or a **member** of the set. For example, the numbers 1, 2, 3, . . . are the elements of a set we call the set of natural numbers. Because the natural numbers never end—we can never reach a *last* number—we refer to this set as an **infinite set.** A set whose elements can be counted is called a **finite set.**

We indicate sets by means of capital letters, such as A, I, or R, or by means of braces, { }, used together with words or symbols.

EXAMPLE 1

a. {natural numbers less than 4} is read "the set whose elements are natural numbers less than 4."

b. If $A = \{\text{natural numbers less than } 7\}$, then $A = \{1, 2, 3, 4, 5, 6\}$. The set A is a finite set.

c. If $B = \{\text{natural numbers greater than } 7\}$, then $B = \{8, 9, \ldots\}$. The set B is an infinite set. ■

We say that two sets are **equal** if they have the same members; thus

$$\{2, 3, 4\} = \{\text{natural numbers between 1 and 5}\},$$

where on the left side of the equality symbol the elements are *listed* and on the right side the elements are *described by a rule.* Of course

$$\{2, 3, 4\} = \{2, 4, 3\} = \{4, 2, 3\}$$

because each set has the same elements.

We can see how the different sets of numbers relate to one another from the diagram in Figure 1.1. Notice how all natural numbers also belong to the set of whole numbers; how all whole numbers also belong to the integers; how all integers are included in the rational numbers; and how all rational and irrational numbers are also real numbers.

Variables and constants

When we want to indicate an unspecified element in a set, we use a lowercase letter, such as a, b, c, x, y, or z. Such symbols are called **variables.** If the given set, called the **replacement set** of the variable, is a set of numbers (as it will be throughout this book), then the variable represents a number. A symbol used to denote a known, or specified, element of a set is called a **constant.**

We use the symbol $\in$ to denote membership in a set.

FIGURE 1.1

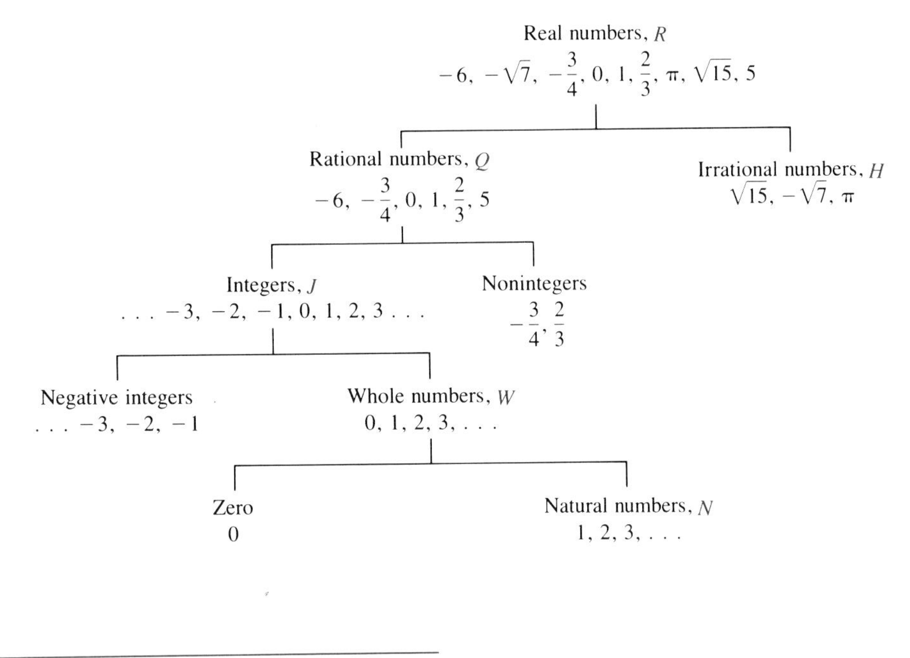

EXAMPLE 2 a. $x \in N$ is read "x is an element of N." The variable is x. The replacement set is N.

b. $x \in \{5\}$. Since x can be only one value, 5, x is a constant. ■

The slash, /, is a handy device for indicating the negation of a symbol.

EXAMPLE 3 a. $\{2, 3, 4\} \neq \{3, 4, 5\}$ is read "the set whose elements are 2, 3, and 4 *is not equal to* the set whose elements are 3, 4, and 5."

b. $2 \notin \{3, 4, 5\}$ is read "2 *is not an element of* the set whose members are 3, 4, and 5." ■

One way we can describe sets is by a convenient notation called **set-builder notation.** Set-builder notation names a variable and states a condition on the variable in order for it to be an element in the set. Notice how, in the following examples, the vertical line is read "such that."

EXAMPLE 4 a. $\{x \mid x \in A \text{ and } x \notin B\}$ is read "the set of all x *such that* x is an element of A and x is not an element of B."

b. $\{x \mid x \in A \text{ and } x \neq 3\}$ is read "the set of all x *such that* x is an element of A and x is not equal to 3." ■

EXAMPLE 5 If $A = \{-5, -\sqrt{17}, -3/4, 0, 2/3, 4.2, 8\}$, then:

a. $\{x \mid x \in A \text{ and } x \text{ an integer}\} = \{-5, 0, 8\}$.

b. $\{x \mid x \in A \text{ and } x \text{ an irrational number}\} = \{-\sqrt{17}\}$.

c. $\{x \mid x \in A \text{ and } x \text{ a real number}\} = A$. ■

Null set

It is sometimes convenient to consider a set with no members. We call such a set containing no members the **empty set** or **null set**, and we regard it as a finite set. The empty set is denoted by $\varnothing$ or $\{\ \}$. Note that the symbol $\varnothing$ (read "the empty set") is *not* enclosed by braces. Thus,

$$\{\text{odd numbers exactly divisible by } 2\} = \varnothing$$

and

$$\{\text{natural numbers less than } 1\} = \varnothing.$$

EXERCISE 1.1

A *Refer to Figure* 1.1 *as necessary.*

■ *Specify each set by listing the members. See Example* 1.

1. {natural numbers between 2 and 6}
2. { integers between -3 and 4}
3. {first three whole numbers}
4. {first three natural numbers}
5. {natural numbers greater than 4}
6. {integers less than 2}
7. {odd natural numbers between 4 and 10}
8. {even integers between -7 and -1}

■ *State whether the given set is finite or infinite. See Example* 1.

9. {whole numbers less than 10,000}
10. {integers less than 10,000}
11. {integers between -3 and 4}
12. {integers greater than -4}
13. {rational numbers between -1 and 0}
14. {rational numbers between 0 and 1}

■ *Replace the question mark with either* $\in$ *or* $\notin$ *to form a true statement. See Examples 2 and 3.*

15. $2 \ ? \ \{\text{rational numbers}\}$
16. $0 \ ? \ \{\text{natural numbers}\}$
17. $4/3 \ ? \ \{\text{real numbers}\}$
18. $-3 \ ? \ \{\text{whole numbers}\}$
19. $-4 \ ? \ \{\text{irrational numbers}\}$
20. $-6 \ ? \ \{\text{integers}\}$

■ *Let* $A = \{-5, -\sqrt{15}, -3.44, -2/3, 0, 1/5, 7/3, 6.1, 8\}$. *List the members of the given set. See Examples 4 and 5.*

21. $\{x \mid x \in A \text{ and } x \text{ a whole number}\}$
22. $\{x \mid x \in A \text{ and } x \text{ a natural number}\}$
23. $\{x \mid x \in A \text{ and } x \text{ an irrational number}\}$
24. $\{x \mid x \in A \text{ and } x \text{ a rational number}\}$
25. $\{x \mid x \text{ is a negative real number in } A\}$
26. $\{x \mid x \text{ is a real number in } A\}$

B

27. If $A = B$ and $4 \notin B$, can 4 be an element of A?
28. If $A \neq B$ and $4 \in B$, must 4 be an element of A?
29. If $A \neq B$ and $B = C$, can $A = C$?
30. If $A \neq B$ and $B \neq C$, must $A \neq C$?

1.2

AXIOMS OF EQUALITY AND ORDER

In mathematics, formal assumptions about numbers or their properties are called **axioms** or **postulates**. Such assumptions are formal statements about properties which we propose to assume as always valid. While we are free to formulate axioms in any way we please, it is clearly desirable that any axioms we adopt lead to useful consequences. A set of useful axioms must not lead to contradictory conclusions.

The words *property, law,* and *principle* are sometimes used to denote assumptions, although these words may also be applied to certain consequences of axioms. In this book we use, in each situation, the word we believe to be the one most frequently encountered. The first such assumptions to be considered concern equality.

Properties of equality

An **equality**, or an "is equal to" assertion, is a mathematical statement that *two symbols, or groups of symbols, are names for the same number.* A number has an infinite variety of names. Thus, 3, 6/2, $4 - 1$, and $2 + 1$ are all names for the same number; hence, the equality

$$4 - 1 = 2 + 1$$

is a statement that "$4 - 1$" and "$2 + 1$" are different names for the same number.

We shall assume that the "is equal to" ($=$) relationship has the following properties.

$a = a$.	**Reflexive property**
If $a = b$, *then* $b = a$.	**Symmetric property**
If $a = b$ *and* $b = c$, *then* $a = c$.	**Transitive property**
If $a = b$, *then b may be replaced by a or a by b in any statement without altering the truthfulness of the statement.*	**Substitution property**

EXAMPLE 1

a. Reflexive property: $x + y = x + y$

b. Symmetric property: If $t = 8$, then $8 = t$.

c. Transitive property: If $x + 2 = y$ and $y = 3$, then $x + 2 = 3$.

d. Substitution property: If $x = 5$, and $x + y = 2$, then $5 + y = 2$.

■

The transitive property can be viewed as a special case of the substitution property. Thus, in Example c above, we can view the equality $x + 2 = 3$ as being obtained from $x + 2 = y$ by *substituting* 3 for y.

We refer to the symbol (or the number it names) to the left of an equals sign as the *left-hand member,* and that to the right as the *right-hand member* of the equality.

Properties of order

The real number b is said to be *less than* the real number a if $a - b$ is positive. The relationship that establishes the order between b and a is called an **inequality**. In symbols, we write

$b < a$ which is read "b is *less than* a,"

or

$a > b$ which is read "a is *greater than* b."

Furthermore, two conditions such as $a < b$ and $b < c$ can be written as the continued inequality $a < b < c$, which is read "a is less than b and b is less than c" or simply "b is between a and c."

It is also possible to combine the concepts of equality and inequality and write two statements at once:

$\leqslant$ is read "is less than or equal to."

and

$\geqslant$ is read "is greater than or equal to."

EXAMPLE 2 a. $y \leqslant 10$ is read "y is less than or equal to 10."

b. $x \geqslant 8$ is read "x is greater than or equal to 8."

c. $-2 < x \leqslant 5$ is read "-2 is less than x and x is less than or equal to 5" or "x is between -2 and 5, or x equals 5." ■

A real number that is not a negative number is called a **nonnegative** number. It can be zero or positive. In symbols, we write

$$a \geqslant 0$$

to indicate that a is nonnegative. Similarly, $a \leqslant 0$ means that a is a **nonpositive** number. It can be zero or negative. For example,

4 and 0 are both nonnegative numbers

and

-4 and 0 are both nonpositive numbers.

Inequalities such as

$$1 < 2 \quad \text{and} \quad 3 < 5$$

are said to be of the *same sense*, because the left-hand member is less than the right-hand member in each case. Inequalities such as

$$1 < 2 \quad \text{and} \quad 5 > 3$$

are said to be of *opposite sense*, because in one case the left-hand member is less than the right-hand member and in the other case the left-hand member is greater than the right-hand member.

We assume the following property concerning the order of real numbers.

If $a < b$ *and* $b < c$, *then* $a < c$.	**Transitive property of inequality**

Thus,

$$\text{if } x < 7 \text{ and } 7 < y, \text{ then } x < y$$

and

$$\text{if } 6 < x \text{ and } x < y + z, \text{ then } 6 < y + z.$$

Number lines

For each real number, there corresponds one and only one point on a line, and vice versa. Hence, a geometric line can be used to visualize relationships between real numbers. For example, to represent 1, 3, and 5 on a line, we scale a straight line in convenient units, with increasing positive direction indicated by an arrow, and mark the required points with dots on the line. This kind of line is called a **number line**, and this particular number line is shown in Figure 1.2. The real numbers corresponding to points on the number line are called the **coordinates** of the points, and the points are called the **graphs** of the numbers.

FIGURE 1.2

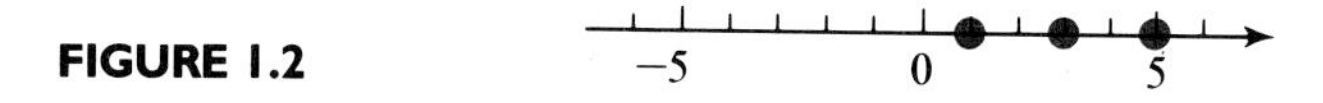

Positive numbers are associated with the points on the line to the right of a point called the **origin** (labeled 0); negative numbers are associated with the points on the line to the left of the origin. The number 0 is neither positive nor negative; it serves as a point of separation for the positive and negative numbers.

EXAMPLE 3 a. $\{-4, -1, 3, 5\}$; graph:

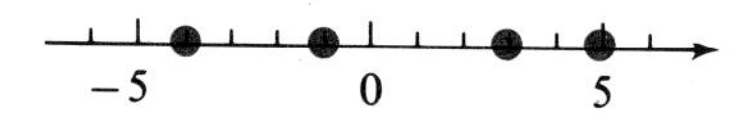

b. $\{x \mid -2 < x \leq 3, x \text{ an integer}\}$; graph:

■

We have noted in the foregoing discussion that certain algebraic statements concerning the order of real numbers can be interpreted geometrically. We summarize some of the more common correspondences in Table 1.1, where in each case a, b, and c are real numbers.

TABLE 1.1

Algebraic statement	Geometric statement
1. $a > 0$; a is positive	1. The graph of a lies to the right of the origin. 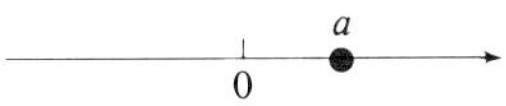
2. $a < 0$; a is negative	2. The graph of a lies to the left of the origin. 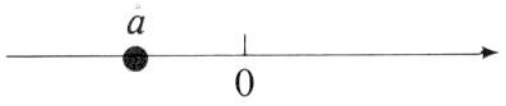
3. $a > b$	3. The graph of a lies to the right of the graph of b. 
4. $a < b$	4. The graph of a lies to the left of the graph of b. 
5. $a < c < b$	5. The graph of c is to the right of the graph of a and to the left of the graph of b. 

Graphs of real numbers

Number lines can be used to display infinite sets of points as well as finite sets. For example, Figure 1.3a is the graph of the set of all integers greater than or equal to 2 *and* less than 5. The graph consists of three points. Figure 1.3b is the graph of the set of all real numbers greater than or equal to 2 and less than 5. The graph is of an *infinite set of points.* The closed dot on the left end of the shaded portion of the graph indicates that the endpoint is part of the graph, while the open dot on the right end indicates that the endpoint is not in the graph.

FIGURE 1.3

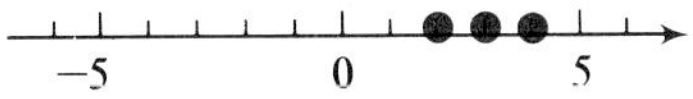

−5 0 5

a.

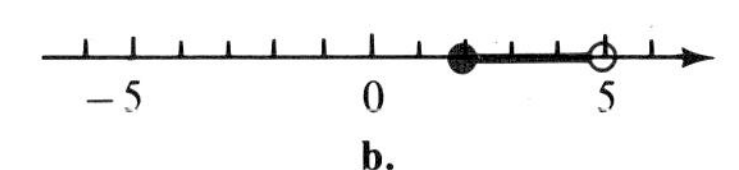

The set of numbers whose graph is shown in Figure 1.3a can be described in set-builder notation by

$$\{x \mid 2 \leq x < 5 \text{ and } x \in J\}$$

(read "the set of all x such that x is greater than or equal to 2 *and* less than 5, and x is an element of the set of integers"). Similarly, the set of numbers whose graph is shown in Figure 1.3b can be described by

$$\{x \mid 2 \leq x < 5 \text{ and } x \in R\}.$$

When no question exists relative to the replacement set for x, the notation $\{x \mid 2 \leq x < 5, x \in R\}$ is abbreviated to $\{x \mid 2 \leq x < 5\}$.

EXAMPLE 4 a. $\{x \mid x \geq -2\}$; graph:

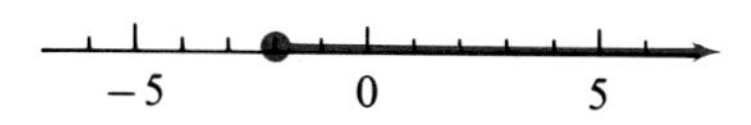

b. $\{y \mid 30 \leq y < 40\}$; graph:

■

EXERCISE 1.2

A

■ *Replace each question mark to make the given statement an application of the given property of equality or order. See Example* 1.

1. $3r = \underline{?}$; reflexive property.
2. If $n = t + 3$, then $\underline{?} = n$; symmetric property.

3. If $r = 6$ and $r - 3 = t$, then $\underline{?} - 3 = t$; substitution property.
4. If $4 = x$ and $x = y$, then $\underline{?} = \underline{?}$; transitive property.
5. If $y = x - 2$ then $x - 2 = \underline{?}$; symmetric property.
6. If $a = c$ and $c = 4$, then $\underline{?} = 4$; transitive property.
7. If $r = n$ and $n + 6 = 8$, then $\underline{?} + 6 = 8$; substitution property.
8. If $t = 4$ and $5 \cdot t = 6s$, then $5 \cdot \underline{?} = 6s$; substitution property.
9. $6 + x = \underline{?}$; reflexive property.
10. If $2 + x = y$, then $y = \underline{?}$; symmetric property.

■ *Express each relation using symbols. See Example 2.*

11. 8 is greater than 5
12. -5 is greater than -8
13. -6 is less than -4
14. -3 is less than 4
15. $x + 1$ is negative
16. $x - 3$ is positive
17. $x - 4$ is nonpositive
18. $x + 2$ is nonnegative
19. y is between -2 and 3
20. y is between -4 and 0
21. x is greater than or equal to 1 and less than 7
22. $3t$ is greater than or equal to 0 and less than or equal to 4

■ *Replace each question mark with an appropriate order symbol to form a true statement. See Example 2.*

23. $-2 \underline{?} 8$
24. $3 \underline{?} 6$
25. $-7 \underline{?} -13$
26. $0 \underline{?} -5$
27. $-6 \underline{?} -3$
28. $\frac{-3}{2} \underline{?} \frac{-3}{4}$
29. $1\frac{1}{2} \underline{?} \frac{3}{2}$
30. $3 \underline{?} \frac{6}{2}$
31. $3 \underline{?} 5 \underline{?} 7$
32. $3 \underline{?} 0 \underline{?} -4$
33. $-7 \underline{?} 0 \underline{?} 2$
34. $-5 \underline{?} -2 \underline{?} 0$

■ *Graph each set on a separate number line. Assume variables are integers. See Example 3.*

35. $\{-3, 0, 1, 4\}$
36. $\{-5, -2, 0, 3\}$
37. $\{x \mid x > 2\}$
38. $\{x \mid x < -3\}$
39. $\{x \mid x \geq -4\}$
40. $\{x \mid x \leq 3\}$
41. $\{x \mid -1 \leq x < 6\}$
42. $\{x \mid -5 < x \leq -1\}$

■ *Graph each set of real numbers on a separate number line. See Example 4.*

43. $\{x \mid x \geq -5\}$
44. $\{y \mid 3 < y \leq 7\}$
45. $\{t \mid t < -3\}$
46. $\{s \mid s \leq 8\}$
47. $\{x \mid -5 \leq x \leq -3\}$
48. $\{y \mid -5 \leq y < 10\}$
49. $\{z \mid 22 < z \leq 28\}$
50. $\{x \mid -25 \leq x < -18\}$
51. $\{r \mid 50 < r < 58\}$
52. $\{y \mid 23 \leq y \leq 28\}$
53. $\{z \mid -30 \leq z\}$
54. $\{x \mid x < 50\}$

1.3

SOME PROPERTIES OF THE REAL NUMBERS

Properties for addition and multiplication

In addition to the axioms of equality and order noted in Section 1.2, we assume the following properties for addition and multiplication of real numbers.

$a + b$ *is a real number.*	Closure for addition
$a + b = b + a$.	Commutative property of addition
$(a + b) + c = a + (b + c)$.	Associative property of addition
ab *is a real number.*	Closure for multiplication
$ab = ba$.	Commutative property of multiplication
$(ab)c = a(bc)$.	Associative property of multiplication
$a(b + c) = (ab) + (ac)$.	Distributive property
There exists a unique number 0 *with the property* $a + 0 = a$ *and* $0 + a = a$.	Identity element for addition
There exists a unique number 1 *with the property* $a \cdot 1 = a$ *and* $1 \cdot a = a$.	Identity element for multiplication
For each real number a, *there exists a unique real number* $-a$ *(opposite of* a*) with the property* $a + (-a) = 0$ *and* $(-a) + a = 0$.	Additive-inverse property
For each real number a *except* 0, *there exists a unique real number* $1/a$ (**reciprocal** *of* a) *with the property* $a\left(\frac{1}{a}\right) = 1$ *and* $\left(\frac{1}{a}\right)a = 1$.	Multiplicative-inverse (reciprocal) property

Parentheses are used in some of the above properties to indicate an order of operations; operations enclosed in parentheses are performed before any others. Note that the distributive property is the only one of the properties that involves the operation of addition *and* the operation of multiplication.

EXAMPLE 1

	For addition	For multiplication
a. Closure property:	$3 + 4$ is a real number	$3 \cdot 4$ is a real number
b. Commutative property:	$3 + 4 = 4 + 3$	$3 \cdot 4 = 4 \cdot 3$
c. Associative property:	$(2 + 3) + 4 = 2 + (3 + 4)$	$(2 \cdot 3) \cdot 4 = 2 \cdot (3 \cdot 4)$
d. Identity element:	$3 + 0 = 3$	$3 \cdot 1 = 3$
e. Inverse property:	$3 + (-3) = 0$	$3\left(\frac{1}{3}\right) = 1$
f. Distributive property:	$3(4 + 5) = 3(4) + 3(5)$	

■

Some important theorems

The axioms listed above imply other properties of the real numbers. Such implications are generally stated as **theorems**, which are logical consequences of the axioms and other theorems.

Theorems generally consist of two parts: an "if" part, called the *hypothesis,* and a "then" part, called the *conclusion.* To simplify the statement of a theorem, we make the following assumption:

In any expression or equation involving variables in this book, unless otherwise stated, the replacement set for the variables will be the set R of real numbers.

The following two theorems will be used in much of our work.

If $a = b$, *then*

$$a + c = b + c. \quad \textbf{Addition property of equality}$$

If $a = b$, *then*

$$ac = bc. \quad \textbf{Multiplication property of equality}$$

EXAMPLE 2 a. If $x = y$, then $x + 3 = y + 3$ and $5x = 5y$.
b. If $x + y = z$, then $x + y + (-y) = z + (-y)$.
c. If $\dfrac{x + y}{4} = z$, then $4\left(\dfrac{x + y}{4}\right) = 4z$. ■

Two other important theorems detail the role of 0 in a product and assert a property of negatives.

$$\boldsymbol{a \cdot 0 = 0}$$ **Zero-factor property**

EXAMPLE 3 a. $3 \cdot 0 = 0$ b. $-8 \cdot 0 = 0$ c. $0 \cdot 0 = 0$ ■

We have given names to the foregoing theorems because they are used frequently throughout the book. In general, theorems are not named.

The following property expresses the fact that *the opposite of the opposite of a number is the number itself.*

$$\boldsymbol{-(-a) = a}$$

EXAMPLE 4 a. $-(-4) = 4$ b. $-(-8) = 8$ c. $-[-(-6)] = -6$ ■

Absolute value of a number

Sometimes we wish to consider only the nonnegative member of a pair of numbers a and $-a$. For example, since the graphs of $-a$ and a are each located the same distance from the origin, when we wish to refer simply to this distance and not to its direction to the left or right of 0, we can use the notation $|a|$ (read "the absolute value of a"). Absolute value is defined more formally as follows.

$$|a| = \begin{cases} a, & \text{if } a \geq 0, \\ -a, & \text{if } a < 0. \end{cases}$$

From this definition, $|a|$ is always nonnegative.

EXAMPLE 5 a. $|3| = 3$ b. $|0| = 0$
c. $|-3| = -(-3) = 3$ d. If $x < 0$, then $|x| = -x$. ■

EXERCISE 1.3

A

■ *In Problems* 1–18, *replace each question mark to make the given statement an application of the given property. See Examples* 1, 2, *and* 3.

1. $12 + x = x + \underline{?}$; commutative property of addition
2. $(2m)n = 2(\underline{?})$; associative property of multiplication
3. $3(x + y) = 3x + \underline{?}$; distributive property
4. $(2 + z) + 3 = 2 + (\underline{?})$; associative property of addition
5. $4 \cdot t = \underline{?}$; commutative property of multiplication
6. $7 + \underline{?} = 0$; additive-inverse property
7. $3 \cdot \frac{1}{3} = \underline{?}$; multiplicative-inverse property
8. $m + \underline{?} = m$; identity element for addition
9. $r \cdot \underline{?} = r$; identity element for multiplication
10. $r + (s + 2) = (s + 2) + \underline{?}$; commutative property of addition
11. $3(x + y) = \underline{?} + \underline{?}$; distributive property
12. $6 \cdot \frac{1}{6} = \underline{?}$; commutative property of multiplication
13. If $z = 4$, then $z + 5 = 4 + \underline{?}$; additive property of equality
14. $(x + y) + z = x + \underline{?}$; associative property of addition
15. If $x = 6$, then $-3x = \underline{?} \cdot 6$; multiplicative property of equality
16. $15 \cdot \underline{?} = 0$; zero-factor property

17. $-4(3x) = (-4 \cdot 3)\underline{?}$; associative property of multiplication
18. If $m = t$, then $m + 3 = \underline{?} + \underline{?}$; additive property of equality

■ *Simplify each expression. See Example 4.*

19. $-(-3)$
20. $-[-(-7)]$
21. $-[-(-x)]$
22. $-[-(x + 2)]$
23. If $x < 0$, does $-x$ represent a positive number or a negative number?
24. If $x < 0$, does $-(-x)$ represent a positive number or a negative number?

■ *Rewrite each expression without using absolute-value notation. See Example 5.*

25. $|-3|$
26. $|-7|$
27. $|4|$
28. $|6|$
29. $-|-2|$
30. $-|-4|$
31. $-|5|$
32. $-|7|$
33. $|x|$
34. $|-x|$
35. $|x - 2|$
36. $|x + 3|$

B

37. Is the set $\{0, 1\}$ closed with respect to addition? With respect to multiplication?
38. Is the set $\{-1, 0, 1\}$ closed with respect to addition? With respect to multiplication?
39. Is the set $\{1, 2\}$ closed with respect to addition? With respect to multiplication?
40. Which of the sets N, W, J, and Q (page 5) are closed with respect to addition? With respect to multiplication?
41. Must $|-x| = |x|$?
42. Can $|x| = -|x + 3|$?
43. Can $|y| > |x| + |y|$?
44. Can $|y| < |x| + |y|$?

1.4

SUMS AND DIFFERENCES

Sums of real numbers

Addition is an operation that associates with each pair of real numbers a and b a third real number $a + b$, called the **sum** of a and b. We first state an obvious result.

> *The sum of two positive numbers is positive.*

EXAMPLE 1

a. If $x > 0$, then $x + 5 > 0$.

b. If $x > 0$ and $y > 5$, then $x + y > 0$. ■

The properties of real numbers imply the familiar laws of signs for sums. For example, consider the sum $-2 + (-3)$. First note that, by the additive-inverse property,

$$2 + (-2) = 0 \quad \text{and} \quad 3 + (-3) = 0;$$

so

$$[2 + (-2)] + [3 + (-3)] = 0.$$

By means of the associative and commutative properties of addition, this can be rewritten as

$$(2 + 3) + [(-2) + (-3)] = 0.$$

Since by the additive-inverse property,

$$(2 + 3) + [-(2 + 3)] = 0,$$

it follows that

$$(-2) + (-3) = -(2 + 3).$$

The above example suggests the following.

> *The sum of two negative numbers is the negative of the sum of their absolute values.*

EXAMPLE 2 a. $(-3) + (-5) = -(3 + 5) = -8$ b. $(-18) + (-6) = -(18 + 6) = -24$ ■

Arguments similar to the one above concerning the sum of a positive number and a negative number lead to the following.

> *The sum of a positive number and a negative number is equal to the nonnegative difference of their absolute values preceded by the sign of the number with the greater absolute value.*

EXAMPLE 3

a. $8 + (-6) = 8 - |-6|$
$= 2$

b. $-3 + 5 = 5 - |-3|$
$= 5 - 3 = 2$

c. $-12 + 10 = -(|-12| - 10)$
$= -(12 - 10)$
$= -2$

d. $5 + (-7) = -(|7| - 5)$
$= -(7 - 5)$
$= -2$ ■

We can usually perform the addition of signed numbers mentally.

Differences of real numbers

We can define the **difference** $a - b$ of two positive numbers a and b as follows: If $a > b > 0$, then $a - b$ is the number d,

$$a - b = d,$$

such that

$$a = b + d.$$

For example,

$$7 - 2 = 5 \quad \text{because} \quad 7 = 2 + 5.$$

This definition for the difference of two *positive numbers* a and b, for $a > b$, can apply to *all numbers*. Thus,

$$2 - 7 = -5 \quad \text{because} \quad 2 = 7 + (-5).$$

We can also define the difference of two numbers in terms of a sum. Note that

$$7 - 2 = 5 \quad \text{and} \quad 7 + (-2) = 5,$$

and that

$$2 - 7 = -5 \quad \text{and} \quad 2 + (-7) = -5.$$

These examples suggest the following.

> *The difference of b subtracted from a equals the sum of a and $-b$,*
>
> $$\boldsymbol{a - b = a + (-b).}$$

EXAMPLE 4

a. $8 - (-3) = 8 + (3)$
$= 8 + 3 = 11$

b. $(-7) - (4) = (-7) + (-4)$
$= -11$

c. $(-5) - (-2) = (-5) + (2)$
$= -5 + 2 = -3$

d. $3 - 8 = 3 + (-8)$
$= -5$ ■

Three uses of the minus sign

Since we have seen that the difference $a - b$ is given by $a + (-b)$, we may consider the symbols $a - b$ as representing either the difference of a and b or, preferably, the sum of a and $(-b)$. Note that we have now used the sign $+$ in two ways and the sign $-$ in three ways. We have used these signs to denote positive and negative numbers and as signs of operation to indicate the sum or difference of two numbers; we also have used the sign $-$ to indicate the "opposite" of a number. For example,

-3; the sign denotes a negative integer.

$7 - 3$; the sign can be viewed as the operation of subtraction where $+3$ is subtracted from $+7$.

$-a$; the sign denotes the "opposite" of a. If a is positive, then $-a$ is negative. If a is negative, then $-a$ is positive.

In discussing the results of operations, it is convenient to use the term **basic numeral**. For example, while the numeral "$3 + 5$" names the sum of the real numbers 3 and 5, we shall refer to "8" as the basic numeral for this number. Similarly, the basic numeral for "$2 - 8$" is "-6."

EXERCISE 1.4

A

■ *Write each sum or difference using a basic numeral. See Examples* 1–4.

1. $4 + 7$
2. $6 + 9$
3. $-2 + 8$
4. $-5 + 7$
5. $6 + (-3)$
6. $5 + (-10)$
7. $-2 + (-5)$
8. $-6 + (-6)$
9. $8 - 3$
10. $6 - 1$
11. $4 - 7$
12. $8 - 14$
13. $-2 - 7$
14. $-5 - 11$
15. $4 - (-2)$
16. $8 - (-4)$
17. $-2 - (-4)$
18. $-8 - (-5)$
19. $6 - 2 + 7$
20. $8 + 3 - 5$
21. $2 - 5 - 3$
22. $6 - 4 - 2$
23. $8 - 7 - (-2)$
24. $4 - (-3) + 2$
25. $7 - (5 - 2)$
26. $8 - (10 - 2)$

27. $(6 - 5) - 11$

28. $(3 - 7) - 1$

29. $(6 - 1 + 8) - 3$

30. $4 - (6 + 2 - 11)$

31. $(7 - 2) + (-3 + 1)$

32. $(5 - 9) + (4 - 2)$

33. $(3 - 5 + 4) - (8 - 13)$

34. $(38 - 25) + (13 - 17 - 2)$

35. $(5 + 24 - 29) + (12 - 8 + 4)$

36. $(-15 + 3 - 1) - (9 - 5 + 12)$

37. $(23 - 15 + 18) - (16 - 3 + 2)$

38. $(-27 + 3 - 6) - (18 + 1 - 12)$

B

39. If subtraction were commutative, we would be able to write $a - b = b - a$. Find a numerical example to show that subtraction is *not* commutative.

40. If subtraction were associative, we would write $(a - b) - c = a - (b - c)$. Find a numerical example to show that subtraction is *not* associative.

41. Which of the following sets are closed under subtraction?
    a. natural numbers **b.** whole numbers **c.** integers

42. For what values of a and b
    a. is the sum $a + b$ positive? **b.** is the difference $a - b$ positive?

1.5

PRODUCTS AND QUOTIENTS

Products of real numbers

Multiplication is an operation that associates with each pair of numbers a and b a third number $a \cdot b$, also written $(a)(b)$ or ab. The number ab is called the **product** of the **factors** a and b. We first state an obvious result.

> *The product of two positive numbers is positive.*

EXAMPLE 1 **a.** If $x > 0$, then $5x > 0$.

b. If $x > 0$ and $y > 3$, then $xy > 0$. ■

Now, what can be said about the product of a positive number and a negative number? Let us consider the product $3(-4)$. First consider the equality

$$4 + (-4) = 0.$$

By the multiplication property of equality,

$$3[4 + (-4)] = 3 \cdot 0;$$

and the distributive and zero-factor properties permit us to write

$$3 \cdot 4 + 3 \cdot (-4) = 0.$$

This implies that $3(-4)$ is the additive inverse of $3 \cdot 4$, which is equal to $-(3 \cdot 4)$.

It can be shown in a similar way that $(-3)(-4)$ is the additive inverse of $-(3 \cdot 4)$; hence $(-3)(-4)$ is equal to $3 \cdot 4$.

Laws of signs for products

In general, as a consequence of properties of the real numbers, we can obtain the following familiar laws of signs for products.

$$(a)(b) = ab, \quad a(-b) = -(ab) = -ab,$$
$$(-a)(b) = -(ab) = -ab, \quad \text{and} \quad (-a)(-b) = ab,$$

where the symbol $-ab$ denotes the additive inverse of ab.

Thus, the product of two numbers with unlike signs is negative and the product of two numbers with like signs is positive.

EXAMPLE 2

a. $(3)(-2) = -6$

b. $(-3)(-2) = 6$

c. $(-2)(3) = -6$

d. $(3)(2) = 6$ ■

Note that since $-1 \cdot a = -(1 \cdot a)$ and $1 \cdot a = a$,

$$-1 \cdot a = -a.$$

Since each *pair* of negative factors yields a positive product, we have the following guideline.

If a product contains

1. *an even number of negative factors, then the product is positive;*
2. *an odd number of negative factors, then the product is negative.*

EXAMPLE 3

a. $(-3)(2)(-4)(5)$
$= (-3)(-4)(2)(5)$
$= 12 \cdot 10 = 120$

b. $(-1)(-5)(3)(-2)(6)$
$= (-1)(-5)(-2)(3)(6)$
$= -10 \cdot 18 = -180$ ■

Prime and composite numbers

If a natural number greater than 1 has no factors that are natural numbers other than itself and 1, it is said to be a **prime number**. Thus, 2, 3, 5, 7, 11, etc., are prime numbers. A natural number greater than 1 that is not a prime number is said to be a **composite number**. Thus, 4, 6, 8, 9, 10, etc., are composite numbers. When a composite number is exhibited as a product of prime factors only, it is said to be **completely factored**. For example, although 30 may be factored into

$$(5)(6), \quad (10)(3), \quad (15)(2), \quad \text{or} \quad (30)(1),$$

if we continue the factorization, we arrive at the set of prime factors 2, 3, and 5 in each case. It is a fact that, except for order, *each composite number has one and only one factorization*. This is known as the **fundamental theorem of arithmetic**.

The words *composite* and *prime* are used in reference to natural numbers only. Any negative integer can, however, be expressed as the product $(-1)a$, where a is a natural number. Hence, if we refer to the completely factored form of a negative integer, we refer to the product of (-1) and the prime factors of the associated natural number. The composite numbers in this section will be less than 100 and hence the factors can be determined by inspection.

EXAMPLE 4 a. $24 = (2)(2)(2)(3)$ b. $-33 = -1(3)(11)$ c. 37 is prime ■

Quotients of real numbers

We can define the *quotient* of two real numbers in terms of multiplication.

> *The* **quotient** *of two numbers a and b is the number q,*
>
> $$\frac{a}{b} = q, \quad \textit{such that} \quad bq = a.$$

EXAMPLE 5

a. $\frac{6}{2} = 3$ because $2 \cdot 3 = 6$.

b. $\frac{-6}{2} = -3$ because $2(-3) = -6$.

c. $\frac{6}{-2} = -3$ because $(-2)(-3) = 6$.

d. $\frac{-6}{-2} = 3$ because $(-2)(3) = -6$. ■

Since the quotient of two numbers a/b is a number q such that $bq = a$, the sign of the quotient of two signed numbers must be consistent with the laws of signs for the product of two signed numbers. Therefore,

$$\frac{+a}{+b} = +q, \quad \frac{-a}{-b} = +q, \quad \frac{+a}{-b} = -q, \quad \frac{-a}{+b} = -q.$$

When a fraction bar is used to indicate that one algebraic expression is to be divided by another, the dividend is the **numerator** and the divisor is the **denominator**. Note that the denominator of a/b is restricted to nonzero numbers, for, if b is 0 and a is not 0, then there exists no q such that

$$0 \cdot q = a.$$

Again, if b is 0 and a is 0, then, for *any* q,

$$0 \cdot q = 0,$$

and the quotient is not unique. Thus: *Division by zero is not defined.*

EXAMPLE 6

a. $\frac{-5}{0}$ is not defined.

b. $\frac{1}{x - 2}$ is not defined for $x = 2$, because $(2) - 2 = 0$. ■

The following alternative definition of a quotient is consistent with the definition above (see Problem 69, Exercise 1.5).

$$\boldsymbol{\frac{a}{b} = a\left(\frac{1}{b}\right) \qquad (b \neq 0).}$$

EXAMPLE 7 a. $\frac{2}{3} = 2\left(\frac{1}{3}\right)$ b. $\frac{a}{3b} = a\left(\frac{1}{3b}\right)$ $(b \neq 0)$ c. $\frac{x}{y - x} = x\left(\frac{1}{y - x}\right)$ $(y \neq x)$ ■

The alternative definition of a quotient also enables us to rewrite some products as quotients. For example,

$$3\left(\frac{1}{5}\right) = \frac{3}{5}, \quad 2 \cdot \frac{3}{7} = \frac{2 \cdot 3}{7}, \quad \text{and} \quad \frac{1}{2} \cdot 3x = \frac{3x}{2}.$$

Note that this alternate definition of a quotient along with the associative law of multiplication can be used to simplify some products.

EXAMPLE 8 a. $\frac{1}{2} \cdot 2x = \left(\frac{1}{2} \cdot 2\right)x$
$= x$

b. $\frac{1}{3} \cdot 3y = \left(\frac{1}{3} \cdot 3\right)y$
$= y$ ■

EXERCISE 1.5

A

■ *Write each product as a basic numeral. See Examples* 1, 2, *and* 3.

1. $(4)(-3)$
2. $(-5)(3)$
3. $(-2)(-6)$
4. $(-8)(-5)$
5. $(3)(-2)(-4)$
6. $(-5)(2)(-3)$
7. $(5)(-1)(4)$
8. $(2)(-2)(6)$
9. $(2)(3)(-1)(-4)$
10. $(5)(-2)(-1)(6)$
11. $(4)(0)(-2)(3)$
12. $(-6)(0)(2)(3)$
13. $-2(3 \cdot 4)$
14. $-2(2 \cdot 6)$
15. $(-4 \cdot 3)(-2)$
16. $(-5 \cdot 2)(-3)$
17. $(2 \cdot 3)(-4 \cdot 1)$
18. $(-2 \cdot 3)(3 \cdot 4)$
19. $-2(-3 \cdot 1 \cdot 4)$
20. $-3(-2 \cdot 2 \cdot 4)$

■ *Express each integer in completely factored form. If the integer is a prime number, so state. See Example* 4.

21. 8
22. 26
23. 49
24. 18
25. 17
26. -16
27. -12
28. 23
29. 56
30. 65
31. -48
32. -52

■ *Write each quotient as a basic numeral, or state if undefined. See Examples 5 and 6.*

33. $\frac{-16}{4}$
34. $\frac{-32}{8}$
35. $\frac{39}{-3}$
36. $\frac{45}{-15}$
37. $\frac{-27}{-9}$
38. $\frac{-54}{-6}$
39. $\frac{0}{-7}$
40. $\frac{0}{-12}$
41. $\frac{-5}{0}$
42. $\frac{-27}{0}$
43. $-\left(\frac{-8}{2}\right)$
44. $-\left(\frac{-12}{-3}\right)$

■ *Rewrite each quotient as a product in which one factor is a natural number and the other factor is the reciprocal (multiplicative inverse) of a natural number. See Example 7.*

45. $\frac{7}{8}$
46. $\frac{27}{7}$
47. $\frac{3}{8}$
48. $\frac{9}{17}$
49. $\frac{82}{11}$
50. $\frac{3}{13}$
51. $\frac{7}{100}$
52. $\frac{3}{1000}$

■ *Rewrite each product as a quotient and simplify if possible. See Examples 7 and 8.*

53. $3\left(\frac{1}{2}\right)$
54. $8\left(\frac{1}{3}\right)$
55. $3\left(\frac{x}{y}\right)$
56. $\left(\frac{2}{z}\right)x$
57. $\frac{1}{4} \cdot 4x$
58. $\frac{1}{5} \cdot 5y$
59. $\frac{1}{2} \cdot 2(x + y)$
60. $\frac{1}{3} \cdot 3(x - y)$
61. Find an example to show that $2(3y) \neq (2 \cdot 3)(2y)$ for all y.
62. Find an example to show that $-3(xy) \neq (-3x)(-3y)$ for all x and y.

B

■ *Under what conditions is each statement in Problems 63–66 a true statement?*

63. $\frac{x}{y} = 0$
64. $\frac{x}{y} \neq 0$
65. $\frac{x}{y} > 0$
66. $\frac{x}{y} < 0$
67. If division were commutative, we would be able to write $a \div b = b \div a$. Find a numerical example to show that division is not commutative.
68. If division were associative, we would write $(a \div b) \div c = a \div (b \div c)$. Find a numerical example to show that division is not associative.
69. Show that if $b \neq 0$, $\frac{a}{b} = a \cdot \frac{1}{b}$. [*Hint:* Start with the fact that $(a/b) = q$ implies that $bq = a$. Then multiply both members by $1/b$.]

1.6

ORDER OF OPERATIONS

Using the associative properties of addition and of multiplication, we obtain the same sum or product regardless of the way that we group terms of a sum or factors of a product. However, if two or more operations are involved in an algebraic expression we can possibly obtain different results depending on the order in which the operations are performed. For example, the expression $2 + 3 \cdot 5$ might be interpreted two ways. We might see it as meaning

$$(2 + 3) \cdot 5 \quad \text{or} \quad 2 + (3 \cdot 5),$$

in which case the result is either

$$5 \cdot 5 = 25 \quad \text{or} \quad 2 + 15 = 17.$$

To avoid such confusion, we make certain agreements about the order of performing operations.

Order of Operations

1. First, any expression within a symbol of inclusion (parentheses, brackets, fraction bars, etc.) is simplified.
2. Next, multiplications and divisions are performed as encountered in order from left to right.
3. Last, additions and subtractions are performed in order from left to right.

Thus in the example above, $2 + 3 \cdot 5$, we first multiply and then add to get

$$2 + 15 = 17.$$

EXAMPLE 1 a. $3 + 2 \cdot 5$
$= 3 + 10$
$= 13$

b. $3(2 + 7) - 3 \cdot (-4)$
$= 3(9) + 12$
$= 27 + 12$
$= 39$

c. $\dfrac{6 + (-15)}{-3} + \dfrac{5}{4 - (-1)}$

$= \dfrac{-9}{-3} + \dfrac{5}{5}$

$= 3 + 1 = 4$

d. $\dfrac{4 - (-2)}{2} - \dfrac{6 + (-12)}{3}$

$= \dfrac{6}{2} - \dfrac{-6}{3}$

$= 3 - (-2) = 5$ ■

Common Error

Note that in Example a, $3 + 2 \cdot 5 \neq (3 + 2) \cdot 5$.

Numerical evaluation

The process of substituting given numbers for variables and simplifying the arithmetic expression (according to the agreed order of operation) is called **numerical evaluation.**

EXAMPLE 2 a. Evaluating $a + (n - 1)d$ for $a = 2$, $n = 5$, and $d = -3$, we obtain

$$\begin{aligned} a + (n - 1)d &= (2) + [(5) - 1)](-3) \\ &= 2 + 4(-3) \\ &= 2 + (-12) = -10. \end{aligned}$$

b. Evaluating $P + Prt$, for $P = 2000$, $r = 0.08$, and $t = 6$, we obtain

$$\begin{aligned} P + Prt &= (2000) + (2000)(0.08)(6) \\ &= 2000 + 960 = 2960. \end{aligned}$$ ■

You may find it helpful to enclose in parentheses each number substituted as shown in the above solutions.

EXERCISE 1.6

A

■ *Write each expression as a basic numeral. See Example* 1.

1. $4 + 4 \cdot 4$
2. $7 - 6 \cdot 2$
3. $4(-1) + 2 \cdot 2$
4. $6 \cdot 3 + (-2)(5)$
5. $-2 \cdot (4 + 6) + 3$
6. $5(6 - 12) + 7$
7. $2 - 3(6 - 1)$
8. $4 - 7(8 + 2)$
9. $\dfrac{7 + (-5)}{2} - 3$

10. $\dfrac{12}{8-2} - 5$

11. $\dfrac{3(6-8)}{-2} - \dfrac{6}{-2}$

12. $\dfrac{5(3-5)}{2} - \dfrac{18}{-3}$

13. $6[3 - 2(4 + 1)] - 2$

14. $6[5 - 3(1 - 4)] + 3$

15. $(4 - 3)[2 + 3(2 - 1)]$

16. $(8 - 6)[5 + 7(2 - 3)]$

17. $64 \div [8(4 - 2[3 + 1])]$

18. $27 \div (3[9 - 3(4 - 2)])$

19. $5[3 + (8 - 1)] \div (-25)$

20. $-3[-2 + (6 - 1)] \div [18 \div (-2)]$

21. $[-3(8 - 2) + 3] \cdot [24 \div (2 - 8)]$

22. $[-2 + 3(5 - 8)] \cdot [-15 \div (5 - 2)]$

23. $\left[\dfrac{7-(-3)}{5-3}\right]\left[\dfrac{4+(-8)}{3-5}\right]$

24. $\left[\dfrac{12+(-2)}{3+(-8)}\right]\left[\dfrac{6+(-15)}{8-5}\right]$

25. $\left(3 - 2\left[\dfrac{5-(-4)}{2+1} - \dfrac{6}{3}\right]\right) + 1$

26. $\left(7 + 3\left[\dfrac{6+(-18)}{4+2}\right] - 5\right) + 3$

27. $\dfrac{8 - 6\left(\dfrac{5+3}{4-8}\right) - 3}{-2 + 4\left(\dfrac{6-3}{1-4}\right) + 6}$

28. $\dfrac{12 + 3\left(\dfrac{12-20}{3-1}\right) - 1}{-8 + 6\left(\dfrac{12-30}{2-5}\right) + 1}$

29. $\dfrac{3(3+2) - 3 \cdot 3 + 2}{3 \cdot 2 + 2(2-1)}$

30. $\dfrac{6 - 2\left(\dfrac{4+6}{5}\right) + 8}{3 - 3 \cdot 2 + 8}$

31. $\dfrac{3|-3| - |-5|}{|-2|} - \dfrac{|-6|}{3}$

32. $\dfrac{2|5-8|}{|-3|} - \dfrac{|-4| + 2}{|-2|}$

■ *Evaluate each expression for the given values of the variables. See Example 2.*

33. $\dfrac{5(F-32)}{9}$; $F = 212$

34. $\dfrac{R+r}{r}$; $R = 12$ and $r = 2$

35. $\dfrac{E-e}{R}$; $E = 18$, $e = 2$, and $R = 4$

36. $\dfrac{a - rs_n}{1-r}$; $r = 2$, $s_n = 12$, and $a = 4$

37. $P + Prt$; $P = 1000$, $r = 0.04$, and $t = 2$

38. $R_0(1 + at)$; $R_0 = 2.5$, $a = 0.05$, and $t = 20$

39. The perimeter of a rectangle of length ℓ and width w is given by $2\ell + 2w$. Determine the perimeter of a rectangle of length 16 centimeters and width 12 centimeters.

40. The area of a trapezoid with bases b and c and height h is given by $\frac{1}{2}h(b + c)$. Determine the area of a trapezoid with bases of 10 and 12 centimeters and height of 14 centimeters.

41. If P dollars is invested at simple rate r, then the amount of money that is accumulated after t years is given by $P + Prt$. Determine the amount accumulated if \$800 is invested at 9% ($r = 0.09$) for 3 years.

42. The net resistance of an electric circuit is given by $\dfrac{rR}{r+R}$, where r and R are two resistors in the circuit. Determine the net resistance if r and R are 10 and 20 ohms, respectively.

CHAPTER SUMMARY

[1.1] A **set** is a collection of objects. The items in the collection are the **members,** or **elements,** of the set. **Natural numbers, whole numbers, integers, rational numbers,** and **irrational numbers** are all **real numbers** (see Figure 1.1).

A **variable** is a symbol used to represent an unspecified element of a given set called the **replacement set** of the variable. A symbol used to denote a single element is called a **constant.**

[1.2] The **equality** $(=)$ and **order** $(<, >)$ relations in the set of real numbers R are governed by a set of assumptions called **axioms.** Order relationships between real numbers can be pictured on a **number line.**

[1.3] The operations of addition and multiplication are governed by the following axioms: The set of real numbers is **closed** under the operations of addition and multiplication. The operations are **commutative** and **associative,** and each operation has an **identity element.** Each number a has an **additive inverse** (**negative**) $-a$, and each nonzero number a has a **multiplicative inverse** (**reciprocal**) $1/a$. Multiplication **distributes** over addition. These properties imply the following properties, called **theorems:**

$$\text{If} \quad a = b, \quad \text{then} \quad a + c = b + c \quad \text{and} \quad ac = bc.$$

Furthermore,

$$a \cdot 0 = 0,$$

and

$$-(-a) = a.$$

The **absolute value** of a number a is defined by

$$|a| = \begin{cases} a, & \text{if} \quad a \geq 0, \\ -a, & \text{if} \quad a < 0. \end{cases}$$

[1.4] The sum of two positive numbers is positive; the sum of two negative numbers is negative; and the sum of a positive number and a negative number is equal to the nonnegative difference of their absolute values preceded by the sign of the number with the greater absolute value.

The difference $a - b$ of two numbers is equal to the sum of a and the additive inverse of b:

$$a - b = a + (-b).$$

[1.5] The product of two positive numbers or two negative numbers is positive; the product of a positive number and a negative number is negative.

A natural number greater than 1 that has no natural-number factors other than itself and 1 is a **prime number.** A natural number greater than 1 that is not a prime number is a **composite number**, and it has one and only one prime factorization.

The quotient of two numbers a/b, with $b \neq 0$, is equal to the number q, such that $b \cdot q = a$. Alternatively,

$$\frac{a}{b} = a\left(\frac{1}{b}\right).$$

Division by 0 is not defined.

[1.6] Operations are performed in the following order:

1. First, any expression within a symbol of inclusion (parentheses, brackets, fraction bars, etc.) is simplified.
2. Next, multiplications and divisions are performed as encountered in order from left to right.
3. Last, additions and subtractions are performed in order from left to right.

REVIEW EXERCISES

A

[1.1]

■ *Let* $A = \left\{-3, -2.55, -\sqrt{2}, -\frac{3}{5}, 0, 1, 5.55\ldots, \frac{13}{2}\right\}$.

1. List the members in set A that are integers.
2. List the members in set A that are whole numbers.

[1.2]

3. Use symbols to show the relationship between a and c if $a < b$ and $b < c$.
4. Use symbols to show that a is equal to or less than 7.
5. Use symbols to show that b is between a and c.

6. Express the relation "y is greater than or equal to 6" using symbols.
7. Graph $\{x \mid -4 \leq x < 5, x \text{ an integer}\}$.
8. Graph $\{x \mid -4 \leq x < 5, x \text{ a real number}\}$.

[1.3]

9. By the commutative property of addition, $x + 2 = \underline{?} + \underline{?}$.
10. By the addition property of equality, if $z = 18$, then $z + t = \underline{?}$.
11. By the multiplicative-inverse property, $3 \cdot \underline{?} = 1$.
12. By the identity property for multiplication, $7 \cdot \underline{?} = 7$.

[1.4]

■ *Write each sum or difference as a basic numeral.*

13. a. $2 - (-4) + 8$ b. $5 - (-6) - 3$
14. a. $3 - |6| - |-3|$ b. $2 + |6 - 12| - |3|$

[1.5]

■ *Write each product as a basic numeral.*

15. a. $3(-2)(6)$ b. $-3(-6)(0)$ c. $7(-3)(-1)$ d. $-1(-2)(-3)(-4)$

■ *Express each integer in completely factored form.*

16. a. 24 b. 28 c. -42 d. -72

■ *Write each quotient as a basic numeral.*

17. a. $\dfrac{-24}{-2}$ b. $\dfrac{0}{-3}$ c. $\dfrac{-16}{2}$ d. $\dfrac{32}{-4}$

■ *Express each quotient as a product.*

18. a. $\dfrac{-7}{5}$ b. $\dfrac{24}{7}$ c. $-\dfrac{4}{5}$ d. $\dfrac{8}{3}$

■ *Express each product as a quotient.*

19. a. $\dfrac{2}{3} \cdot x$ b. $\dfrac{3}{4} \cdot y$ c. $x \cdot \dfrac{1}{3}$ d. $y \cdot \dfrac{1}{4}$

[1.6]

■ *Write each expression as a basic numeral.*

20. a. $5 - \dfrac{3 + 9}{6}$ b. $\dfrac{6 + 2 \cdot 3}{10 - 4 \cdot 2}$
21. a. $\dfrac{6 \cdot 8 \div 4 - 2}{7 - 5}$ b. $\dfrac{6 - 4 \cdot 3 + 2}{4 - 4 \cdot 2}$
22. a. $(4 - 2)[3 + (6 - 2)] + 2$ b. $(8 + 1)[4 - (8 - 3)] - 2$

■ *Evaluate each algebraic expression for the given values of the variables.*

23. $\frac{9C}{5} + 32$ for $C = 100$.

24. $\frac{E - e}{R}$ for $E = 20$, $e = 4$, and $R = 2$.

■ *Simplify each expression.*

25. $\left[\frac{4 - (-2)}{-3}\right] - 2[8 - 3 \cdot 4]$

26. $2\left[4 - \frac{2 - (-6)}{4}\right] - 5\left[3 - \frac{6 \cdot 3}{-4 - 2}\right]$

27. $\frac{3|-2| - (-4)}{|-5|} - \frac{|-4| - 4}{|-3|}$

28. $\frac{2|4 - 6| - |6 - 4|}{|4 - 6|} + 2[|5 - 6| - 3]$

B

29. For what values of x is $|x| = x$?

30. For what values of x and y will the following inequalities be true statements?
a. $x - y > 0$ b. $x - y < 0$

2 Polynomials

2.1

DEFINITIONS

In Chapter 1 we saw that the product of two numbers a and b is expressed by ab. Sometimes, to avoid any possible confusion, we used a centered dot to indicate multiplication; for example, $a \cdot b$, $x \cdot y$, $2 \cdot 3$. In other cases, we used parentheses around one or both of the symbols, as in $(2)(3)$, $2(3)$, and $(x)(x)$.

Powers

If the factors in a product are identical, the number of such factors is indicated by means of a natural-number superscript called an **exponent**. Thus, in the expression 3^4, which denotes $3 \cdot 3 \cdot 3 \cdot 3$, the exponent is 4. The expression 3^4 is called a **power**, and the number 3 is called the **base** of the power. In general, a power is defined as follows.

> *If n is a natural number,*
>
> $$\boldsymbol{a^n = a \cdot a \cdot a \cdot \cdots \cdot a \qquad (n \text{ factors}).}$$

For example,

$$x^2 = x \cdot x, \qquad y^3 = y \cdot y \cdot y, \quad \text{and} \quad z^5 = z \cdot z \cdot z \cdot z \cdot z.$$

If no exponent appears on a variable, as in y for example, the exponent 1 is assumed. Thus, $y = y^1$.

Algebraic expressions

We have used "algebraic expression" or simply "expression" for any meaningful collection of numerals, variables, and signs of operation. In an expression of the form $A + B + C + \cdots$, A, B, and C are called **terms** of the expression. For example:

> $3x + 4yz$ contains two terms; the first term, $3x$, contains two factors and the second term, $4yz$, contains three factors.
>
> $3(x + 4y^2)$ contains only one term; however, the factor $x + 4y^2$ contains two terms.

Differences as sums

From the discussion on page 20 we know that $a - b$ is equal to $a + (-b)$. Thus,

$$2x^2 - 3y = 2x^2 + (-3y)$$

consists of the terms $2x^2$ and $-3y$, and

$$3(x + 4y^2) - 2x^3y - y^3 = 3(x + 4y^2) + (-2x^3y) + (-y^3)$$

consists of the terms $3(x + 4y^2)$, $-2x^3y$, and $-y^3$.

In our work it will be helpful to view any difference as a sum.

Coefficient

Any factor or group of factors in a term is said to be the coefficient of the product of the remaining factors in the term. Thus, in the term $3xyz$, the product $3x$ is the coefficient of yz; y is the coefficient of $3xz$; 3 is the coefficient of xyz; and so on. Hereafter, the word "coefficient" will refer to a number unless otherwise indicated. For example, the coefficient of the term $4xy$ is 4; the coefficient of $-2a^2b$ is -2. The term **numerical coefficient** is sometimes used to emphasize this point. If no coefficient appears in a term, the coefficient is understood to be 1. Thus, x is viewed as $1x$. For example:

> In the expression $3x^2 - 5x$, the coefficient of x^2 is 3 and the coefficient of x is -5;
>
> in the expression $x^2 - y$, the coefficient of x^2 is 1 and the coefficient of y is -1.

EXAMPLE 1

a. The terms of $x^2 - 3x$ are x^2 and $-3x$ with coefficients of 1 and -3, respectively.

b. The terms of $2x^3 - xy + y^2$ are $2x^3$, $-xy$, and y^2 with coefficients of 2, -1, and 1, respectively. ■

Names for algebraic expressions

An algebraic expression of the form cx^n, where c is a constant and n is a *whole number*, or a product of such expressions, is called a **monomial**. For example,

$$y^3, \quad -3x^4, \quad \text{and} \quad 2x^2y^3$$

are monomials. A **polynomial** is an algebraic expression that contains only terms that are monomials. Thus,

$$x^2, \quad \frac{1}{5}x - 2, \quad 3x^2 - 2x + 1, \quad \text{and} \quad x^3 - 2x^2 + 1$$

are polynomials.

If the polynomial contains two or three terms, we refer to it as a **binomial** or **trinomial**, respectively. Polynomials with four or more terms do not have special names.

EXAMPLE 2

a. 3, $5x^3$, and x^2y are monomials.

b. $x + y$, $2x^2 - 3x$, and $4x + 3y$ are binomials.

c. $4x + 3y + 2z$, $x^2 + 2x + 1$, and $5xy + 2x - 3y$ are trinomials. ■

The degree of a monomial in one variable is given by the exponent on the variable. Thus, $3x^4$ is of fourth degree, and $7x^5$ is of fifth degree. The degree of a polynomial in one variable is the same as the degree of its term of largest degree.

EXAMPLE 3

a. $3x^2 + 2x + 1$ is a second-degree polynomial.

b. $x^5 - x - 1$ is a fifth-degree polynomial.

c. $2x - 3$ is a first-degree polynomial. ■

Order of operations

Recall the agreement that was made in Section 1.6 concerning the order in which operations are to be performed. Because a power is a product, we make the additional agreement.

> *Simplify any powers of ungrouped bases in an expression before performing other multiplication or division operations.*

EXAMPLE 4

a. $5 + 3 \cdot 4^2$
$= 5 + 3 \cdot 16$
$= 5 + 48 = 53$

b. $5 + (3 \cdot 4)^2$
$= 5 + 12^2$
$= 5 + 144 = 149$

c. $(5 + 3 \cdot 4)^2$
$= (5 + 12)^2$
$= 17^2 = 289$ ■

Particular care should be taken when evaluating powers that involve minus signs. For example, note that

$$(-2)^2 = (-2)(-2) = 4$$

and that

$$-2^2 = -(2)^2 = -(2)(2) = -4.$$

Thus, $(-2)^2 \neq -2^2$. In general $(-x)^2 \neq -x^2$ for all $x \neq 0$.

We can now numerically evaluate algebraic expressions involving powers by simply following the appropriate order of operations.

EXAMPLE 5 Given $x = 3$ and $y = -2$:

a. $$2xy - y^2 = 2(3)(-2) - (-2)^2 = -12 - 4 = -16$$

b. $$\frac{x^2 - y}{3x + 2} + x(-y) = \frac{(3)^2 - (-2)}{3(3) + 2} + 3[-(-2)] = \frac{9 + 2}{9 + 2} + 3(2) = \frac{11}{11} + 6 = 1 + 6 = 7$$ ■

Names for polynomials

Polynomials are frequently represented by symbols such as

$$P(x), \quad D(y), \quad \text{and} \quad Q(z),$$

where the symbol in parentheses designates the variable. The symbols are read "P of x," "D of y," and "Q of z," respectively. For example, we might write

$$P(x) = x^2 - 2x + 1,$$
$$D(y) = y^6 - 2y^2 + 3y - 2,$$
$$Q(z) = 8z^4 + 3z^3 - 2z^2 + z - 1.$$

The notation $P(x)$ can be used to denote values of the polynomial for specific values of x. Thus, $P(2)$ represents the value of the polynomial $P(x)$ when x is replaced by 2.

EXAMPLE 6 If $P(x) = x^2 - 2x + 1$, then

$$P(2) = (2)^2 - 2(2) + 1 = 1,$$
$$P(3) = (3)^2 - 2(3) + 1 = 4,$$
$$P(-4) = (-4)^2 - 2(-4) + 1 = 25. \quad ■$$

If $P(x)$ and $R(x)$ name two polynomials, then $P[R(x)]$ is the value of the polynomial $P(x)$ at $R(x)$.

EXAMPLE 7 If $P(x) = x^2 + 1$ and $R(x) = 2x - 3$, find the value of each expression.

a. $P[R(4)]$

b. $R[P(3)]$

Solutions

a. $R(4) = 2(4) - 3$
$= 5;$
$P[R(4)] = P(5)$
$= (5)^2 + 1$
$= 26$

b. $P(3) = (3)^2 + 1$
$= 10;$
$R[P(3)] = R(10)$
$= 2(10) - 3$
$= 17$ ■

EXERCISE 2.1

A

■ *Identify each polynomial as a monomial, binomial, or trinomial. Give the degree of the polynomial and name the coefficient of each nonconstant term. See Examples* 1, 2, *and* 3.

1. $2x^3 - x^2$ 2. $x^2 - 2x + 1$ 3. $5n^4$ 4. $3n + 1$
5. $3r^2 - r + 2$ 6. r^3 7. $y^3 - 2y^2 - y$ 8. $3y^2 + 1$

■ *Simplify. See Example 4.*

9. -5^2 10. $(-5)^2$ 11. $(-3)^2$
12. -3^2 13. $4^2 - 2 \cdot 3^2$ 14. $4^2 - (2 \cdot 3)^2$
15. $\frac{4 \cdot 2^3}{16} + 3 \cdot 4^2$ 16. $\frac{4 \cdot 3^2}{6} + (3 \cdot 4)^2$ 17. $\frac{3^2 - 5}{6 - 2^2} - \frac{6^2}{3^2}$
18. $\frac{3^2 \cdot 2^2}{4 - 1} + \frac{(-3)(2)^3}{6}$ 19. $\frac{(-5)^2 - 3^2}{4 - 6} + \frac{(-3)^2}{2 + 1}$ 20. $\frac{7^2 - 6^2}{10 + 3} - \frac{8^2 \cdot (-2)}{(-4)^2}$

■ *Given $x = 3$ and $y = -2$, evaluate each expression. See Example 5.*

21. $3x + y$ 22. $x - y^2$ 23. $x^2 - 2y$
24. $(x + 2y)^2$ 25. $x^2 - y^2$ 26. $(3y)^2 - 3x$
27. $\frac{4x}{y} - xy$ 28. $\frac{-xy^2}{6} + 2xy$ 29. $\frac{(x - y)^2}{-5} + \frac{(xy)^2}{6}$
30. $(x + y)^2 + (x - y)^2$

■ *Evaluate each expression for the given values of the variables.*

31. $\frac{1}{2}gt^2$; $g = 32$ and $t = 2$ 32. $\frac{1}{2}gt^2 - 12t$; $g = 32$ and $t = 3$
33. $\frac{Mv^2}{g}$; $M = 64$, $v = 2$, and $g = 32$ 34. $\frac{32(V - v)^2}{g}$; $V = 12$, $v = 4$, and $g = 32$
35. ar^{n-1}; $a = 2$, $r = 3$, and $n = 4$ 36. $\frac{a - ar^n}{1 - r}$; $a = 4$, $r = 2$, and $n = 3$

■ *Find the values of each expression for the specified values of the variable. See Example 6.*

37. If $P(x) = x^3 - 3x^2 + x + 1$, find $P(2)$ and $P(-2)$.
38. If $P(x) = 2x^3 + x^2 - 3x + 4$, find $P(3)$ and $P(-3)$.
39. If $D(x) = 3x^2 - 16x + 16$, find $D(2)$ and $D(0)$.
40. If $D(x) = 11x^2 - 6x + 1$, find $D(4)$ and $D(0)$.
41. If $P(x) = x^2 + 3x + 1$ and $Q(x) = x^3 - 1$, find $P(3)$ and $Q(2)$.
42. If $P(x) = -3x^2 + 1$ and $Q(x) = 2x^2 - x + 1$, find $P(0)$ and $Q(-1)$.
43. If $P(x) = x^5 + 3x - 1$ and $Q(x) = x^4 + 2x - 1$, find $P(2)$ and $Q(-2)$.
44. If $P(x) = x^6 - x^5$ and $Q(x) = x^7 - x^6$, find $P(-1)$ and $Q(-1)$.

B

■ *Given that $P(x) = x + 1$, $Q(x) = x^2 - 1$, and $R(x) = x^2 + x - 1$, find the value for each expression. See Examples 6 and 7.*

45. $P(2) + R(3) - Q(1)$ 46. $P(1) + R(-2) \cdot Q(-1)$ 47. $P(-2) - R(1) \cdot Q(2)$
48. $P(3)[R(-1) + Q(1)]$ 49. $P[Q(2)]$ 50. $R[P(-1)]$

51. $Q[P(3)]$
52. $Q[R(-2)]$
53. $R[Q(0)]$
54. $Q[P(0)]$
55. $P[R(-2)]$
56. $P[Q(-2)]$
57. Which axiom(s) from Chapter 1 justifies the assertion that if the variables in a monomial represent real numbers, the monomial represents a real number? That the same is true of any polynomial with real coefficients?
58. Argue that, for each real number $x > 0$, $(-x)^n = -x^n$ for n an odd natural number, and $(-x)^n = x^n$ for n an even natural number.
[*Hint:* Write $(-x)^n$ as $(-x)(-x)\ldots(-x) = (-1)(x)(-1)(x)\ldots(-1)(x)$.]

2.2

SUMS AND DIFFERENCES

Rewriting sums

We can use the distributive property in the form

$$ba + ca = (b + c)a$$

to rewrite sums such as

$$2x + 3x, \quad 5y + 3y + 6y, \quad \text{and} \quad 3x + 2y + x + 3y.$$

EXAMPLE 1

a. $3x + 2y + x + 3y$
$= 3x + x + 2y + 3y$
$= 4x + 5y$

b. $x + 2y + z + 3x + 5z$
$= x + 3x + 2y + z + 5z$
$= 4x + 2y + 6z$ ■

Equivalent expressions

Terms that differ only in their numerical coefficients are commonly called **like terms.** When we apply the distributive property to like terms, as in the examples above, we are *combining like terms.* The expression obtained from combining like terms represents the same number as the original expression for all real-number replacements of the variable or variables involved.

Expressions that represent (or name) *the same number for all replacements of the variables* are called **equivalent expressions.** Thus, in the examples above, we obtained equivalent expressions when we simplified each polynomial.

We can show that two expressions are *not* equivalent by simply finding replacement(s) for the variable(s) for which the two expressions have different values. For example, we can show that $-(xy)^2$ is not equivalent to $-xy^2$ by substituting numbers for x and y. Let us try 2 for x and 3 for y. Thus,

$$-(xy)^2 = -(2 \cdot 3)^2 = -6^2 = -36$$

and

$$-xy^2 = -2 \cdot 3^2 = -2 \cdot 9 = -18.$$

Hence, $-(xy)^2 \neq -xy^2$. We have used what is known as a **counterexample** to show that the two expressions are not equivalent.

Rewriting differences

In the discussion on page 38 we noted that the signs in any polynomial are viewed as indications of positive or negative coefficients, and the operations involved are understood to be addition.

EXAMPLE 2

a. $3x - 5x + 4x$
$= 3x + (-5x) + 4x$
$= (3 - 5 + 4)x$
$= 2x$

b. $3x^2 + 2y^2 - x^2 + y^2$
$= 3x^2 + (-x^2) + 2y^2 + y^2$
$= (3 - 1)x^2 + (2 + 1)y^2$
$= 2x^2 + 3y^2$ ■

We can write an expression such as $a + (b + c)$ without parentheses as $a + b + c$. However, for an expression such as

$$a - (b + c),$$

in which the parentheses are preceded by a negative sign, we first write

$$a - (b + c) = a + [-(b + c)].$$

Then, since

$$-(b + c) = -b - c,$$

we have

$$\begin{aligned} a - (b + c) &= a + [-b - c] \\ &= a - b - c. \end{aligned}$$

Thus:

An expression in parentheses preceded by a negative sign can be written equivalently without parentheses by replacing each term within the parentheses with its negative (additive inverse).

EXAMPLE 3 a. $(x^2 + 2x) - (2x^2 - 3x)$
$= x^2 + 2x - 2x^2 + 3x$
$= -x^2 + 5x$

b. $(5x^2 - 2) - (x^2 - 2x) - (2x^2 - 4x + 3)$
$= 5x^2 - 2 - x^2 + 2x - 2x^2 + 4x - 3$
$= 2x^2 + 6x - 5$ ■

Common Error

Note that in Example a, $-(2x^2 - 3x) \neq -2x^2 - 3x$.

Sums and differences can also be obtained in vertical form, with like terms vertically aligned. For example the sum

$$(2x^2 - 3x + 1) + (-x + 3) + (x^2 + 5x - 2)$$

can be obtained by writing the sum in a vertical arrangement in which like terms are aligned.

$$\begin{array}{r} 2x^2 - 3x + 1 \\ -\ x + 3 \\ \underline{x^2 + 5x - 2} \\ 3x^2 + \ x + 2 \end{array}$$

If a difference is written in vertical form, we can simply replace each term in the subtrahend with its negative, and then add. For example, the difference

$$\begin{array}{r} 3t^2 - 4t + 3 \\ \underline{(-)\ -5t^2 + 2t - 2} \end{array} \quad \text{can be written as the sum} \quad \begin{array}{r} 3t^2 - 4t + 3 \\ \underline{(+)\ 5t^2 - 2t + 2} \\ 8t^2 - 6t + 5 \end{array}.$$

Simplifying expressions

The rewriting of a polynomial by combining like terms is a process called **simplifying** the polynomial, since the result is a polynomial with fewer terms than the original. If a polynomial contains no like terms or grouping symbols, we shall say that the polynomial is in **simple form.** When simplifying expressions where grouping devices occur within grouping devices, a great deal of difficulty can be avoided by removing the inner devices first and working outward. Furthermore, it is usually helpful to simplify an expression inside a grouping device before removing the parentheses or brackets.

EXAMPLE 4

a. $3x - [2 - (3x + 1)]$
$= 3x - [2 - 3x - 1]$
$= 3x - [1 - 3x]$
$= 3x - 1 + 3x$
$= 6x - 1$

b. $x^2 - [3x - (x^2 - 2)]$
$= x^2 - [3x - x^2 + 2]$
$= x^2 - 3x + x^2 - 2$
$= 2x^2 - 3x - 2$ ■

EXERCISE 2.2

A

■ *Write each expression in simplest form. See Examples* 1, 2, *and* 3.

1. $3x^2 + 4x^2$
2. $7x^3 - 3x^3$
3. $-6y^2 + 3y^2$
4. $-5y^2 - 6y^2$
5. $8z^2 - 8z^2 + z^2$
6. $-6z^3 + 6z^2 - z^2$
7. $3x^2y + 4x^2y - 2x$
8. $6xy^2 - 4xy^2 + 3y$
9. $3r^2 + (3r^2 + 4r)$
10. $r^2 - (2r^2 + r)$
11. $(s^2 - s) - 3s$
12. $(s^2 + s) - 3s^2$
13. $(2t^2 + 3t - 1) - (t^2 + t)$
14. $(t^2 - 4t + 1) - (2t^2 - 2)$
15. $(u^2 - 3u - 2) - (3u^2 - 2u + 1)$
16. $(2u^2 + 4u + 2) - (u^2 - 4u - 1)$
17. $(2x^2 - 3x + 5) - (3x^2 + x - 2)$
18. $(4y^2 - 3y - 7) - (6y^2 - y + 2)$
19. $(4t^3 - 3t^2 + 2t - 1) - (5t^3 + t^2 + t - 2)$
20. $(4s^3 - 3s^2 + 2s - 1) - (s^3 - s^2 + 2s - 1)$
21. $(4a^2 + 6a - 7) - (a^2 + 5a + 2) + (-2a^2 - 7a + 6)$
22. $(7c^2 - 10c + 8) - (8c + 11) + (-6c^2 - 3c - 2)$
23. $(4x^2y - 3xy + xy^2) - (-5x^2y + xy - 2xy^2) - (x^2y - 3xy)$
24. $(m^2n^2 - 2mn + 7) - (-2m^2n^2 + mn - 3) - (3m^2n^2 - 4mn + 2)$
25. Subtract $4x^2 - 3x + 2$ from the sum of $x^2 - 2x + 3$ and $x^2 - 4$.
26. Subtract $2t^2 + 3t - 1$ from the sum of $2t^2 - 3t + 5$ and $t^2 + t + 2$.
27. Subtract the sum of $2b^2 - 3b + 2$ and $b^2 + b - 5$ from $4b^2 + b - 2$.
28. Subtract the sum of $7c^2 + 3c - 2$ and $3 - c - 5c^2$ from $2c^2 + 3c + 1$.

■ *Simplify. See Example* 4.

29. $y - [2y + (y + 1)]$
30. $3a + [2a - (a + 4)]$
31. $3 - [2x - (x + 1) + 2]$
32. $5 - [3y + (y - 4) - 1]$
33. $(3x + 2) - [x + (2 + x) + 1]$
34. $-(x - 3) + [2x - (3 + x) - 2]$
35. $[x^2 - (2x + 3)] - [2x^2 + (x - 2)]$
36. $[2y^2 - (4 - y)] + [y^2 - (2 + y)]$
37. Show by a counterexample that $-(x + 1)$ is not equivalent to $-x + 1$.
38. Show by a counterexample that $-(x - y)$ is not equivalent to $-x - y$.

B

■ *Simplify.*

39. $3y - (2x - y) - (y - [2x - (y - 2x)] + 3y)$
40. $[x - (y + x)] - (2x - [3x - (x - y)] + y)$
41. $[x - (3x + 2)] - (2x - [x - (4 + x)] - 1)$
42. $-(2y - [2y - 4y + (y - 2)] + 1) + [2y - (4 - y) + 1]$

■ *Given that* $P(x) = x - 1$, $Q(x) = x^2 + 1$, *and* $R(x) = x^2 - x + 1$, *write each expression in terms of x and simplify.*

43. $P(x) + Q(x) - R(x)$
44. $P(x) - Q(x) + R(x)$
45. $R(x) - [Q(x) - P(x)]$
46. $Q(x) - [R(x) + P(x)]$
47. $Q(x) - [R(x) - P(x)]$
48. $P(x) - [R(x) - Q(x)]$
49. $R(x) + [P(x) - Q(x)]$
50. $Q(x) + [R(x) - P(x)]$

2.3

PRODUCTS OF MONOMIALS

Consider the product

$$a^m a^n,$$

where m and n are natural numbers. Since

$$a^m = aaa \cdot \cdots \cdot a \qquad (m \text{ factors}),$$

and

$$a^n = aaa \cdot \cdots \cdot a \qquad (n \text{ factors}),$$

it follows that

$$\begin{aligned} a^m a^n &= \overbrace{(aaa \cdot \cdots \cdot a)}^{m \text{ factors}}\overbrace{(aaa \cdot \cdots \cdot a)}^{n \text{ factors}} \\ &= \overbrace{aaa \cdot \cdots \cdot a}^{m + n \text{ factors}}. \end{aligned}$$

Hence, we have the following property on page 48.

> *For all natural numbers m and n,*
> $$a^m a^n = a^{m+n}. \quad (1)$$

We shall refer to this property as the **first law of exponents.** Thus, we can simplify an expression for the product of two natural-number powers of the same base simply by adding the exponents and using the sum as an exponent on the same base.

EXAMPLE 1 a. $x^2x^3 = x^{2+3} = x^5$ b. $xx^3x^4 = x^{1+3+4} = x^8$ c. $y^3y^4y^2 = y^{3+4+2} = y^9$ ■

We can use the commutative and associative properties of multiplication with the first law of exponents to multiply two or more monomials.

EXAMPLE 2 a. $(3x^2y)(2xy^2) = 3 \cdot 2x^2xyy^2 = 6x^3y^3$ b. $(-2xy^4)(x^2y)(4y^2) = -2 \cdot 4xx^2y^4yy^2 = -8x^3y^7$ ■

Now consider the powers $(a^m)^n$ and $(ab)^n$, where m and n are natural numbers:

$$\begin{aligned}(a^m)^n &= a^m \cdot a^m \cdot a^m \cdot \cdots \cdot a^m \quad (n \text{ factors})\\ &= a^{m+m+m+\cdots+m} \quad (n \text{ terms})\\ &= a^{mn}\end{aligned}$$

and

$$\begin{aligned}(ab)^n &= (ab)(ab)(ab) \cdot \cdots \cdot (ab) \quad (n \text{ factors})\\ &= \overbrace{(a \cdot a \cdot \cdots \cdot a)}^{n \text{ factors}}\overbrace{(b \cdot b \cdot \cdots \cdot b)}^{n \text{ factors}}\\ &= a^n b^n.\end{aligned}$$

Hence, we have the following properties.

For all natural numbers m and n,

$$(a^m)^n = a^{mn} \qquad (2)$$

and

$$(ab)^n = a^n b^n. \qquad (3)$$

We shall refer to these properties as the **second** and **third laws of exponents,** respectively.

EXAMPLE 3 a. From (2),

$$(x^2)^3 = x^6 \quad \text{and} \quad (x^5)^2 = x^{10}.$$

b. From (3),

$$(xy)^3 = x^3y^3 \quad \text{and} \quad (3x)^4 = 3^4x^4 = 81x^4.$$

c. From (2) and (3),

$$(x^2y^3)^3 = x^6y^9 \quad \text{and} \quad (2x^3y^2z)^2 = 2^2x^6y^4z^2 = 4x^6y^4z^2.$$ ■

The following example illustrates the three laws of exponents above in a more general way.

EXAMPLE 4

a. $x^n \cdot x^{n+3}$
$= x^{n+n+3}$
$= x^{2n+3}$

b. $(y^{2n+1})^3$
$= y^{3(2n+1)}$
$= y^{6n+3}$

c. $(x^{n+1} \cdot y^{n-1})^2$
$= x^{2(n+1)} \cdot y^{2(n-1)}$
$= x^{2n+2} \cdot y^{2n-2}$ ■

EXERCISE 2.3

A

■ *Write each product as a polynomial in simplest form. See Examples 1 and 2.*

1. $(7t)(-2t^2)$
2. $(4c^3)(3c)$
3. $(4a^2b)(-10ab^2c)$
4. $(-6r^2s^2)(5rs^3)$
5. $(11x^2yz)(4xy^3z)$
6. $(-8abc)(-b^2c^3)$
7. $2(3x^2y)(x^3y^4)$
8. $-5(ab^3)(-3a^2bc)$
9. $(-r^3)(-r^2s^4)(-2rt^2)$
10. $(-5mn)(2m^2n)(-n^3)$
11. $(y^2z)(-3x^2z^2)(-y^4z)$
12. $(-3xy)(2xz^4(3x^3y^2z)$
13. $(2rt)(-3r^2t)(-t^2)$
14. $-a^2(ab^2)(2a)(-3b^2)$
15. $-(2x)(x^2)(-3x)$
16. $(-2y^2)(y^2)(y)$
17. $z^3(-2z)(3z)(-1)$
18. $(-t)(2t^2)(-t)(0)$

■ *Write each expression as a polynomial in simplest form. See Example 3.*

19. $(x^3)^2$
20. $(x^4)^3$
21. $(xy)^4$
22. $(xy)^2$
23. $(y^2z)^3$
24. $(yz^4)^2$
25. $(2xz^2)^3$
26. $(3x^2z)^2$
27. $(2xy^2z)^2$
28. $(3x^2yz^2)^3$
29. $(-2xy^3z)^3$
30. $(-3x^2yz^3)^3$
31. $(x^2y)^2 + (xy)^3$
32. $(2xy^3)^4 + (3xy)^2$
33. $(2xy^2z)^2 - (xyz^2)^2$
34. $(3x^2yz)^3 - (4x^2y^2z^2)^2$
35. $(-xy)^2 - xy^2(xy)^2$
36. $(x^2y)(xy)^2 + (xy^2)^2$
37. $(2x^2y)^2(xy) + (xy^2)$
38. $(xy)^2 + (-x^2y)^2(-xy^2)$
39. $(2xy)^2 - 3x(x^2y)^2 + 4x(xy^2)$
40. $3(x^2y)^2 + x(x^2y) - x^2(x^2y^2)$
41. $(x^2y)^2 - 2x^2(x^2y^2) + 3xy(x^3y)$
42. $2x^2(y^3) + 4y(xy)^2 - xy(xy^2)$

B

■ *Use the laws of exponents to simplify each expression. (Assume that exponents are natural numbers.) See Example 4.*

43. $a^{2n} \cdot a^{n-3}$
44. $b^{-n} \cdot b^{2n+1}$
45. $x^{n^2-n} \cdot x^{2n-n^2}$
46. $y^{2n+6} \cdot y^{4-n}$
47. $a^{2n-2} \cdot a^{n+3}$
48. $b^{n+2} \cdot b^{2n-1}$
49. $(x^{2n}y)^3$
50. $(xy^{3n})^2$
51. $(x^{n-2}y)^3$
52. $(xy^{n+2})^2$
53. $(x^{2n+1}y^{n-1})^3$
54. $(x^{n-2}y^{2n+1})^2$

2.4

PRODUCTS OF POLYNOMIALS

The associative property of addition can be used to extend the distributive property to cases where the right-hand factor contains more than two terms.

Thus,

$$\begin{aligned} a[b+c+d] &= a[(b+c)+d] \\ &= a(b+c)+ad \\ &= ab+ac+ad, \\ a[b+c+d+e] &= a([b+c+d]+e) \\ &= ab+ac+ad+ae, \qquad \text{etc.} \end{aligned}$$

We refer to this as the **generalized distributive property.**

EXAMPLE 1 a. $3x(x+y+z) = 3x(x)+3x(y)+3x(z)$
$= 3x^2+3xy+3xz$

b. $-2ab^2(3a^2b-ab+2ab^2)$
$= (-2ab^2)(3a^2b)+(-2ab^2)(-ab)+(-2ab^2)(2ab^2)$
$= -6a^3b^3+2a^2b^3-4a^2b^4$ ■

The distributive property can also be applied to simplify expressions for products of polynomials containing more than one term. For example,

$$\begin{aligned} (x-y)(3x+2y) &= 3x(x-y)+2y(x-y) && (1) \\ &= 3x^2-3xy+2xy-2y^2 && (2) \\ &= 3x^2-xy-2y^2. && (3) \end{aligned}$$

The four products in step (2) can be obtained more directly by writing the products in the following form.

$$\begin{aligned} (3x+2y)(x-y) &= \overset{①}{3x^2} - \overset{②}{3xy} + \overset{③}{2xy} - \overset{④}{2y^2} \\ &= 3x^2-xy-2y^2. \end{aligned}$$

This process is sometimes called the "FOIL" method, where "FOIL" represents:

the product of the First terms;
the product of the Outer terms;
the product of the Inner terms;
the product of the Last terms.

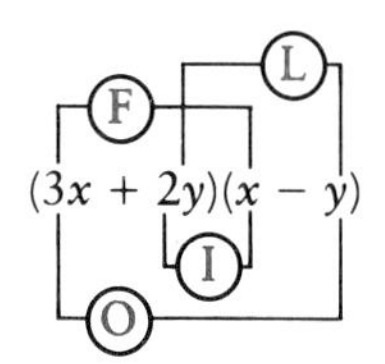

EXAMPLE 2 a. $(2x - 1)(x + 3)$
$= 2x^2 + 6x - x - 3$
$= 2x^2 + 5x - 3$

b. $(3x + 1)(2x - 1)$
$= 6x^2 - 3x + 2x - 1$
$= 6x^2 - x - 1$ ■

Special products of binomials

The following products are special cases of the multiplication of binomials. Since they occur so frequently you should learn to recognize them on sight.

> **I.** $(x + a)^2 = (x + a)(x + a) = x^2 + 2ax + a^2$
>
> **II.** $(x - a)^2 = (x - a)(x - a) = x^2 - 2ax + a^2$
>
> **III.** $(x + a)(x - a) = x^2 - a^2$

Common Error

Note that in I, $(x + a)^2 \neq x^2 + a^2$ *and in* II, $(x - a)^2 \neq x^2 - a^2$.

EXAMPLE 3 a. $(z - 3)^2$
$= (z - 3)(z - 3)$
$= z^2 - 2 \cdot 3z + 3^2$
$= z^2 - 6z + 9$

b. $(x + 4)^2$
$= (x + 4)(x + 4)$
$= x^2 + 2 \cdot 4x + 16$
$= x^2 + 8x + 16$

c. $(y + 5)(y - 5)$
$= y^2 - 5^2$
$= y^2 - 25$ ■

More about products

The distributive property can also be used to simplify expressions involving products of polynomials containing more than two terms by multiplying each term of one factor by each of the terms in the other factor.

EXAMPLE 4 a. $(x + 2)(x^2 - x + 1) = x(x^2 - x + 1) + 2(x^2 - x + 1)$
$= x^3 - x^2 + x + 2x^2 - 2x + 2$
$= x^3 + x^2 - x + 2$

b. $(2x - 3)(x^2 + 2x - 1) = 2x(x^2 + 2x - 1) - 3(x^2 + 2x - 1)$
$= 2x^3 + 4x^2 - 2x - 3x^2 - 6x + 3$
$= 2x^3 + x^2 - 8x + 3$ ■

The distributive property frequently plays a role in the simplification of expressions involving grouping devices. For example, we can begin to simplify

$$2[x - 3y + 3(y - x)] - 2(2x + y)$$

by applying the distributive property to $3(y - x)$, that is, to the *inner* set of parentheses, and then combining like terms.

EXAMPLE 5

$$\begin{aligned} 2[x - 3y + 3(y - x)] - 2(2x + y) &= 2[x - 3y + 3y - 3x] - 2(2x + y) \\ &= 2[-2x] - 2(2x + y) \\ &= -4x - 4x - 2y \\ &= -8x - 2y \end{aligned}$$

■

In Section 1.4 we used the definition of a difference to write expressions of the form

$$a - (b + c) \quad \text{and} \quad a - (b - c)$$

without parentheses as

$$a - b - c \quad \text{and} \quad a - b + c,$$

respectively. You may find it helpful to first express the differences as

$$a - 1 \cdot (b + c) \quad \text{and} \quad a - 1 \cdot (b - c)$$

and then use the distributive property as shown to obtain the same results.

EXAMPLE 6

a. $3[2x - (x - 2) + 3]$

$$\begin{aligned} &= 3[2x - 1 \cdot (x - 2) + 3] \\ &= 3[2x - x + 2 + 3] \\ &= 3[x + 5] \\ &= 3x + 15 \end{aligned}$$

b. $4[x - (x + 2)^2 + 4]$

$$\begin{aligned} &= 4[x - 1 \cdot (x^2 + 4x + 4) + 4] \\ &= 4[x - x^2 - 4x - 4 + 4] \\ &= 4[-x^2 - 3x] \\ &= -4x^2 - 12x \end{aligned}$$

■

The following examples illustrate the application of laws of exponents along with the distributive property in a general way.

EXAMPLE 7

a. $2a^{2n}(3a^n - 2)$
$= (2a^{2n})3a^n - (2a^{2n})2$
$= 6a^{2n+n} - 4a^{2n}$
$= 6a^{3n} - 4a^{2n}$

b. $(2a^n + 1)(a^n - 2)$
$= 2a^n(a^n - 2) + 1(a^n - 2)$
$= 2a^{2n} - 4a^n + a^n - 2$
$= 2a^{2n} - 3a^n - 2$ ■

EXERCISE 2.4

A

■ *Write each expression as a polynomial in simple form. See Examples* 1, 2, *and* 3.

1. $4y(x - 2y)$
2. $3x(2x + y)$
3. $-6x(2x^2 - x + 1)$
4. $-2y(y^2 - 3y + 2)$
5. $-(x^2 - 4x - 1)$
6. $-(2y^2 - y + 3)$
7. $(x + 3)^2$
8. $(y - 4)^2$
9. $(2y - 5)^2$
10. $(3x + 2)^2$
11. $(x + 3)(x - 3)$
12. $(x - 7)(x + 7)$
13. $(n + 2)(n + 8)$
14. $(r - 1)(r - 6)$
15. $(r + 5)(r - 2)$
16. $(y - 2)(y + 3)$
17. $(y - 6)(y - 1)$
18. $(z - 3)(z - 5)$
19. $(2z + 1)(z - 3)$
20. $(3t - 1)(2t + 1)$
21. $(4r + 3)(2r - 1)$
22. $(2z - 1)(3z + 5)$
23. $(2x - a)(2x + a)$
24. $(3t - 4s)(3t + 4s)$

■ *Simplify each product. See Example* 4.

25. $(y + 2)(y^2 - 2y + 3)$
26. $(t + 4)(t^2 - t - 1)$
27. $(x - 3)(x^2 + 5x - 6)$
28. $(x - 7)(x^2 - 3x + 1)$
29. $(x - 2)(x - 1)(x + 3)$
30. $(y + 2)(y - 2)(y + 4)$
31. $(z - 3)(z + 2)(z + 1)$
32. $(z - 5)(z + 6)(z - 1)$
33. $(2x + 3)(3x^2 - 4x + 2)$
34. $(3x - 2)(4x^2 + x - 2)$
35. $(2a^2 - 3a + 1)(3a^2 + 2a - 1)$
36. $(b^2 - 3b + 5)(2b^2 - b + 1)$

■ *Simplify each expression. See Examples* 5 *and* 6.

37. $2[a - (a - 1) + 2]$
38. $3[2a - (a + 1) + 3]$
39. $a[a - (2a + 3) - (a - 1)]$
40. $-2a[3a + (a - 3) - (2a + 1)]$
41. $-[a - 3(a + 1) - (2a + 1)]$
42. $-[(a + 1) - 2(3a - 1) + 4]$
43. $2(a - [a - 2(a + 1) + 1] + 1)$
44. $-4(4 - [3 - 2(a - 1) + a] + a)$
45. $-x(x - 3[2x - 3(x + 1)] + 2)$
46. $x(4 - 2[3 - 4(x + 1)] - x)$
47. $2[4x + (x + 1)^2]$
48. $3[2x + (x + 2)^2]$
49. $-x + 2x[4 - (x - 3)^2]$
50. $-2x + x[3 - (x + 4)^2]$

51. $-2x[x + (2x - 1)^2 - 4]$

52. $-x[2x - (2x + 1)^2 + 3]$

53. $-4[2x^2 - 2(x + 1)(x - 2) - 4x]$

54. $-3[2x^2 - 3(x - 2)(x + 3) + 3x]$

■ *Show that the left-hand member is equivalent to the right-hand member.*

55. $(x + a)(x + b) = x^2 + (a + b)x + ab$

56. $(x + a)^2 = x^2 + 2ax + a^2$

57. $(x + a)(x - a) = x^2 - a^2$

58. $(ax + by)(cx + dy) = acx^2 + (ad + bc)xy + bdy^2$

59. $(x + a)(x^2 - ax + a^2) = x^3 + a^3$

60. $(x - a)(x^2 + ax + a^2) = x^3 - a^3$

61. Use a counterexample to show that $(x + y)^2$ is not equivalent to $x^2 + y^2$.

62. Use a counterexample to show that $(x - y)^2$ is not equivalent to $x^2 - y^2$.

B

■ *In Problems 63–74, assume that all variables in exponents denote natural numbers. See Example 7.*

63. $x^n(2x^n - 1)$

64. $3t^n(2t^n + 3)$

65. $a^{n+1}(a^n - 1)$

66. $b^{n-1}(b + b^n)$

67. $a^{2n+1}(a^n + a)$

68. $b^{2n+2}(b^{n-1} + b^n)$

69. $(1 + a^n)(1 - a^n)$

70. $(a^n - 3)(a^n + 2)$

71. $(a^{3n} + 2)(a^{3n} - 1)$

72. $(a^{3n} - 3)(a^{3n} + 3)$

73. $(2a^n - b^n)(a^n + 2b^n)$

74. $(a^{2n} - 2b^n)(a^{3n} + b^{2n})$

CHAPTER SUMMARY

[2.1] Expressions of the form a^n, where

$$a^n = aaa \cdot \cdots \cdot a \qquad (n \text{ factors}),$$

are called **powers**; a is the **base** and n is the **exponent** of the power.

Any meaningful collection of numerals, variables, and signs of operation is called an **expression.** In an expression of the form $A + B + C + \cdots$, A, B, and C are called **terms.** Any factor or group of factors in a term is the **coefficient** of the product of the remaining factors.

An algebraic expression of the form cx^n, where c is a constant and n is a whole number, or a product of such expressions, is called a **monomial.** A **polynomial** is an algebraic expression that contains only terms that are monomials. Polynomials of two and three terms are called **binomials** and **trinomials**, respectively.

The degree of a monomial in one variable is given by the exponents on the variable. The degree of a polynomial is the degree of its term of highest degree.

Polynomials are represented by symbols such as $P(x)$, $Q(z)$, etc., and the values of these polynomials for some specific value a are represented by $P(a)$, $Q(a)$, etc.

[2.2] Terms that differ only in their numerical coefficients are called **like terms.** Two expressions that are equal for all real-number replacements of any variable or variables involved are **equivalent expressions.**

[2.3] The following laws of exponents are useful in rewriting powers:

$$a^m \cdot a^n = a^{m+n}, \qquad (a^m)^n = a^{mn}, \qquad (ab)^m = a^m b^m.$$

[2.4] The distributive property in the form

$$a(b + c + d + \cdots) = ab + ac + ad + \cdots$$

is used to rewrite products as equivalent expressions without parentheses.

The following products are special cases of products of binomials.

$$\text{I.} \quad (x + a)^2 = x^2 + 2ax + a^2$$

$$\text{II.} \quad (x - a)^2 = x^2 - 2ax - a^2$$

$$\text{III.} \quad (x + a)(x - a) = x^2 - a^2$$

REVIEW EXERCISES

A

[2.1]

■ *In Exercises 1 and 2, identify each polynomial as a monomial, binomial, or trinomial. State the degree of the polynomial.*

1. a. $2y^3$ b. $3x^2 - 2x + 1$
2. a. $3x^2 - x^5$ b. $5y^4 - y^3 - y^2$
3. Find the value of $\dfrac{2x^2 - 3y}{x - y}$ for $x = -2$ and $y = 3$.
4. Find the value of $x^2 - 2xy^2 - xy + y^4$ for $x = 3$ and $y = -1$.
5. If $P(x) = 2x^2 - 3x - 1$, find a. $P(3)$ b. $P(-2)$
6. If $Q(x) = x^3 - 2x^2 - x$, find a. $Q(-1)$ b. $Q(-2)$

[2.2]

■ *Simplify each expression.*

7. a. $(2x - y) - (x - 2y + z)$ b. $(2x^2 - 3z^2) - (x - 2y) + (z^2 + y)$
8. a. $(x - 3y) + (2x + y) - (y - z)$ b. $(y - 3z) - (x + 2y) - (x + 3z)$
9. a. $(x^2 - 2y - z^2) - (3x^2 + 2y + z)$ b. $(2x - y^2 + z) - (2x^2 + y - z)$
10. a. $2x - [x - (x - 3) + 2]$ b. $x^2 - (x + 1) - [2x^2 + (x - 1)]$
11. a. $y^2 + y - [y - (y^2 - 1)]$ b. $y^2 - [y^2 - (y + 1) - y]$
12. a. $z - [z^2 - (z^2 - z) + z]$ b. $z - (z^2 - 1) - [z - (z^2 + 1)]$

[2.3]

■ *Simplify each expression.*

13. a. $(2x^2y)(-3xy^3)$ b. $(3xy)(2xz^2)(-y^2z)$
14. a. $(-2x^2y^3z)^3$ b. $(xy^2)^3 - (2x^3y)^2$
15. a. $(x^2y)(3xy^2)^2$ b. $(xy^2)^2(2x^2y)$
16. a. $(2xy)(3yz^2)(xy)^2$ b. $(yz)^2(3xy^2)(2xz)$
17. a. $3x^2(xy^2) + 2x(xy)^2 - x^2y(xy)^2$ b. $x(x^2y^3) - 3x(xy)^2 + 4(xy)^3$
18. a. $(-xy)^2 + 2x(x^2y) + 4x^2y^2$ b. $(2x)^2y^3 + 3xy^2(xy) - x^2y^2(y)$

[2.4]

■ *Write each expression as a polynomial in simple form.*

19. a. $2x(x^2 - 2x + 1)$ b. $2(2x - 1)(x + 3)$
20. a. $(y - 2)(y^2 - y + 1)$ b. $(z - 1)(z + 1)(z + 2)$
21. a. $3[x - 2(x + 1) - 3]$ b. $-x\{1 - 2[x + 3(x - 1)] + x\}$
22. a. $y[y - 2(y - 1) + 4]$ b. $-y[4y - (2y + 1) - 3]$
23. a. $-[x^2 - (x - 3)^2 + 4x]$ b. $-[2x^2 - (x - 2)(x + 3) - 4]$
24. a. $2[x - 4(x - 2)(x - 1) - 3]$ b. $-3[x - 2(x + 2)^2 - 5]$

B

■ *Given that* $P(x) = x^2 - 1$ *and* $Q(x) = x^2 + 3x - 1$, *find the value of each expression.*

25. $Q[P(2)]$
26. $P[Q(-3)]$

■ *Given that* $P(x) = x + 1$, $Q(x) = x^2 - x$, *and* $R(x) = 2x^2 - 1$, *find the value of each expression.*

27. $P(2) - [Q(1) + R(0)]$
28. $Q(-3) - [P(-2) - R(1)]$

■ *Write each expression without using parentheses. (Assume that variables in exponents denote natural numbers.)*

29. $(x^{n-1}y^n)^3$
30. $(2x^n + 1)(x^n - 3)$

3 Equations and Inequalities

The development of skills that are used to solve narrative exercises, commonly called word problems, is an important objective in the study of algebra. There are two distinct skills involved that we will consider in this chapter. One skill involves writing equations or inequalities that are mathematical models for conditions stated in the word problems. The second skill involves solving the equations or inequalities.

We will first review properties and procedures that are used to solve equations that involve polynomials.

3.1

SOLVING EQUATIONS

Equivalent equations

Any number that satisfies an equation is called a **solution** of the equation, and the set of all solutions is called the **solution set**. The process of finding solution sets of equations involves generating equations that have the same solution set. Equations that have identical solution sets are called **equivalent equations**. For example,

$$5x - 4 = 16,$$
$$5x = 20,$$

and

$$x = 4$$

are equivalent equations because $\{4\}$ is the solution set of each.

Simplifying members of an equation

There are several properties that enable us to form equivalent equations. One way is to simplify a member of an equation by writing the sum or difference of two or more terms as a single term. For example, the solution of the equation $9x - 8x = 12$ becomes evident when it is written in the equivalent form $x = 12$.

Addition and multiplication laws

Another way to generate equivalent equations follows from the equality axioms of real numbers (see Section 1.2).

1. *The addition of the same expression to each member of an equation produces an equivalent equation.*
2. *The multiplication of each member of an equation by the same nonzero number produces an equivalent equation.*

Any application of these properties is called an **elementary transformation.** An elementary transformation always produces an equivalent equation.

The application of Properties 1 and 2 above enables us to transform an equation whose solution may not be obvious, through a series of equivalent equations, until we obtain an equation that has an obvious solution. If necessary, we may obtain an equation in which the variable with coefficient of 1 stands alone as one member.

EXAMPLE 1 To solve

$$3x + 5 = 11,$$

we first add -5, the additive inverse of 5, to each member to obtain

$$3x + 5 - 5 = 11 - 5,$$
$$3x = 6.$$

Now, by inspection, we can readily determine that 2 is the solution. However, if necessary, we can multiply each member by $1/3$, the multiplicative inverse of 3, in order to obtain the variable x with a coefficient of 1. Thus,

$$\frac{1}{3}(3x) = \frac{1}{3}(6),$$
$$x = 2.$$

In either case, the solution set is $\{2\}$. ■

We can always determine whether an apparent solution of an equation is really a solution by substituting it into the original equation and verifying that the resulting statement is true. If each of the equations in a sequence is simply obtained by means of an application of Properties 1 and 2 above, the sole purpose for such a check is to detect arithmetic errors.

Let us consider several other examples.

EXAMPLE 2 To solve

$$3y - 7 = y + 5,$$

we first add $-y$ and $+7$ to each member, where $-y$ is the additive inverse of y and 7 is the additive inverse of -7, to obtain

$$3y - 7 + (-y) + 7 = y + 5 + (-y) + 7,$$
$$3y - y = 5 + 7.$$

Simplifying each member, we have

$$2y = 12.$$

We now have an equation in which one member includes only terms containing the variable and the other member includes only constants. By inspection, we can determine that 6 is the solution. Alternatively, we can multiply each member by $\frac{1}{2}$, the multiplicative inverse of 2, to obtain the variable y with a coefficient of 1. Thus,

$$\frac{1}{2}(2y) = \frac{1}{2}(12),$$
$$y = 6.$$

The solution set is $\{6\}$. ■

Equations containing parentheses

When solving an equation that contains grouping symbols, it may be necessary to use the distributive property to simplify one or both members.

EXAMPLE 3 To solve

$$3(x + 2) - 2(x + 1) = 8$$

we first apply the distributive property to obtain

$$3x + 6 - 2x - 2 = 8,$$

from which

$$x = 4.$$

Hence, the solution set is $\{4\}$. ■

EXAMPLE 4 To solve

$$2(x - 3)^2 - 2x^2 = 9,$$

we first apply the distributive property to obtain

$$2(x^2 - 6x + 9) - 2x^2 = 9,$$

and then apply it again to obtain

$$\begin{aligned} 2x^2 - 12x + 18 - 2x^2 &= 9, \\ -12x + 18 &= 9, \\ -12x &= -9, \\ x &= \frac{9}{12} = \frac{3}{4}. \end{aligned}$$

Hence, the solution set is $\{3/4\}$. ■

Sometimes grouping symbols are nested within a second set of grouping symbols and we have to apply the distributive law more than once.

EXAMPLE 5 To solve

$$4[x - 3(x + 2) - 4] = 3,$$

we first simplify $-3(x + 2)$ to obtain

$$\begin{aligned} 4[x - 3x - 6 - 4] &= 3, \\ 4[-2x - 10] &= 3. \end{aligned}$$

Applying the distributive law a second time, we get

$$\begin{aligned} -8x - 40 &= 3, \\ -8x &= 43, \\ x &= -\frac{43}{8}. \end{aligned}$$

Hence, the solution set is $\{-43/8\}$. ■

The procedures that we used to solve equations in the previous examples are also applicable to solve equations that involve decimals.

EXAMPLE 6 To solve

$$0.10(x - 2) + 0.20(x + 3) = 2.50,$$

we first apply the distributive property to get

$$\begin{aligned} 0.10x - 0.20 + 0.20x + 0.60 &= 2.50, \\ 0.30x + 0.40 &= 2.50. \end{aligned}$$

Now we add -0.40 to each member to obtain

$$\begin{aligned} 0.30x + 0.40 - 0.40 &= 2.50 - 0.40, \\ 0.30x &= 2.10. \end{aligned}$$

Dividing each by 0.30, we obtain

$$\begin{aligned} \frac{0.30x}{0.30} &= \frac{2.10}{0.30}, \\ x &= 7. \end{aligned}$$

The solution set is $\{7\}$. ■

It is usually easier to work with nondecimal numbers. In the above example we could first multiply each member of the original equation by 100 to obtain the equation

$$\begin{aligned} 100[0.10(x - 2) + 0.20(x + 3)] &= 100(2.50) \\ 0.10(x - 2) + 0.20(x + 3) &= 2.50, \end{aligned}$$

or

$$10(x - 2) + 20(x + 3) = 250,$$

and then proceed to solve the equation.

Solving formulas

Equations that express relationships between quantities of practical interest are called **formulas.** We can use Properties 1 and 2 above to solve a formula for a specific variable if we are given values for the other variables in the formula.

EXAMPLE 7 The perimeter P of a rectangle is given by the formula

$$P = 2l + 2w,$$

where l is the length and w is the width. Find l given that P is 38 centimeters and w is 4 centimeters.

Solution We substitute the given values for P and w in the formula to obtain

$$38 = 2l + 2(4).$$

Adding -8, the additive inverse of 2(4), to each member, we get

$$38 - 8 = 2l + 8 - 8,$$
$$30 = 2l.$$

Then multiplying each member by ½, the multiplicative inverse of 2, we obtain

$$\frac{1}{2}(30) = \frac{1}{2}(2l),$$
$$15 = l.$$

The solution set is {15}. Hence, the length of the rectangle is 15 centimeters. ■

Solving equations for specified symbols

An equation containing more than one variable or symbols representing constants can be solved for one of the symbols in terms of the remaining symbols. In general, we apply the properties listed on page 60 until the desired symbol stands alone as one member of an equation. The following suggestions may be helpful.

To Solve a First-degree Equation for a Specified Variable:

1. Transform the equation to a form where all terms containing the specified variable are in one member and all terms not containing that variable are in the other member.
2. Combine like terms in each member.
3. If unlike terms containing the specified variable remain, factor that variable from the terms.
4. Divide each member by the coefficient of the specified variable.

EXAMPLE 8 To solve the formula

$$P = 2l + 2w \quad \text{for } w,$$

we first add $-2l$ to each member to obtain

$$\begin{aligned} P - 2l &= 2l + 2w - 2l, \\ P - 2l &= 2w. \end{aligned}$$

Then we divide each member by 2, the coefficient of w, (or multiply by ½) to get

$$\frac{P - 2l}{2} = w \quad \text{or} \quad w = \frac{P - 2l}{2}. \qquad ■$$

EXAMPLE 9 To solve the formula

$$S = lw + lh + 2wh \quad \text{for } w,$$

we first obtain

$$S - lh = lw + 2wh$$

in which all terms containing the specified variable w are in one member and all terms not containing w are in the other member. Then we factor w from $lw + 2wh$ to obtain

$$S - lh = w(l + 2h),$$

and finally divide each member by $(l + 2h)$, the coefficient of w, to get

$$\frac{S - lh}{l + 2h} = w \quad \text{or} \quad w = \frac{S - lh}{l + 2h}. \qquad ■$$

The method that we used to solve an equation for one of the variables in terms of other variables and constants is also applicable when the equation contains parentheses. When parentheses are involved, usually there are several ways to proceed that may lead to equations that have different forms. However, they will be equivalent.

EXAMPLE 10 Solve $2A = h(b + c)$ for b.

Solution (1)
Dividing each member by h, the coefficient of $b + c$, yields

$$\frac{2A}{h} = \frac{h(b + c)}{h},$$

$$\frac{2A}{h} = b + c.$$

Adding $-c$ to each member, we get

$$\frac{2A}{h} - c = b \quad \text{or} \quad b = \frac{2A}{h} - c.$$

Solution (2)
Applying the distributive law in the right-hand member yields

$$2A = hb + hc.$$

Adding $-hc$ to each member, we get

$$2A - hc = hb + hc - hc,$$

$$2A - hc = hb.$$

Finally, dividing each member by h, the coefficient of b, we get

$$\frac{2A - hc}{h} = b \quad \text{or} \quad b = \frac{2A - hc}{h}.$$

■

In our later work we will develop methods to verify that the different equations obtained are equivalent.

EXERCISE 3.1

A

■ *Solve each equation. See Examples 1–5.*

1. $3x + 5 = 26$
2. $2 + 5x = 37$
3. $4x - 6 = 22$
4. $3x - 5 = 7$
5. $y + (y + 140) = 620$
6. $y + (y - 160) = 830$
7. $y + (y - 210) + (y - 490) = 7970$
8. $y + (y - 620) + (y - 810) = 8620$
9. $3(z + 2) = 14$
10. $2(z - 3) = 15$
11. $3z - (z - 4) = 12$
12. $5z - (z + 1) = 14$
13. $2[x - 3(x + 2) - 4] = 6$
14. $3[3x - 2(x - 3) + 1] = 8$
15. $3[4 + 2(x - 3) - x] = 10$
16. $2[1 - 3(x + 1) - 2x] = 12$
17. $-[4 - (y - 2) + 2y] = 0$
18. $-[7 - (y - 3) - 4y] = 0$
19. $-2[y - (y + 1)] = 3(y - 2)$
20. $-3[2y - (y - 2)] = 2(y + 3)$
21. $(z - 3)^2 - z^2 = 6$
22. $(z + 2)^2 - z^2 = 8$
23. $4 + (x - 1)(x + 2) = x^2 + 8$
24. $2 + (x - 2)(x + 3) = x^2 + 12$
25. $-2[y - (y - 1)^2] = 2y^2 - 6$
26. $-3[y - (y + 2)^2] = 3y^2 - 10$
27. $3[x + 2(x + 2)^2 - x^2] = 3x^2 + 2$
28. $2[2x + 3(x + 1)^2 - x^2] = 4x^2 + 1$
29. $-[(2x - 1)^2 - 2x] = 2x - (2x + 1)^2$
30. $-[(2x - 1)^2 + 2x] = 2x - (2x - 1)^2$

■ *Solve each equation. See Example 6.*

31. $0.40(y - 4) = 2.80$
32. $0.60(y + 2) = 3.60$
33. $0.25y + 0.10(y + 32) = 11.60$
34. $0.12y + 0.08(y + 10{,}000) = 12{,}000$
35. $0.10x + 0.80(20) = 0.50(x + 20)$
36. $0.10x + 0.12(x + 4{,}000) = 920$

■ *Solve each formula for the specified variable. See Example 7.*

37. $P = 2l + 2w$ for l, given that $w = 6$ and $P = 48$
38. $A = \frac{1}{2}bh$ for b, given that $A = 18$ and $h = 9$
39. $A + B + C = 180$ for B, given that $A = 62$ and $C = 44$
40. $V = lwh$ for w, given that $V = 48$, $l = 4$, and $h = 2$
41. $A = P + Prt$ for t, given that $A = 1320$, $P = 1000$, and $r = 0.08$
42. $A = P + Prt$ for t, given that $A = 2320$, $P = 1000$, and $r = 0.11$

■ *Solve each formula for the specified variable. See Examples 8–10.*

43. $f = ma$ for m
44. $pv = k$ for v
45. $I = prt$ for p
46. $V = lwh$ for h
47. $P = 2l + 2w$ for w
48. $S = 3\pi d + \pi d$ for d
49. $v = k + gt$ for g
50. $v = k + gt$ for t
51. $S = 2\pi(r + h)$ for h
52. $S(l - r) = a$ for r
53. $l = a + (n - l)d$ for n
54. $2A = h(b + c)$ for c
55. $A = P(1 + rt)$ for t
56. $S = 2[l(w + h) + wh]$ for l

3.2

CONSTRUCTING MATHEMATICAL MODELS; APPLICATIONS

In this section we solve a variety of word problems by constructing mathematical models in the form of equations.

Translating sentences

In the first example we translate several simple word sentences into equations where we have represented a number by x.

EXAMPLE 1 a. Four more than three times a number is equal to 25.

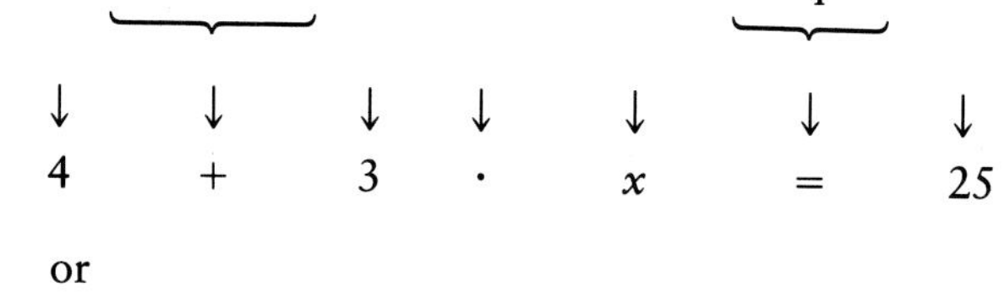

or

$$4 + 3x = 25$$

b. Two times the sum of a number and six equals 18.

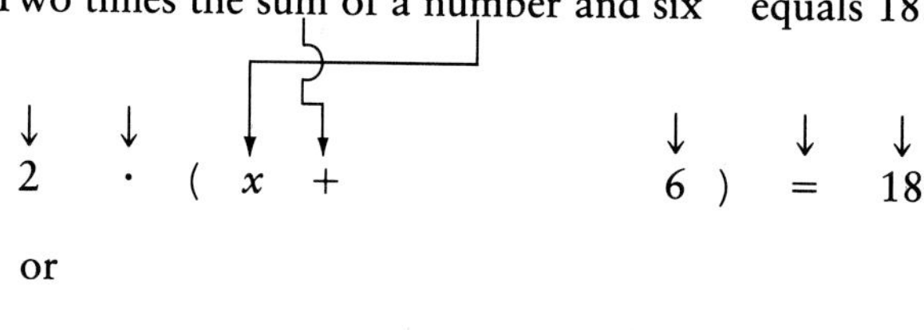

or

$$2(x + 6) = 18$$ ■

In the above examples the word sentences were simple, and translating each sentence into an equation was direct. Mathematical models are not always so easy to construct.

Constructing mathematical models

In problems where the relationship between given quantities and one or more quantities that we wish to find is not evident from a reading of the problem, *we can at least start the process of constructing a mathematical model by first writing in words the quantity or quantities we want to find.* Then these quantities can be represented by variables (in this section one variable), or by expressions containing the variable. These first two steps will help us write an appropriate equation that describes the condition on the variable stated in the problem.

We first look at several simple examples where only one quantity is to be found. In each example we proceed in the following order:

1. Represent the unknown quantity using words.
2. Assign a variable to represent the unknown quantity.
3. Write an equation expressing the condition on the variable.
4. Solve the equation.
5. Interpret the solution of the equation in terms of the quantity represented by the variable in Step 2. (Does your answer make sense?)

EXAMPLE 2 Six percent (0.06) of the students in all algebra sections received an "A" grade. How many students were enrolled if nine students received A's?

Solution

Step 1 → Number of students enrolled: n ← Step 2

Step 3 Write an equation expressing the fact that 0.06 of the number of students enrolled equals 9.

$$0.06n = 9.$$

Step 4 Solve the equation.

$$n = \frac{9}{0.06} = 150.$$

Step 5 Hence, 150 students are enrolled. ■

EXAMPLE 3 A camera is on sale at a 25% (0.25) discount. What was the original cost if the discount is \$45?

Solution

Step 1 → Original cost: c ← Step 2

Step 3 Write an equation expressing the fact that 0.25 of the original cost is \$45.

$$0.25c = 45.$$

Step 4 Solve the equation.

$$c = \frac{45}{0.25} = 180.$$

Step 5 Hence, the original cost is \$180. ■

In each of the above examples only one quantity was to be found. *If two or more quantities are to be found, separate words or word phrases should be used for each quantity.* After assigning a variable to represent one of the quantities, the conditions in the problem are used to determine the algebraic expressions involving the variable, which are then assigned to the other quantities.

In our later work we will sometimes use different variables for different quantities; however, at this time we will use one variable.

EXAMPLE 4 A small business calculator sells for \$5 less than a scientific model. If the cost to buy both is \$41, how much does each cost?

Solution

Step 1 First write *two* word phrases for the quantities to be found.

Step 2 Assign a variable, say C, for the *cost* of the scientific calculator.

Step 2a Then, because the business calculator is \$5 less than the scientific calculator, the *cost* of the business calculator can be expressed as $C - 5$.

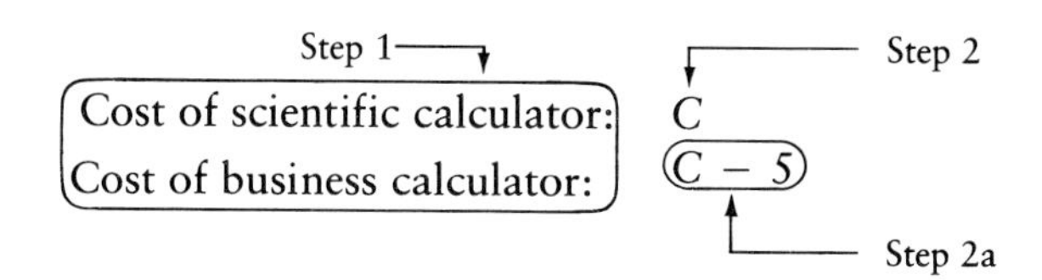

Step 3 Write an equation expressing the fact that the total cost is \$41.

$$C + (C - 5) = 41.$$

Step 4 Solve the equation.

$$2C = 46$$
$$C = 23.$$

Step 5 Hence, the cost of the scientific calculator is \$23 and the cost of the business calculator is $C - 5 = 23 - 5$ or \$18. ■

It is sometimes possible to write more than one equation that would be correct for the conditions of a problem, depending on the quantity to which we assign the variable. In the above example, if we had let C represent the cost of the business calculator, then $C + 5$ would represent the cost of the scientific calculator and an appropriate equation would be

$$C + (C + 5) = 41.$$

The procedure used in Example 4 can also be used when three or more quantities are to be obtained. In the following example we will let the variable represent the *largest* number we wish to obtain and then express each of the other smaller quantities in terms of this variable.

EXAMPLE 5 One thousand sixty students voted to elect a candidate to represent their college at a national convention. One candidate received 60 votes more than a second candidate and 104 votes more than a third candidate. How many votes did each candidate receive?

Solution

Step 1 →	Step 2 →
Number of votes of winning candidate:	x
Number of votes of second candidate:	$x - 60$
Number of votes of third candidate:	$x - 104$
	← Step 2a

Step 3 Write an equation expressing the fact that there was a total of 1060 votes.

$$x + (x - 60) + (x - 104) = 1060.$$

Step 4 Solve the equation.

$$3x - 164 = 1060$$
$$3x = 1224$$
$$x = 408.$$

Step 5 Hence, the winning candidate received 408 votes, the second candidate received $x - 60 = 408 - 60$ or 348 votes, and the third candidate received $x - 104 = 408 - 104$ or 304 votes. ■

Using formulas and figures Solving problems involving geometric figures requires a knowledge of some simple formulas. Formulas and other geometric relationships that are needed for problems in this book are given in Appendix C. Sometimes it is helpful to make a sketch and label parts of the figure before trying to write an equation.

EXAMPLE 6 The vertex angle of an isosceles triangle measures 30° more than the sum of the measures of the other two equal angles. What is the measure of each angle in the triangle? (Recall: The sum of the measures of the angles in a triangle is always 180°.)

Solution

Step 1 Use the fact that in an isosceles triangle, the two angles opposite the two sides of equal length have equal measure. Express these quantities in a simple phrase.

Step 2 Represent the measure of each base angle by a variable, say x.

Step 2a Represent the measure of the third angle (vertex angle) by $x + x + 30$.

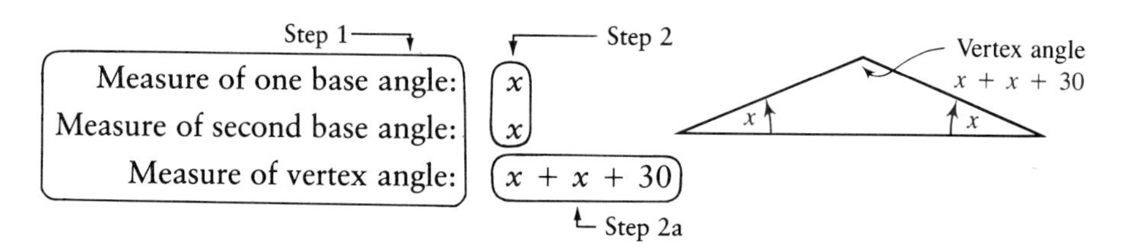

Step 3 Write an equation using the fact that the sum of the measures of the angles in any triangle is 180°.

$$(x) + (x) + (2x + 30) = 180.$$

Step 4 Solve the equation.

$$4x = 150$$
$$x = 37\tfrac{1}{2}.$$

Step 5 Hence, each base angle equals $37\tfrac{1}{2}°$ and the vertex angle equals

$$x + x + 30 = 37\tfrac{1}{2} + 37\tfrac{1}{2} + 30 \quad \text{or} \quad 105°.$$ ■

The step-by-step procedure to construct mathematical models shown above may seem to be very detailed for these simple examples. However, if you take the simple approach suggested, you will be prepared to use the same procedures when the problems are more difficult and writing an appropriate equation is not so easy.

Sometimes it is helpful to summarize information in a tabular form to assist us in constructing a mathematical model.

Coin problems The basic idea of problems involving coins (or bills) is that the value of a number of coins (or bills) of the same denomination is equal to the product of the value of a single coin (or bill) and the total number of coins (or bills).

$$\begin{bmatrix}\text{value of}\\ n\\ \text{coins}\end{bmatrix} = \begin{bmatrix}\text{value of}\\ 1\\ \text{coin}\end{bmatrix} \times \begin{bmatrix}\text{number}\\ \text{of}\\ \text{coins}\end{bmatrix}$$

EXAMPLE 7 A collection of coins consisting of dimes and quarters has a value of \$11.60. How many dimes and quarters are in the collection if there are 32 more dimes than quarters?

Solution

Step 1 Express the *two* quantities asked for in *two* simple phrases.

Step 2 Represent the number of quarters using a variable, say q.

Step 2a Represent the number of dimes by $q + 32$ (there are 32 more dimes than quarters).

Step 1 → Number of quarters: q ← Step 2

Number of dimes: $q + 32$ ← Step 2a

Step 3 A table showing the total value of the quarters and the total value of the dimes will be helpful to construct a mathematical model.

Denomination	Value of 1 coin	Number of coins	Value of coins
Quarters	0.25	q	$0.25q$
Dimes	0.10	$q + 32$	$0.10(q + 32)$

Write an equation relating the value of the quarters and the value of the dimes to the value of the entire collection.

$$\begin{bmatrix}\text{value of}\\ \text{quarters}\end{bmatrix} + \begin{bmatrix}\text{value of}\\ \text{dimes}\end{bmatrix} = \begin{bmatrix}\text{value of}\\ \text{collection}\end{bmatrix}$$
$$0.25q + 0.10(q + 32) = 11.60$$

Step 4 Solve the equation. Generally it is easier to first multiply each member by 100 in order to obtain an equivalent equation in which none of the terms contain decimals.

$$25q + 10q + 320 = 1160$$
$$35q = 840$$
$$q = 24.$$

Step 5 Therefore, there are 24 quarters and $24 + 32$ or 56 dimes in the collection. ■

Interest problems

Simple interest problems involve the fact that the interest (I) earned during a single year is equal to the amount (A) invested times the annual rate (r) of interest. Thus, $I = Ar$.

EXAMPLE 8 A man has an annual income of $12,000 from two investments. He has $10,000 more invested at 8% than he has invested at 12%. How much does he have invested at each rate?

Solution

Step 1 Express the *two* quantities asked for in *two* simple phrases.

Step 2 Represent the amount invested at 12% using a variable, say A.

Step 2a Represent the amount invested at 8% by $A + 10{,}000$.

Step 1 → Amount invested at 12%: A ← Step 2

Amount invested at 8%: $A + 10{,}000$ ← Step 2a

Step 3 A table showing the total interest received on each investment will be helpful to construct the mathematical model.

Investment Rate		Amount Invested	Interest
12%	0.12	A	$0.12A$
8%	0.08	$A + 10{,}000$	$0.08(A + 10{,}000)$

Write an equation relating the interest from each investment and the total interest received.

$$\begin{bmatrix}\text{interest from}\\ \text{12\% investment}\end{bmatrix} + \begin{bmatrix}\text{interest from}\\ \text{8\% investment}\end{bmatrix} = [\text{total interest}]$$

$$0.12A \quad + 0.08(A + 10{,}000) = \quad 12{,}000$$

Step 4 Solve the equation. First multiply each member by 100.

$$\begin{aligned} 12A + 8(A + 10{,}000) &= 1{,}200{,}000 \\ 12A + 8A + 80{,}000 &= 1{,}200{,}000 \\ 20A &= 1{,}120{,}000 \\ A &= 56{,}000. \end{aligned}$$

Step 5 Therefore, \$56,000 is invested at 12% and \$56,000 + 10,000 or \$66,000 is invested at 8%. ■

Mixture problems

The key to solving mixture problems is to recognize that the amount of a given substance in a mixture is obtained by multiplying the amount of the mixture by the percent (or rate) of the given substance in the mixture.

$$\begin{bmatrix}\text{amount}\\ \text{of}\\ \text{substance}\end{bmatrix} = \begin{bmatrix}\text{percent (rate)}\\ \text{of substance}\\ \text{in mixture}\end{bmatrix} \times \begin{bmatrix}\text{amount}\\ \text{of}\\ \text{mixture}\end{bmatrix}$$

Tables are especially helpful in mixture problems.

EXAMPLE 9 How many liters of a 10% solution of acid should be added to 20 liters of a 60% solution of acid to obtain a 50% solution?

Solution

Step 1 Express the quantity asked for in a simple phrase.

Step 2 Represent the quantity using a variable, say n.

Step 1 → Number of liters of 10% solution: Step 2 → n

Step 3 Set up a table showing the amount of *pure* acid in each solution.

Percent acid in mixture		Number of liters	Amount of acid
10%	0.10	n	$0.10n$
60%	0.60	20	$0.60(20)$
50%	0.50	$n + 20$	$0.50(n + 20)$

A figure may also be helpful. Here, the amount of pure acid in the solution is shown in red; the remainder is water.

n liters [$.10n$] + 20 liters [$.60(20)$] = [$.50(n + 20)$] $n + 20$ liters

Write an equation relating the amount of *pure* acid.

$$\begin{bmatrix}\text{pure acid in}\\ \text{10\% solution}\end{bmatrix} + \begin{bmatrix}\text{pure acid in}\\ \text{60\% solution}\end{bmatrix} = \begin{bmatrix}\text{pure acid in}\\ \text{50\% solution}\end{bmatrix}$$

$$0.10n + 0.60(20) = 0.50(n + 20)$$

Step 4 Solve the equation. First multiply each member by 100.

$$\begin{aligned} 10n + 60(20) &= 50(n + 20) \\ 10n + 1200 &= 50n + 1000 \\ -40n &= -200 \\ n &= 5. \end{aligned}$$

Step 5 Therefore, 5 liters of a 10% solution are needed. ■

You may wish to refer to the following summary of the steps we have suggested as you proceed to solve the word problems in this section and others.

Suggestions for Solving Word Problems

Step 1. Read the problem and note particularly what is asked for. *Write* a short word phrase to describe each unknown quantity.

Step 2a. Represent one unknown quantity by a variable.

b. Represent other unknown quantities in terms of this variable.

(Where applicable: Draw a sketch and label the parts.)

Step 3. Write an equation that involves the variable and is a "model" for the conditions on the variable stated in the problem.

Step 4. Solve the equation.

Step 5. Interpret the solution of the equation in terms of the quantity or quantities that were represented by the variable in Step 2.

EXERCISE 3.2

A

■ *Write each sentence as an equation using x as the variable. See Example* 1.

1. The sum of three times a number and 5 is 26.
2. Two plus five times a number equals 37.
3. Four times a number less 6 equals 22.
4. Three times a number minus 5 is 7.
5. The sum of two times a number and the number itself equals 21.
6. The sum of three times a number and two times the number equals 40.
7. The sum of an integer and 2, divided by 4, equals 5.
8. An integer less 2, divided by 5, equals 3.

■ *Solve each problem. Use the five steps suggested on page 69. See Examples 2 and 3 for Problems 9–18.*

9. One year the Dean of Admissions at City College selected 75% of all applicants for the freshman class. How many students applied for entrance if the college selected 600 students?
10. A part-time secretary takes home 80% of her total salary. What is her total salary if she takes home $120 per week?

11. A saleswoman earns a commission of 2% (0.02) on her sales. How much must she sell to have an income of $300 per week?
12. A company contracts to produce 1200 gears that meet certain specifications. If 3% (0.03) of the gears manufactured are defective, what is the total number of gears that must be manufactured?
13. Thirty percent (0.30) of all the students enrolled in a mathematics class are women. How many students are enrolled if there are 12 women in the class?
14. A basketball player made 40% (0.40) of all his shooting attempts. How many shots did he attempt if he made 8 baskets?
15. A company contracts to manufacture 1710 baseball bats. If 5% (0.05) of the bats are discarded because of defects, what is the total number of bats that must be manufactured to yield the 1710 bats?
16. A company that manufactures transistors always ships 2% (0.02) additional transistors over any amount ordered to cover any defective ones that might be included. How many transistors were ordered if 22,400 were shipped?
17. An employer finds that 94% (0.94) of his employees were present on Monday. If 9 employees were absent, how many employees did he have?
18. A computer was on sale for a 40% (0.40) discount. What was the original cost if the sale price was $384?

■ *See Examples* 4 *and* 5 *for Problems* 19–24.

19. A 24-foot rope is cut into two pieces so that one piece is 6 feet longer than the other. How long is each piece?
20. A 36-foot chain is cut into two pieces so that one piece is twice as long as the other piece. How long is each piece?
21. Six hundred twenty students voted in an election for two candidates for student body president. The winner received 140 more votes than the loser. How many votes did each candidate receive?
22. Eight hundred thirty students voted in an election for student body treasurer. The loser received 160 fewer votes than the winner. How many votes did each candidate receive?
23. Three candidates ran for city council member in an election in which 6560 votes were cast. The winner received 210 votes more than the second candidate and 490 votes more than the third candidate. How many votes did each candidate receive?
24. One candidate in an election received 620 votes more than a second candidate and 810 votes more than a third candidate. How many votes did each candidate receive if there were 8650 votes cast?

■ *See Example* 6 *for Problems* 25–34.

25. One angle of a triangle measures two times another angle, and the third angle measures 12° more than the sum of the measures of the other two. Find the measure of each angle.
26. The measure of the smallest angle of a triangle is 25° less than the measure of another angle, and 50° less than the measure of the third angle. Find the measure of each angle.

27. One angle of a triangle measures 10° more than another and the measure of the third angle equals the sum of the measures of the first two angles. Find the measures of each angle.

28. The measure of one angle of a triangle is 20° more than that of another and the measure of the third angle is six times the measure of the smaller. Find the measures of each angle.

29. The length of each of two equal sides of an isosceles triangle is 15 centimeters greater than the length of the third side. Find the length of the three sides if the perimeter is 66 centimeters.

30. The length of the longest side of a triangle is three times that of the shortest side and the length of the third side is 72 centimeters more than that of the shortest side. Find the length of the three sides if the perimeter is 272 centimeters.

31. When the length of each side of a square is increased by 5 centimeters, the area is increased by 85 square centimeters. Find the length of a side of the original square.

32. The length of each side of a square is decreased by 6 inches. If the area is decreased by 84 square inches, find the length of the side of the original square.

33. The length of a table tennis table is four feet longer than its width, and its perimeter is 28 feet. Find its dimensions.

34. The length of a tennis court for singles is 24 feet longer than twice its width, and its perimeter is 210 feet. Find its dimensions.

■ *See Example 7 for Problems 35–44.*

35. A collection of dimes and quarters has a value of $3.60. There are twice as many dimes as quarters. How many of each kind are there?

36. A collection of nickels and dimes has a value of $4.90. There are 22 more dimes than nickels. How many of each kind are there?

37. A woman who has 12 more quarters than dimes has a total of $12.45 in quarters and dimes. How many of each kind of coin does she have?

38. A savings bank containing $1.75 in dimes, quarters, and nickels contains 3 more dimes than quarters, and 3 times as many nickels as dimes. How many coins of each kind are in the bank?

39. On an airplane flight for which first-class fare is $80 and tourist fare is $64 there were 42 passengers. If receipts for the flight totaled $2880, how many first-class and how many tourist passengers were on the flight?

40. An ice cream vendor bought 50 ice cream bars—chocolate-covered bars at 18¢ each and sandwich bars at 15¢ each. If the total cost was $8.10, how many bars of each kind did the vendor purchase?

41. One brand of coffee sells for $1.20 per pound and a second brand sells for $1.40 per pound. A restaurant buyer paid $28.80 for some of each brand. How many pounds of each brand did he buy if he bought 2 pounds more of the expensive brand than the less expensive brand?

42. A hardware clerk sold some 12-cent and 16-cent bolts for a total of $7.80. How many of each were sold if there were three times as many 12-cent bolts as there were 16-cent bolts?

43. A theater charged \$4.00 for an adult ticket and \$2.50 for children under 12 years old. One evening 82 people viewed a movie. If the receipts totaled \$310, how many of each kind of ticket were sold?

44. The first-class fare on an airplane flight is \$240 and the tourist fare is \$150. If 48 passengers paid a total of \$8100 for the flight, how many of each ticket were sold?

■ *See Example* 8 *for Problems* 45–52.

45. Two investments produce an annual income of \$920. One investment earns 10% and the other earns 12%. How much is invested at each rate if the amount invested at 12% is \$4000 more than the amount invested at 10%?

46. An amount of money is invested at 8% and \$3000 more than that amount is invested at 10%. How much is invested at each rate if the total income is \$1110?

47. An amount of \$42,000 is invested, part at 11% and the remainder at 9%. Find the amount invested at each rate if the yearly incomes on the investments are equal.

48. An amount of money is invested at 8% and twice that amount is invested at 12%. How much is invested at each rate if the total income is \$1280?

49. A woman has invested \$8000, part in a bank at 10% and part in a savings and loan association at 12%. If her annual return is \$844, how much has she invested at each rate?

50. A sum of \$2400 is split between an investment in a mutual fund paying 14% and one in corporate bonds paying 11%. If the return on the 14% investment exceeds that on the 11% investment by \$111 per year, how much is invested at each rate?

51. If \$3000 is invested in bonds at 8%, how much additional money must be invested in stocks paying 13% to make the earnings on the total investment 10%?

52. If \$6000 is invested in bonds at 9%, how much additional money must be invested in stocks at 12% to earn a return of 11% on the total investment?

■ *See Example* 9 *for Problems* 53–60.

53. How many quarts of a 10% solution of acid should be added to 20 quarts of a 40% solution of acid to obtain a 30% solution of acid?

54. How many quarts of a 30% salt solution must be added to 50 quarts of a 10% salt solution to obtain a 25% salt solution?

55. How many ounces of an alloy containing 40% aluminum must be melted with an alloy containing 60% aluminum to obtain 60 ounces of an alloy containing 50% aluminum?

56. How many pounds of an alloy containing 30% copper must be melted with an alloy containing 10% copper to obtain 12 pounds of an alloy containing 25% copper?

57. How many liters of a 30% salt solution must be added to 40 liters of a 12% salt solution to obtain a 20% solution?

58. How many grams of an alloy containing 45% silver must be melted with an alloy containing 60% silver to obtain 40 grams of an alloy containing 48% silver?

59. How much pure alcohol should be added to 12 liters of a 45% solution to obtain a 60% solution?

60. How many liters of a 20% sugar solution should be added to 40 liters of a 32% solution to obtain a 28% solution?

3.3

INEQUALITIES; INTERVAL NOTATION

Solution of an inequality

Expressions such as

$$x + 3 \geqslant 10 \quad \text{and} \quad \frac{-2y - 3}{3} < 5$$

are called **inequalities.** For appropriate values of the variable, one member of an inequality represents a real number that is less than ($<$), less than or equal to ($\leqslant$), greater than or equal to ($\geqslant$), or greater than ($>$) the real number represented by the other member.

Any element of the replacement set of the variable for which an inequality is true is called a **solution**, and the set of all solutions of an inequality is called the **solution set** of the inequality.

Equivalent inequalities

As in the case of equations, we shall solve a given inequality by generating a series of equivalent inequalities (inequalities that have the same solution set) until we arrive at one whose solution set is obvious. To do this we shall need some fundamental properties of inequalities. Notice that

$$2 < 3,$$
$$2 + 5 < 3 + 5,$$

and

$$2 - 5 < 3 - 5.$$

Figures 3.1 a and b (page 82) demonstrate that the addition of 5 or -5 to each member of $2 < 3$ simply shifts the graphs of the members the same number of units to the right or left on the number line, with the order of the members left unchanged. This will be the case for the addition of any real number to each member of an inequality. Since for any real value of the variable for which an expression is defined the expression represents a real number, we generalize this idea and assert the following.

FIGURE 3.1

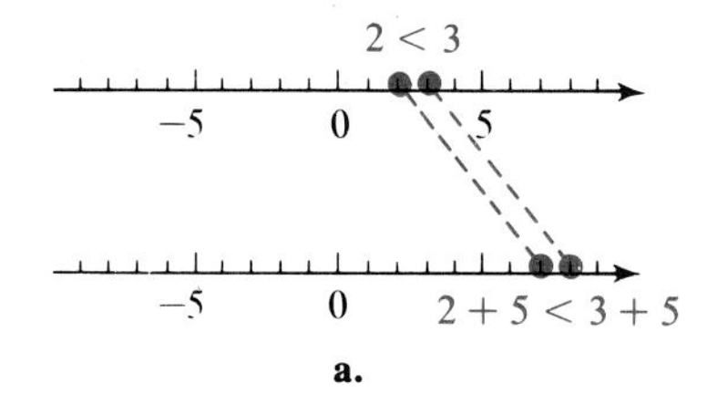

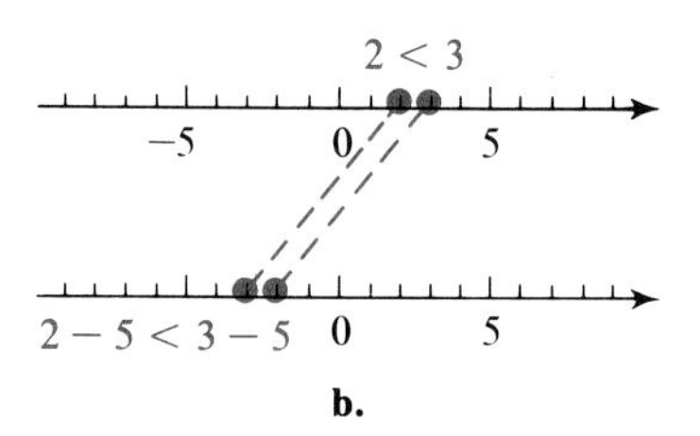

> 1. *The addition of the same expression representing a real number to each member of an inequality produces an equivalent inequality in the* **same sense.**

Next, if we multiply each member of

$$2 < 3$$

by 2, we have

$$4 < 6,$$

where the products form an inequality in the same sense. If, however, we multiply each member of

$$2 < 3$$

by -2, we have

$$-4 > -6,$$

where the inequality is in the opposite sense. Figure 3.2 illustrates this. Multiplying each member of $2 < 3$ by 2 simply moves the graph of each member out twice as far in a positive direction (part a). Multiplying by -2 also doubles

FIGURE 3.2

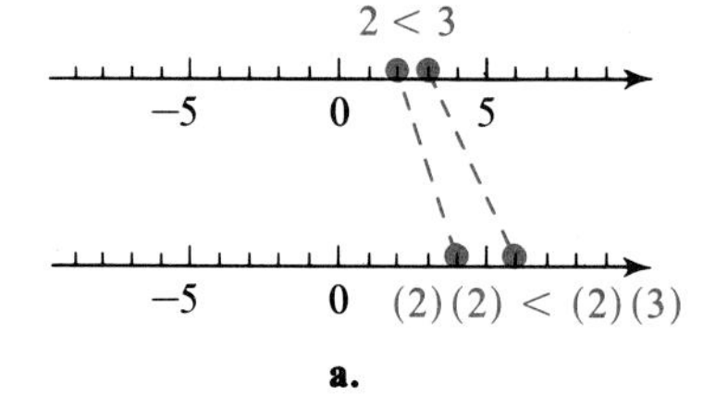

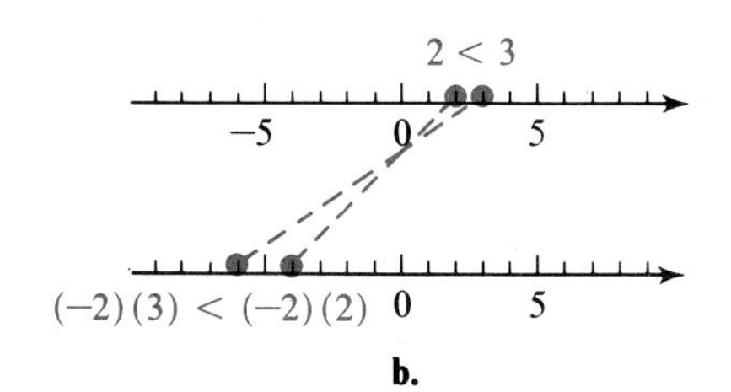

the absolute value of each member, but the products are negative and the sense of the inequality is reversed as shown in part b. In general, we assert the following.

2. *If each member of an inequality is multiplied by the same expression representing a positive number, the result is an equivalent inequality in the* **same sense**.
3. *If each member of an inequality is multiplied by the same expression representing a negative number, the result is an equivalent inequality in the* **opposite sense.**

Statements 1, 2, and 3 above can be expressed in symbols as follows.

The inequality

$$a < b$$

is equivalent to

$$a + c < b + c \qquad (1)$$
$$a \cdot c < b \cdot c \quad \text{for} \quad c > 0, \qquad (2)$$
$$a \cdot c > b \cdot c \quad \text{for} \quad c < 0. \qquad (3)$$

These relationships are also true with $<$ replaced by $\leqslant$ and $>$ replaced by $\geqslant$.

Note that none of these assertions permits multiplying by 0 and that variables in expressions used as multipliers are restricted from values for which the expression vanishes or is not defined.

These properties can be applied to solve first-degree inequalities in the same way the equality properties are applied to solve first-degree equations.

EXAMPLE 1 To solve the inequality

$$3x - 9 < 8,$$

we add 9 to each member, giving us

$$3x < 17.$$

Then we multiply each member by $\frac{1}{3}$ to obtain

$$x < \frac{17}{3}.$$

The solution set is $\{x \mid x < 17/3\}$. ■

Graphs of solution sets

The solution set of the inequality in the above example is a half-line and can be graphed as shown in Figure 3.3, where the red line represents the points whose coordinates are in the solution set. The end-point is shown as an open dot to show that $17/3$ is not included in the set.

EXAMPLE 2 To solve the inequality

$$x + 4 \leq 18 + 3x$$

we first add -4 and $-3x$ to each member to obtain

$$-2x \leq 14.$$

Then we multiply each member by $-\frac{1}{2}$ and *reverse* the sense of the inequality to get

$$x \geq -7.$$

The solution set is $\{x \mid x \geq -7\}$. The end-point in the graph is shown as a solid dot to show that -7 is included in the set.

■

We can use the properties above to solve continued inequalities of the form shown in the following example.

EXAMPLE 3 To solve

$$4 < x + 4 \leq 6,$$

we add -4 to each member to obtain

$$4 + (-4) < x + 4 + (-4) \leq 6 + (-4)$$
$$0 < x \leq 2.$$

The solution set is $\{x \mid 0 < x \leq 2\}$. The left end-point of its graph is an open dot and the right end-point is a closed dot.

■

Intersection of two sets

Sometimes two conditions of inequality are placed on a variable. For example, we may want to consider $\{x \mid -2 < x \text{ and } x \leq 5\}$, that is, the numbers in $\{x \mid -2 < x\}$ and also in $\{x \mid x \leq 5\}$. We call the resulting set the **intersection** of the two given sets. In general, we have the following.

> If A and B are sets, the **intersection** of A and B consists of all those numbers that are in set A and also in set B. We designate this set by $\boldsymbol{A \cap B}$, read "the intersection of sets A and B."

The set $\{x \mid -2 < x \text{ and } x \leq 5\}$ may be written as

$$\{x \mid -2 < x\} \cap \{x \mid x \leq 5\}.$$

The graph of the intersection is shown in Figure 3.4.

FIGURE 3.4

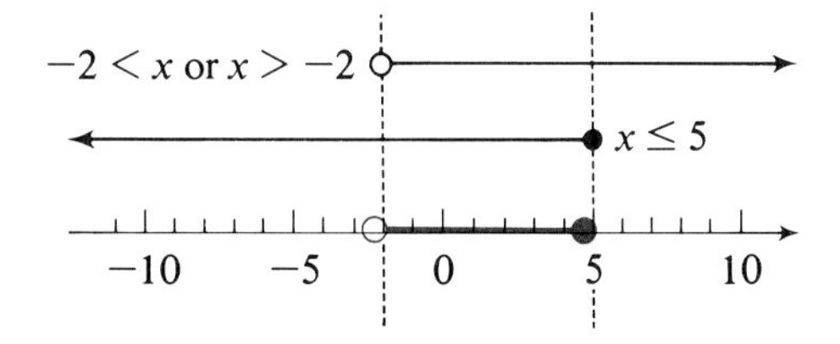

EXAMPLE 4 To graph

$$\{x \mid x > -4\} \cap \{x \mid x + 1 \leqslant 4\}$$

we first graph each inequality. Then we locate the interval common to both number lines.

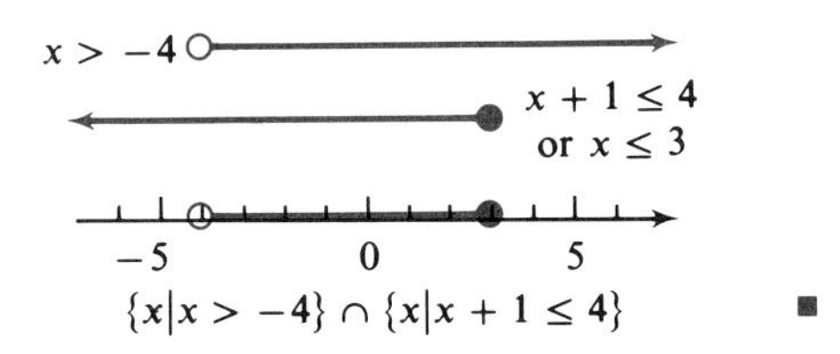

■

Note that in Example 4 the intersection of the two sets can be written as the single set

$$\{x \mid -4 < x \leqslant 3\}.$$

Interval notation

A special notation is used sometimes to describe an interval of real numbers. The symbols (a, b) are used for the interval that includes all real numbers *between* a and b, where $a < b$. The symbol "[" is used instead of "(" if a is included in the interval, and "]" is used instead of ")" if b is included in the interval. That is, (a, b) means all x such that $a < x < b$ and $[a, b]$ means all x such that $a \leqslant x \leqslant b$. For example:

a. $\{x \mid 3 < x < 5\} = (3, 5)$
b. $\{y \mid 6 \leqslant y \leqslant 7\} = [6, 7]$
c. $\{z \mid 1 < z \leqslant 3\} = (1, 3]$
d. $\{x \mid -2 \leqslant x < 4\} = [-2, 4)$

In the event an interval includes all real numbers greater than a given real number or less than a given real number, we incorporate the symbols $+\infty$ and $-\infty$ in the notation. These symbols are read "positive infinity" and "negative infinity," respectively. They do not denote real numbers. For example:

a. $\{x \mid x \geqslant 2\} = [2, +\infty)$
b. $\{y \mid y > -4\} = (-4, +\infty)$
c. $\{z \mid z \leqslant 0\} = (-\infty, 0]$
d. $\{t \mid t < 6\} = (-\infty, 6)$

Because an interval such as $[-5, 8]$ contains both of its endpoints, it is called a **closed interval.** Intervals such as $(3, 18]$ and $[-5, -2)$ are called **half-open intervals,** and intervals such as $(3, 8)$ or $(-\infty, 2)$ are called **open intervals.**

We have now used three different ways to represent an inequality, as shown in the chart below.

Set	Interval	Graph
$\{x \mid -2 < x \leqslant 3\}$	$(-2, 3]$	−5 0 5
$\{x \mid x < 4\}$	$(-\infty, 4)$	−5 0 5
$\{x \mid x \geqslant -2\}$	$[-2, +\infty)$	−5 0 5

Parentheses and brackets used in interval notation can also be used when constructing number lines. For example, the three graphs above would appear as

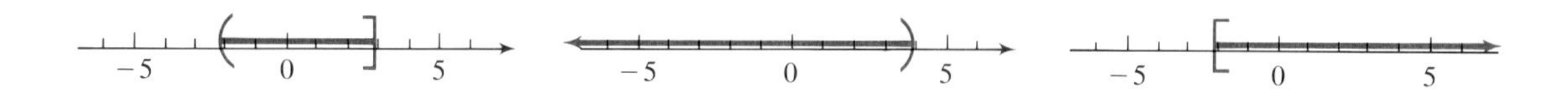

Disjoint sets

Sets that do not have any numbers in common are called **disjoint sets.** If A and B are sets, and

$$A \cap B = \varnothing,$$

then A and B are disjoint. For example:

a. $\{2, 3\} \cap \{4, 5\} = \varnothing$

b. $\{x \mid x < 1\} \cap \{x \mid x > 3\} = \varnothing$

c. $(-5, -1) \cap (1, 5) = \varnothing$

d. $(-\infty, 3) \cap (3, +\infty) = \varnothing$

Word problems

The suggestions for solving word problems in Section 3.2, where the mathematical model is an equation, are also applicable for problems where the mathematical model is an inequality, except that symbols such as $\leqslant$ and $<$ replace the symbol $=$.

EXAMPLE 5 A student must have an average of 80% or more and less than 90% on five tests in a course to receive a B. Her grades on the first four tests were 98%, 76%, 86%, and 92%. What grade on the fifth test would give her a B in the course?

Solution

Step 1 We first express the quantity asked for in a simple phrase.

Step 2 Then we represent the quantity symbolically.

Step 1 → Grade (in percent) on the fifth test: x ← Step 2

Step 3 We then write an inequality expressing the fact that the average of the five tests is greater than or equal to 80 and less than 90.

$$80 \leqslant \frac{98 + 76 + 86 + 92 + x}{5} < 90.$$

Step 4 Solving the inequality we obtain

$$400 \leqslant 352 + x < 450$$
$$48 \leqslant x < 98.$$

Step 5 The solution set is $\{x \mid 48 \leqslant x < 98\}$. Therefore, any grade equal to or greater than 48 and less than 98 would give the student a B in the course. ■

EXERCISE 3.3

A

■ *Solve and graph each solution set. See Examples* 1 *and* 2.

1. $3x < 6$
2. $x + 7 > 8$
3. $x - 5 \leqslant 7$
4. $2x - 3 < 4$
5. $3x - 2 > 1 + 2x$
6. $2x + 3 \leqslant x - 1$
7. $\dfrac{2x - 6}{3} > 0$
8. $\dfrac{2x - 3}{2} \leqslant 5$
9. $\dfrac{5x - 7x}{3} > 4$
10. $\dfrac{x - 3x}{5} \leqslant 6$
11. $\dfrac{2x - 5x}{2} \leqslant 7$
12. $\dfrac{x - 6x}{2} < -20$
13. $5x + 10 > 3x - 12$
14. $6x + 3 \geqslant 5x - 6$
15. $8x + 4 - 4x \geqslant 4$
16. $8 - 2x < 5x + 2$

■ *Solve and graph each solution set. See Example 3.*

17. $4 < x - 2 < 8$
18. $0 \leq 2x \leq 12$
19. $-3 < 2x + 1 \leq 7$
20. $2 \leq 3x - 4 \leq 8$
21. $6 < 4 - x < 10$
22. $-3 < 3 - 2x < 9$

■ *Graph each set. See Example 4.*

23. $\{x \mid x < 2\} \cap \{x \mid x > -2\}$
24. $\{x \mid x \leq 5\} \cap \{x \mid x \geq 1\}$
25. $\{x \mid x + 1 \leq 3\} \cap \{x \mid -x - 1 \leq 3\}$
26. $\{x \mid 2x - 3 < 5\} \cap \{x \mid -2x + 3 < 5\}$
27. $\{x \mid x - 7 < 2x\} \cap \{x \mid -3x \geq 4x + 28\}$
28. $\{x \mid 9 \leq 3x < 21\} \cap \{x \mid -10 < 5x \leq 15\}$

■ *Solve each word problem. Use the five steps suggested on page 69. See Example 5.*

29. A student must have an average of 80% or more and less than 90% on 5 tests to receive a B in a course. What grade on the fifth test would give the student a B if her grades on the first four tests were 78%, 64%, 88%, and 76%?
30. In the preceding example, what grade on the final examination would give a student a B if her grades on the first four hourly tests were 72%, 68%, 84%, and 70%, and the final examination counted for two hourly tests?
31. The Fahrenheit and Celsius temperature scales are related by the formula

$$C = \frac{5F - 160}{9}.$$

Within what range must the temperature be in degrees Fahrenheit for the temperature to lie between 30°C and 40°C?

32. Within what range must the temperature be in degrees Fahrenheit for the temperature to lie between −10°C and 20°C?

B

■ *Graph the given intervals on the same line graph, and represent the numbers involved using set-builder notation.*

33. $[-8, 2]$; $(3, 7]$
34. $(-4, 0]$; $(2, +\infty)$
35. $[-7, -3]$; $(0, 4]$
36. $(-\infty, 0)$; $(0, +\infty)$
37. $(-5, -3]$; $(-2, 0]$; $(1, 3)$
38. $(-\infty, -4]$; $(-2, 0]$; $(2, +\infty)$

■ *Write each non-empty interval as a single interval. If the sets are disjoint, so state.*

39. $[-8, -4] \cap [-6, 2)$
40. $(-3, 0] \cap [-1, 4]$
41. $[-5, -3) \cap [-2, 0)$
42. $[-6, 4] \cap (6, 7]$
43. $(0, 3) \cap \varnothing$
44. $(-4, 0] \cap \varnothing$
45. $(-3, 2) \cap (-6, 4)$
46. $(-5, -3) \cap (-7, 4)$

3.4

EQUATIONS AND INEQUALITIES INVOLVING ABSOLUTE VALUE

Equations involving absolute value

In Section 1.4, we defined the absolute value of a real number by

$$|x| = \begin{cases} x, & \text{if } x \geq 0 \\ -x, & \text{if } x < 0 \end{cases}$$

and interpreted it in terms of distance on a number line. For example, by definition $|-5| = 5$; but 5 also denotes the distance between the graph of -5 and the origin. More generally, we have

$$|x - a| = \begin{cases} x - a, & \text{if } x - a \geq 0, \text{ or equivalently, if } x \geq a \\ -(x - a), & \text{if } x - a < 0, \text{ or equivalently, if } x < a \end{cases}$$

and $|x - a|$ can be interpreted on a line graph as denoting the distance the graph of x is located from the graph of a, as shown in Figure 3.5. Since distance is nonnegative, the smaller value is subtracted from the larger.

FIGURE 3.5

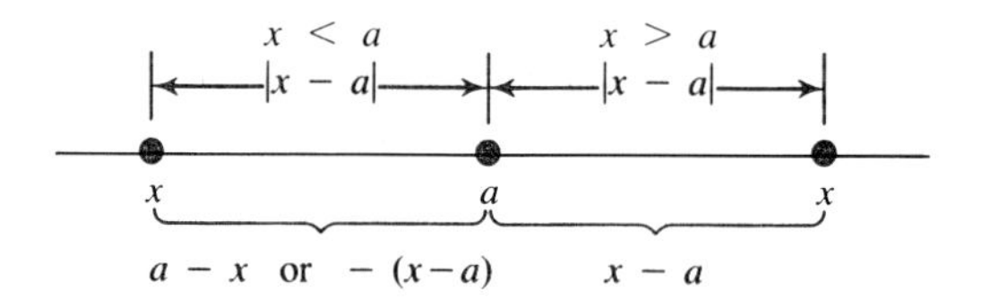

We can solve equations of the form

$$|x - a| = b$$

or, more generally, of the form

$$|ax - b| = c$$

by appealing to the definition of absolute value. Formally, we have the following.

> $|ax - b| = c$ *is equivalent to the joint statement*
>
> $$ax - b = c \quad \textit{or} \quad -(ax - b) = c. \qquad (1)$$

EXAMPLE 1 Solve $|2x - 3| = 5$.

Solution This equation implies that

$$\begin{aligned} 2x - 3 &= 5 \\ 2x &= 8 \\ x &= 4 \end{aligned} \quad \text{or} \quad \begin{aligned} -(2x - 3) &= 5 \\ 2x - 3 &= -5 \\ 2x &= -2 \\ x &= -1 \end{aligned}$$

Hence, the solution set is $\{4, -1\}$. ■

Union of two sets

Note that this set is really a combination of the member of $\{4\}$ and the member of $\{-1\}$. We call such a combination the **union** of the two sets.

> If A and B are sets, the **union** of A and B consists of all those members that are either in set A or in set B or in both. We denote this set by $A \cup B$, read "the union of sets A and B."

In the above example, we have

$$\{4\} \cup \{-1\} = \{4, -1\}.$$

Inequalities involving absolute value

Consider the graph of the solution set of $|x| < 5$ in Figure 3.6. In this simple case, the fact that this is indeed the graph of the inequality can be verified by inspection. Note that if $x \geqslant 0$, then $x < 5$, and if $x < 0$, then $x > -5$. These two conditions can be expressed as the continued inequality $-5 < x < 5$.

FIGURE 3.6

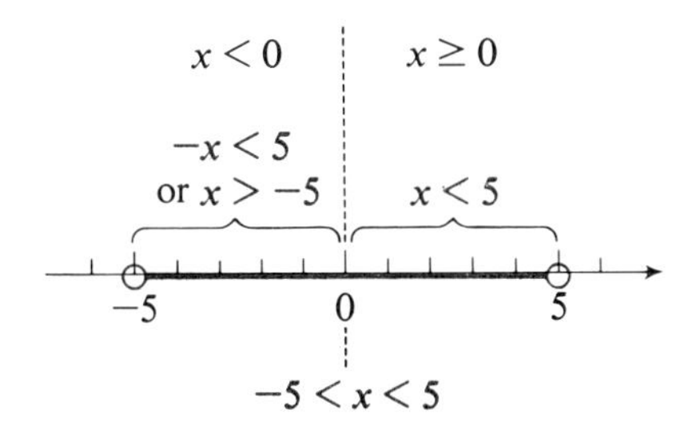

Now consider the graph of the solution set of $|x| > 5$ in Figure 3.7, which can also be verified by inspection.

FIGURE 3.7

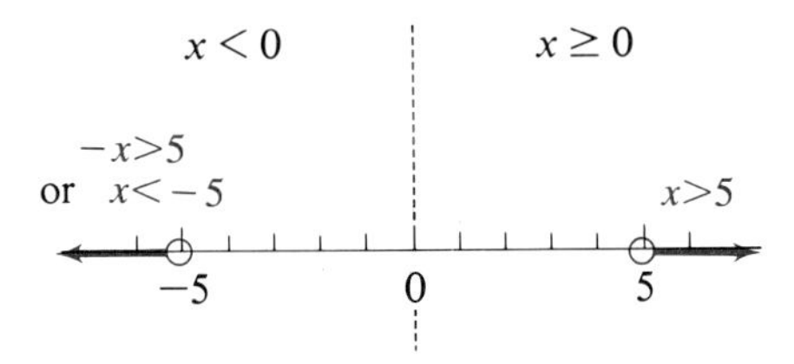

Note that in this case the two conditions $x < -5$ and $x > 5$ cannot be written as a continued inequality.

The above examples suggest the following properties which can be used to solve inequalities that involve absolute value.

$$|ax - b| < c \quad \textit{is equivalent to} \quad -c < ax - b < c, \tag{2}$$

and

$$|ax - b| > c \quad \textit{is equivalent to} \quad ax - b > c \quad \textit{or} \quad -(ax - b) > c. \tag{3}$$

EXAMPLE 2 Solve $|x + 1| < 3$ and graph the solution set.

Solution From Property 2 above, this inequality is equivalent to

$$-3 < x + 1 < 3,$$

from which we have

$$-4 < x < 2.$$

Hence, the solution set is $(-4, 2)$, with the graph as shown.

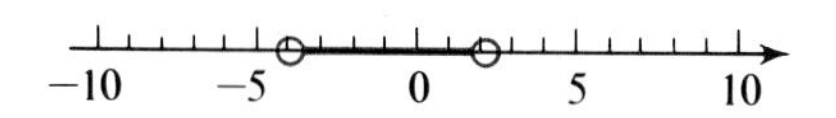

■

EXAMPLE 3 Solve $|x + 1| \geqslant 3$ and graph the solution set.

Solution From Property 3 above, the inequality is equivalent to the two inequalities

$$x + 1 \geqslant 3 \quad \text{or} \quad -(x + 1) \geqslant 3,$$

from which we have

$$x \geqslant 2 \quad \text{or} \quad x \leqslant -4.$$

Hence, the solution set is $(-\infty, -4] \cup [2, +\infty)$, with the graph as shown.

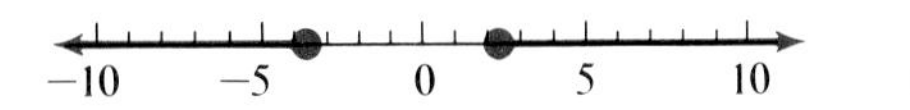

■

EXERCISE 3.4

A

■ *Solve. See Example* 1.

1. $|x| = 5$
2. $|x| = 7$
3. $|x - 4| = 9$
4. $|x - 3| = 7$
5. $|2x + 1| = 13$
6. $|3x - 1| = 5$
7. $|4 - 3x| = 1$
8. $|6 - 5x| = 4$
9. $|4x + 3| = 0$
10. $|3x - 7| = 0$
11. $|2x - 3| = 3$
12. $|3x - 2| = 2$

■ *Solve and graph each solution set. See Example* 2.

13. $|x| < 2$
14. $|x| < 5$
15. $|x + 3| \leqslant 4$
16. $|x + 1| \leqslant 8$
17. $|2x - 5| < 3$
18. $|2x + 4| < 6$
19. $|4 - x| \leqslant 8$
20. $|5 - 2x| \leqslant 15$

■ *Solve and graph each solution set. See Example* 3.

21. $|x| > 3$
22. $|x| \geqslant 5$
23. $|x - 2| > 5$
24. $|x + 5| > 2$
25. $|3 - 2x| \geqslant 7$
26. $|4 - 3x| > 10$

■ *Solve each equation or inequality. See Examples* 2 *and* 3.

27. $|x + 1| \leqslant 7$
28. $|x - 2| < 5$
29. $|2x - 6| > 3$
30. $|3x + 5| \geqslant 2$
31. $|5x + 2| = 0$
32. $|2x - 6| = 0$
33. $|4x - 1| \leqslant 6$
34. $|3x + 2| > 5$

B

■ *Write each interval as a single interval.*

35. $([-5, -2] \cup [-3, 4]) \cap [0, 6]$
36. $([-6, -1] \cup [-1, 2]) \cap [-3, 5]$
37. $((-4, 2) \cup (-2, 3)) \cap [-3, 4]$
38. $((-7, 0) \cup (-2, 1)) \cap [-3, 2]$
39. $((-3, -1) \cup (-4, 0)) \cap [-2, 0]$
40. $((-1, 5) \cup (-3, -1]) \cap [-4, 4]$

CHAPTER SUMMARY

[3.1] A replacement for the variable in an equation that results in a true statement is called a **solution** of the equation; the set of all solutions is called the **solution set**.

Equations that have identical solution sets are called **equivalent equations.** A conditional equation can be solved by applying elementary transformations to generate a sequence of equivalent equations until an equation is obtained that can be solved by inspection.

The *equation*

$$a = b$$

is equivalent to

$$a + c = b + c,$$
$$a \cdot c = b \cdot c.$$

[3.2] The first step in a systematic approach in constructing a mathematical model for a word problem is to write separate word phrases for each quantity that is to be found. Then one quantity is represented by a variable and any other quantities are represented in terms of this variable.

[3.3] An inequality can be solved by producing equivalent inequalities until one is obtained that can be solved by inspection.

The inequality

$$a < b$$

is equivalent to

$$a + c < b + c$$
$$a \cdot c < b \cdot c \quad \text{for} \quad c > 0,$$
$$a \cdot c > b \cdot c \quad \text{for} \quad c < 0.$$

The **intersection** of two sets, $A \cap B$, is the set containing all members that are in both set A and set B. If the intersection is $\varnothing$, then A and B are **disjoint sets.**

An **interval** of real numbers is an infinite set. An interval may be **closed** (two endpoints), **half-open** (one endpoint), or **open** (no endpoints).

[3.4] $|ax - b| = c$ is equivalent to $ax - b = c$ or $-(ax - b) = c$;
$|ax - b| < c$ is equivalent to $-c < ax - b < c$;
$|ax - b| > c$ is equivalent to $ax - b > c$ or $-(ax - b) > c$.

The **union** of two sets, $A \cup B$, is the set containing all members that are in either set A or set B or both.

REVIEW EXERCISES

A

[3.1]

■ *Solve.*

1. $2x - 6 = 4x - 8$
2. $0.40x = 240$
3. $2(x - 3) + 4(x + 2) = 6$
4. $2x - (x + 3) = 2(x + 1)$
5. $2[2x - (x + 3) + 4] = 12$
6. $-[x - (3x - 2) - 2] = 16$
7. $-[x^2 - (x - 3)^2] = 0$
8. $-[2(x + 1)^2 - 2x^2] = 0$
9. $0.30(y + 2) = 2.10$
10. $0.06y + 0.04(y + 1000) = 60$
11. Solve $3N = 5t - 3c$ for t, given that $N = 10$ and $c = 5$.
12. Solve $C = 10 + 2p - 2t$ for p, given that $C = 38$ and $t = 3$.
13. Solve $l = a + nd - d$ for n.
14. Solve $2s = at + k$ for a.
15. Solve $S = 2\pi(R - r)$ for R.
16. Solve $9C = 5(F - 32)$ for F.

[3.2]

■ *Solve each problem. Use the five steps suggested on page 69.*

17. A tennis player made 26% (0.26) of her first serves in a match. If her first serve was good 52 times, how many times did she serve?
18. A part-time secretary has a take-home pay of \$180 per week. What is her total salary if 20% has been deducted for income taxes?
19. Computer A sells for \$120 more than Computer B. How much does each computer cost if their total cost is \$460?
20. One candidate in a student body election for president received 120 votes more than a second candidate and 160 votes more than a third candidate. How many votes did each candidate receive if there were 440 votes cast?

21. A hardware clerk sold some 10-cent and 13-cent bolts for a total of \$2.58. There were three times as many 10-cent bolts as 13-cent bolts sold. How many of each were sold?

22. A postal clerk sells some 2-cent and 22-cent stamps for a total of \$7.32. If there are six more 2-cent than 22-cent stamps, how many of each were sold?

23. Two investments produce an annual interest of \$320. An amount of \$1000 more is invested at 11% than at 10%. How much is invested at each rate?

24. An amount of \$3600 is invested, part at 8% and the remainder at 10%. How much is invested at each rate if the early income on each investment is the same?

25. How many quarts of a 20% solution of acid should be added to 10 quarts of a 30% solution to obtain a 25% solution?

26. How many pounds of an alloy containing 60% copper must be melted with an alloy containing 20% copper to obtain 4 pounds of an alloy containing 35% copper?

[3.3]

■ *Solve and graph each solution set. Represent the solution set in set-builder notation and interval notation.*

27. $x - 3 \leq 24$

28. $6x - 6 \leq 2x$

29. $1 < 2x - 3 \leq 7$

30. $-5 < 4 - 3x < 1$

■ *Graph.*

31. $\{x \mid x > -5\} \cap \{x \mid x < 2\}$

32. $\{x \mid x + 2 \geq -1\} \cap \{x \mid 0 < x < 4\}$

[3.4]

■ *Solve.*

33. a. $|2x + 1| = 7$ b. $|3 - 4x| = 8$

■ *Solve and graph each solution set.*

34. a. $|x + 3| \geq 5$ b. $|x - 2| < 5$

B

■ *Represent the numbers involved using set-builder notation.*

35. a. $[-4, 5)$ b. $(-2, 6]$

36. a. $[-5, -3]$ b. $(4, 8)$

■ *Write each interval as a single interval. If the sets are disjoint, so state.*

37. $[-4, 5) \cap [0, 8]$

38. $(-2, 6] \cap [-6, 0)$

■ *Write each interval as a single interval.*

39. $([-3, 0] \cup [-2, 3]) \cap [-1, 5)$

40. $((-4, 2] \cup [-1, 5)) \cap (-6, 0]$

4 Factoring Polynomials

In Chapter 3 we used the distributive property to write equivalent expressions without parentheses for products of polynomials containing parentheses. In this chapter we will consider methods to reverse this process. This reverse process, called **factoring**, will enable us to solve a wide variety of problems. In this chapter the factoring process will enable us to solve some second-degree equations and hence to solve some applied problems that can be modeled by such equations.

4.1

FACTORING MONOMIALS FROM POLYNOMIALS

In Section 1.5, we factored natural numbers that were not prime as the product of prime factors. Now we shall see how the distributive property in the form

$$ax + bx + cx + dx = (a + b + c + d)x$$

furnishes us a means of writing a polynomial as a single term comprised of two or more factors. By the distributive property,

$$3x^2 + 6x = 3x(x + 2).$$

Of course, we can also write

$$3x^2 + 6x = 3(x^2 + 2x)$$

or

$$3x^2 + 6x = 3x^2\left(1 + \frac{2}{x}\right) \qquad (x \neq 0)$$

or any other of an infinite number of such expressions. We are, however, primarily interested in factoring a polynomial into a unique form (except for signs and order of factors) referred to as the **completely factored form.** A polynomial with integral coefficients is in completely factored form if:

1. It is written as a product of polynomials with integral coefficients.
2. No polynomial—other than a monomial—can be further factored.

The restriction that the factors be polynomials means that all the variables involved have exponents from $\{1, 2, 3, \ldots\}$. Restricting the coefficients to integers prohibits such factorizations as

$$x + 3 = 3\left(\frac{1}{3}x + 1\right).$$

Because the choice of signs and order of factors is arbitrary, the factored form of an expression that seems most "natural" should be used, although this is admittedly not always easy to determine. For instance, the forms

$$a(1 - x - x^2) \quad \text{and} \quad -a(x^2 + x - 1)$$

are equivalent, but it is difficult to decide which is more "natural."

Common monomial factors can be factored from a polynomial by first identifying such common factors and then writing the resultant factored expression. For example, observe that the polynomial

$$6x^3 + 9x^2 - 3x$$

contains the monomial $3x$ as a factor of each term. We therefore write

$$6x^3 + 9x^2 - 3x = 3x(\qquad\qquad)$$

and insert within the parentheses the appropriate polynomial factor. This factor can be determined by inspection. We ask ourselves for the monomials that multiply $3x$ to yield $6x^3$, $9x^2$, and $-3x$. The final result appears as

$$6x^3 + 9x^2 - 3x = 3x(2x^2 + 3x - 1).$$

Checking the result of factoring

We can always check the result of factoring an expression by multiplying the factors. In the example above,

$$3x(2x^2 + 3x - 1) = 6x^3 + 9x^2 - 3x.$$

EXAMPLE 1

a. $18x^2y - 24xy^2$
$= 6xy(\underline{?} - \underline{?})$
$= 6xy(3x - 4y);$
because
$6xy(3x - 4y)$
$= 18x^2y - 24xy^2$

b. $y(x - 2) + z(x - 2)$
$= (x - 2)(\underline{?} + \underline{?})$
$= (x - 2)(y + z);$
because
$(x - 2)(y + z)$
$= y(x - 2) + z(x - 2)$

c. $2(x + 4)^2 - 2(x + 4)$
$= 2(x + 4)(\underline{?} - \underline{?})$
$= 2(x + 4)(x + 4 - 1)$
$= 2(x + 4)(x + 3)$

d. $3x(x - 2)^2 + (x - 2)^2$
$= (x - 2)^2(\underline{?} + \underline{?})$
$= (x - 2)^2(3x + 1)$ ■

Notice that complete factorization of monomial factors is not required. Thus, in Example 1a above, it is not necessary that the factor $6xy$ be written $2 \cdot 3xy$ in order for the expression to be considered completely factored.

Factoring $a - b$

One particularly useful factorization is of the form

$$\begin{aligned} a - b &= (-1)(-a + b) \\ &= (-1)(b - a) \\ &= -(b - a). \end{aligned}$$

Hence, we have the following important relationship.

$$\boldsymbol{a - b = -(b - a).}$$

That is, $a - b$ and $b - a$ are negatives of each other.

EXAMPLE 2 a. $3x - y = -(y - 3x)$ b. $a - 2b = -(2b - a)$ ■

The following examples show the factorization of polynomials in which exponents include variables.

EXAMPLE 3

a. $x^{3n} + x^n$
$= x^n(\underline{?} + \underline{?})$
$= x^n(x^{2n} + 1)$

b. $x^{n+2} - 2x^2$
$= x^n \cdot x^2 - 2x^2$
$= x^2(\underline{?} - \underline{?})$
$= x^2(x^n - 2)$

c. $x^{2n} + x^{n+1} - x^n$
$= x^{2n} + x^n x - x^n$
$= x^n(\underline{?} + \underline{?} - \underline{?})$
$= x^n(x^n + x - 1)$

■

EXERCISE 4.1

A

■ *Factor completely. Check by multiplying factors. See Example 1.*

1. $2x + 6$
2. $3x - 9$
3. $4x^2 + 8x$
4. $3x^2y + 6xy$
5. $3x^2 - 3xy + 3x$
6. $x^3 - x^2 + x$
7. $24a^2 + 12a - 6$
8. $15r^2s + 18rs^2 - 3rs$
9. $2x^4 - 4x^2 + 8x$
10. $3n^4 - 6n^3 + 12n^2$
11. $12z^4 + 15z^3 - 9z^2$
12. $2x^2y^2 - 3xy + 5x^2$
13. $ay^2 + aby + ab$
14. $x^2y^2z^2 + 2xyz - xz$
15. $3m^2n - 6mn^2 + 12mn$
16. $6x^2y - 9xy^2 + 12x$
17. $15a^2c^2 - 12ac + 6ac^3$
18. $14xy + 21x^2y^2 - 28xyz$
19. $a(a + 3) + b(a + 3)$
20. $b(a - 2) + a(a - 2)$
21. $2x(x + 3) - y(x + 3)$
22. $y(y - 2) - 3x(y - 2)$
23. $2y(a + b) - x(a + b)$
24. $3x(2a - b) + 4y(2a - b)$
25. $(x + 3) + (x + 3)^2$
26. $(x - 6)^2 + (x - 6)$
27. $4(x - 2)^2 - 8(x - 2)$
28. $6(x + 1) - 3(x + 1)^2$
29. $x(2x - 1)^2 + (2x - 1)^2$
30. $(2x + 3)^2 + x(2x + 3)^2$
31. $x(x - 5)^2 - x^2(x - 5)$
32. $x^2(x + 3) - x(x + 3)^2$
33. $4(x - 1)(x + 3) + 2(x + 1)(x + 3)$
34. $3(x + 2)(x - 4) + 6(x + 2)(x + 1)$
35. $(x - 1)^2 - (x - 1)(x + 3)$
36. $(x + 2)^2 - (x + 2)(x - 1)$

■ *Supply the missing factors or terms. See Example 2.*

37. $7 - r = -(\underline{?} - \underline{?})$
38. $3m - 2n = -(\underline{?} - \underline{?})$
39. $2a - b = -(\underline{?} - \underline{?})$
40. $r^2 - s^2t^2 = -(\underline{?} - \underline{?})$
41. $-2x + 2 = -2(\underline{?})$
42. $-6x - 9 = -3(\underline{?})$
43. $-ab - ac = \underline{?}(b + c)$
44. $-a^2 + ab = \underline{?}(a - b)$
45. $2x - 1 = -(\underline{?})$
46. $x^2 - 3 = -(\underline{?})$
47. $x - y + z = -(\underline{?})$
48. $3x + 3y - 2z = -(\underline{?})$

B

■ *Factor completely. (Assume that variables in the exponents denote natural numbers.) See Example 3.*

49. $x^{2n} - x^n$
50. $x^{4n} + x^{2n}$
51. $x^{3n} - x^{2n} - x^n$
52. $y^{4n} + y^{3n} + y^{2n}$
53. $x^{n+2} + x^n$
54. $x^{n+2} + x^{n+1} + x^n$
55. $-x^{2n} - x^n = \underline{?}(x^n + 1)$
56. $-x^{5n} + x^{2n} = \underline{?}(x^{3n} - 1)$
57. $-x^{a+1} - x^a = -x^a(\underline{?} + \underline{?})$
58. $-y^{a+2} + y^2 = -y^2(\underline{?} - \underline{?})$

4.2

FACTORING QUADRATIC POLYNOMIALS

A common type of factoring involves one of the following quadratic (second-degree) binomials or trinomials.

$$x^2 + (a + b)x + ab = (x + a)(x + b) \tag{1}$$
$$x^2 + 2ax + a^2 = (x + a)^2 \tag{2}$$
$$x^2 - a^2 = (x + a)(x - a) \tag{3}$$
$$acx^2 + (ad + bc)xy + bdy^2 = (ax + by)(cx + dy) \tag{4}$$
$$(ax)^2 - (by)^2 = (ax - by)(ax + by) \tag{5}$$

The factorization of the polynomials in (1) to (5) above can be easily verified by using the distributive property on each right-hand member.

Again, we shall require integral coefficients and positive integral exponents on the variables when factoring these polynomials.

As an example of the application of form (1) above, consider the trinomial

$$x^2 + 6x - 16.$$

We desire, if possible, to find two binomial factors,

$$(x + a)(x + b),$$

whose product is the given trinomial. We see from form (1) that a and b are two integers such that $a + b = 6$ and $ab = -16$; that is, their sum must be the coefficient of the linear term $6x$ and their product must be -16. By inspection, or by trial and error, we determine that the two numbers are 8 and -2, so that

$$x^2 + 6x - 16 = (x + 8)(x - 2).$$

Checking, we note that $(x + 8)(x - 2) = x^2 + 6x - 16$.

EXAMPLE 1 Factor:

a. $x^2 - 7x + 12$ b. $x^2 - x - 12$

Solutions a. We want to find two numbers whose product is 12 and whose sum is -7. Since the product is positive and the sum is negative, the two numbers must both be negative. By inspection, or trial and error, the two numbers are -4 and -3. Hence,

$$x^2 - 7x + 12 = (x - 4)(x - 3).$$

b. We want to find two numbers whose product is -12 and whose sum is -1. Since the product is negative, the two numbers must be of opposite sign and their sum must be -1. By inspection, or trial and error, the two numbers are -4 and 3. Hence,

$$x^2 - x - 12 = (x - 4)(x + 3). \quad ■$$

Although we have not specifically noted the check in the above examples, the check should be done mentally for each factorization.

Perfect-square trinomials Form (2), the square of a binomial, is simply a special case of (1).

EXAMPLE 2 Factor:

a. $x^2 + 8x + 16$ b. $x^2 - 10x + 25$

Solutions a. Two numbers whose product is 16 and whose sum is 8 are 4 and 4. Hence,

$$\begin{aligned} x^2 + 8x + 16 &= (x + 4)(x + 4) \\ &= (x + 4)^2. \end{aligned}$$

b. Two numbers whose product is 25 and whose sum is -10 are -5 and -5. Hence,

$$\begin{aligned} x^2 - 10x + 25 &= (x - 5)(x - 5) \\ &= (x - 5)^2. \end{aligned}$$ ■

Equations of form (2) are sometimes called **perfect-square trinomials** because they are the squares of binomials.

Difference of two squares

Form (3) is another special case of (1), in which the coefficient of the first-degree term in x is 0. For example,

$$x^2 - 25 = x^2 + 0x - 25 = (x - 5)(x + 5).$$

In particular, form (3) states:

> *The difference of the squares of two numbers is equal to the product of the sum and the difference of the two numbers.*

The factors $x - 5$ and $x + 5$ in the above example are called **conjugates** of each other. In general, any binomials of the form $\boldsymbol{a - b}$ and $\boldsymbol{a + b}$ are called **conjugate pairs.**

EXAMPLE 3 Factor:

a. $x^2 - 81$ **b.** $x^2 + 81$

Solutions

a. $x^2 - 81$ can be written as the difference of two squares, $(x)^2 - (9)^2$, which can be factored as a conjugate pair:

$$x^2 - 81 = x^2 - 9^2 = (x - 9)(x + 9).$$

b. $x^2 + 81$, equivalent to $x^2 + 0x + 81$, is *not* of form (3). It is *not* factorable, because no two real numbers have a product of 81 and a sum of 0. ■

Form (4) is a generalization of (1)—that is, in (4) we are confronted with a quadratic trinomial where the coefficient of the term of second degree in x is other than 1.

EXAMPLE 4 To factor $8x^2 - 9 - 21x$:

1. Write in decreasing powers of x.

$$8x^2 - 21x - 9$$

2. Consider possible combinations of first-degree factors of the first term.

$$(8x \quad)(x \quad)$$
$$(4x \quad)(2x \quad)$$

3. Consider combinations of the factors ① of the last term.

$$(8x \quad 9)(x \quad 1)$$
$$(8x \quad 1)(x \quad 9)$$
$$(8x \quad 3)(x \quad 3)$$
$$(4x \quad 9)(2x \quad 1)$$
$$(4x \quad 1)(2x \quad 9)$$
$$(4x \quad 3)(2x \quad 3)$$

4. Select the combination(s) of products ② and ③ whose sum(s) could be the second term $(-21x)$.

$$(8x \quad 3)(x \quad 3)$$

5. Insert the proper signs.

$$(8x + 3)(x - 3) \quad ■$$

Although the process in the above example can normally be done mentally, it was written in detail for the purposes of illustration.

Signs on factors

Factoring trinomials of the form $Ax^2 + Bx + C$, where A is positive, can be facilitated by making use of the following considerations:

1. If both B and C are positive, both signs in the factored form are positive. For example, as a first step in factoring $6x^2 + 11x + 4$, we could write

$$(\quad + \quad)(\quad + \quad).$$

2. If B is negative and C is positive, both signs in the factored form are negative. Thus, as the first step in factoring $6x^2 - 11x + 4$, we could write

$$(\quad - \quad)(\quad - \quad).$$

3. If C is negative, the signs in the factored form are opposite. Thus, as a first step in factoring $6x^2 - 5x - 4$, we could write

$$(\quad + \quad)(\quad - \quad) \text{ or } (\quad - \quad)(\quad + \quad).$$

EXAMPLE 5

a. $6x^2 + 5x + 1$
$= (\quad + \quad)(\quad + \quad)$
$= (3x + 1)(2x + 1)$

b. $6x^2 - 5x + 1$
$= (\quad - \quad)(\quad - \quad)$
$= (3x - 1)(2x - 1)$

c. $6x^2 - x - 1$
$= (\quad + \quad)(\quad - \quad)$
$= (3x + 1)(2x - 1)$

d. $6x^2 + x - 2$
$= (\quad + \quad)(\quad - \quad)$
$= (3x + 2)(2x - 1)$ ■

Difference of two squares

Form (5) on page 103 is a special case of form (3), in which the coefficient of the xy term is 0.

EXAMPLE 6

a. $16y^2 - 1$
$= (4y)^2 - (1)^2$
$= (4y - 1)(4y + 1)$

b. $4x^2 - 9y^2$
$= (2x)^2 - (3y)^2$
$= (2x - 3y)(2x + 3y)$ ■

If a polynomial of more than one term contains a common monomial factor in each of its terms, this monomial should be factored from the polynomial before seeking other factors.

EXAMPLE 7

a. $32x^2 - 84x - 36$
$= 4(8x^2 - 21x - 9)$
$= 4(8x + 3)(x - 3)$

b. $4x^2 - 100$
$= 4(x^2 - 25)$
$= 4(x - 5)(x + 5)$ ■

Suggestions for Factoring Polynomials

1. Write a polynomial in one variable in descending powers of the variable.
2. Factor out any factors that are common to each term in the polynomial.
3. Factor any binomial that is the difference of two squares as the product of conjugate pairs.
4. Factor any trinomials that can be factored.
5. Check the result of the factoring by multiplying the factors.

The examples and suggestions above are primarily concerned with factoring second-degree polynomials. However, the suggestions are also applicable to polynomials of higher degree.

EXAMPLE 8

a. $x^4 + 2x^2 + 1$
$= (x^2 + 1)(x^2 + 1)$
$= (x^2 + 1)^2$

b. $x^4 - 3x^2 - 4$
$= (x^2 - 4)(x^2 + 1)$
$= (x - 2)(x + 2)(x^2 + 1)$ ■

The following examples show the factorization of polynomials in which exponents include variables.

EXAMPLE 9

a. $9 - x^{2n}$
$= (3 - x^n)(3 + x^n)$

b. $y^{2n} - y^n - 6$
$= (y^n - 3)(y^n + 2)$ ■

EXERCISE 4.2

A

■ *Factor completely. See Examples 1–6.*

1. $x^2 + 5x + 6$
2. $x^2 + 5x + 4$
3. $y^2 - 7y + 12$
4. $y^2 - 7y + 10$
5. $x^2 - x - 6$
6. $x^2 - 2x - 15$
7. $y^2 - 3y - 10$
8. $y^2 + 4y - 21$
9. $2x^2 + 3x - 2$

10. $3x^2 - 7x + 2$
11. $4x^2 + 7x - 2$
12. $6x^2 - 5x + 1$
13. $3x^2 - 4x + 1$
14. $4x^2 - 5x + 1$
15. $9x^2 - 21x - 8$
16. $10x^2 - 3x - 18$
17. $10x^2 - x - 3$
18. $8x^2 + 5x - 3$
19. $4x^2 + 12x + 9$
20. $4y^2 + 4y + 1$
21. $3x^2 - 7ax + 2a^2$
22. $9x^2 + 9ax - 10a^2$
23. $9x^2y^2 + 6xy + 1$
24. $4x^2y^2 + 12xy + 9$
25. $x^2 - 25$
26. $x^2 - 36$
27. $(xy)^2 - 1$
28. $(xy)^2 - 4$
29. $y^4 - 9$
30. $y^4 - 49$
31. $x^2 - 4y^2$
32. $9x^2 - y^2$
33. $4x^2 - 25y^2$
34. $16x^2 - 9y^2$
35. $16x^2y^2 - 1$
36. $64x^2y^2 - 1$

■ *Factor completely. See Example 7.*

37. $3x^2 + 12x + 12$
38. $2x^2 + 6x - 20$
39. $2a^3 - 8a^2 - 10a$
40. $2a^3 + 15a^2 + 7a$
41. $4a^2 - 8ab + 4b^2$
42. $20a^2 + 60ab + 45b^2$
43. $4x^2y - 36y$
44. $x^2 - 4x^2y^2$
45. $12x - x^2 - x^3$
46. $x^2 - 2x^3 + x^4$
47. $x^4y^2 - x^2y^2$
48. $x^3y - xy^3$

B

■ *Factor completely. See Example 8.*

49. $y^4 + 3y^2 + 2$
50. $a^4 + 5a^2 + 6$
51. $3x^4 + 7x^2 + 2$
52. $4x^4 - 11x^2 - 3$
53. $x^4 + 3x^2 - 4$
54. $x^4 - 6x^2 - 27$
55. $x^4 - 5x^2 + 4$
56. $y^4 - 13y^2 + 36$
57. $2a^4 - a^2 - 1$
58. $3x^4 - 11x^2 - 4$
59. $x^4 + a^2x^2 - 2a^4$
60. $4x^4 - 33a^2x^2 - 27a^4$

■ *Factor completely. See Example 9.*

61. $x^{4n} - 1$
62. $16 - y^{4n}$
63. $x^{4n} - y^{4n}$
64. $x^{4n} - 2x^{2n} + 1$
65. $3x^{4n} - 10x^{2n} + 3$
66. $6y^{2n} + 30y^n - 900$

4.3

FACTORING OTHER POLYNOMIALS

A few other polynomials occur frequently enough to justify a study of their factorization.

Factoring by grouping

Sometimes a polynomial is factorable by grouping. For example, we can factor a from the first two terms of

$$ax + ay + bx + by$$

and b from the last two terms to obtain

$$a(x + y) + b(x + y).$$

Now, since the factor $x + y$ is a common factor of both terms, we can write the expression as

$$(x + y)(a + b).$$

Thus,

$$\mathbf{ax + ay + bx + by = (a + b)(x + y).}$$

EXAMPLE 1 To factor $3x^2y + 2y + 3xy^2 + 2x$:

We first rewrite the expression in the form

$$3x^2y + 2x + 3xy^2 + 2y.$$

We then factor the common monomial x from the first group of two terms and the common monomial y from the second group of two terms to obtain

$$x(3xy + 2) + y(3xy + 2).$$

If we now factor the common binomial $(3xy + 2)$ from each term, we have

$$(3xy + 2)(x + y). \quad ■$$

The polynomials in the following examples are factored in a similar way.

EXAMPLE 2 a. $yb - ya + xb - xa$

$$= y(b - a) + x(b - a)$$
$$= (b - a)(y + x)$$

b. $x^2 + xb - ax - ab$

$$= x(x + b) - a(x + b)$$
$$= (x + b)(x - a) \quad ■$$

In the above examples we accomplished the factorization of the expression by grouping the four terms as the sum or difference of two binomials. Sometimes it is necessary to view an expression of four terms as a trinomial and monomial in order to factor the expression.

EXAMPLE 3

a. $x^2 - 2x + 1 - y^2$
$= (x - 1)^2 - y^2$
$= (x - 1 - y)(x - 1 + y)$

b. $x^2 - y^2 + 4yz - 4z^2$
$= x^2 - (y^2 - 4yz + 4z^2)$
$= x^2 - (y - 2z)^2$
$= (x - y + 2z)(x + y - 2z)$ ■

Sum and difference of cubes

Two other factorizations are of special interest. These are the sum of two cubes, $x^3 + y^3$, and the difference of two cubes, $x^3 - y^3$. Our ability to factor these binomials is a result of observing that

$$(x + y)(x^2 - xy + y^2) = x^3 - x^2y + xy^2 + yx^2 - xy^2 + y^3$$
$$= x^3 + y^3$$

and that

$$(x - y)(x^2 + xy + y^2) = x^3 + x^2y + xy^2 - yx^2 - xy^2 - y^3$$
$$= x^3 - y^3.$$

Viewing the two equations from right to left, we have the following special factorizations.

$$\mathbf{x^3 + y^3 = (x + y)(x^2 - xy + y^2)}$$

and

$$\mathbf{x^3 - y^3 = (x - y)(x^2 + xy + y^2).}$$

EXAMPLE 4

a. $8a^3 + b^3$
$= (2a)^3 + b^3$
$= (2a + b)[(2a)^2 - 2ab + b^2]$
$= (2a + b)(4a^2 - 2ab + b^2)$

b. $x^3 - 27y^3$
$= x^3 - (3y)^3$
$= (x - 3y)[x^2 + 3xy + (3y)^2]$
$= (x - 3y)(x^2 + 3xy + 9y^2)$ ■

EXERCISE 4.3

A

■ *Factor. See Examples 1 and 2.*

1. $ax + a + bx + b$
2. $5a + ab + 5b + b^2$
3. $ax^2 + x + a^2x + a$
4. $a + ab + b + b^2$
5. $x^2 + ax + xy + ay$
6. $x^3 - x^2y + xy - y^2$
7. $3ab - cb - 3ad + cd$
8. $2ac - bc + 2ad - bd$
9. $3x + y - 6x^2 - 2xy$
10. $5xz - 5yz - x + y$
11. $a^3 + 2ab^2 - 2a^2b - 4b^3$
12. $6x^3 - 4x^2 + 3x - 2$
13. $x^2 - x + 2xy - 2y$
14. $2a^2 + 3a - 2ab - 3b$
15. $2a^2b + 6a^2 - b - 3$
16. $2ab^2 + 5a - 8b^2 - 20$
17. $x^3y^2 + x^3 - 3y^2 - 3$
18. $12 - 4y^3 - 3x^2 + x^2y^3$
19. $x^3 + 2x^2 + 4x + 8$
20. $x^3 + 3x^2 + 3x + 9$
21. $2x^3 - 3x^2 + 2x - 3$
22. $2x^3 + 7x^2 + 4x + 14$
23. $x^3 - x^2 + x - 1$
24. $x^3 - 2x^2 - 3x + 6$

■ *Factor. See Example 3.*

25. $(x + 2)^2 - y^2$
26. $x^2 - (y - 3)^2$
27. $x^2 + 2x + 1 - y^2$
28. $x^2 - 6x + 9 - y^2$
29. $y^2 - x^2 + 2x - 1$
30. $y^2 - x^2 + 4x - 4$
31. $4x^2 + 4x + 1 - 4y^2$
32. $9x^2 - 6x + 1 - 9y^2$

■ *Factor. See Example 4.*

33. $x^3 - 1$
34. $y^3 + 27$
35. $(2x)^3 + y^3$
36. $y^3 - (3x)^3$
37. $a^3 - 8b^3$
38. $27a^3 + b^3$
39. $(xy)^3 - 1$
40. $8 + x^3y^3$
41. $27a^3 + 64b^3$
42. $a^3 - 125b^3$
43. $64a^3b^3 - 1$
44. $8a^3b^3 + 1$

B

■ *Factor.*

45. $x^3 + (x - y)^3$
46. $(x + y)^3 - z^3$
47. $(x + 1)^3 - 1$
48. $x^6 + (x - 2y)^3$
49. $(x + 1)^3 - (x - 1)^3$
50. $(2y - 1)^3 + (y - 1)^3$
51. Show that $ac - ad + bd - bc$ can be factored as $(a - b)(c - d)$ and as $(b - a)(d - c)$.
52. Show that $a^2 - b^2 - c^2 + 2bc$ can be factored as $(a - b + c)(a + b - c)$.

53. Consider the polynomial $x^4 + x^2y^2 + y^4$. If x^2y^2 is both added to and subtracted from this expression (thus producing an equivalent expression in which the first three terms form a "perfect square"), we have

$$\begin{aligned} x^4 + x^2y^2 + y^4 &= x^4 + x^2y^2 + y^4 + x^2y^2 - x^2y^2 \\ &= x^4 + 2x^2y^2 + y^4 - x^2y^2 \\ &= (x^2 + y^2)^2 - (xy)^2 \\ &= (x^2 + y^2 - xy)(x^2 + y^2 + xy). \end{aligned}$$

By adding and subtracting an appropriate monomial, factor $x^4 + 3x^2y^2 + 4y^4$.

54. Use the method of Problem 53 to factor $x^4 - 8x^2y^2 + 4y^4$.

55. Use the method of Problem 53 to factor $a^4 + 6a^2b^2 + 25b^4$.

56. Use the method of Problem 53 to factor $4a^4 - 5a^2b^2 + b^4$.

4.4

SOLUTION OF QUADRATIC EQUATIONS BY FACTORING

Standard form

A second-degree equation in one variable is called a **quadratic equation** in that variable. We shall designate as **standard form** for such equations

$$\mathbf{ax^2 + bx + c = 0,}$$

where a, b, and c are constants representing real numbers and $a \neq 0$.

If $b = 0$ or $c = 0$, then the equations are of the form

$$ax^2 + c = 0 \quad \text{or} \quad ax^2 + bx = 0$$

and are called **incomplete quadratic equations.**

Writing equivalent equations

In Chapter 3 we solved first-degree equations by performing certain elementary transformations. These transformations are equally applicable to equations of higher degree and, in particular, to quadratic equations. For example, adding $8x^2 - 6$ to each member of

$$2x = 6 - 8x^2,$$

we obtain

$$8x^2 + 2x - 6 = 0.$$

Then, dividing each member by 2, we have

$$4x^2 + x - 3 = 0.$$

The three equations are equivalent.

Solution by factoring

If the left-hand member of a quadratic equation in standard form is factorable, we may solve the equation by making use of the following principle.

> *The product of two factors equals* 0 *if and only if one or both of the factors equals* 0.

Thus:

$$\mathbf{ab = 0} \quad \textit{if and only if} \quad \mathbf{a = 0} \quad \textit{or} \quad \mathbf{b = 0.}$$

EXAMPLE 1 $(x - 1)(2x + 3) = 0$ if and only if

$$x - 1 = 0 \qquad (x = 1)$$

or

$$2x + 3 = 0 \qquad (x = -3/2)$$ ■

The word "or" in the above statement and in the example is used in an inclusive sense to mean either one *or* the other *or* both.

The above principle enables us to solve quadratic equations if the left-hand member of an equation in standard form is factorable.

EXAMPLE 2 Solve $x^2 + 2x = 15$. (1)

Solution First write the equation in standard form and then factor the left-hand member to obtain

$$x^2 + 2x - 15 = 0 \qquad (2)$$

$$(x + 5)(x - 3) = 0,$$

which will be true if and only if

$$x + 5 = 0 \quad \text{or} \quad x - 3 = 0$$
$$x = -5 \qquad\qquad x = 3.$$

Thus, either -5 or 3, when substituted for x in (2) or (1), will make the left-hand member 0. The solution set is $\{-5, 3\}$. ■

EXAMPLE 3 Solve $3x(x + 1) = 2x + 2$.

Solution Write in standard form.

$$3x^2 + x - 2 = 0$$

Factor the left-hand member.

$$(3x - 2)(x + 1) = 0$$

Set each factor equal to 0 and solve the equations.

$$3x - 2 = 0 \quad \text{or} \quad x + 1 = 0$$
$$x = \frac{2}{3} \qquad\qquad x = -1$$

The solution set is $\{2/3, -1\}$. ■

In general, the solution set of a quadratic equation can be expected to contain two elements. However, if the left-hand member of a quadratic equation in standard form is the square of a binomial, we find that the solution set contains only one member. For example, to solve

$$x^2 - 2x + 1 = 0,$$

we factor the left-hand member to obtain

$$(x - 1)(x - 1) = 0,$$

and the solution set is $\{1\}$, which contains only one element.

For reasons of convenience and consistency in more advanced work, the unique solution obtained in the above example is said to be of **multiplicity two**.

Writing equations from given solutions

Notice that the solution set of the quadratic equation

$$(x - r_1)(x - r_2) = 0 \tag{3}$$

is $\{r_1, r_2\}$*. Therefore, if r_1 and r_2 are given as solutions of a quadratic equation, the equation can be written directly as (3). By completing the indicated multiplication, the equation can be transformed to standard form.

EXAMPLE 4 If 2 and -3 are given as solutions of a quadratic equation, then

$$[x - (2)][x - (-3)] = 0.$$

That is,

$$(x - 2)(x + 3) = 0,$$

from which

$$x^2 + x - 6 = 0.$$ ■

EXERCISE 4.4

A

■ *Solve. Determine your answer by inspection or follow the procedure in Example* 1.

1. $(x + 2)(x - 5) = 0$
2. $(x + 3)(x - 4) = 0$
3. $(2x + 5)(x - 2) = 0$
4. $(x + 1)(3x - 1) = 0$
5. $x(2x + 1) = 0$
6. $x(3x - 7) = 0$
7. $4(x - 6)(2x + 3) = 0$
8. $5(2x - 7)(x + 1) = 0$
9. $3(x - 2)(2x + 1) = 0$
10. $7(x + 5)(3x - 1) = 0$
11. $4(2x - 5)(3x + 2) = 0$
12. $2(3x - 4)(2x + 5) = 0$

■ *Solve. See Examples 2 and 3.*

13. $x^2 - 3x = 0$
14. $x^2 + 5x = 0$
15. $2x^2 = 6x$
16. $3x^2 = 3x$
17. $x^2 - 9 = 0$
18. $x^2 - 4 = 0$
19. $2x^2 - 18 = 0$
20. $3x^2 - 3 = 0$
21. $9x^2 = 4$

*The subscripts 1 and 2 used in r_1 and r_2 in this equation are used to identify constants. The symbols are read "r sub one" and "r sub two" or simply "r one" and "r two."

22. $25x^2 = 4$
23. $4x^2 - 9 = 0$
24. $9x^2 - 25 = 0$
25. $x^2 - 5x + 4 = 0$
26. $x^2 + 5x + 6 = 0$
27. $x^2 - 5x - 14 = 0$
28. $x^2 - x - 42 = 0$
29. $3x^2 - 6x = -3$
30. $12x^2 = 8x + 15$
31. $x(2x - 3) = -1$
32. $2x(x - 2) = x + 3$
33. $(x - 2)(x + 1) = 4$
34. $x(3x + 2) = (x + 2)^2$
35. $(x - 1)^2 = 2x^2 + 3x - 5$
36. $x(x + 1) = 4 - (x + 2)^2$
37. $t(t + 4) - 1 = 4$
38. $z(z + 5) + 18 = 4(1 - z)$
39. $2z(z + 3) = 3 + z$
40. $(n - 3)(n + 2) = 6$
41. $(t + 2)(t - 5) = 8$
42. $(x + 1)(2x - 3) = 3$
43. $(y - 1)^2 = 4(y - 2)$
44. $6y = (y + 1)^2 + 3$
45. $4x - [(x + 1)(x - 2) + 6] = 0$
46. $2x - [(x + 2)(x - 3) + 8] = 0$
47. $[(x + 1)^2 - 2x] = 10$
48. $3[(x + 2)^2 - 4x] = 15$

■ *Given the solutions of a quadratic equation, r_1 and r_2, write the equation in standard form with integral coefficients. See Example 4.*

49. -2 and 1
50. -4 and 3
51. 0 and -5
52. 0 and 5
53. -3 and -3
54. 4 and 4
55. a and b
56. $-a$ and $-b$

B

■ *Solve each equation for x in terms of a and b.*

57. $x^2 - 4b^2 = 0$
58. $x^2 - (a + b)^2 = 0$
59. $x^2 - 3ax - 4a^2 = 0$
60. $x^2 - 4bx - 12b^2 = 0$
61. $x^2 - (a + b)x + ab = 0$
62. $x^2 + (2a - b)x - 2ab = 0$
63. $x^2 + \left(a + \frac{b}{2}\right)x + \frac{ab}{2} = 0$
64. $x^2 - \frac{1}{2}(a + b)x + \frac{ab}{4} = 0$

4.5

APPLICATIONS

In some cases, the mathematical model we obtain for a physical situation is a quadratic equation and hence may have two solutions. It may be that one but not both of the solutions of the equation fits the physical situation. For example, if we were asked to find two consecutive *natural numbers* whose product is 72, we would write the equation

$$x(x + 1) = 72$$

as our model. Solving this equation, we have

$$x^2 + x - 72 = 0$$
$$(x + 9)(x - 8) = 0,$$

where the solution set is $\{8, -9\}$. Since -9 is not a natural number, we must reject it as a possible answer to the original question; however, the solution 8 leads to the consecutive natural numbers 8 and 9. As additional examples, observe that we would not accept -6 feet as the height of a man or $27/4$ for the number of people in a room.

A quadratic equation used as a model for a physical situation may have two, one, or no meaningful solutions—meaningful, that is, in a physical sense. Answers to word problems should always be checked in the original problem.

We will continue to follow the step-by-step procedure that we used in Chapter 3 to solve word problems, giving careful attention to the construction of the mathematical model.

EXAMPLE 1 The length of a rectangle is 4 centimeters greater than the width, and the area is 77 square centimeters. Find the dimensions of the rectangle.

Solution

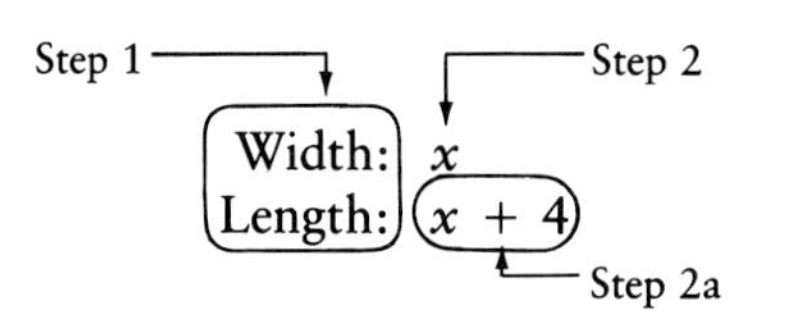

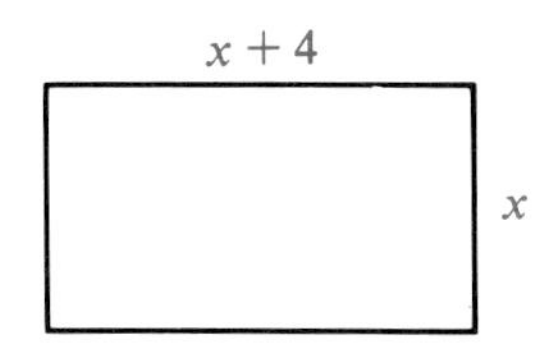

3. Write a model for the conditions on the variable. Since the area (77 square centimeters) of a rectangle is the product of its width and length, the model is

$$x(x + 4) = 77.$$

4. Solve the equation:

$$x^2 + 4x - 77 = 0,$$
$$(x + 11)(x - 7) = 0.$$

5. The solution set is $\{-11, 7\}$. Since a dimension cannot be negative, the only acceptable value for x is 7. If $x = 7$, then $x + 4 = 11$, and the dimensions are 7 centimeters and 11 centimeters. ■

EXAMPLE 2 A ball thrown vertically upward reaches a height h in feet, given by the equation $h = 64t - 16t^2$, where t is the time in seconds after the throw. How long will it take the ball to reach a height of 48 feet on its way up?

Solution Step 1 → Time to reach 48 feet on way up: t ← Step 2

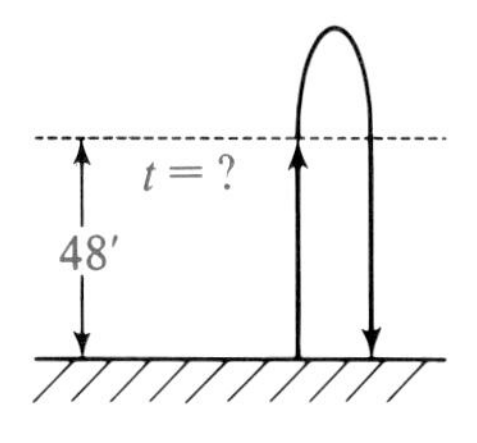

3. In this case, the model is given:

$$h = 64t - 16t^2.$$

4. Substitute 48 for h and solve for t:

$$48 = 64t - 16t^2$$
$$16t^2 - 64t + 48 = 0$$
$$16(t^2 - 4t + 3) = 0$$
$$16(t - 1)(t - 3) = 0$$
$$t = 1 \quad \text{or} \quad t = 3$$

5. Use the figure to interpret the two solutions: it takes 1 second to reach 48 feet *on the way up*. (In 3 seconds the ball is also at 48 feet on its way down.) ■

EXAMPLE 3 A deck of uniform width was constructed around a 20 by 25-foot rectangular pool. Find the width of the deck if 196 square feet of brick was used.

Solution

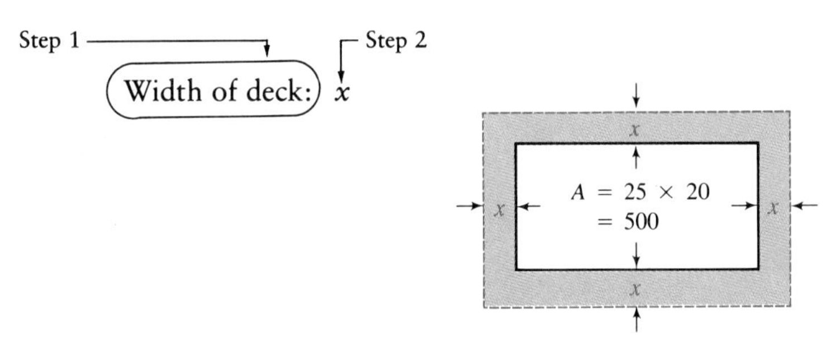

3. Since the area of the deck is 196 square feet:

$$\text{enlarged area} - \text{area of pool} = 196$$
$$(25 + 2x)(20 + 2x) - 500 = 196$$

4. Solve the equation:

$$500 + 90x + 4x^2 - 500 = 196$$
$$4x^2 + 90x - 196 = 0$$
$$2x^2 + 45x - 98 = 0$$
$$(2x + 49)(x - 2) = 0$$
$$x = -49/2 \quad \text{or} \quad x = 2$$

5. The width cannot be $-49/2$. Hence, the width of the deck is 2 feet. ■

The above examples show that when solving quadratic equations that are models for stated conditions in word problems, solutions are sometimes obtained that do not apply. As we have noted, it is important to always check solutions to see if they are meaningful for the stated conditions.

EXERCISE 4.5

A

■ *Solve each problem. Use the five steps suggested in Examples* 1, 2, *and* 3.

1. Find two positive numbers that differ by 3 and have a product of 40.
2. Find two negative numbers that differ by 7 and have a product of 60.
3. Find two consecutive negative integers such that the sum of their squares is 61.
4. Find two consecutive positive integers such that the sum of their squares is 85.
5. The sum of the squares of three consecutive positive integers is 149. Find the integers.
6. The sum of the squares of three consecutive negative integers is 434. Find the integers.
7. A rectangular lawn is 2 meters longer than it is wide. If its area is 63 square meters, what are the dimensions of the lawn?
8. The length of a rectangular steel plate is 2 centimeters greater than 2 times its width. If the area of the plate is 40 square centimeters, find its dimensions.
9. The area of a triangle is 27 square inches. Find the lengths of the base and the altitude if the base is 3 inches shorter than the altitude.
10. The area of a triangle is 70 square centimeters. Find the lengths of the base and the altitude if the base is 13 centimeters longer than the altitude.
11. The hypotenuse of a right triangle is 10 centimeters. A second side is 2 centimeters longer than the third side. What are the lengths of the two shorter sides?

12. One side of a right triangle is 7 inches smaller than a second side. If the length of the hypotenuse is 13 inches, how long are the two shorter sides?
13. Using the information in Example 2 on page 118, determine how long after the ball was thrown it will take for the ball to reach a height of 28 feet on the way down.
14. Using the information in Example 2, determine how long after the ball was thrown it will return to the ground.
15. The position of a particle moving in a straight line is given by the formula

$$s = t^2 - 5t,$$

where s is the distance from the starting point (in centimeters) and t is the time (in seconds) that the particle has been in motion. How long will it take for the particle to move 24 centimeters in a positive direction?
16. Using the motion of the particle in Problem 15, where will the particle be in 2 seconds? 3 seconds? When will the particle be back at its starting point?
17. A photographer uses 48 square inches for a picture on a mat whose length is twice its width. If the margin around the picture is to be 2 inches uniformly, what are the dimensions of the mat?
18. A book designer wants to use 35 square inches of type on a page and to make the height of the page 2 inches longer than its width. If the margin is to be 1 inch uniformly, what are the dimensions of the page?
19. A rectangular garden measuring 25 by 50 meters has its area increased by 318 square meters by a border of uniform width along both shorter sides and one longer side. Find the width of the border.
20. A rectangular garden measuring 12 by 18 meters has its area increased by 216 square meters by a border of uniform width on all sides. Find the width of the border.
21. A tray is formed from a rectangular piece of metal whose length is 2 centimeters greater than its width by cutting a square with sides 2 centimeters in length from each corner, and then bending up the sides. Find the dimensions of the tray if the volume is 160 cubic centimeters.
22. A tray is formed from a rectangular piece of metal whose length is twice its width by cutting a square with sides 1 centimeter in length from each corner, and then bending up the sides. Find the dimensions of the tray if the volume is 144 cubic centimeters.

CHAPTER SUMMARY

[4.1–4.3] The process of using the distributive property to rewrite a polynomial as a single term comprised of two or more factors is called **factoring.** If a polynomial of more than one term contains a common factor in each of its terms, this common factor should be factored from the polynomial first. Three special cases of factoring are:

$$x^2 - a^2 = (x - a)(x + a)$$
$$x^3 + a^3 = (x + a)(x^2 - ax + a^2)$$
$$x^3 - a^3 = (x - a)(x^2 + ax + a^2)$$

Some trinomials of the form $ax^2 + bx + c$ can be factored by trial and error. Some polynomials that consist of more than three terms can be factored by first grouping terms with like factors. Thus:

$$\begin{aligned} ax + ay + bx + by &= a(x + y) + b(x + y) \\ &= (a + b)(x + y). \end{aligned}$$

[4.4] The **standard form** for a quadratic equation in one variable is

$$ax^2 + bx + c = 0,$$

where a, b, and c are constants with $a \neq 0$.

To solve an equation by **factoring**, we use the fact that the product of two factors is 0 if and only if one or both of the factors equals 0. That is,

$$ab = 0 \quad \text{if and only if} \quad a = 0 \quad \text{or} \quad b = 0.$$

[4.5] Quadratic equations are sometimes useful as mathematical models in physical situations. In such cases, it may be that not all the solutions of an equation are meaningful.

REVIEW EXERCISES

A

[4.1]

■ *Factor each polynomial.*

1. a. $12x^2 - 8x + 4$ b. $x^3 - 3x^2 - x$
2. a. $4y^3 - 8y^2$ b. $4x^3y^2 - 2x^2y^2 + 6xy^2$
3. a. $3x - y = -(\underline{?})$ b. $2x - y + z = -(\underline{?})$
4. a. $-x^2 + 2x = -x(\underline{?})$ b. $-6x^2y + 3xy - 3xy^2 = -3xy(\underline{?})$
5. a. $x(a + 2) - (a + 2)$ b. $2a(x - y) + b(x - y)$
6. a. $x + 2y - x(x + 2y)$ b. $y(x + y) + (x + y)$

[4.2]

■ *Factor each polynomial.*

7. a. $x^2 - 2x - 35$ b. $y^2 + 4y - 32$
8. a. $(xy)^2 - 36$ b. $a^2 - 49b^2$
9. a. $3y^2 + 11y - 4$ b. $x^3 + 3x^2 - 10x$
10. a. $9x^2 - 36$ b. $12x^2 - 3y^2$
11. a. $2x^2 + 3xy - 2y^2$ b. $6x^2 - xy - y^2$
12. a. $15a^2 + 28ab + 12b^2$ b. $12a^2 - 18ab + 6b^2$

[4.3]

■ *Factor each polynomial.*

13. a. $2xy + 2x^2 + y + x$ b. $xy - 3x - y + 3$
14. a. $ax - 2bx + ay - 2by$ b. $2ax - 4ay + bx - 2by$
15. a. $2a^3 - a^2 - 8a + 4$ b. $a^3 + 2a^2 - 4a - 8$
16. a. $(2y - 3)^2 - 4x^2$ b. $4y^2 - 4y + 1 - x^2$
17. a. $y^2 - 4x^2 - 4x - 1$ b. $x^2 - y^2 + 8y - 16$
18. a. $(2x)^3 - y^3$ b. $x^3 + (4y)^3$
19. a. $27y^3 + z^3$ b. $x^3 - 8a^3$
20. a. $8x^3y^3 - 125$ b. $1 + 64x^3y^3$

[4.4]

■ *Solve for x by factoring.*

21. a. $x^2 - 2x = 0$ b. $x^2 - 5x + 6 = 0$
22. a. $(x - 6)(x + 4) = -9$ b. $x(x + 1) = 6$
23. a. $x - 1 = \frac{1}{4}x^2$ b. $x + \frac{3}{x} = 4$

■ *Given the solutions of a quadratic equation, write the equation in standard form.*

24. a. -2 and 5 b. 0 and -3

[4.5]

■ *Solve.*

25. The base of a triangle is 1 inch longer than 2 times the length of its altitude, and its area is 18 square inches. Find the length of the base and the altitude.
26. The width of a rectangle is 4 centimeters less than its length, and the area is 60 square centimeters. Find the dimensions of the rectangle.
27. The sum of the squares of three consecutive positive integers is 194. Find the integers.

28. The distance d (in meters) that an object travels in t seconds when given an initial velocity of 15 meters per second is given by

$$d = 15t + 2t^2.$$

Find the time necessary for an object to travel 63 meters.

B

■ *Factor each expression completely. (Assume that variables in exponents denote natural numbers.)*

29. $x^{4n} + x^{2n}$

30. $x^{n+2} - x^n$

31. $(x + y)^3 + (x - y)^3$

32. $(x + y)^3 - (x - y)^3$

PYTAGORAS
8
16
6
4
12
2
PHYLOLAVS

5 Rational Expressions

Recall from Section 1.1 that integers and quotients of integers (divisor not equal to zero) are called rational numbers. In a similar way, polynomials and quotients of polynomials are called **rational expressions.** For example,

$$x^2 + 2x, \qquad 3y, \qquad \frac{y}{y+1}, \quad \text{and} \quad \frac{x^2 - 2x + 1}{2x}$$

are rational expressions. For each replacement of the variable(s) for which the numerator and denominator of a fraction represent real numbers and for which the denominator is not zero, a rational expression represents a real number. Of course, for any value of the variable(s) for which the denominator vanishes (is equal to zero), the fraction does not represent a real number and is said to be undefined.

5.1

REDUCING FRACTIONS

Fundamental principle of fractions

There are infinitely many fractions that correspond to a given quotient. Thus, for example,

$$\frac{1}{2} = \frac{2}{4} = \frac{3}{6} = \frac{4}{8} \cdots \quad \text{and} \quad \frac{3}{5} = \frac{6}{10} = \frac{9}{15} = \frac{12}{20} \cdots.$$

The properties of real numbers can be used to establish the following property which enables us to write such equivalent fractions.

An equivalent fraction is obtained if the numerator and the denominator of a fraction are each multiplied or divided by the same nonzero number.

This property is commonly called **the fundamental principle of fractions** and is expressed in symbols as follows:

$$\frac{a}{b} = \frac{ac}{bc} \qquad (b,c \neq 0).$$

Signs on fractions

There are three signs associated with a fraction: a sign for the numerator, a sign for the denominator, and a sign for the fraction itself. Although there are eight different possible symbols associated with the symbol "a/b" and the two signs "+" and "−," these symbols represent only two real numbers, a/b and its additive inverse $-(a/b)$. The property below follows from the definition of a quotient and the fundamental principle of fractions.

$$\frac{a}{b} = \frac{-a}{-b} = -\frac{a}{-b} = -\frac{-a}{b} \qquad (b \neq 0) \qquad (1)$$

and

$$\frac{-a}{b} = \frac{a}{-b} = -\frac{a}{b} = -\frac{-a}{-b} \qquad (b \neq 0). \qquad (2)$$

EXAMPLE 1 a. $\frac{-2}{3} = \frac{2}{-3} = -\frac{2}{3} = -\frac{-2}{-3}$ b. $\frac{2}{3} = \frac{-2}{-3} = -\frac{2}{-3} = -\frac{-2}{3}$ ■

We can use (1) and (2) to write a given fraction as an equivalent fraction by replacing any two of the fraction's three elements—the fraction itself, the numerator, and the denominator—with their negatives.

The forms $\frac{a}{b}$ and $\frac{-a}{b}$, in which the sign of the fraction and the sign of the denominator are both positive, are generally the most convenient representations and will be referred to as **standard forms**. Thus,

$$\frac{-3}{5}, \quad \frac{3}{5}, \quad \text{and} \quad \frac{7}{10}$$

are in standard form, while

$$\frac{3}{-5}, \quad -\frac{-3}{5}, \quad \text{and} \quad -\frac{7}{-10}$$

are not.

If the numerator or denominator of a fraction is an expression containing more than one term, there are alternative standard forms. For example, since

$$a - b = -(b - a),$$

we have

$$\frac{-b}{a - b} = \frac{-b}{-(b - a)} = \frac{b}{b - a},$$

and either

$$\frac{-b}{a - b} \quad \text{or} \quad \frac{b}{b - a}$$

may be taken as standard form, as convenience dictates. Observe that the quotient is not defined for $a - b = 0$ or $b - a = 0$; so we must make the restriction $a \neq b$.

Particular care should be taken when writing a fraction in standard form when the numerator contains more than one term. For example,

$$-\frac{a - b}{c} = \frac{-(a - b)}{c},$$

where the minus sign in the right member precedes *the entire numerator* $a - b$. This fraction can now be written as

$$\frac{-a + b}{c} \quad \text{or} \quad \frac{b - a}{c}.$$

Common Error

In particular, note that $-\dfrac{a - b}{c} \neq \dfrac{-a - b}{c}.$

EXAMPLE 2 Write each fraction in standard form and specify any values of the variables for which the fraction is undefined.

a. $-\dfrac{5}{-y}$ b. $-\dfrac{a}{a-2}$ c. $-\dfrac{x-1}{3}$

Solutions a. $-\dfrac{5}{-y} = \dfrac{5}{y} \quad (y \neq 0)$ b. $-\dfrac{a}{a-2} = \dfrac{-a}{a-2}$ or $\dfrac{a}{2-a} \quad (a \neq 2)$ c. $-\dfrac{x-1}{3} = \dfrac{-(x-1)}{3}$ or $\dfrac{-x+1}{3}$ ■

Common Error *In Example c, note that* $-\dfrac{x-1}{3} \neq \dfrac{-x-1}{3}$.

EXAMPLE 3 Write each fraction on the left as an equal fraction in standard form with the denominator shown on the right.

a. $\dfrac{-1}{2-x}; \dfrac{\quad}{x-2}$ b. $-\dfrac{a}{b-a}; \dfrac{\quad}{a-b}$ c. $\dfrac{3}{3x-2y}; \dfrac{\quad}{2y-3x}$

Solutions a. $\dfrac{-1}{2-x} = \dfrac{-1}{-(x-2)} = \dfrac{1}{x-2}$

b. $-\dfrac{a}{b-a} = -\dfrac{a}{-(a-b)} = \dfrac{a}{a-b}$

c. $\dfrac{3}{3x-2y} = \dfrac{3}{-(2y-3x)} = \dfrac{-3}{2y-3x}$ ■

Reducing fractions

A fraction is said to be in lowest terms if the numerator and denominator do not contain common factors. The arithmetic fraction a/b, where a and b are integers and $b \neq 0$, is in lowest terms providing a and b are relatively prime—that is, providing they contain no common integral factor other than 1 or -1. If the numerator and denominator of a fraction are polynomials with integral coefficients, then the fraction is said to be in lowest terms if the numerator and denominator do not contain a common polynomial factor with integral coefficients other than 1 or -1.

To express a given fraction in lowest terms (to **reduce** the fraction), we factor the numerator and denominator, and then apply the fundamental principle of fractions.

EXAMPLE 4

a. $$\frac{yz^2}{y^3z} = \frac{z \cdot yz}{y^2 \cdot yz} = \frac{z}{y^2} \quad (y,z \neq 0)$$

b. $$\frac{8x^3y}{6x^2y^3} = \frac{4x \cdot 2x^2y}{3y^2 \cdot 2x^2y} = \frac{4x}{3y^2} \quad (x,y \neq 0)$$ ■

Slash lines are sometimes used to abbreviate the procedure in the above examples. For example, instead of writing

$$\frac{y}{y^2} = \frac{1 \cdot y}{y \cdot y} = \frac{1}{y} \qquad (y \neq 0),$$

we can write

$$\frac{y}{y^2} = \frac{\overset{1}{\cancel{y}}}{\underset{y}{\cancel{y^2}}} = \frac{1}{y} \qquad (y \neq 0).$$

or simply

$$\frac{\cancel{y}}{\underset{y}{\cancel{y^2}}} = \frac{1}{y} \qquad (y \neq 0).$$

The division of a polynomial containing more than one term by a monomial may also be considered a special case of changing a fraction to lowest terms, providing the monomial is contained as a factor in each term of the polynomial.

EXAMPLE 5

a. $$\frac{6y - 3}{3} = \frac{\cancel{3}(2y - 1)}{\cancel{3}} = 2y - 1$$

b. $$\frac{9x^3 - 6x^2 + 3x}{3x} = \frac{\cancel{3x}(3x^2 - 2x + 1)}{\cancel{3x}} = 3x^2 - 2x + 1 \quad (x \neq 0)$$ ■

Common binomial factors in a numerator and denominator of a fraction can also be "divided out" by using the fundamental principle of fractions. It is usually necessary to factor polynomials in the numerator and denominator in order to see the common factors.

EXAMPLE 6

a. $\dfrac{x^2 - 7x + 6}{x^2 - 36}$

$= \dfrac{\cancel{(x - 6)}(x - 1)}{\cancel{(x - 6)}(x + 6)}$

$= \dfrac{x - 1}{x + 6} \qquad (x \neq 6, -6)$

b. $\dfrac{a - b}{b^2 - a^2}$

$= \dfrac{-1\cancel{(b - a)}}{(b + a)\cancel{(b - a)}}$

$= \dfrac{-1}{b + a} \qquad (a \neq b, -b)$

Notice that in Example b it was necessary to write $a - b$ in the numerator as $-1(b - a)$ so that the common factors $b - a$ could be "divided out." ■

If the divisor is contained as a factor in the dividend, then the division of one polynomial by another polynomial, where each contains more than one term, can also be considered an example of reducing a fraction to lowest terms.

EXAMPLE 7

a. $\dfrac{y^2 - 5y + 4}{y - 1}$

$= \dfrac{(y - 4)\cancel{(y - 1)}}{\cancel{y - 1}}$

$= y - 4 \qquad (y \neq 1)$

b. $\dfrac{2x^2 + x - 15}{x + 3}$

$= \dfrac{(2x - 5)\cancel{(x + 3)}}{\cancel{x + 3}}$

$= 2x - 5 \qquad (x \neq -3)$ ■

Common Errors

Note that the fundamental principle of fractions enables us to obtain an equivalent expression by dividing out any nonzero factors *that are common to both numerator and denominator of a fraction.* The fundamental principle of fractions does not apply to common terms. *For example,*

$$\frac{2xy}{3y} = \frac{2x}{3} \qquad (y \neq 0)$$

because y is a common factor in the numerator and denominator of the left member. However,

$$\frac{2x + y}{3 + y} \neq \frac{2x}{3}$$

because y is a common term but is not a common factor *of the numerator and the denominator. Furthermore,*

$$\frac{5x + 3}{5y} \neq \frac{x + 3}{y}$$

because 5 is not *a common factor of the* entire *numerator.*

EXAMPLE 8 a. $\dfrac{2x + 4}{4}$ b. $\dfrac{3x + 6}{3}$ c. $\dfrac{9x^2 + 3}{6x + 3}$

Solutions a. $\dfrac{2x + 4}{4}$

$$= \frac{\cancel{2}(x + 2)}{\cancel{2}(2)}$$

$$= \frac{x + 2}{2}$$

b. $\dfrac{3x + 6}{3}$

$$= \frac{\cancel{3}(x + 2)}{\cancel{3}}$$

$$= x + 2$$

c. $\dfrac{9x^2 + 3}{6x + 3}$

$$= \frac{\cancel{3}(3x^2 + 1)}{\cancel{3}(2x + 1)}$$

$$= \frac{3x^2 + 1}{2x + 1} \quad \left(x \neq -\frac{1}{2}\right)$$

■

Common Errors

Note that, in Example 8a *above,*

$$\frac{2x + 4}{4} \neq \frac{2x + \cancel{4}}{\cancel{4}};$$

in Example b,

$$\frac{9x^2 + 3}{6x + 3} \neq \frac{9x^2 + \cancel{3}}{6x + \cancel{3}};$$

and in Example c,

$$\frac{3x + 6}{3} \neq \frac{\cancel{3}x + 6}{\cancel{3}}.$$

To avoid the necessity of always having to note restrictions on divisors (denominators), we shall assume in the remaining exercise sets in this chapter that no denominator is 0. However, we will continue to show restrictions in the examples in the text.

EXERCISE 5.1

A

■ *Write in standard form and specify any values of the variables for which the fraction is undefined. See Examples 1 and 2.*

1. $\dfrac{1}{-4}$
2. $-\dfrac{1}{3}$
3. $-\dfrac{3}{-5}$
4. $-\dfrac{-3}{4}$

5. $\frac{-2}{-5}$
6. $\frac{-6}{-7}$
7. $-\frac{-3}{-7}$
8. $-\frac{-4}{-5}$
9. $-\frac{2x}{y}$
10. $\frac{x}{-3y}$
11. $-\frac{-3x}{4y}$
12. $-\frac{x}{-2y}$
13. $\frac{x+1}{-x}$
14. $\frac{x+3}{-x}$
15. $-\frac{x-y}{y+2}$
16. $-\frac{y-x}{y-1}$

■ *Write each fraction on the left as an equal fraction in standard form with the denominator shown on the right. (Assume that no denominator equals 0.) See Example 3.*

17. $-\frac{4}{3-y}; \frac{}{y-3}$
18. $\frac{-3}{2-x}; \frac{}{x-2}$
19. $\frac{1}{x-y}; \frac{}{y-x}$
20. $\frac{-6}{x-y}; \frac{}{y-x}$
21. $\frac{x-2}{3-x}; \frac{}{x-3}$
22. $\frac{2x-5}{3-y}; \frac{}{y-3}$
23. $\frac{x+1}{x-y}; \frac{}{y-x}$
24. $\frac{x+3}{y-x}; \frac{}{x-y}$
25. $-\frac{x-2}{x-y}; \frac{}{y-x}$
26. $-\frac{x-4}{x-2y}; \frac{}{2y-x}$
27. $\frac{-a+1}{-3a-b}; \frac{}{3a+b}$
28. $\frac{-a-1}{2b-3a}; \frac{}{3a-2b}$
29. Show by a counterexample that $-\frac{x+3}{2}$ is not equivalent to $\frac{-x+3}{2}$.
30. Show by a counterexample that $-\frac{2x-y}{3}$ is not equivalent to $\frac{-2x-y}{3}$.

■ *Use the fundamental principle of fractions to reduce each fraction to lowest terms in standard form. (Assume no denominator is 0.) See Example 4 for Exercises 31–42.*

31. $\frac{6x^3y^2}{3xy^3}$
32. $\frac{12a^4b^2}{4a^2b^4}$
33. $\frac{14tr^4}{7t^2r^2}$
34. $\frac{22a^2bc^3}{11a^4c^2}$
35. $\frac{14cd}{-7c^2d^3}$
36. $\frac{100mn}{-5m^2n^3}$
37. $\frac{abc}{a^4bc^3}$
38. $\frac{xyz}{xy^2z^3}$
39. $\frac{m^2np}{-6m^2np^3}$
40. $\frac{abc}{-7abc^3}$
41. $\frac{-12r^2st}{-6rst^2}$
42. $\frac{-15xy^3z}{-3y^2z^4}$

■ *See Examples 5–8 for Exercises 43–74.*

43. $\frac{4x+6}{6}$
44. $\frac{2y-8}{8}$
45. $\frac{9x-3}{9}$
46. $\frac{5y-10}{5}$
47. $\frac{a^3-3a^2+2a}{-a}$
48. $\frac{3x^3-6x^2+3x}{-3x}$

49. $\dfrac{y^2 + 5y - 14}{y - 2}$

50. $\dfrac{x^2 + 5x + 6}{x + 3}$

51. $\dfrac{5a - 10}{3a - 6}$

52. $\dfrac{6x + 9}{10x + 15}$

53. $\dfrac{(a - b)^2}{2a - 2b}$

54. $\dfrac{x^2 - 16}{2x + 8}$

55. $\dfrac{6t^2 - 6}{(t - 1)^2}$

56. $\dfrac{4x^2 - 4}{(x + 1)^2}$

57. $\dfrac{2y^2 - 8}{2y + 4}$

58. $\dfrac{5y^2 - 20}{2y - 4}$

59. $\dfrac{6 - 2y}{y^2 - 9}$

60. $\dfrac{4 - 2y}{y^2 - 4}$

61. $\dfrac{y^2 - x^2}{(x - y)^2}$

62. $\dfrac{(2x - y)^2}{y^2 - 4x^2}$

63. $\dfrac{x^2 - 5x + 4}{x^2 - 1}$

64. $\dfrac{y^2 - 2y - 3}{y^2 - 9}$

65. $\dfrac{2y^2 + y - 6}{y^2 + y - 2}$

66. $\dfrac{6x^2 - x - 1}{2x^2 + 9x - 5}$

67. $\dfrac{x^2 + xy - 2y^2}{x^2 - y^2}$

68. $\dfrac{4x^2 - 9y^2}{2x^2 + xy - 6y^2}$

69. $\dfrac{x^2 + ax + xy + ay}{2x + 2a}$

70. $\dfrac{ax^2 + x + a^2x + a}{3x + 3a}$

71. $\dfrac{ax - 2bx + ay - 2by}{a^2 - 4b^2}$

72. $\dfrac{2ax - 4ay + bx - 2by}{x^2 - 4y^2}$

73. $\dfrac{8y^3 - 27}{4y^2 - 9}$

74. $\dfrac{8x^3 - 1}{4x^2 - 1}$

75. Show by a counterexample that $\dfrac{2x + y}{y}$ is not equivalent to $2x$.

76. Show by a counterexample that $\dfrac{4x - y}{4}$ is not equivalent to $x - y$.

5.2

DIVIDING POLYNOMIALS

In Section 5.1 we used the fundamental principle of fractions to reduce fractions in which common factors occurred in numerators and denominators. However, we can use other methods to rewrite fractions if the polynomial in the numerator and the polynomial in the denominator do not have common factors.

Monomial denominators

If the divisor in a quotient of polynomials is a monomial, we can first rewrite the quotient as the sum of two or more fractions; that is,

$$\frac{a + b + c}{g} = \frac{a}{g} + \frac{b}{g} + \frac{c}{g}.$$

Once the quotient is written as separate fractions, each can be reduced as before.

EXAMPLE 1 a. $\dfrac{6y^2 + 4y + 1}{2}$

$$= \frac{6y^2}{2} + \frac{4y}{2} + \frac{1}{2}$$

$$= 3y^2 + 2y + \frac{1}{2}$$

b. $\dfrac{9x^3 - 6x^2 + 4}{3x}$

$$= \frac{9x^3}{3x} - \frac{6x^2}{3x} + \frac{4}{3x}$$

$$= 3x^2 - 2x + \frac{4}{3x} \quad (x \neq 0)$$

■

Long division If the denominator in a quotient of polynomials is not a monomial, we can use a method similar to the long division process used in arithmetic to rewrite the quotient as the sum of a polynomial and a fraction.

EXAMPLE 2 Rewrite $\dfrac{2x^2 + x - 7}{x + 3}$ as the sum of a polynomial and a rational expression.

Solution We first write

$$x + 3 \overline{\big)\, 2x^2 + x - 7}$$

and then divide $2x^2$ by x $(2x^2/x)$ to obtain $2x$; subtract the product of $2x$ and $x + 3$ from $2x^2 + x$, and "bring down" -7:

$$\begin{array}{rl} & \,2x \\ x + 3 & \overline{\big)\, 2x^2 + x - 7} \\ \text{Change signs and add} \rightarrow & \,\underline{2x^2 + 6x} \\ & \; - 5x - 7 \end{array}$$

We then divide $-5x$ by x $(-5x/x)$ to obtain -5; subtract the product of -5 and $x + 3$ from $-5x - 7$.

$$\begin{array}{rl} & \,2x - 5 \\ x + 3 & \overline{\big)\, 2x^2 + x - 7} \\ \text{Change signs and add} \rightarrow & \,\underline{2x^2 + 6x} \\ & \; - 5x - 7 \\ \text{Change signs and add} \rightarrow & \; \underline{- 5x - 15} \\ & \; 8 \end{array}$$

Hence,

$$\frac{2x^2 + x - 7}{x + 3} = 2x - 5 + \frac{8}{x + 3} \qquad (x \neq -3). \qquad ■$$

When using this process of long division, it helps to write the dividend in descending powers of the variable. Furthermore, it is sometimes helpful to insert a term with a zero coefficient so that like terms will be aligned for convenient computation.

EXAMPLE 3 Rewrite $\dfrac{3x - 1 + 4x^3}{2x - 1}$ using the long division process.

Solution We first write $3x - 1 + 4x^3$ in descending powers as $4x^3 + 3x - 1$. Then we insert $0x^2$ between $4x^3$ and $3x$ and divide.

$$\begin{array}{rl}
 & 2x^2 + x + 2 \\
2x - 1 & \overline{)\,4x^3 + 0x^2 + 3x - 1} \\
\text{Change signs and add} \rightarrow & \underline{4x^3 - 2x^2} \\
 & 2x^2 + 3x \\
\text{Change signs and add} \rightarrow & \underline{2x^2 - x} \\
 & 4x - 1 \\
\text{Change signs and add} \rightarrow & \underline{4x - 2} \\
 & 1
\end{array}$$

Hence,

$$\frac{3x - 1 + 4x^3}{2x - 1} = 2x^2 + x + 2 + \frac{1}{2x - 1} \qquad \left(x \neq \frac{1}{2}\right). \qquad ■$$

We call an expression such as

$$2x^2 + x + 2 + \frac{1}{2x - 1}$$

a **mixed expression,** just as we call symbols such as $3\frac{1}{2}$ mixed numbers.

The procedure followed in the examples above can also be used to rewrite quotients with denominators that are polynomials of degree greater than one.

EXAMPLE 4 Divide: $\dfrac{z^4 - 3z^3 + 2z^2 - 3z + 1}{z^2 + 2z - 1}$.

Solution

$$\begin{array}{r|l}
 & z^2 - 5z + 13 \\
\hline
z^2 + 2z - 1 & z^4 - 3z^3 + 2z^2 - 3z + 1 \\
 & \underline{z^4 + 2z^3 - z^2} \\
 & \quad -5z^3 + 3z^2 - 3z \\
 & \quad \underline{-5z^3 - 10z^2 + 5z} \\
 & \qquad\qquad 13z^2 - 8z + 1 \\
 & \qquad\qquad \underline{13z^2 + 26z - 13} \\
 & \qquad\qquad\qquad -34z + 14
\end{array}$$

Hence,

$$\frac{z^4 - 3z^3 + 2z^2 - 3z + 1}{z^2 + 2z - 1} = z^2 - 5z + 13 + \frac{-34z + 14}{z^2 + 2z - 1}.$$ ■

EXERCISE 5.2

A

■ *Divide. See Example 1.*

1. $\dfrac{8a^2 + 4a + 1}{2}$
2. $\dfrac{15t^3 - 12t^2 + 5t}{3t^2}$
3. $\dfrac{7y^4 - 14y^2 + 3}{7y^2}$
4. $\dfrac{21n^4 + 14n^2 - 7}{7n^2}$
5. $\dfrac{18r^2s^2 - 15rs + 6}{3rs}$
6. $\dfrac{12x^3 - 8x^2 + 3x}{4x}$
7. $\dfrac{8a^2x^2 - 4ax^2 + ax}{2ax}$
8. $\dfrac{9a^2b^2 + 3ab^2 + 4a^2b}{ab^2}$
9. $\dfrac{25m^6 - 15m^3 + 7}{-5m^3}$
10. $\dfrac{36t^5 + 24t^3 - 12t}{-12t^2}$
11. $\dfrac{40m^4 - 25m^2 + 7m}{5m^2}$
12. $\dfrac{15s^{10} - 21s^5 + 6}{3s^2}$

■ *Divide. See Examples 2 and 3.*

13. $\dfrac{4y^2 + 12y + 7}{2y + 1}$
14. $\dfrac{2n^2 + 13n - 6}{2n - 1}$
15. $\dfrac{4t^2 - 4t - 5}{2t - 1}$
16. $\dfrac{2x^2 - 3x - 15}{2x + 5}$
17. $\dfrac{x^3 + 2x^2 + x + 1}{x - 2}$
18. $\dfrac{2x^3 - 3x^2 - 2x + 4}{x + 1}$
19. $\dfrac{2a - 3a^2 + a^4 - 1}{a + 3}$
20. $\dfrac{3 + 2b^2 + 2b^4}{b - 4}$
21. $\dfrac{4z^2 + 5z + 8z^4 + 3}{2z + 1}$
22. $\dfrac{7 - 3t^3 - 23t^2 + 10t^4}{2t + 3}$
23. $\dfrac{x^4 - 1}{x - 2}$
24. $\dfrac{y^5 + 1}{y - 1}$

B

■ *Divide. See Example* 4.

25. $\dfrac{x^3 - 3x^2 + 2x + 5}{x^2 - 2x + 7}$

26. $\dfrac{2y^3 + 5y^2 - 3y + 2}{y^2 - y - 3}$

27. $\dfrac{4a^4 + 3a^3 - 2a + 1}{a^2 + 3a - 1}$

28. $\dfrac{2b^4 - 3b^2 + b + 2}{b^2 + b - 3}$

29. $\dfrac{t^4 - 3t^3 + 2t^2 - 2t + 1}{t^3 - 2t^2 + t + 2}$

30. $\dfrac{r^4 + r^3 - 2r^2 + r + 5}{r^3 + 2r + 3}$

31. Determine k so that the polynomial $x^3 - 3x + k$ has $x - 2$ as a factor.

32. Determine k so that the polynomial $x^3 + 2x^2 + k$ has $x + 3$ as a factor.

5.3

SYNTHETIC DIVISION

In Section 5.2, we rewrote quotients of polynomials of the form $P(x)/D(x)$ using a long division algorithm (process). If the divisor $D(x)$ is of the form $(x - a)$—*a first-degree polynomial where the coefficient of x is 1 and* $a \neq 0$—this algorithm may be simplified by a procedure known as **synthetic division.** Consider the quotient

$$\frac{x^4 + x^2 + 2x - 1}{x + 3}.$$

The division can be accomplished by long division as follows:

$$\begin{array}{r|l}
 & x^3 - 3x^2 + 10x - 28 \\
\hline
x + 3 & x^4 + 0x^3 + x^2 + 2x - 1 \\
 & \underline{x^4 + 3x^3} \\
 & \quad -3x^3 + x^2 \\
 & \quad \underline{-3x^3 - 9x^2} \\
 & \qquad 10x^2 + 2x \\
 & \qquad \underline{10x^2 + 30x} \\
 & \qquad\quad -28x - 1 \\
 & \qquad\quad \underline{-28x - 84} \\
 & \qquad\qquad 83 \ \text{(remainder).}
\end{array}$$

We see that, for $x \neq -3$,

$$\frac{x^4 + x^2 + 2x - 1}{x + 3} = x^3 - 3x^2 + 10x - 28 + \frac{83}{x + 3}$$

or, clearing for fractions,

$$x^4 + x^2 + 2x - 1 = (x^3 - 3x^2 + 10x - 28)(x + 3) + 83.$$

If we omit the variables, writing only the coefficients of the terms, and use 0 for the coefficient of any missing power, we have

$$\begin{array}{r|l}
 & \underline{1 - 3 + 10 - 28} \\
1 + 3 & 1 + 0 + 1 + 2 - 1 \\
 & \underline{1 + 3} \\
 & \quad - 3 + (1) \\
 & \quad \underline{- 3 - 9} \\
 & \qquad\quad 10 + (2) \\
 & \qquad\quad \underline{10 + 30} \\
 & \qquad\qquad - 28 - (1) \\
 & \qquad\qquad \underline{- 28 - 84} \\
 & \qquad\qquad\qquad 83 \quad \text{(remainder).}
\end{array}$$

Now, observe that the numbers shown in color are repetitions of the numbers written immediately above and are also repetitions of the coefficients of the associated variable in the quotient. The numbers shown in parentheses are repetitions of the coefficients of the dividend. Therefore, the whole process can be written in compact form as

$$\begin{array}{r|rrrrrll}
3 & 1 & 0 & 1 & 2 & -1 & & (1) \\
 & & 3 & -9 & 30 & -84 & & (2) \\
\hline
 & 1 & -3 & 10 & -28 & 83 & \text{(remainder: 83),} & (3)
\end{array}$$

where the repetitions are omitted and where 1, the coefficient of x in the divisor, has also been omitted.

The numbers in line (3) which are coefficients of variables in the quotient, and the remainder, have been obtained by *subtracting* the **detached coefficients** in line (2) from the detached coefficients of terms of the same degree in line (1). We could obtain the same result by replacing 3 with -3 in the divisor and *adding* instead of subtracting at each step. This is what is done in the *synthetic division* process. The final form then is

$$\begin{array}{r|rrrrrll}
-3 & 1 & 0 & 1 & 2 & -1 & & (1) \\
 & & -3 & 9 & -30 & 84 & & (2) \\
\hline
 & 1 & -3 & 10 & -28 & 83 & \text{(remainder: 83)} & (3)
\end{array}$$

Note that the numbers in line (2) are obtained by multiplying the preceding number to the left in line (3) by -3.

Comparing the results of using synthetic division with the same process using long division, we observe that the numbers in line (3) are the coefficients of the polynomial

$$x^3 - 3x^2 + 10x - 28,$$

and that there is a remainder of 83.

As another example, let us write the quotient

$$\frac{3x^3 - 4x - 1}{x - 2}$$

in the form $Q(x) + r/(x - a)$. Using synthetic division, we begin by writing

$$\begin{array}{r|rrrr} 2 & 3 & 0 & -4 & -1 \\ \hline \end{array}$$

where 0 has been inserted in the position corresponding to the coefficient of a second-degree term. The divisor is the negative of -2, or 2. Then, we have

$$\begin{array}{r|rrrrl} 2 & 3 & 0 & -4 & -1 & \quad (1) \\ & & 6 & 12 & 16 & \quad (2) \\ \hline & 3 & 6 & 8 & 15 & \text{(remainder: 15)} \quad (3) \end{array}$$

This process employs these steps:

1. 3 is "brought down" from line (1) to line (3).
2. 6, the product of 2 and 3, is written in the next position on line (2).
3. 6, the sum of 0 and 6, is written on line (3).
4. 12, the product of 2 and 6, is written in the next position on line (2).
5. 8, the sum of -4 and 12, is written on line (3).
6. 16, the product of 2 and 8, is written in the next position on line (2).
7. 15, the sum of -1 and 16, is written on line (3).

We now use the first three numbers on line (3) as coefficients to write a polynomial of degree one less than the degree of the dividend. This polynomial is the quotient lacking the remainder. The last number is the remainder. Thus, for $x - 2 \neq 0$, the quotient of $3x^3 - 4x - 1$ divided by $(x - 2)$ is

$$3x^2 + 6x + 8$$

with a remainder of 15; that is,

$$\frac{3x^3 - 4x - 1}{x - 2} = 3x^2 + 6x + 8 + \frac{15}{x - 2} \qquad (x \neq 2).$$

We will use synthetic division in the following examples to rewrite each quotient in the form $Q(x) + r/(x - a)$, where r is a constant.

EXAMPLE 1 $\dfrac{2x^4 + x^3 - 1}{x + 2}$

Solution We begin by writing

$$\underline{-2|} \quad 2 \quad 1 \quad 0 \quad 0 \quad -1,$$

where 0 is the coefficient of any missing power. The divisor is the negative of 2, i.e., -2. We proceed by using the following steps.

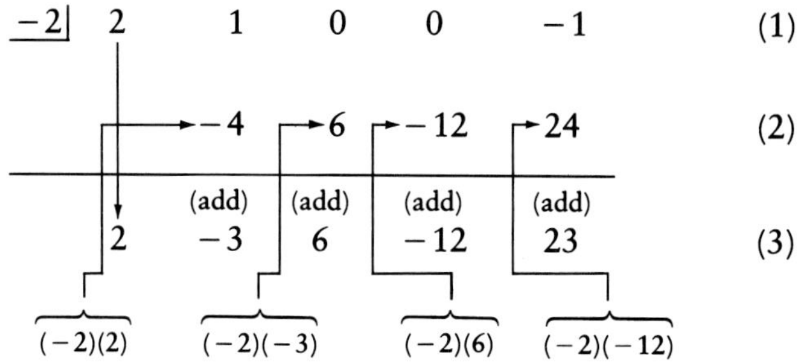

$\underline{-2\vert}$	2	1	0	0	−1	(1)
		−4	6	−12	24	(2)
		(add)	(add)	(add)	(add)	
	2	−3	6	−12	23	(3)
		(−2)(2)	(−2)(−3)	(−2)(6)	(−2)(−12)	

We use the first four numbers on line (3) as the coefficients of $Q(x)$. The last number is the remainder:

$$2x^3 - 3x^2 + 6x - 12 + \frac{23}{x + 2} \qquad (x \neq -2) \qquad ■$$

EXAMPLE 2 $\dfrac{x^4 - 2x^2 + 3}{x - 2}$

Solution We begin by writing

$$\underline{2|} \quad 1 \quad 0 \quad -2 \quad 0 \quad 3,$$

where 0 is the coefficient of any missing power. The divisor is the negative of -2, i.e., 2. We proceed by using the following steps.

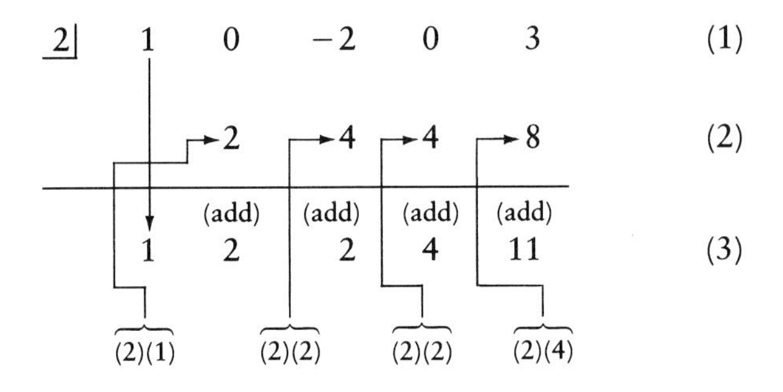

We use the first four numbers on line (3) as the coefficients of $Q(x)$. The last number is the remainder.

$$x^3 + 2x^2 + 2x + 4 + \frac{11}{x - 2} \qquad (x \neq 2)$$ ■

EXERCISE 5.3

A

■ *Use synthetic division to write each quotient* $P(x)/(x - a)$ *in the form* $Q(x)$ *or* $Q(x) + r/(x - a)$, *where r is a constant. See Examples 1 and 2.*

1. $\dfrac{x^2 - 8x + 12}{x - 6}$
2. $\dfrac{a^2 + a - 6}{a + 3}$
3. $\dfrac{x^2 + 4x + 4}{x + 2}$
4. $\dfrac{x^2 + 6x + 9}{x + 3}$
5. $\dfrac{x^4 - 3x^3 + 2x^2 - 1}{x - 2}$
6. $\dfrac{x^4 + 2x^2 - 3x + 5}{x - 3}$
7. $\dfrac{2x^3 + x - 5}{x + 1}$
8. $\dfrac{3x^3 + x^2 - 7}{x + 2}$
9. $\dfrac{2x^4 - x + 6}{x - 5}$
10. $\dfrac{3x^4 - x^2 + 1}{x - 4}$
11. $\dfrac{x^3 + 4x^2 + x - 2}{x + 2}$
12. $\dfrac{x^3 - 7x^2 - x + 3}{x + 3}$
13. $\dfrac{x^4 - 2x^3 + x^2 - 1}{x - 2}$
14. $\dfrac{x^4 + x^3 - 2x + 3}{x + 1}$
15. $\dfrac{x^5 + x^4 + 2x^2 - 1}{x - 1}$
16. $\dfrac{x^5 - 2x^3 + x^2 - 1}{x + 2}$
17. $\dfrac{x^5 - 3x^2 - 1}{x + 1}$
18. $\dfrac{x^5 - 2x^2 + 1}{x - 2}$
19. $\dfrac{x^6 + x^4 - x}{x - 1}$
20. $\dfrac{x^6 + 3x^3 - 2x - 1}{x - 2}$
21. $\dfrac{x^5 - 1}{x - 1}$
22. $\dfrac{x^5 + 1}{x + 1}$
23. $\dfrac{x^6 - 1}{x - 1}$
24. $\dfrac{x^6 + 1}{x + 1}$

5.4

BUILDING FRACTIONS

Just as we changed fractions to equivalent fractions in lowest terms in Section 3.2 by applying the fundamental principle in the form

$$\frac{ac}{bc} = \frac{a}{b} \qquad (b, c \neq 0),$$

we can also change fractions to equivalent fractions in higher terms by applying the fundamental principle in the form

$$\boldsymbol{\frac{a}{b} = \frac{ac}{bc} \qquad (b, c \neq 0).}$$

For example, $\frac{1}{2}$ can be changed to an equivalent fraction with a denominator of 8 by multiplying the numerator by 4 and the denominator by 4. Thus,

$$\frac{1}{2} = \frac{1 \cdot 4}{2 \cdot 4} = \frac{4}{8}.$$

This process is called **building** a fraction, and the number 4 is said to be a **building factor.** The fraction $\frac{4}{8}$ is said to be in **higher terms** than $\frac{1}{2}$.

In general, when building a/b to an equivalent fraction with bc as a denominator (i.e., $a/b = ?/bc$) we can usually determine the building factor c by inspection, and then multiply the numerator and the denominator of the original fraction by this building factor. If the building factor cannot be obtained by inspection, the desired denominator (bc) can be divided by the denominator of the given fraction (b) to determine the building factor (c).

EXAMPLE 1 Express each fraction as an equivalent fraction with the given denominator.

a. $\dfrac{5x}{3y} = \dfrac{?}{12y^2}$ b. $\dfrac{2}{y-1} = \dfrac{?}{y^2-1}$

Solutions a. By inspection we note that the building factor is $4y$. Alternatively,

$$12y^2 \div 3y = 4y.$$

Hence,

$$\frac{5x(4y)}{3y(4y)} = \frac{20xy}{12y^2}.$$

b. We first factor $y^2 - 1$ as $(y - 1)(y + 1)$. Then by inspection we note that the building factor is $y + 1$. Alternatively,

$$(y^2 - 1) \div (y - 1) = y + 1.$$

Hence,

$$\frac{2(y + 1)}{(y - 1)(y + 1)} = \frac{2y + 2}{y^2 - 1}. \quad ■$$

Least common denominator

Sometimes it is necessary to change two or more fractions with unlike denominators into fractions with a common denominator. In particular, the common denominator that is the most useful is the **least common multiple** (**LCM**) of the denominators, called the **least common denominator** (**LCD**). The LCM of two or more natural numbers is the smallest natural number that is exactly divisible by each of the given numbers. Thus, 24 is the LCM of 6 and 8, because 24 is the smallest natural number each will divide into without a remainder.

In the example above it was easy to find the LCM of 6 and 8 (24) by inspection. Sometimes it is necessary to factor each number first in order to find the LCM.

To find the LCM of a Set of Natural Numbers:

1. Express each number in completely factored form.
2. Write as factors of a product each *different* prime factor occurring in any of the numbers, including each factor the greatest number of times it occurs in any one of the given numbers.

EXAMPLE 2 Find the LCM of 12, 9, and 15.

Solution

The numbers:	12	9	15
Appear in prime factor form as:	$2 \cdot 2 \cdot 3$	$3 \cdot 3$	$3 \cdot 5$
The LCM contains the factors:	$2 \cdot 2 \cdot 3 \cdot 3 \cdot 5$		
The LCM is:	180		

The factors 2 and 3 are each used twice because 2 appears twice as a factor of 12 and 3 appears twice as a factor of 9. ■

We define the LCM of a set of polynomials as the polynomial of lowest degree yielding a polynomial quotient upon division by each of the given polynomials. We can find the LCM of a set of polynomials with integral coefficients in a manner comparable to that used with a set of natural numbers.

EXAMPLE 3 Find the LCM of $x^2 - 9$ and $x^2 - x - 6$.

Solution We first factor each polynomial.

$$x^2 - 9 \qquad\qquad x^2 - x - 6$$
$$(x - 3)(x + 3) \qquad\qquad (x - 3)(x + 2)$$

Hence, the LCM is $(x - 3)(x + 3)(x + 2)$. ■

We can now write two or more fractions with unlike denominators as equivalent fractions with a common denominator. In particular, with the least common denominator (LCD).

EXAMPLE 4 Write the fractions $\frac{1}{12}$, $\frac{2}{9}$, and $\frac{4}{15}$ as equivalent fractions with the LCD of the fractions.

Solution From Example 2 above, the LCM of 12, 9, and 15 is 180. Hence, the LCD is 180 and we seek numerators such that

$$\frac{1}{12} = \frac{?}{180}, \quad \frac{2}{9} = \frac{?}{180}, \quad \text{and} \quad \frac{4}{15} = \frac{?}{180}.$$

Since $180 \div 12 = 15$, $180 \div 9 = 20$, and $180 \div 15 = 12$, we have

$$\frac{1(15)}{12(15)} = \frac{15}{180}, \quad \frac{2(20)}{9(20)} = \frac{40}{180}, \quad \text{and} \quad \frac{4(12)}{15(12)} = \frac{48}{180}.$$ ■

A similar procedure is used if the fractions involve variables.

EXAMPLE 5 Write the fractions $\frac{3}{x^2 - 9}$ and $\frac{4x}{x^2 - x - 6}$ as equivalent fractions with the LCD of the fractions.

Solution From Example 3 above, the LCM of $x^2 - 9$ and $x^2 - x - 6$ is the product $(x - 3)(x + 3)(x + 2)$. Hence, the LCD is $(x - 3)(x + 3)(x + 2)$ and we seek numerators such that

$$\frac{3}{x^2 - 9} = \frac{?}{(x - 3)(x + 3)(x + 2)}$$

and

$$\frac{4x}{x^2 - x - 6} = \frac{?}{(x - 3)(x + 3)(x + 2)}.$$

Since

$$(x - 3)(x + 3)(x + 2) \div (x - 3)(x + 3) = x + 2$$

and

$$(x - 3)(x + 3)(x + 2) \div (x - 3)(x + 2) = x + 3,$$

we have

$$\frac{3(x + 2)}{(x^2 - 9)(x + 2)} = \frac{3x + 6}{(x - 3)(x + 3)(x + 2)}$$

and

$$\frac{4x(x + 3)}{(x^2 - x - 6)(x + 3)} = \frac{4x^2 + 12x}{(x - 3)(x + 3)(x + 2)}.$$ ■

EXERCISE 5.4

A

■ *Express each fraction as an equivalent fraction with the given denominator. See Example* 1.

1. $\frac{2}{3} = \frac{?}{9}$
2. $\frac{3}{4} = \frac{?}{8}$
3. $\frac{-15}{7} = \frac{?}{14}$
4. $\frac{-12}{5} = \frac{?}{20}$
5. $4 = \frac{?}{5}$
6. $6 = \frac{?}{7}$
7. $\frac{2}{6x} = \frac{?}{18x}$
8. $\frac{5}{3y} = \frac{?}{21y}$

9. $\dfrac{-a^2}{b^2} = \dfrac{?}{b^3}$

10. $\dfrac{-a}{b} = \dfrac{?}{ab^2}$

11. $y = \dfrac{?}{xy}$

12. $x = \dfrac{?}{xy^3}$

13. $\dfrac{3}{a-b} = \dfrac{?}{a^2-b^2}$

14. $\dfrac{5}{2a+b} = \dfrac{?}{4a^2-b^2}$

15. $\dfrac{3x}{y+2} = \dfrac{?}{y^2-y-6}$

16. $\dfrac{5x}{y+3} = \dfrac{?}{y^2+y-6}$

17. $\dfrac{-2}{x+1} = \dfrac{?}{x^2+3x+2}$

18. $\dfrac{-3}{a+2} = \dfrac{?}{a^2+3a+2}$

■ *Find the LCM. See Examples 2 and 3.*

19. 4, 6, 10
20. 3, 4, 5
21. 6, 8, 15
22. 4, 15, 18
23. 14, 21, 36
24. 4, 11, 22
25. $2ab, 6b^2$
26. $12xy, 24x^3y^2$
27. $6xy, 8x^2, 3xy^2$
28. $7x, 8y, 6z$
29. $(a-b), a(a-b)^2$
30. $6(x+y)^2, 4xy^2$
31. $a^2-b^2, a-b$
32. $x+2, x^2-4$
33. $a^2+5a+4, (a+1)^2$
34. $x^2-3x+2, (x-1)^2$
35. $x^2+3x-4, (x-1)^2$
36. $x^2-x-2, (x-2)^2$
37. $x^2-x, (x-1)^3$
38. $y^2+2y, (y+2)^2$
39. $4a^2-4, (a-1)^2, 2$
40. $3x^2-3, (x-1)^2, 4$
41. $x^3, x^2-x, (x-1)^2$
42. $y, y^3-y, (y-1)^3$

■ *Write each fraction equivalently with the LCD of the given fractions. See Examples 4 and 5.*

43. $\dfrac{2}{3x}$ and $\dfrac{1}{6xy}$

44. $\dfrac{1}{4y}$ and $\dfrac{3}{8xy}$

45. $\dfrac{3}{4y^2}$ and $\dfrac{2}{3y}$

46. $\dfrac{1}{5x}$ and $\dfrac{3}{2x^2}$

47. $\dfrac{1}{2x+2}$ and $\dfrac{1}{2}$

48. $\dfrac{3}{4x-4}$ and $\dfrac{1}{4}$

49. $\dfrac{3}{y-3}$ and $\dfrac{2}{y+2}$

50. $\dfrac{1}{y+4}$ and $\dfrac{3}{y-1}$

51. $\dfrac{5}{3y-6}$ and $\dfrac{1}{6y+3}$

52. $\dfrac{3}{2y+4}$ and $\dfrac{5}{4y-6}$

53. $\dfrac{2}{a^2-1}$ and $\dfrac{3}{a-1}$

54. $\dfrac{4}{(a-1)^2}$ and $\dfrac{2}{a-1}$

55. $\dfrac{1}{a^2-5a+4}$ and $\dfrac{2}{a^2-2a+1}$

56. $\dfrac{3}{a^2-1}$ and $\dfrac{1}{a^2+a-2}$

57. $\dfrac{3y}{y;+3y+2}$ and $\dfrac{y}{y^2+4y+4}$

58. $\dfrac{y}{y^2+5y+4}$ and $\dfrac{y}{y^2-16}$

59. $\dfrac{2x}{4x^2-y^2}$ and $\dfrac{3x}{(2x-y)^2}$

60. $\dfrac{x}{(3x+y)^2}$ and $\dfrac{2x}{9x^2-y^2}$

61. $\dfrac{5y}{x^2+2xy+y^2}$ and $\dfrac{3}{2x+2y}$

62. $\dfrac{2x}{x^2-6xy+9y^2}$ and $\dfrac{4}{3x-9y}$

63. $\dfrac{1}{x^3-1}$ and $\dfrac{3}{x-1}$

64. $\dfrac{3}{x^3+8}$ and $\dfrac{1}{x+2}$

65. $\dfrac{4x}{x^3 + y^3}$ and $\dfrac{y}{x + y}$

66. $\dfrac{y}{x^3 - y^3}$ and $\dfrac{2x}{x - y}$

67. $\dfrac{3x}{(x - y)^2}$ and $\dfrac{2x}{y - x}$

68. $\dfrac{2y}{y - 2x}$ and $\dfrac{5y}{(x - 2y)^2}$

69. $\dfrac{1}{x - 2y}$ and $\dfrac{3}{4y - 2x}$

70. $\dfrac{4}{6x - 3y}$ and $\dfrac{2}{y - 2x}$

5.5

SUMS AND DIFFERENCES

Fractions with common denominators

The sum or difference of two fractions with common denominators can be written as a single fraction as follows.

$$\frac{a}{c} + \frac{b}{c} = \frac{a + b}{c}$$

and

$$\frac{a}{c} - \frac{b}{c} = \frac{a - b}{c} \qquad (c \neq 0).$$

EXAMPLE 1

a. $\dfrac{2x}{9} + \dfrac{5x}{9}$

$= \dfrac{2x + 5x}{9}$

$= \dfrac{7x}{9}$

b. $\dfrac{a + 1}{b} - \dfrac{a - 1}{b}$

$= \dfrac{a + 1 - (a - 1)}{b}$

$= \dfrac{a + 1 - a + 1}{b} = \dfrac{2}{b}$ ■

Common Error

Note that in Example b, $-(a - 1) \neq -a - 1$.

Fractions with unlike denominators

If the fractions in a sum or difference have unlike denominators, we can build the fractions to equivalent fractions that have common denominators and then rewrite the sum as above.

EXAMPLE 2 $\frac{2y}{9} + \frac{5y}{6} - \frac{y}{4}$; the LCD is $3 \cdot 3 \cdot 2 \cdot 2 = 36$.

We build each fraction to a fraction with this LCD.

$$\frac{(4)2y}{(4)9} = \frac{8y}{36}; \qquad \frac{(6)5y}{(6)6} = \frac{30y}{36}; \qquad \frac{(9)y}{(9)4} = \frac{9y}{36}$$

Then we obtain

$$\frac{8y}{36} + \frac{30y}{36} - \frac{9y}{36} = \frac{8y + 30y - 9y}{36} = \frac{29y}{36}. \qquad ■$$

EXAMPLE 3 $\frac{1}{3}x - \frac{3}{8}x + \frac{1}{2}x$; the LCD is $3 \cdot 2 \cdot 2 \cdot 2 = 24$.

We build each fraction to a fraction with this LCD.

$$\frac{(8)}{(8)} \cdot \frac{1}{3}x = \frac{8}{24}x; \qquad \frac{(3)}{(3)} \cdot \frac{3}{8}x = \frac{9}{24}x; \qquad \frac{(12)}{(12)} \cdot \frac{1}{2}x = \frac{12}{24}x.$$

Then we obtain

$$\frac{8}{24}x - \frac{9}{24}x + \frac{12}{24}x = \frac{8 - 9 + 12}{24}x = \frac{11}{24}x. \qquad ■$$

EXAMPLE 4 $\frac{3x}{x + 2} - \frac{2x}{x - 3}$; the LCD is $(x + 2)(x - 3)$.

We build each fraction to a fraction with this LCD.

$$\frac{3x}{x + 2} = \frac{3x(x - 3)}{(x + 2)(x - 3)}; \qquad \frac{2x}{x - 3} = \frac{2x(x + 2)}{(x - 3)(x + 2)}$$

Then we obtain

$$\frac{3x}{x+2} - \frac{2x}{x-3} = \frac{3x^2 - 9x - (2x^2 + 4x)}{(x+2)(x-3)}$$
$$= \frac{3x^2 - 9x - 2x^2 - 4x}{(x+2)(x-3)}$$
$$= \frac{x^2 - 13x}{(x+2)(x-3)}. \quad ■$$

A similar procedure is used if the fractions involve denominators that contain polynomials that must be factored.

EXAMPLE 5 $\dfrac{2}{x^2 - 9} - \dfrac{1}{x^2 - x - 6}$; the LCD is $(x - 3)(x + 3)(x - 2)$.

We build each fraction to a fraction with this LCD.

$$\frac{2}{(x-3)(x+3)} = \frac{2(x+2)}{(x-3)(x+3)(x+2)};$$
$$\frac{1}{(x-3)(x+2)} = \frac{1(x+3)}{(x-3)(x+2)(x+3)}$$

Then, we obtain

$$\frac{2(x+2)}{(x-3)(x+3)(x+2)} - \frac{(x+3)}{(x-3)(x+2)(x+3)} = \frac{2x + 4 - (x+3)}{(x-3)(x+3)(x+2)}$$
$$= \frac{x+1}{(x-3)(x+3)(x+2)}. \quad ■$$

Since polynomials are easier to work with in factored form, it is usually advantageous to leave the LCM of a set of polynomials in factored form rather than carry out the indicated multiplication. Sometimes it is also convenient to leave numerators or denominators of fractions in factored form.

EXAMPLE 6 $\dfrac{2}{x^2 - 4} - \dfrac{3}{x^2 - 5x + 6} + \dfrac{1}{x - 3}$; the LCD is $(x + 2)(x - 2)(x - 3)$.

We build each fraction to a fraction with this LCD.

$$\frac{2}{(x+2)(x-2)} = \frac{2(x-3)}{(x+2)(x-2)(x-3)};$$
$$\frac{3}{(x-2)(x-3)} = \frac{3(x+2)}{(x-2)(x-3)(x+2)};$$
$$\frac{1}{x-3} = \frac{1(x-2)(x+2)}{(x-3)(x-2)(x+2)}$$

Then we obtain

$$\frac{2(x-3)}{(x+2)(x-2)(x-3)} - \frac{3(x+2)}{(x-2)(x-2)(x-3)} + \frac{1(x-2)(x+2)}{(x-3)(x-2)(x+2)}$$
$$= \frac{2x-6-3x-6+x^2-4}{(x-3)(x-2)(x+2)} = \frac{x^2-x-16}{(x-3)(x-2)(x+2)}.$$ ■

EXERCISE 5.5

A

■ *Write each sum or difference as a single fraction in lowest terms. See Example* 1.

1. $\frac{x}{2} - \frac{3}{2}$
2. $\frac{y}{7} - \frac{5}{7}$
3. $\frac{1}{6}a + \frac{1}{6}b - \frac{1}{6}c$
4. $\frac{1}{3}x - \frac{2}{3}y + \frac{1}{3}z$
5. $\frac{x-1}{2y} + \frac{x}{2y}$
6. $\frac{y+1}{b} + \frac{y-1}{b}$
7. $\frac{3}{x+2y} - \frac{x+3}{x+2y} - \frac{x-1}{x+2y}$
8. $\frac{2}{a-3b} - \frac{b-2}{a-3b} + \frac{b}{a-3b}$
9. $\frac{a+1}{a^2-2a+1} + \frac{5-3a}{a^2-2a+1}$
10. $\frac{x+4}{x^2-x+2} + \frac{2x-3}{x^2-x+2}$

■ *See Examples* 2 *and* 3.

11. $\frac{x}{2} + \frac{2x}{3}$
12. $\frac{3y}{4} + \frac{y}{3}$
13. $\frac{y}{3} - \frac{2y}{5}$
14. $\frac{3x}{5} - \frac{x}{2}$
15. $\frac{2x}{3} - \frac{3x}{4} + \frac{x}{2}$
16. $\frac{y}{2} + \frac{2y}{3} - \frac{3y}{4}$
17. $\frac{1}{2}x + \frac{2}{3}x$
18. $\frac{3}{4}y + \frac{2}{3}y$
19. $\frac{5}{6}y - \frac{3}{4}y$
20. $\frac{3}{4}x - \frac{1}{6}x$
21. $\frac{2}{3}y - \frac{1}{6}y + \frac{1}{4}y$
22. $\frac{3}{4}y + \frac{1}{3}y - \frac{5}{6}y$

■ *See Example 4.*

23. $\frac{x+1}{x}+\frac{2y-1}{y}$

24. $\frac{y-2}{y}+\frac{2x-3}{x}$

25. $\frac{1}{2y}-\frac{2}{3x}$

26. $\frac{2}{3x}-\frac{1}{2y}$

27. $\frac{5}{x+1}-\frac{3}{y-1}$

28. $\frac{2}{y+2}-\frac{3}{x-1}$

29. $\frac{x}{3x+2}+\frac{x}{x-1}$

30. $\frac{2y}{y-1}+\frac{y}{2y+1}$

31. $\frac{y}{2y-1}-\frac{2y}{y+1}$

32. $\frac{2x}{3x+1}-\frac{x}{x-2}$

33. $\frac{x+1}{x+2}+\frac{x+2}{x+3}$

34. $\frac{y-2}{y+1}+\frac{y+3}{y-2}$

35. $\frac{y-1}{y+1}-\frac{y-2}{2y-3}$

36. $\frac{x-2}{2x+1}-\frac{x+1}{x-1}$

■ *See Example 5.*

37. $\frac{3}{2x+4}+\frac{4}{3x+6}$

38. $\frac{5}{4y-8}+\frac{3}{5y-10}$

39. $\frac{7}{5x-10}-\frac{5}{3x-6}$

40. $\frac{2}{3y+6}-\frac{3}{2y+4}$

41. $\frac{2}{x^2-x-2}+\frac{2}{x^2+2x+1}$

42. $\frac{1}{y^2-1}+\frac{1}{y^2+2y+1}$

43. $\frac{y}{y^2-16}-\frac{y+1}{y^2-5y+4}$

44. $\frac{x}{x^2-5x+6}-\frac{x-1}{x^2-9}$

45. $\frac{x}{x^2-1}-\frac{2x+1}{x^2-2x+1}$

46. $\frac{y}{y^2+3y-10}-\frac{2y-1}{y^2-4}$

47. $\frac{y-1}{y^2-3y}-\frac{y+1}{y^2+2y}$

48. $\frac{x+1}{x^2+2x}-\frac{x-1}{x^2-3x}$

49. $\frac{2x+1}{x^2-4}-\frac{3x-2}{x^2-4x+4}$

50. $\frac{3y-1}{y^2-4y+3}-\frac{y+2}{(y-3)^2}$

■ *See Example 6.*

51. $\frac{1}{z^2-7z+12}+\frac{2}{z^2-5z+6}-\frac{3}{z^2-6z+8}$

52. $\frac{4}{a^2-4b^2}+\frac{2}{a^2+3ab+2b^2}+\frac{4}{a^2-ab-2b^2}$

53. $\frac{y+2}{y^2-6y+8}+\frac{3y-8}{y^2-5y+6}+\frac{2y-5}{y^2-7y+12}$

54. $\frac{2y+5}{y^2+5y+4}+\frac{y+13}{y^2-y-20}+\frac{y+7}{y^2-4y-5}$

55. $\frac{2y+1}{y^2+y-20}-\frac{3y-10}{y^2-6y+8}+\frac{3y+8}{y^2+3y-10}$

56. $\frac{3y}{y^2+y-2}-\frac{3y-4}{y^2-y-6}+\frac{2y-4}{y^2-4y+3}$

57. $x+\frac{1}{x-1}-\frac{1}{(x-1)^2}$ $\left[\textit{Hint}: \text{Write } x \text{ as } \frac{x}{1}.\right]$

58. $y - \frac{2}{y^2 - 1} + \frac{3}{y + 1}$ $\left[\textit{Hint}\text{: Write } y \text{ as } \frac{y}{1}.\right]$

59. $y - \frac{y^2}{y - 1} + \frac{y^2}{y + 1}$

60. $x + \frac{2x^2}{x + 2} - \frac{3x^2}{x - 1}$

61. $1 - \frac{x - 2}{(x + 1)^2} + \frac{2x - 1}{x^2 - 1}$

62. $1 + \frac{y + 1}{y^2 - 4} - \frac{y - 1}{(y + 2)^2}$

B

■ a. *Write each expression as a single fraction.*

b. *Use long division to rewrite the result obtained for* a *in the form*

$$P(x) + \frac{r}{x - a}.$$

63. $x - 1 + \frac{3}{x + 2}$

64. $x + 3 + \frac{1}{x - 1}$

65. $x + 2 - \frac{2}{x + 3}$

66. $x - 3 - \frac{3}{x - 2}$

67. $x^2 - x + 3 - \frac{1}{x + 1}$

68. $x^2 + x - 4 - \frac{1}{x - 1}$

■ *Show that each pair of expressions is equivalent, given that $c^2 + s^2 = 1$. Assume that the variables represent numbers for which each expression is defined. [Hint: Simplify the first expression.]*

69. $\frac{c}{s} + \frac{s}{c}$; $\frac{1}{sc}$

70. $\frac{1}{c^2} + \frac{1}{s^2}$; $\frac{1}{c^2s^2}$

71. $c + \frac{s^2}{c}$; $\frac{1}{c}$

72. $\left(\frac{c}{s}\right)^2 + 1$; $\frac{1}{s^2}$

73. $\frac{(c - 1)(c + 1)}{s^2} + 1$; 0

74. $s + \frac{c(c - s)}{s}$; $\frac{1 - cs}{s}$

75. $\frac{s - c}{s + c} + \frac{2sc}{s^2 - c^2}$; $\frac{1}{s^2 - c^2}$

76. $\frac{c^3 + c^2s}{c^2 - s^2} - s$; $\frac{1 - cs}{c - s}$

5.6

PRODUCTS AND QUOTIENTS

Products of fractions

In earlier mathematics courses, we learned that, for example,

$$\frac{2}{3} \cdot \frac{4}{5} = \frac{2 \cdot 4}{3 \cdot 5} = \frac{8}{15}.$$

In general:

$$\frac{a}{b} \cdot \frac{c}{d} = \frac{ac}{bd} \qquad (b, d \neq 0). \qquad (1)$$

Thus, the product of two fractions equals the product of their numerators divided by the product of their denominators. For example,

$$\frac{6x^2}{y} \cdot \frac{xy}{2} = \frac{6x^2 \cdot xy}{y \cdot 2} = \frac{6x^3y}{2y}$$
$$= \frac{3x^3(2y)}{1(2y)} = 3x^3 \qquad (y \neq 0).$$

We can arrive at the same result for this example by using slash lines:

$$\frac{6x^2}{y} \cdot \frac{xy}{2} = \frac{\overset{3}{\cancel{6}}x^2}{\underset{1}{\cancel{y}}} \cdot \frac{x\overset{1}{\cancel{y}}}{\underset{1}{\cancel{2}}} = 3x^3 \qquad (y \neq 0).$$

In the above example we have shown the factors "1." Although you may find it helpful to continue to do so, we will omit showing such factors in the following examples.

If any of the fractions, or any of the factors of the numerators or denominators of the fractions, have negative signs, it is advisable to proceed as if all the signs were positive and then attach the appropriate sign to the simplified product. If there is an even number of negative signs involved, the result has a positive sign; if there is an odd number of negative signs involved, the result has a negative sign.

EXAMPLE 1

a. $$\frac{\cancel{-x}}{\cancel{y^2}} \cdot \frac{-2\overset{y}{\cancel{y^3}}}{\underset{x}{\cancel{x^2}}} = \frac{2y}{x} \qquad (x, y \neq 0)$$

b. $$\frac{\overset{-2x}{\cancel{-4x^2}}}{\cancel{3y}} \cdot \frac{\cancel{y}}{\cancel{2x}} \cdot \frac{\cancel{3}}{5} = \frac{-2x}{5} \qquad (x, y \neq 0)$$

c. $$\frac{-\cancel{4}}{\underset{3}{\cancel{9}}}x^2 \cdot \frac{\cancel{3}}{\cancel{4}}x = \frac{-x^3}{3}$$

d. $$-\frac{\cancel{2}}{\cancel{5}}y \cdot \frac{\cancel{5}}{\underset{4}{\cancel{8}}}y^2 = \frac{-y^3}{4}$$ ■

Polynomials in a fraction should be factored and common factors in numerators and denominators divided out before Property (1) above is applied.

EXAMPLE 2 a. $\dfrac{x^2 - 3x + 4}{3x} \cdot \dfrac{x}{x^2 - 1} = \dfrac{(x - 4)\cancel{(x - 1)}}{3\cancel{x}} \cdot \dfrac{\cancel{x}}{(x + 1)\cancel{(x - 1)}}$

$$= \frac{x - 4}{3(x + 1)} \qquad (x \neq -1,0,1)$$

b. $\dfrac{4y^2 - 1}{y^2 - 4} \cdot \dfrac{y^2 + 2y}{4y + 2} = \dfrac{(2y - 1)\cancel{(2y + 1)}}{(y - 2)\cancel{(y + 2)}} \cdot \dfrac{y\cancel{(y + 2)}}{2\cancel{(2y + 1)}}$

$$= \frac{y(2y - 1)}{2(y - 2)} \qquad (y \neq -2,2) \quad ■$$

We can use the distributive property to rewrite products in which one or more of the factors is a polynomial that includes fractions.

EXAMPLE 3 a. $\dfrac{2}{3}x\left(\dfrac{1}{2}x - 9\right) = \dfrac{\cancel{2}}{3}x\left(\dfrac{1}{\cancel{2}}x\right) - \dfrac{2}{\cancel{3}}x(\overset{3}{\cancel{9}})$

$$= \frac{1}{3}x^2 - 6x$$

b. $\left(x - \dfrac{1}{2}\right)^2 = \left(x - \dfrac{1}{2}\right)\left(x - \dfrac{1}{2}\right)$

$$= x^2 - \frac{1}{2}x - \frac{1}{2}x + \frac{1}{4}$$

$$= x^2 - x + \frac{1}{4} \quad ■$$

Quotients of fractions

When rewriting quotients

$$\frac{a}{b} \div \frac{c}{d} \qquad (b, c, d \neq 0),$$

we seek a quotient q such that

$$\left(\frac{c}{d}\right)q = \frac{a}{b} \qquad (b, c, d \neq 0).$$

To obtain q in terms of the other variables, we use the multiplication law of equality and multiply each member of this equality by d/c:

$$\left(\frac{d}{c}\right)\left(\frac{c}{d}\right)q = \left(\frac{d}{c}\right)\left(\frac{a}{b}\right),$$

from which

$$q = \left(\frac{d}{c}\right)\left(\frac{a}{b}\right) = \left(\frac{a}{b}\right)\left(\frac{d}{c}\right),$$

and we have the following result.

$$\boldsymbol{\frac{a}{b} \div \frac{c}{d} = \frac{a}{b} \cdot \frac{d}{c} \qquad (b, c, d \neq 0).} \qquad (2)$$

We often refer to this property by the familiar phrase "invert the divisor and multiply."

EXAMPLE 4 a. $\dfrac{2x^3}{3y} \div \dfrac{4x}{5y^2}$

$$= \frac{\cancelto{x^2}{2x^3}}{3\cancel{y}} \cdot \frac{\cancelto{y}{5y^2}}{\cancelto{2}{4x}}$$

$$= \frac{5x^2y}{6} \qquad (x, y \neq 0)$$

b. $\dfrac{x^2 - 1}{x + 3} \div \dfrac{x^2 - x - 2}{x^2 + 5x + 6}$

$$= \frac{(x-1)\cancel{(x+1)}}{\cancel{x+3}} \cdot \frac{\cancel{(x+3)}(x+2)}{\cancel{(x+1)}(x-2)}$$

$$= \frac{(x-1)(x+2)}{(x-2)}$$

$$= \frac{x^2 + x - 2}{x - 2} \qquad (x \neq -3, -1, 2)$$ ■

As special cases of (2), observe that

$$a \div \frac{c}{d} = \frac{a}{1} \cdot \frac{d}{c} = \frac{ad}{c},$$

$$\frac{a}{b} \div c = \frac{a}{b} \cdot \frac{1}{c} = \frac{a}{bc},$$

and

$$1 \div \frac{a}{b} = 1 \cdot \frac{b}{a} = \frac{b}{a}.$$

EXAMPLE 5

a. $2x \div \frac{x}{y}$

$= 2\cancel{x} \cdot \frac{y}{\cancel{x}}$

$= 2y$

$(x, y \neq 0)$

b. $\frac{x}{y} \div 2x$

$= \frac{\cancel{x}}{y} \cdot \frac{1}{2\cancel{x}}$

$= \frac{1}{2y}$

$(x, y \neq 0)$

c. $1 \div \frac{x}{y}$

$= 1 \cdot \frac{y}{x}$

$= \frac{y}{x}$

$(x, y \neq 0)$ ■

The order of operations that we have established for the basic operations are applicable when simplifying algebraic expressions involving fractions. Multiplications are performed before additions and subtractions.

EXAMPLE 6

$$\frac{3}{y-2} + \frac{4y-2}{y^2-5y+4} \cdot \frac{y^2-y-12}{2y^2+5y-3} = \frac{3}{y-2} + \frac{2\cancel{(2y-1)}}{\cancel{(y-4)}(y-1)} \cdot \frac{\cancel{(y-4)}\cancel{(y+3)}}{\cancel{(2y-1)}\cancel{(y+3)}}$$

$$= \frac{3}{y-2} + \frac{2}{y-1}$$

$$= \frac{3(y-1)}{(y-2)(y-1)} + \frac{2(y-2)}{(y-1)(y-2)}$$

$$= \frac{3y-3+2y-4}{(y-1)(y-2)}$$

$$= \frac{5y-7}{(y-1)(y-2)}$$ ■

EXERCISE 5.6

A

■ *Write each product as a single fraction in lowest terms. See Example 1.*

1. $\frac{16}{38} \cdot \frac{19}{12}$
2. $\frac{4}{15} \cdot \frac{3}{16}$
3. $\frac{21}{4} \cdot \frac{2}{15}$
4. $\frac{7}{8} \cdot \frac{48}{64}$
5. $\frac{24}{3} \cdot \frac{20}{36} \cdot \frac{3}{4}$
6. $\frac{3}{10} \cdot \frac{16}{27} \cdot \frac{30}{36}$
7. $\frac{7x}{12} \cdot \frac{3}{14x^2}$
8. $\frac{4a^2}{3} \cdot \frac{6b}{2a}$
9. $\frac{15n^2}{3p} \cdot \frac{5p^2}{n^3}$
10. $\frac{21t^2}{5s} \cdot \frac{15s^3}{7st}$
11. $\frac{-4}{3n} \cdot \frac{6n^2}{16}$
12. $\frac{14a^3b}{3b} \cdot \frac{-6}{7a^2}$

13. $\frac{1}{3}x^2 \cdot \frac{6}{7}x^3$

14. $\frac{2}{3}y \cdot \frac{9}{10}y^2$

15. $\frac{3}{4}x^2y \cdot \frac{2}{3}xy^2$

16. $\frac{1}{4}x^3y \cdot \frac{2}{5}xy$

17. $-\frac{1}{2}xyz^2 \cdot \frac{2}{3}x^2y$

18. $-\frac{3}{5}x^2y \cdot \frac{5}{6}xy^2z$

19. $\frac{-12a^2b}{5c} \cdot \frac{10b^2c}{24a^3b}$

20. $\frac{a^2}{xy} \cdot \frac{3x^3y}{4a}$

21. $\frac{-2ab}{7c} \cdot \frac{3c^2}{4a^3} \cdot \frac{-6a}{15b^2}$

22. $\frac{10x}{12y} \cdot \frac{3x^2z}{5x^3z} \cdot \frac{6y^2x}{3yz}$

23. $5a^2b^2 \cdot \frac{1}{a^3b^3}$

24. $15x^2y \cdot \frac{3}{45xy^2}$

■ *See Example 2.*

25. $\frac{5x+25}{2x} \cdot \frac{4x}{2x+10}$

26. $\frac{3y}{4xy-6y^2} \cdot \frac{2x-3y}{12x}$

27. $\frac{4a^2-1}{a^2-16} \cdot \frac{a^2-4a}{2a+1}$

28. $\frac{9x^2-25}{2x-2} \cdot \frac{x^2-1}{6x-10}$

29. $\frac{x^2-x-20}{x^2+7x+12} \cdot \frac{(x+3)^2}{(x-5)^2}$

30. $\frac{4x^2+8x+3}{2x^2-5x+3} \cdot \frac{6x^2-9x}{1-4x^2}$

31. $\frac{x^2-6x+5}{x^2+2x-3} \cdot \frac{x^2-4x-21}{x^2-10x+25}$

32. $\frac{x^2-x-2}{x^2+4x+3} \cdot \frac{x^2-4x-5}{x^2-3x-10}$

33. $\frac{2x^2-x-6}{3x^2-4x+1} \cdot \frac{3x^2+7x+2}{2x^2+7x+6}$

34. $\frac{3x^2-7x-6}{2x^2-x-1} \cdot \frac{2x^2-9x-5}{3x^2-13x-10}$

35. $\frac{7a+14}{14a-28} \cdot \frac{4-2a}{a+2} \cdot \frac{a-3}{a+1}$

36. $\frac{5x^2-5x}{3} \cdot \frac{x^2-9x-10}{4x-40} \cdot \frac{y^2}{2-2x^2}$

■ *Write each product as a polynomial. See Example 3.*

37. $\frac{1}{2}x\left(\frac{2}{5}x-6\right)$

38. $\frac{3}{4}y\left(\frac{1}{6}y+8\right)$

39. $\left(x+\frac{1}{3}\right)\left(x+\frac{1}{3}\right)$

40. $\left(y-\frac{1}{3}\right)\left(y-\frac{1}{3}\right)$

41. $\left(y-\frac{1}{4}\right)^2$

42. $\left(y+\frac{1}{4}\right)^2$

■ *Write each quotient as a single fraction in lowest terms. See Example 4.*

43. $\frac{3}{4} \div \frac{9}{16}$

44. $\frac{2}{3} \div \frac{9}{15}$

45. $\frac{xy}{a^2b} \div \frac{x^3y^2}{ab}$

46. $\frac{9ab^3}{x} \div \frac{3}{2x^3}$

47. $\frac{28x^2y^3}{a^2} \div \frac{21x^2y^2}{5a}$

48. $\frac{24a^3b}{x^2} \div \frac{3a^2b}{7x}$

49. $\frac{4x-8}{3y} \div \frac{6x-12}{y}$

50. $\frac{6y-27}{5x} \div \frac{4y-18}{x}$

51. $\frac{a^2-a-6}{a^2+2a-15} \div \frac{a^2-4}{a^2+6a+5}$

52. $\frac{a^2+2a-15}{a^2+3a-10} \div \frac{a^2-9}{a^2-9a+14}$

53. $\frac{x^2+x-2}{x^2+2x-3} \div \frac{x^2+7x+10}{x^2-2x-15}$

54. $\frac{x^2+6x-7}{x^2+x-2} \div \frac{x^2+5x-14}{x-3x-10}$

55. $\dfrac{10x^2 - 13x - 3}{2x^2 - x - 3} \div \dfrac{5x^2 - 9x - 2}{3x^2 + 2x - 1}$

56. $\dfrac{9x^2 + 3x - 2}{12x^2 + 5x - 2} \div \dfrac{9x^2 - 6x + 1}{8x^2 + 10x - 3}$

■ *See Example 5.*

57. $1 \div \dfrac{x^2 - 1}{x + 2}$

58. $1 \div \dfrac{x^2 + 3x + 1}{x - 2}$

59. $(x^2 - 5x + 4) \div \dfrac{x^2 - 1}{x^2}$

60. $(x^2 - 9) \div \dfrac{x^2 - 6x + 9}{3x}$

61. $\dfrac{x^2 + 3x}{2y} \div 3x$

62. $\dfrac{2y^2 + y}{3x} \div 2y$

■ *See Example 6.*

63. $\dfrac{4}{2y - 1} - \dfrac{3y + 6}{3y^2 - y - 2} \cdot \dfrac{4y^2 - 3y - 1}{4y^2 + 9y + 2}$

64. $\dfrac{3}{2y - 1} - \dfrac{4y + 12}{3y^2 + 13y + 4} \cdot \dfrac{2y^2 + 9y + 4}{2y^2 + 7y + 3}$

65. $\dfrac{3}{2y - 3} - \dfrac{2y - 8}{3y^2 + 10y - 8} \div \dfrac{y^2 - 6y + 8}{y^2 + 2y - 8}$

66. $\dfrac{2}{y - 4} + \dfrac{6y - 3}{12y^2 - 25y + 12} \div \dfrac{6y^2 + 5y - 4}{9y^2 - 16}$

67. $\left(\dfrac{1}{y - 2} + \dfrac{1}{y + 2}\right)\left(\dfrac{2}{y^2} + \dfrac{1}{y}\right)$

68. $\left(\dfrac{3}{y - 3} + \dfrac{4}{y + 4}\right)\left(\dfrac{1}{3} - \dfrac{1}{y}\right)$

B

■ *Write as a single fraction in lowest terms.*

69. $\dfrac{x^3 + y^3}{x} \div \dfrac{x + y}{3x}$

70. $\dfrac{8x^3 - y^3}{x + y} \div \dfrac{2x - y}{x^2 - y^2}$

71. $\dfrac{xy - 3x + y - 3}{x - 2} \div \dfrac{x + 1}{x^2 - 4}$

72. $\dfrac{2xy + 4x + 3y + 6}{2x + 3} \div \dfrac{y + 2}{y - 1}$

73. $\dfrac{x^3 - 1}{x} \cdot \dfrac{x^2 - 1}{x^2 + x + 1} \div \dfrac{(x - 1)^3}{x}$

74. $\dfrac{x}{x^3 - 8} \div \dfrac{1}{x^2 + 2x + 4} \cdot \dfrac{(x - 2)^2}{2x}$

■ *Let* $P(x) = \dfrac{x}{x - 1}$, $Q(x) = \dfrac{x^2}{x^2 - 1}$, *and* $R(x) = \dfrac{x^3}{(x - 1)^2}$. *Write each expression as a single fraction in lowest terms.*

75. $P(x) \div Q(x)$

76. $Q(x) \div R(x)$

77. $P(x) \cdot Q(x) \div R(x)$

78. $P(x) \cdot R(x) \div Q(x)$

5.7

COMPLEX FRACTIONS

A fraction that contains one or more fractions in either its numerator or denominator or both is called a **complex fraction**. For example,

$$\frac{\frac{2}{3}}{\frac{1}{6}} \quad \text{and} \quad \frac{x + \frac{3}{4}}{x - \frac{1}{2}}$$

are complex fractions. Like simple fractions, complex fractions represent quotients. For example,

$$\frac{\frac{2}{3}}{\frac{5}{6}} = \frac{2}{3} \div \frac{5}{6} \tag{1}$$

and

$$\frac{x + \frac{3}{4}}{x - \frac{1}{2}} = \left(x + \frac{3}{4}\right) \div \left(x - \frac{1}{2}\right). \tag{2}$$

In cases like Equation (1), in which the denominator of the complex fractions does not contain sums or differences, we can simply multiply the numerator by the inverse of the denominator.

EXAMPLE 1 a. $\dfrac{\frac{2}{3}}{\frac{5}{6}} = \dfrac{2}{3} \div \dfrac{5}{6}$

$= \dfrac{2}{\cancel{3}} \cdot \dfrac{\overset{2}{\cancel{6}}}{5} = \dfrac{4}{5}$

b. $\dfrac{\frac{x}{8}}{\frac{y}{4}} = \dfrac{x}{8} \div \dfrac{y}{4}$

$= \dfrac{x}{\cancel{8}} \cdot \dfrac{\cancel{4}}{y} = \dfrac{x}{2y}$ ■

In a complex fraction of the Form (2), in which the numerator or denominator of the complex fraction contains sums or differences, it is more convenient to use the fundamental principle of fractions to simplify the fractions. In fact, we can also use the fundamental principle of fractions to simplify complex fractions of Form (1).

EXAMPLE 2 Simplify $\dfrac{\frac{2}{3}}{\frac{5}{6}}$ by using the fundamental principle of fractions.

Solution Multiplying the numerator ⅔ and the denominator ⅚ by 6, the LCD of the two fractions, we obtain

$$\frac{\frac{2}{3}}{\frac{5}{6}} = \frac{\frac{2}{\cancel{3}}\overset{2}{\cancel{(6)}}}{\frac{5}{\cancel{6}}\cancel{(6)}} = \frac{4}{5},$$

a simple fraction equivalent to the original fraction. ■

EXAMPLE 3 Simplify $\dfrac{x + \frac{3}{4}}{x - \frac{1}{2}}$ [the fraction in (2) above].

Solution Multiplying the numerator and denominator by 4, the LCD of the fractions ¾ and ½, we obtain

$$\frac{4\left(x + \frac{3}{4}\right)}{4\left(x - \frac{1}{2}\right)} = \frac{4(x) + 4\left(\frac{3}{4}\right)}{4(x) - 4\left(\frac{1}{2}\right)} = \frac{4x + 3}{4x - 2} \quad \left(x \neq \frac{1}{2}\right). \quad ■$$

Alternatively, the fraction in Example 3 may be simplified by first representing the numerator as a single fraction and the denominator as a single fraction, and then writing the quotient as the product of the numerator and the reciprocal of the denominator:

$$\frac{x + \frac{3}{4}}{x - \frac{1}{2}} = \frac{\frac{4 \cdot x}{4 \cdot 1} + \frac{3}{4}}{\frac{2 \cdot x}{2 \cdot 1} - \frac{1}{2}} = \frac{\frac{4x + 3}{4}}{\frac{2x - 1}{2}}$$

$$= \frac{4x + 3}{4} \div \frac{2x - 1}{2} = \frac{4x + 3}{4} \cdot \frac{2}{2x - 1}$$

$$= \frac{4x + 3}{2(2x - 1)} = \frac{4x + 3}{4x - 2} \quad \left(x \neq \frac{1}{2}\right).$$

Sometimes it is difficult to use the fundamental principle directly to simplify a complex fraction. In such cases, we can simplify the fraction in small steps.

EXAMPLE 4 Simplify $\dfrac{a}{a + \dfrac{3}{3 + \dfrac{1}{2}}}$.

Solution We can first simplify part of the complex fraction:

$$\frac{(3) \cdot 2}{\left(3 + \dfrac{1}{2}\right) \cdot 2} = \frac{6}{6 + 1} = \frac{6}{7}.$$

Then substituting $6/7$ and applying the fundamental principle of fractions again, we have

$$\frac{a}{a + \dfrac{6}{7}} = \frac{(a) \cdot 7}{\left(a + \dfrac{6}{7}\right) \cdot 7} = \frac{7a}{7a + 6}.$$ ■

EXERCISE 5.7

A

■ *Write each complex fraction as a single fraction in lowest terms.*
See Examples 1 and 2.

1. $\dfrac{\dfrac{3}{7}}{\dfrac{2}{7}}$

2. $\dfrac{\dfrac{4}{5}}{\dfrac{7}{5}}$

3. $\dfrac{\dfrac{2}{9}}{\dfrac{7}{3}}$

4. $\dfrac{\dfrac{5}{2}}{\dfrac{21}{4}}$

5. $\dfrac{\dfrac{b}{c}}{\dfrac{b^2}{a}}$

6. $\dfrac{\dfrac{5x}{6y}}{\dfrac{4x}{5y}}$

7. $\dfrac{\dfrac{2x}{5y}}{\dfrac{3x}{10y^2}}$

8. $\dfrac{\dfrac{3ab}{4}}{\dfrac{3b}{8a^2}}$

■ *See Example 3.*

9. $\dfrac{\frac{3}{4}}{4 - \frac{1}{4}}$ 10. $\dfrac{\frac{1}{3}}{4 + \frac{2}{3}}$ 11. $\dfrac{1 - \frac{2}{3}}{3 + \frac{1}{3}}$ 12. $\dfrac{\frac{1}{2} + \frac{3}{4}}{\frac{1}{2} - \frac{3}{4}}$

13. $\dfrac{\frac{2}{3}}{\frac{1}{3} + \frac{3}{4}}$ 14. $\dfrac{\frac{1}{4}}{\frac{2}{3} + \frac{1}{2}}$ 15. $\dfrac{\frac{1}{5} + \frac{1}{6}}{\frac{1}{2} + \frac{2}{3}}$ 16. $\dfrac{\frac{3}{4} + \frac{1}{3}}{\frac{1}{2} + \frac{5}{6}}$

17. $\dfrac{\frac{2}{a} + \frac{3}{2a}}{5 + \frac{1}{a}}$ 18. $\dfrac{1 + \frac{2}{a}}{1 - \frac{4}{a^2}}$ 19. $\dfrac{x + \frac{x}{y}}{1 + \frac{1}{y}}$ 20. $\dfrac{1 + \frac{1}{x}}{1 - \frac{1}{x}}$

21. $\dfrac{1}{1 - \frac{1}{x}}$ 22. $\dfrac{4}{\frac{2}{x} + 2}$ 23. $\dfrac{y - 2}{y - \frac{4}{y}}$ 24. $\dfrac{y + 3}{\frac{9}{y} - y}$

25. $\dfrac{\frac{1}{y + 1}}{1 - \frac{1}{y^2}}$ 26. $\dfrac{\frac{1}{y - 1}}{\frac{1}{y^2} + 1}$ 27. $\dfrac{x - \frac{x}{y}}{y + \frac{y}{x}}$ 28. $\dfrac{y + \frac{x}{y}}{x - \frac{y}{x}}$

■ *Write each quotient as a complex fraction and then as a single fraction in lowest terms.*

29. $\left(\frac{1}{y^2} - \frac{1}{4}\right) \div \left(\frac{1}{y} + \frac{1}{2}\right)$ 30. $\left(\frac{4}{y} - \frac{1}{3}\right) \div \left(\frac{16}{y^2} - \frac{1}{9}\right)$

31. $\left(\frac{9}{y^2} - 1\right) \div \left(\frac{3}{y} + 1\right)$ 32. $\left(1 + \frac{1}{y^3}\right) \div \left(1 + \frac{1}{y}\right)$

33. $\left(y + 3 + \frac{10}{2y - 3}\right) \div \left(y - \frac{2}{2y - 3}\right)$ 34. $\left(y - 1 - \frac{4}{3y - 2}\right) \div \left(y - \frac{1}{3y - 2}\right)$

B

■ *Write each complex fraction as a single fraction in lowest terms. See Example 4.*

35. $\dfrac{1 + \dfrac{1}{1 - \frac{a}{b}}}{1 - \dfrac{3}{1 - \frac{a}{b}}}$ 36. $\dfrac{1 - \dfrac{1}{\frac{a}{b} + 2}}{1 + \dfrac{3}{\frac{a}{2b} + 1}}$ 37. $\dfrac{a + 2 - \dfrac{12}{a + 3}}{a - 5 + \dfrac{16}{a + 3}}$

38. $\dfrac{a + 4 - \dfrac{7}{a - 2}}{a - 1 + \dfrac{2}{a - 2}}$ 39. $\dfrac{\frac{1}{ab} + \frac{2}{bc} + \frac{3}{ac}}{\frac{2a + 3b + c}{abc}}$ 40. $\dfrac{\frac{a}{bc} - \frac{b}{ac} + \frac{c}{ab}}{\frac{1}{a^2b^2} - \frac{1}{a^2c^2} + \frac{1}{b^2c^2}}$

■ *Show that each pair of expressions is equivalent, given that* $c^2 + s^2 = 1$. *Assume that the variables represent numbers for which each expression is defined. [Hint: Simplify the first expression.]*

41. $\dfrac{(c+s)^2}{c-s} - \dfrac{1}{\dfrac{c-s}{2cs}}$; $\dfrac{1}{c-s}$

42. $\dfrac{s}{\dfrac{s}{s+c}} - \dfrac{2sc}{s+c}$; $\dfrac{1}{s+c}$

43. $\dfrac{\dfrac{1}{s-c}}{\dfrac{s}{s^2-c^2}} + \dfrac{s-c}{s}$; 2

44. $\dfrac{\dfrac{1}{s^2}+\dfrac{1}{c^2}}{\dfrac{1}{s \cdot c}} - \dfrac{c}{s}$; $\dfrac{s}{c}$

5.8

SOLUTION OF EQUATIONS

The properties that we used in previous chapters to solve equations can also be used to solve equations that contain fractions.

Very often it is helpful to first generate an equivalent equation that does not contain fractions.

EXAMPLE 1 Solve $\dfrac{x}{3} + 2 = \dfrac{x}{2}$.

Solution We first multiply each member by 6, the LCD of the denominators 3 and 2, to generate an equivalent equation free of fractions and obtain

$$6\left(\frac{x}{3} + 2\right) = 6\left(\frac{x}{2}\right),$$

$$2x + 12 = 3x.$$

Now adding $-2x$ to each member we obtain

$$2x + 12 + (-2x) = 3x + (-2x),$$

$$12 = x.$$

Hence, the solution set is $\{12\}$. ■

Multiplying by zero Care must be exercised in an application of the multiplication property, for we have specifically excluded *multiplication by* 0. For example, the equation $x = 3$ with solution set $\{3\}$ is not equivalent to $0 \cdot x = 0 \cdot 3$, with solution set $\{x \mid x \in R\}$.

EXAMPLE 2 Solve $\dfrac{x}{x-3} = \dfrac{3}{x-3} + 2.$ (1)

Solution We might first multiply each member by $(x - 3)$ to attempt to produce an equivalent equation that is free of fractions. We have

$$(x-3)\frac{x}{x-3} = (x-3)\frac{3}{x-3} + (x-3)2$$

or

$$x = 3 + 2x - 6, \tag{2}$$

from which

$$x = 3,$$

and 3 *appears* to be a solution of (1). But, upon substituting 3 for x in (1), we have

$$\frac{3}{0} = \frac{3}{0} + 2,$$

where neither member is defined. In obtaining Equation (2), each member of Equation (1) was multiplied by $(x - 3)$, but if x is 3, then $(x - 3)$ is 0, and Equation (2) is not equivalent to Equation (1). Equation (1) has no solution. Its solution set is $\varnothing$. ■

We can always determine whether an apparent solution of an equation is really a solution by substituting the suggested solution in the original equation and verifying that the resulting statement is true. If the equations in a sequence are equivalent, the sole purpose for such a check is to detect arithmetic errors. We shall dispense with checking solutions in the examples that follow *except in cases where we multiply or divide by an expression containing a variable.*

Proportions

A **proportion** is a special kind of fractional equation of the form $\dfrac{a}{b} = \dfrac{c}{d}$, in which each member is a ratio. This proportion can be read "a is to b as c is to d." The numbers a, b, c, and d are called the first, second, third, and fourth **terms** of the proportion, respectively. The first and fourth terms are called the **extremes** of the proportion and the second and third terms are called the **means** of the proportion.

Means $\quad \dfrac{a}{b} = \dfrac{c}{d} \quad$ Extremes

If each member of the proportion is multiplied by bd, we have

$$(\not{b}d)\frac{a}{\not{b}} = (b\not{d})\frac{c}{\not{d}},$$
$$ad = bc.$$

Thus:

$$\text{If} \quad \frac{a}{b} = \frac{c}{d}, \quad \text{then} \quad ad = bc. \tag{3}$$

In any proportion, the product of the extremes is equal to the product of the means.

This property provides a "shortcut" way to solve an equation that is a proportion.

EXAMPLE 3 Solve $\dfrac{x + 6}{3} = \dfrac{x + 1}{2}$.

Solution Using Property (3) we set the product of the extremes equal to the product of the means to obtain

$$2(x + 6) = 3(x + 1),$$

from which,

$$2x + 12 = 3x + 3,$$
$$9 = x.$$

Hence, the solution set is {9}. *Note*: We would of course obtain the same result if we had multiplied each member of the original equation by the LCD 6. ■

Common Error *Property (3) can only be used when solving a proportion. It is* not *applicable for other kinds of fractional equations. For example, equations such as*

$$\frac{x}{3} + 2 = \frac{x}{2} \quad \textit{and} \quad \frac{x + 2}{3} = \frac{x}{2} - 5$$

are not proportions and Property (3) cannot be used. These equations can be solved using the general method of first multiplying each member by its respective least common denominator as shown in Example 1.

All the equations that we solved in the above examples were first-degree equations. The fractional equation in the next example is quadratic. After we obtain an equivalent equation without fractions, we use the procedure we followed in Section 4.4 to complete the solution.

EXAMPLE 4 Solve $x^2 - \frac{17}{3}x = 2$.

Solution Write in standard form.

$$x^2 - \frac{17}{3}x - 2 = 0$$

Multiply each member by 3 and simplify.

$$3(x^2) - 3\left(\frac{17}{3}x\right) - 3(2) = 3(0)$$

$$3x^2 - 17x - 6 = 0$$

Factor the left-hand member.

$$(3x + 1)(x - 6) = 0$$

Set each factor equal to 0 and solve the equations.

$$3x + 1 = 0 \quad \text{or} \quad x - 6 = 0$$

$$x = -\frac{1}{3} \qquad\qquad x = 6$$

The solution set is $\{-1/3, 6\}$. ■

Solving for specified variables

The suggestions given in Section 3.1 for solving an equation for a specified variable in terms of other variables are also applicable to equations that contain fractions. In such an equation it usually is helpful to first write the equation as an equivalent equation without fractions.

EXAMPLE 5 Temperature is commonly measured in Fahrenheit (F) or Celsius (C) units. These units are related by the formula $F = \frac{9}{5}C + 32$. Solve for C.

Solution Multiply each member by 5.

$$5(F) = 5\left(\frac{9}{5}C + 32\right)$$
$$5F = 9C + 160$$

Add -160 to each member.

$$5F - 160 = 9C + 160 - 160$$
$$5F - 160 = 9C$$

Multiply each member by $\frac{1}{9}$ (or divide by 9).

$$\frac{5F - 160}{9} = \frac{9C}{9}$$
$$C = \frac{5F - 160}{9}$$ ■

EXERCISE 5.8

A

■ *Solve each equation. See Example 1.*

1. $7 + \frac{5x}{3} = x - 2$
2. $x + 4 = \frac{2}{5}x - 3$
3. $1 + \frac{x}{9} = \frac{4}{3}$
4. $4 + \frac{x}{5} = \frac{5}{3}$
5. $\frac{1}{5}x - \frac{1}{2}x = 9$
6. $\frac{1}{4}x = 2 - \frac{1}{3}x$
7. $\frac{2x - 1}{5} - \frac{x + 1}{2} = 0$
8. $\frac{2x}{3} - \frac{2x + 5}{6} = \frac{1}{2}$

■ *Solve each equation. See Example 2.*

9. $\frac{x}{x - 2} = \frac{2}{x - 2} + 7$
10. $\frac{x}{x - 3} + \frac{9}{x + 3} = 1$

11. $\frac{2}{y + 1} + \frac{1}{3y + 3} = \frac{1}{6}$

12. $\frac{5}{x - 3} = \frac{x + 2}{x - 3} + 3$

13. $\frac{4}{2x - 3} + \frac{4x}{4x^2 - 9} = \frac{1}{2x + 3}$

14. $\frac{y}{y + 2} - \frac{3}{y - 2} = \frac{y^2 + 8}{y^2 - 4}$

■ *Solve each proportion. See Examples* 1 *or* 3.

15. $\frac{2}{3} - \frac{x}{x + 2}$

16. $\frac{7}{5} = \frac{x}{x - 2}$

17. $\frac{y + 3}{y + 5} - \frac{1}{3}$

18. $\frac{y}{6 - y} = \frac{1}{2}$

19. $\frac{3}{4} = \frac{y + 2}{12 - y}$

20. $\frac{-3}{4} = \frac{y - 7}{y + 14}$

21. $\frac{50}{r} = \frac{75}{r + 20}$

22. $\frac{30}{r} = \frac{20}{r - 10}$

■ *Solve each equation. See Example* 4.

23. $\frac{2x^2}{3} + \frac{x}{3} - 2 = 0$

24. $2x - \frac{5}{3} = \frac{x^2}{3}$

25. $\frac{x^2}{6} + \frac{x}{3} = \frac{1}{2}$

26. $\frac{x}{4} - \frac{3}{4} = \frac{1}{x}$

27. $3 = \frac{10}{x^2} - \frac{7}{x}$

28. $\frac{4}{3x} + \frac{3}{3x + 1} + 2 = 0$

■ *Solve.*

29. The normal weight w, in pounds, of a male between 60 and 70 inches tall is related to his height h, in inches, by

$$w = \frac{11}{2}h - 220.$$

What should be the height of a boy who weighs 143 pounds?

30. Celsius (C) and Fahrenheit (F) temperatures are related by

$$F = \frac{9}{5}C + 32.$$

What is the Celsius temperature when the Fahrenheit temperature is $-40°$?

31. The pressure p, in pounds per square inch, is related to the depth d, in feet below the surface of an ocean, by

$$p = \frac{5}{11}d + 15.$$

At what depth is the pressure 75 pounds per square inch?

32. The horsepower (H) generated by water flowing over a dam of height s, in feet, is given by

$$H = \frac{62.4N \cdot s}{33{,}000},$$

where N is the number of cubic feet of water flowing over the dam per minute. Find the number of cubic feet of water that flowed over a 165-foot dam if 468 hp were generated.

33. The net resistance R_n of an electrical circuit is given by

$$\frac{1}{R_n} = \frac{1}{R_1} + \frac{1}{R_2},$$

where R_1 and R_2* are individual resistors in parallel. If one of the individual resistors is 40 ohms, what must the other be if the net resistance is 30 ohms?

34. The net resistance R_n of an electrical circuit containing resistors R_1, R_2, and R_3* in parallel is given by

$$\frac{1}{R_n} = \frac{1}{R_1} + \frac{1}{R_2} + \frac{1}{R_3}.$$

If two of the individual resistors are 20 ohms and 40 ohms, what must the third resistor be if the net resistance is 10 ohms?

35. A manufacturer plans to depreciate a large lathe that cost \$12,000. The value V in t years is given by the formula

$$V = C\left(1 - \frac{t}{15}\right),$$

where C is the original cost. In how many years will the value be \$5,000? In how many years will the lathe be completely depreciated?

36. The force F necessary to raise a weight w using a special pulley system is given by

$$F = \frac{w}{2}\left(\frac{D_1 - D_2}{D_1}\right).$$

Find the weight that can be lifted using a force of 45 pounds, where $D_1 = 25$ and $D_2 = 22$.

37. Three units are normally used as a measure of temperature, namely, Celsius (C), Fahrenheit (F), and Kelvin (K). These units are related by the formulas

$$C = \frac{5F - 160}{9} \quad \text{and} \quad K = C + 273.$$

What Fahrenheit temperature corresponds to a temperature of 290° Kelvin?

38. Use the formulas in Problem 37 to find the Kelvin temperature that corresponds to a temperature of 98.6° Fahrenheit.

39. The number of diagonals, D, of a polygon of n sides is given by

$$D = \frac{n}{2}(n - 3).$$

How many sides does a polygon with 90 diagonals have?

*The subscripts 1, 2, and 3 used in R_1, R_2, and R_3 in the equations are read "R sub one," "R sub two," and "R sub three."

40. The formula

$$s = \frac{n}{2}(n + 1)$$

gives the sum of the first n natural numbers 1, 2, 3, ... How many consecutive natural numbers starting with 1 will give a sum of 406?

■ *Solve each formula for the specified variable. See Example 5.*

41. $S = \dfrac{a}{1 - r}$, for r

42. $A = \dfrac{h}{2}(b + c)$, for c

43. $\dfrac{1}{r} = \dfrac{1}{s} + \dfrac{2}{t}$, for t

44. $S = \dfrac{n}{2}(a + s_n)$, for s_n

B

■ *Solve each equation for the specified variable in terms of the other variables.*

45. $V = C\left(1 - \dfrac{t}{15}\right)$, for t

46. $\dfrac{1}{R_n} = \dfrac{1}{R_1} + \dfrac{1}{R_2} + \dfrac{1}{R_3}$, for R_3

47. $B = \dfrac{2}{5}w\left(1 + \dfrac{2n}{100}\right)$, for n

48. $C = \dfrac{5F - 160}{9}$ and $K = C + 273$, for F in terms of K

5.9

APPLICATIONS

Suggestions were made in Section 3.2 on how to solve word problems where the mathematical models were equations whose members were polynomials. In this section we will solve several kinds of problems where the equations we obtain involve fractions.

We will continue to follow the step-by-step procedure that we used in Chapters 3 and 4 to solve word problems, giving careful attention to the construction of the mathematical models.

EXAMPLE 1 If two-thirds of a certain number is added to three-fourths of the number, the result is 51. Find the number.

Solution Step 1 → (The number:) x ← Step 2

3. Construct a mathematical model for the conditions on the variable:

$$\frac{2}{3}x + \frac{3}{4}x = 51$$

4. Solve the equation. The LCD is 12.

$$12\left(\frac{2}{3}x\right) + 12\left(\frac{3}{4}x\right) = 12(51)$$
$$8x + 9x = 612$$
$$17x = 612$$
$$x = 36$$

5. The number we seek is 36. ■

The problem in the next example leads to a quadratic equation which has two solutions. Again, we must check each solution to determine whether it is a meaningful solution for the stated conditions.

EXAMPLE 2 The sum of an integer and two times its reciprocal is $11/3$. What is the integer?

Solution Step 1 → (The integer:) x ← Step 2

3. Construct a mathematical model for the conditions on the variable.

$$x + 2 \cdot \frac{1}{x} = \frac{11}{3}.$$

4. Solve. Multiply each member by $3x$.

$$3x\left(x + \frac{2}{x}\right) = 3x\left(\frac{11}{3}\right)$$
$$3x^2 + 6 = 11x$$

Write in standard form and solve by the factoring method.

$$3x^2 - 11x + 6 = 0$$
$$(3x - 2)(x - 3) = 0$$
$$3x - 2 = 0 \quad \text{or} \quad x - 3 = 0$$
$$x = \frac{2}{3} \qquad\qquad x = 3$$

5. Since ⅔ is not an integer, the number 3 is the solution:

$$(3) + 2 \cdot \frac{1}{(3)} = \frac{11}{3}. \qquad ■$$

Ratio and proportion problems

Proportions can be used as mathematical models in the solution of a wide variety of word problems. A table, such as the one in the following example, can be helpful to set up a proportion.

EXAMPLE 3 A car uses 8 gallons of gas to travel 140 miles. How many gallons would be required to travel 450 miles?

Solution

Step 1 → Number of gallons required for 450 miles: x ← Step 2

3. Set up a table as shown in (a). Then enter x and 450 as the numerators in the appropriate columns and 8 and 140 as the denominators in the appropriate columns to obtain the proportion in (b).

Gallons		Miles
———	=	———

(a)

Gallons		Miles
$\frac{x}{8}$	=	$\frac{450}{140}$

(b)

4. Solve the proportion in (b) by first equating the product of the means to the product of the extremes to obtain

$$140 \cdot x = 8 \cdot 450,$$

from which

$$x = \frac{8 \cdot 450}{140}$$
$$= \frac{180}{7} = 25\frac{5}{7}.$$

5. Hence, $25\frac{5}{7}$ gallons of gas are required to travel 450 miles. ■

Different models that are also correct would be obtained in the above example by entering the given data in different ways. Thus,

$$\frac{450}{140} = \frac{x}{8}, \quad \frac{8}{x} = \frac{140}{450}, \quad \text{and} \quad \frac{140}{450} = \frac{8}{x}$$

are also correct equations. You can verify that in each equation the products of the means and the products of the extremes are the same as those we obtained above.

Uniform-motion problems

Solving uniform-motion problems requires applying the fact that the distance traveled at a uniform rate is equal to the product of the rate and the time traveled. This relationship may be expressed by any one of the three equations

$$d = rt, \qquad r = \frac{d}{t}, \qquad t = \frac{d}{r}.$$

It is frequently helpful to use a table and a figure to solve these problems.

EXAMPLE 4 An express train travels 150 miles in the same time that a freight train travels 100 miles. If the express goes 20 miles per hour faster than the freight, find the rate of each.

Solution

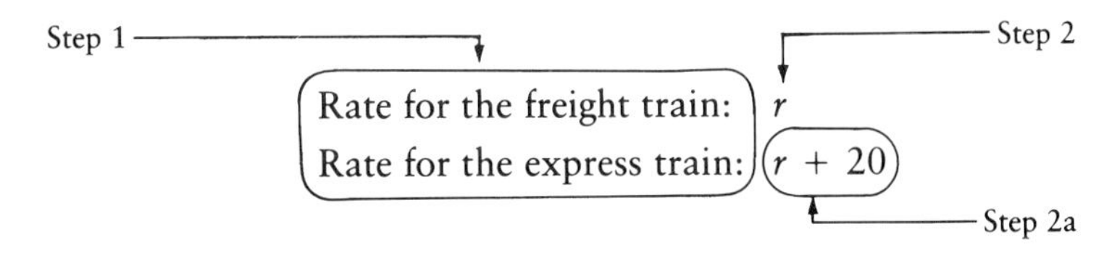

3. Construct a mathematical model. A table and/or a figure can be helpful.

	d	r	$t = d/r$
Freight	100	r	$\frac{100}{r}$
Express	150	$r + 20$	$\frac{150}{r + 20}$

Freight $d = 100$ rate: r

Express $d = 150$ rate: $r + 20$

The fact that the times are the same provides the significant equality in the problem.

$$[t \text{ freight}] = [t \text{ express}]$$
$$\frac{100}{r} = \frac{150}{r + 20}$$

4. Solve the equation. Because the equation is a proportion, set the product of the extremes equal to the product of the means.

$$\begin{aligned} 100(r + 20) &= 150(r) \\ 100r + 2000 &= 150r \\ -50r &= -2000 \\ r &= 40 \\ r + 20 &= 60. \end{aligned}$$

5. Thus, the freight train's rate is 40 miles per hour and the express train's rate is 60 miles per hour. ■

Work problems

Some problems involve the accomplishment of a task in some fixed time when a steady rate of work is assumed. For example, if it takes 6 hours for a man to paint a room, then in one hour he can paint $1/6$ of the room, in two hours he can paint $(1/6)(2)$ of the room and in t hours he can paint $(1/6)(t)$ of the room. In general in such problems, the part of the work done in t units of time equals the work done in one unit of time multiplied by t.

EXAMPLE 5 One pipe can empty a tank in 6 hours, and a second pipe can empty the same tank in 9 hours. How long will it take both pipes to empty the tank?

Solution

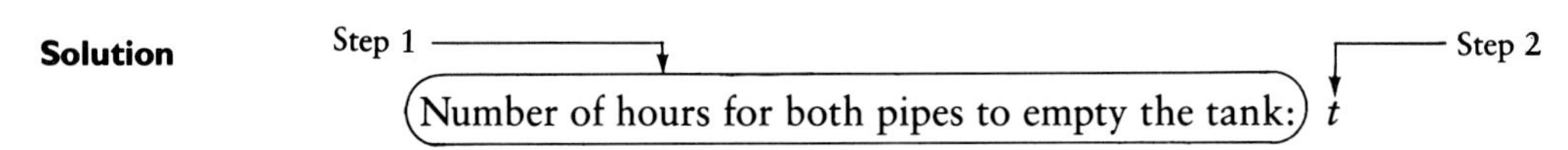

3. Construct a mathematical model. A table can be helpful.

$$\begin{bmatrix}\text{part of tank}\\ \text{emptied in}\\ \text{1 hour}\end{bmatrix} \times [\text{ hours }] = \begin{bmatrix}\text{part of tank}\\ \text{emptied by}\\ \text{each pipe}\end{bmatrix}$$

Pipe 1	$\frac{1}{6}$	t	$\left(\frac{1}{6}\right)t$
Pipe 2	$\frac{1}{9}$	t	$\left(\frac{1}{9}\right)t$

$$\begin{bmatrix}\text{part of tank}\\ \text{emptied by pipe 1}\end{bmatrix} + \begin{bmatrix}\text{part of tank}\\ \text{emptied by pipe 2}\end{bmatrix} = \begin{bmatrix}\text{entire tank}\\ \text{emptied by both}\end{bmatrix}$$

$$\left(\frac{1}{6}\right)t \quad + \quad \left(\frac{1}{9}\right)t \quad = \quad 1$$

4. Solve the equation. The LCD is 18.

$$\begin{aligned}(18)\left(\frac{1}{6}\right)t + (18)\left(\frac{1}{9}\right)t &= (18)1\\ 3t + 2t &= 18\\ 5t &= 18\\ t &= \frac{18}{5}.\end{aligned}$$

5. The tank will be emptied in $3\frac{3}{5}$ hours when both pipes are open. ■

EXERCISE 5.9

A

■ *Solve each problem. Use the five steps suggested in Examples* 1 *and* 2.

1. If one-half of a certain number is added to three times the number, the result is $35/2$. Find the number.
2. If two-thirds of a certain number is subtracted from twice the number, the result is 20. Find the number.

3. Find two consecutive integers such that the sum of one-half the first and two-thirds of the next is 17.
4. Find two consecutive integers such that twice the second less one-half of the first is 14.
5. The sum of a number and its reciprocal is $17/4$. What is the number?
6. The difference of a number and 2 times its reciprocal is $17/3$. What is the number?
7. The positive numerator of a fraction is 1 less than the denominator. The sum of the fraction and 2 times its reciprocal is $41/12$. Find the numerator and the denominator.
8. The positive numerator of a fraction is 2 less than the denominator. The sum of the fraction and 3 times its reciprocal is $28/5$. Find the numerator and the denominator.
9. A partner receives $2/3$ of the profits of the partnership. How much must the business make in profits if this partner is to receive \$160 per week from the partnership?
10. A man owns a three-eighths interest in a lot that was purchased for \$4000. What should the lot sell for if the man is to obtain a \$600 profit on his investment?
11. A woman saves one-sixth of her weekly wages. How much more than \$100 per week does she earn if she saves \$25 per week?
12. After a man used two-thirds of his inheritance as a down payment on a house and one-sixth as a down payment on a car he had \$10,000 left. How much was his inheritance?

■ *Solve. See Example* 3.

13. The denominator of a certain fraction is 6 more than the numerator, and the fraction is equivalent to $3/4$. Find the numerator.
14. The denominator of a certain fraction is 8 more than the numerator, and the fraction is equivalent to $3/5$. Find the denominator.
15. A mason can lay 450 bricks in 3 hours. At the same rate, how many bricks can he lay in a 40-hour week?
16. An automobile uses 6 gallons of gasoline to travel 102 miles. How many gallons are required to make a trip of 459 miles?
17. If 2.5 pounds of tin are required to make 12 pounds of a certain alloy, how many pounds of tin are needed to make 90 pounds of the alloy?
18. A man walks approximately 6.4 miles in 2 hours. At the same rate, how far can he walk in 5 hours?
19. It takes 3 hours to address 144 envelopes. At the same rate, how many envelopes can be addressed in 5 hours?
20. If $3/4$ of a centimeter on a map represents 10 kilometers, how many kilometers does 9 centimeters represent?
21. A secretary earns \$6200 in 20 weeks. At the same rate of earnings, how much would she earn in one year (52 weeks)?
22. A typist can type 200 words in 5 minutes. Typing at the same rate, how long would it take this typist to type 10 pages of a manuscript that averages 240 words per page?

■ *Solve. See Example* 4.

23. An airplane travels 1260 miles in the same time that an automobile travels 420 miles. If the rate of the airplane is 120 miles per hour greater than the rate of the automobile, find the rate of each.

24. Two planes leave an airport at the same time and travel in opposite directions. If one plane averages 440 miles per hour over the ground and the other 560 miles per hour, in how long will they be 2500 miles apart?

25. A ship traveling at 20 knots is 5 nautical miles out from a harbor when another ship leaves the harbor at 30 knots sailing on the same course. How long does it take the second ship to catch up to the first?

26. A boat sails due west from a harbor at 36 knots. An hour later, another boat leaves the harbor on the same course at 45 knots. How far out at sea will the second boat overtake the first?

27. A woman drives 120 miles in the same time that a man drives 80 miles. If the speed of the woman is 20 miles per hour greater than the speed of the man, find the speed of each.

28. Two men drive from town A to town B, a distance of 400 miles. If one man drives twice as fast as the other and arrives at town B 5 hours ahead of the other, how fast was each driving?

29. A riverboat that travels at 18 miles per hour in still water can go 30 miles up a river in 1 hour less time than it can go 63 miles down the river. What is the speed of the current in the river?

30. If a boat travels 20 miles per hour in still water and takes 3 hours longer to go 90 miles up a river than it does to go 75 miles down the river, what is the speed of the current in the river?

31. A commuter takes a train 10 miles to his job in a city. The train returns him home at a rate 10 miles per hour greater than the rate of the train that takes him to work. If he spends a total of 50 minutes a day commuting, what is the rate of each train?

32. On a 50 mile trip, a woman traveled 10 miles in heavy traffic and then 40 miles in less congested traffic. If her average rate in heavy traffic was 20 miles per hour less than her average rate in light traffic, what was each rate if the trip took 1 hour and 30 minutes?

■ *Solve. See Example* 5.

33. It takes one pipe 30 hours to fill a tank, while a second pipe can fill the same tank in 45 hours. How long will it take both pipes running together to fill the tank?

34. One pipe can fill a tank in 4 hours and another can empty it in 6 hours. If both pipes are open, how long will it take to fill the empty tank?

35. A new billing machine can process a firm's monthly billings in 10 hours. By using an older machine together with the new machine, the billings can be completed in 6 hours. How long would it take the older machine to do the job alone?

36. One printing machine can run the necessary copies for the daily circulation of a

newspaper in 5 hours. A second printer can run the same number of copies in 4 hours. How long does it take to run the copies when both machines are working?

37. One tractor operator takes $4/5$ as long as another operator to plow a field. If by working together they can do the job in 20 hours, how long will it take the slow operator to do the job alone?

38. A tractor plows $5/9$ of a field in 10 hours. By adding another tractor the job is finished in another 3 hours. How long would it take the second tractor to do the job alone?

CHAPTER SUMMARY

[5.1] A **fraction** is an expression that denotes a quotient. If the numerator and denominator are polynomials, then the fraction is a **rational expression.**

The **fundamental principle of fractions** states:

If the numerator and the denominator of a fraction are multiplied or divided by the same nonzero number, the result is a fraction ***equivalent*** *to the given fraction:*

$$\frac{a}{b} = \frac{ac}{bc} \qquad (b, c \neq 0).$$

For a given fraction, the replacement of any two of the three elements of the fraction—the fraction itself, the numerator, the denominator—by their negatives results in an equivalent fraction:

$$\frac{a}{b} = \frac{-a}{-b} = -\frac{a}{-b} = -\frac{-a}{b} \qquad (b \neq 0),$$

$$\frac{-a}{b} = \frac{a}{-b} = -\frac{a}{b} = -\frac{-a}{-b} \qquad (b \neq 0).$$

Standard forms for fractions are $\dfrac{a}{b}$ and $\dfrac{-a}{b}$.

A fraction is in **lowest terms** if the numerator and the denominator do not contain factors in common. We can reduce a fraction to lowest terms by using the fundamental principle of fractions to divide the numerator and the denominator by their common nonzero factors.

[5.2] Some quotients of polynomials can be rewritten as equivalent **mixed expressions.** A method similar to the long division process used in arithmetic can be used for this purpose.

[5.3] **Synthetic division** is a condensation of the division algorithm using only coefficients.

[5.4] The **least common denominator** (LCD) of a set of fractions with natural-number denominators is the smallest natural number that is exactly divisible by each of the denominators.

[5.5] The operations of addition and subtraction of fractions are governed by the following properties:

$$\frac{a}{c} + \frac{b}{c} = \frac{a+b}{c} \quad \text{and} \quad \frac{a}{c} - \frac{b}{c} = \frac{a-b}{c} \qquad (c \neq 0).$$

If the fractions in a sum or difference have unlike denominators, we can build each fraction to higher terms by using the fundamental principle of fractions in order to obtain the fractions with a common denominator.

[5.6] The operations of multiplication and division of fractions are governed by the following properties:

$$\frac{a}{b} \cdot \frac{c}{d} = \frac{ac}{bd} \quad (b, d \neq 0) \qquad \frac{a}{b} \div \frac{c}{d} = \frac{a}{b} \cdot \frac{d}{c} \quad (b, c, d \neq 0)$$

[5.7] A **complex fraction** is a fraction containing other fractions in its numerator or denominator or both. It is often convenient to use the fundamental principle to simplify such fractions.

[5.8–5.9] Equations that contain fractions can be solved by using the properties introduced in previous chapters.

A **proportion** is a special type of equation of the form $\frac{a}{b} = \frac{c}{d}$. In such equations, the product of the extremes a and d equals the product of the means b and c. Thus, $ad = bc$.

REVIEW EXERCISES

A

■ *Assume that no denominator is* 0.

[5.1]

1. a. Write six fractions equal to $\frac{1}{a-b}$ by changing the sign or signs of the numerator, the denominator, or the fraction itself.

b. What are the conditions on a and b for the fraction in part a to represent a positive number? A negative number?

2. Express $-\dfrac{1}{1-a}$ in standard form in two ways.

■ *Reduce each fraction to lowest terms.*

3. a. $\dfrac{4x^2y^3}{10xy^4}$ b. $\dfrac{2x^2-8}{2x+4}$

4. a. $\dfrac{4x^2-1}{1-2x}$ b. $\dfrac{x^2-y^2}{2y^2-2x^2}$

[5.2]

■ *Divide.*

5. a. $\dfrac{12x^2-6x+3}{2x}$ b. $\dfrac{6y^3+3y^2-y}{3y^2}$

6. a. $\dfrac{2y^2-3y+1}{2y+3}$ b. $\dfrac{6x^4-3x^3+2x+2}{x-1}$

[5.3]

■ *Use synthetic division to write each quotient as a polynomial in simple form.*

7. $\dfrac{y^3+3y^2-2y-4}{y+1}$

8. $\dfrac{y^7-1}{y-1}$

[5.4]

■ *Express each given fraction as an equivalent fraction with the given denominator.*

9. a. $\dfrac{-3}{4}=\dfrac{?}{24}$ b. $\dfrac{x}{2y}=\dfrac{?}{2xy^2}$

10. a. $\dfrac{2}{x+3y}=\dfrac{?}{x^2-9y^2}$ b. $\dfrac{y}{3-y}=\dfrac{?}{y^2-4y+3}$

[5.5]

■ *Write each expression as a single fraction in lowest terms.*

11. a. $\dfrac{3x+y}{3}+\dfrac{x-y}{3}$ b. $\dfrac{x-2y}{4x}-\dfrac{2x-y}{4x}$

12. a. $\dfrac{2}{5x}-\dfrac{3}{4y}+\dfrac{7}{20xy}$ b. $\dfrac{y}{2y-6}+\dfrac{2}{3y-9}$

13. a. $\dfrac{5}{x-2}-\dfrac{3}{x+3}$ b. $\dfrac{1}{x+2y}+\dfrac{3}{x^2-4y^2}$

14. a. $\dfrac{2}{x^2-1}+\dfrac{3}{x^2-2x+1}$ b. $\dfrac{y}{y^2-16}-\dfrac{y+1}{y^2-5y+4}$

[5.6]

■ *Write each expression as a single fraction in lowest terms.*

15. a. $\frac{2x}{y^2} \cdot \frac{3y^3}{8x} \cdot \frac{4x^2}{9y}$ b. $\frac{4x}{2xy + 8y^2} \cdot \frac{x + 4y}{x}$

16. a. $\frac{x^2 + 2x - 3}{x^2 + 6x + 9} \cdot \frac{2x + 6}{2x - 2}$ b. $\frac{y^3 - y}{2y + 1} \cdot \frac{4y + 2}{y^2 + 2y + 1}$

17. a. $\frac{10x^2y^3}{9} \div \frac{2xy^2}{3}$ b. $\frac{2y - 6}{y + 2x} \div \frac{4y - 12}{2y + 4x}$

18. a. $\frac{y^2 + 4y + 3}{y^2 - y - 2} \div \frac{y^2 - 4y - 5}{y^2 - 3y - 10}$ b. $\frac{x^2}{y - x} \div \frac{x^3 - x^2}{x - y}$

[5.7]

■ *Write each expression as a single fraction in lowest terms.*

19. a. $\frac{\frac{3x}{2y}}{\frac{9x}{10y}}$ b. $\frac{\frac{2}{3}}{4 - \frac{1}{3}}$

20. a. $\frac{2 + \frac{3}{x}}{1 - \frac{2}{3x}}$ b. $\frac{y - \frac{1}{y}}{y + \frac{1}{y}}$

21. a. $\frac{2}{x - \frac{1}{x}}$ b. $\frac{\frac{y}{2} + \frac{y}{3}}{\frac{y}{3} - \frac{y}{4}}$

22. a. $\frac{y - \frac{1}{y}}{1 - \frac{1}{y + 1}}$ b. $\frac{x + \frac{1}{x}}{1 - \frac{1}{x - 1}}$

[5.8]

■ *Solve.*

23. a. $2 - \frac{x}{3} = \frac{11}{6}$ b. $\frac{1}{4}x + \frac{2}{3}x = 11$

24. a. $\frac{3}{y - 1} + \frac{2}{3y - 3} = \frac{11}{9}$ b. $\frac{1}{y + 1} - \frac{3}{y + 1} = \frac{14}{3}$

25. a. $\frac{2}{5} = \frac{x}{x - 2}$ b. $\frac{x + 2}{x - 3} = \frac{3}{4}$

26. a. $\frac{1}{x} + 2x = \frac{33}{4}$ b. $\frac{1}{x} + \frac{2}{x - 2} = \frac{2}{3}$

[5.9]

27. An enlargement is made of a 4 by 5-inch photograph. What is the dimension of the larger side if the smaller side of the enlargement is 11 inches?

28. A car and a light plane start together and travel in the same direction. The plane travels three times as fast as the car. How fast is each traveling if they are 100 miles apart at the end of one hour?

29. A passenger train and a freight train leave a station at the same time and travel in opposite directions. The passenger train averages 60 miles per hour and the

freight train averages 20 miles per hour. How long after leaving the station will they be 200 miles apart?

30. A boy got 17 hits on his first 60 times at bat. How many hits does he need in the next 20 times at bat to have an average of 0.300?

B

■ *Reduce each fraction.*

31. $\dfrac{x^3 - 1}{x^2 - 1}$

32. $\dfrac{8y^3 + 1}{4y^2 - 1}$

■ *Divide.*

33. $\dfrac{x^3 - 2x^2 + x + 1}{x^2 - x + 2}$

34. $\dfrac{x^4 - 3x^2 + 4}{x^3 - 2x + 1}$

35. $\dfrac{x^4 + 3x^3 - x^2 - 6x - 2}{x^2 - 2}$

36. $\dfrac{x^4 - x^2 - 2}{x^2 + 4}$

■ *Simplify each expression.*

37. $\dfrac{x^3 - y^3}{x^2 + xy} \cdot \dfrac{x + y}{x^2 - xy}$

38. $\dfrac{2x + xy + y^2 + 2y}{2x + 2y} \cdot \dfrac{y^3 - 8}{y^2 - 4}$

39. $1 + \dfrac{1 - \dfrac{1}{2}}{1 + \dfrac{1}{1 + \dfrac{1}{2}}}$

40. $\dfrac{3 - \dfrac{1}{\dfrac{x}{y} - 2}}{2 + \dfrac{3}{\dfrac{x}{2y} - 1}}$

41. $\dfrac{\dfrac{2}{ab} + \dfrac{1}{bc} - \dfrac{3}{ac}}{\dfrac{a - 3b + 2c}{abc}}$

42. $\dfrac{a - 3 - \dfrac{2}{a - 1}}{a + 2 - \dfrac{3}{a - 1}}$

43. $\dfrac{\dfrac{1}{a - b}}{\dfrac{a}{a^2 - b^2}} + \dfrac{ab - b^2}{ab}$

44. $\dfrac{a^2 - b^2}{\dfrac{a - b}{b}} - \dfrac{a - b}{\dfrac{1}{b}}$

■ *Solve.*

45. A student store sold \$96 worth of notebooks in one month. The store could sell 20 fewer notebooks to get the same income if the price on each was raised \$0.40. How much did each notebook sell for originally?

46. A theater now has an income of \$1200 for each performance that is sold out. The management could obtain the same income by adding 60 seats and lowering the ticket price \$1.00. How many seats does the theater have now?

Cumulative Review Chapters 1-5

The numbers in brackets refer to the sections where such problems are first considered.

A

1. Graph the set $\{x \mid -7 \leq x < -3,\ x \text{ a real number}\}$. [1.2]
2. Rewrite $|3 - x|$ without absolute value notation. [1.3]
3. Evaluate $2x^2(x - y)^2 - (2x^2)(x - y^2)$ for $x = -2$ and $y = 3$. [2.1]
4. If $P(x) = 2x^2 - 2x + 2$, find $P(-2)$. [2.1]
5. Evaluate $ar^{n-1} - br^{n+1}$ for $a = 2$, $b = -3$, $r = 2$, and $n = 3$. [2.1]
6. Multiply $(2x - 1)(x + 3)(x - 2)$. [2.4]

■ *Factor completely.*

7. $3x^2(x - 3)^4(x + 2) - 12x(x + 2)^2(x - 3)^2$ [4.1]
8. $8x^6 - y^9$ [4.2]
9. $x^3 + 2x^2y^2 - xy - 2y^3$ [4.3]

■ *Simplify.*

10. $-6 - |-3| - |6 - 3|$ [1.6]
11. $\dfrac{-16 - 6(4)}{10(6 - 4)} - \left[\dfrac{3(-6) - 6}{-3(-4) - 6}\right]\left[5 - 3\left(\dfrac{6 + 9}{5}\right)\right]$ [1.6]
12. $\dfrac{-3^2 - 3(-2)^3}{3 - 8} - \dfrac{1 - 2^4(-2)^2}{(4 - 7)^2}$ [2.1]

13. $2x - (x^2 - 4x) - [(1 - 2x^2) - (3x - 4)]$ [2.2]

14. $(-2b)^3(a^2b)^2 + ab(3a^2b)(-ab^3) - a^2(ab^2)^2$ [2.3]

15. $x(x[x(x - 3) + 5] - 2) + 6$ [2.4]

■ *Solve.*

16. $0.30(x - 2) = 1.80$ [3.1]

17. $|-3x - 5| = -7$ [3.4]

18. $0.08x + 0.12(x - 3000) = 1240$ [3.1]

19. $6x - [(x - 2)(x + 5) - 8] = 0$ [4.4]

20. $\dfrac{x + 1}{2x + 1} = \dfrac{x - 3}{x - 2} - \dfrac{1}{2}$ [5.8]

21. $-8 \leqslant \dfrac{-6 - 2x}{4} \leqslant -2$ [3.3]

22. $|6 - 4x| \geqslant 2$ [3.4]

23. Solve $A = P + Prt$ for P. [3.1]

24. Solve $S = 2\pi r(r + h)$ for h. [3.1]

25. Reduce $\dfrac{2x^2 - 9x + 9}{18x - 2x^3}$. [5.1]

26. Use synthetic division to divide $\dfrac{x^5 - 3x^3 - 1}{x - 2}$. [5.3]

27. Find the LCM of $6x^3 - 12x^2$, $3x^2 - 12$, and $9x(x - 2)^2$. [5.4]

28. Write as a single fraction in lowest terms:

$$\frac{2x - 1}{x^2 - 4x + 3} - \frac{x + 4}{x^2 - x - 6} + \frac{x}{x^2 + x - 2} \quad [5.5]$$

29. Write as a single fraction in lowest terms:

$$\frac{2x}{x - 1} - \frac{x + 2}{x + 1} \div \left(\frac{1}{x} - \frac{2}{x + 1}\right). \quad [5.6]$$

30. To accumulate an amount A of money in t years at simple interest rate r, one must invest $\dfrac{A}{1 + rt}$ dollars. How much must be invested at 8% in order to accumulate \$1680 in 5 years? [1.6]

31. Find a quadratic equation with integral coefficients whose solutions are $-\frac{1}{2}$ and $\frac{2}{3}$. [4.4]

32. 32% of the freshman class are from out-of-state. If 493 freshmen are in-state students, how large is the freshman class? [3.2]

33. Ernest has three accounts: checking, savings, and Christmas club. He divides his paycheck of \$1484 between them. He puts \$200 more into his savings account than into the Christmas club, and twice as much into checking as into savings. How much does he deposit into each account? [3.2]

34. If \$16,000 is invested at 9%, how much must be invested at 14% to earn a return of 10% on the total investment? [3.2]

35. How many liters of 80% acid must be added to 40 liters of 30% acid to obtain a mixture which is 60% acid? [3.2]

36. The length of a rectangular garden is twice its width. The garden is bordered by a 3-foot wide path. The area of the garden including the path is 972 square feet more than the area of the garden alone. Find the dimensions of the garden. [3.2]

37. The hypotenuse of a right triangle is 3 inches shorter than twice the length of the shortest leg. The longer leg is 12 inches long. Find the lengths of the other two sides. [4.5]

38. The height h in feet of a ball thrown off a building is given by the equation $h = 500 - 28t - 16t^2$, where t is the time in seconds after the throw. How long will it take the ball to reach a height of 380 feet? [4.5]

39. If $1\frac{1}{4}$ inches on a map represents 20 miles, how many miles does $1\frac{5}{8}$ inches represent? [5.9]

40. A car travels 20 miles per hour less than twice as fast as a train. On a 700-mile trip, the train takes 6 hours longer than the car. Find the speed of the train. [5.9]

B

41. a. Find values for x and y so that $|x + y| = |x| + |y|$.
    b. Find values for x and y so that $|x + y| < |x| + |y|$. [1.3]

42. Write $([-5, -1] \cup (3, 5)) \cap (-2, 3]$ as a single interval. [3.4]

43. Given that $P(x) = 2 - 2x^2$, $Q(x) = x^2 - 3x$, and $R(x) = x^2 + x - 2$, find $Q(0) - [R(-2) - P(2)]$. [2.2]

44. Simplify $(x^{n-1}y^{2n})^3$. [2.3]

45. Factor $x^{3n} - x^{n+3}$. [4.1]

46. Factor $2y^6 - y^3 - 10$. [4.2]

47. Divide $\dfrac{y^4 - 2y^2}{y^2 + 3y - 1}$. [5.2]

48. Write as a single fraction in lowest terms: $x + \dfrac{1}{x} + \dfrac{1}{x - 1}$. [5.5]

49. Simplify $\dfrac{\dfrac{a}{\frac{1}{a} + \frac{1}{b}}}{\dfrac{b}{\frac{1}{a} - \frac{1}{b}}}$. [5.7]

50. Solve for b: $\dfrac{1}{a + b} + \dfrac{1}{b + c} = \dfrac{1}{q}$. [5.8]

TC 267

TV
PVS A·
RITH
ME
TICAE·
BOETIVS

6 Exponents, Roots, and Radicals

6.1

POSITIVE INTEGRAL EXPONENTS

In Section 3.1 the expression a^n, where n is a natural number, was defined as follows.

$$a^n = a \cdot a \cdot a \cdot \cdots \cdot a \qquad (n \text{ factors})$$

The following three laws were developed from this definition in Section 3.3.

Laws of exponents for products

$$\text{I.} \quad a^m \cdot a^n = a^{m+n}$$
$$\text{II.} \quad (a^m)^n = a^{mn}$$
$$\text{III.} \quad (ab)^m = a^m b^m$$

EXAMPLE 1 a. $x^5 \cdot x^2 = x^{5+2} = x^7$ b. $(x^5)^2 = x^{5 \cdot 2} = x^{10}$ c. $(x^2y^3)^2 = (x^2)^2(y^3)^2 = x^4y^6$ ■

Laws of exponents for quotients

Note that if $m > n$,

$$\frac{a^m}{a^n} = \frac{\overbrace{(a \cdot a \cdot a \cdot \cdots \cdot a)}^{(m-n) \text{ factors}}\overbrace{(a \cdot a \cdot a \cdot \cdots \cdot a)}^{n \text{ factors}}}{\underbrace{a \cdot a \cdot a \cdot \cdots \cdot a}_{n \text{ factors}}}$$

$$= a^{m-n}.$$

Furthermore, if $m < n$,

$$\frac{a^m}{a^n} = \frac{\overbrace{a \cdot a \cdot \cdots \cdot a}^{m \text{ factors}}}{\underbrace{(a \cdot a \cdot \cdots \cdot a)}_{m \text{ factors}}\underbrace{(a \cdot a \cdot \cdots \cdot a)}_{(n-m) \text{ factors}}}$$

$$= \frac{1}{a^{n-m}}.$$

The following laws follow from these arguments.

IV. $\dfrac{a^m}{a^n} = a^{m-n} \qquad (m > n, a \neq 0)$

IVa. $\dfrac{a^m}{a^n} = \dfrac{1}{a^{n-m}} \qquad (m < n, a \neq 0)$

EXAMPLE 2

a. $\dfrac{x^4y^6}{x^2y} = x^{4-2}y^{6-1}$

$= x^2y^5 \qquad (x, y \neq 0)$

b. $\dfrac{x^2y}{x^3y^2} = \dfrac{1}{x^{3-2}y^{2-1}}$

$= \dfrac{1}{xy} \qquad (x, y \neq 0)$ ■

Note that the fourth law of exponents is consistent with the process of reducing fractions by using the fundamental principle of fractions. The same results would be obtained in the above examples by using the fundamental principle of fractions.

Now, since

$$\left(\frac{a}{b}\right)^n = \overbrace{\frac{a}{b} \cdot \frac{a}{b} \cdot \frac{a}{b} \cdot \cdots \cdot \left(\frac{a}{b}\right)}^{n \text{ factors}} = \frac{\overbrace{a \cdot a \cdot a \cdot \cdots \cdot a}^{n \text{ factors}}}{\underbrace{b \cdot b \cdot b \cdot \cdots \cdot b}_{n \text{ factors}}}$$

$$= \frac{a^n}{b^n},$$

we have another useful law for powers.

$$\textbf{V.} \quad \left(\frac{a}{b}\right)^n = \frac{a^n}{b^n} \qquad (b \neq 0)$$

EXAMPLE 3 a. $\left(\dfrac{x}{y}\right)^3 = \dfrac{x^3}{y^3} \quad (y \neq 0)$ b. $\left(\dfrac{2x^2}{y}\right)^4 = \dfrac{(2x^2)^4}{y^4} \quad (y \neq 0)$ ■

Ordinarily, two or more of the laws of exponents are required to simplify expressions containing exponents. In the last example, we can further simplify the result: From Law III,

$$\frac{(2x^2)^4}{y^4} = \frac{2^4(x^2)^4}{y^4} \qquad (y \neq 0),$$

and from Law II,

$$\frac{2^4(x^2)^4}{y^4} = \frac{16x^8}{y^4} \qquad (y \neq 0).$$

EXAMPLE 4 a. $\left(\dfrac{2x^3}{y}\right)^2 = \dfrac{2^2 \cdot x^6}{y^2} = \dfrac{4x^6}{y^2} \quad (y \neq 0)$ b. $\left(\dfrac{2 + x}{3y^2}\right)^3 = \dfrac{(2 + x)^3}{(3y^2)^3} = \dfrac{(2 + x)^3}{27y^6} \quad (y \neq 0)$ ■

Common Error *Note that in Example b,*

$$(2 + x)^3 \neq 2^3 + x^3; \quad (2 + x)^3 = (2 + x)(2 + x)(2 + x).$$

The laws of exponents can be applied to simplify powers with variable exponents.

EXAMPLE 5

a. $\dfrac{x^n \cdot x^{n+1}}{x^{n-1}}$
$= x^{n+(n+1)-(n-1)}$
$= x^{n+2} \quad (x \neq 0)$

b. $\dfrac{(y^{n-1})^2}{y^{n-2}}$
$= y^{(2n-2)-(n-2)}$
$= y^n \quad (y \neq 0)$

c. $(x^{n-1} \cdot x^{2n+3})^2$
$= (x^{3n+2})^2$
$= x^{6n+4}$ ■

In the exercise sets in this chapter we shall assume that no denominator equals 0 unless otherwise stated.

EXERCISE 6.1

A

■ *Using one or more of the laws of exponents, write each expression as a product or quotient in which each variable occurs only once, and all exponents are positive. See Example 1.*

1. $x^2 \cdot x^3$
2. $y \cdot y^4$
3. $a^3 \cdot a^5$
4. $b^5 \cdot b^4$
5. $(a^2)^3$
6. $(b^3)^4$
7. $(x^2)^3$
8. $(y^4)^2$
9. $(xy^2)^3$
10. $(x^2y^3)^2$
11. $(abc^2)^4$
12. $(a^2b^3c)^2$
13. $(2x)(-2x)^3$
14. $(-3x^2)^2(-5x)$
15. $(ab^2)^3(-2a^2)^2$
16. $(a^2b^2)^3(-ab^2)^3$

■ *See Examples 1–4.*

17. $\dfrac{x^5}{x^3}$
18. $\dfrac{y^2}{y^6}$
19. $\dfrac{x^2y^4}{xy^2}$
20. $\dfrac{x^4y^3}{x^2y}$
21. $\left(\dfrac{x}{y^2}\right)^3$
22. $\left(\dfrac{y^2}{z^3}\right)^2$
23. $\left(\dfrac{2x}{y^2}\right)^3$
24. $\left(\dfrac{3y^2}{x}\right)^2$
25. $\left(\dfrac{-2x}{3y^2}\right)^3$
26. $\left(\dfrac{-x^2}{2y}\right)^4$
27. $\dfrac{(4x)^3}{(2x^2)^2}$
28. $\dfrac{(5x)^2}{(3x^2)^3}$
29. $\dfrac{(xy^2)^3}{(x^2y)^2}$
30. $\dfrac{(-xy^2)^2}{(x^2y)^3}$
31. $\dfrac{(xy)^2(x^2y)^3}{(x^2y^2)^2}$
32. $\dfrac{(-x)^2(-x^2)^4}{(x^2)^3}$
33. $\left(\dfrac{2x}{y^2}\right)^3\left(\dfrac{y^2}{3x}\right)^2$
34. $\left(\dfrac{x^2z}{2}\right)^2\left(-\dfrac{2}{x^2z}\right)^3$
35. $\left(\dfrac{-3}{y^2}\right)^2(2y^3)^2$
36. $\left(\dfrac{y}{x}\right)^2\left(-\dfrac{3}{4xy}\right)^3$

37. $\left[\left(\frac{r^2s^3t}{xy}\right)^3\left(\frac{x^2y}{r^3st^2}\right)^2\right]^2$

38. $\left[\left(\frac{a^3bc}{x^2y}\right)^4\left(\frac{x^2yz}{ab^2c^3}\right)^2\right]^2$

39. $\left(\frac{x^2}{a^2b}\right)^2\left(-\frac{ab}{x^3}\right)^3\left(\frac{x}{ab}\right)^2$

40. $\left(\frac{m^3n^2p}{r^2s}\right)^2\left(\frac{rs}{mn^2p^2}\right)^3\left(-\frac{mnp}{rs}\right)^2$

41. $\left(\frac{y+z}{x^2}\right)^3\left(\frac{x}{y+z}\right)^2$

42. $\left(\frac{y^2+z}{y}\right)^2\left(\frac{y^2}{x^2+z}\right)^3$

43. Use a counterexample to show that $(x^2 + y^2)^3$ is not equivalent to $x^6 + y^6$.

44. Rewrite $(x^2 + y^2)^3$ as an equivalent expression without using parentheses.

B

■ *Write each expression as a product or quotient in which each variable occurs only once. See Example 5.*

45. $x^n \cdot x^n$

46. $\frac{x^{2n}x^n}{x^{n+1}}$

47. $\frac{(x^{n+1}x^{2n-1})^2}{x^{3n}}$

48. $\left(\frac{y^2 \cdot y^3}{y}\right)^{2n}$

49. $\left(\frac{x^{3n}x^{2n}}{x^{4n}}\right)^2$

50. $\frac{(y^{n+1})^n}{y^n}$

6.2

ZERO AND NEGATIVE INTEGRAL EXPONENTS

Zero exponent

Note that if the second law of exponents holds for the case where $m = n$, we have

$$\frac{a^n}{a^n} = a^{n-n} = a^0 \qquad (a \neq 0). \tag{1}$$

By the definition of a quotient,

$$\frac{a^n}{a^n} = 1 \qquad (a \neq 0). \tag{2}$$

Since $a^n/a^n = a^0$ by (1) above and $a^n/a^n = 1$ by (2) above, a^0 must be defined as 1.

$$\mathbf{a^0 = 1 \qquad (a \neq 0)}$$

EXAMPLE 1 a. $3^0 = 1$ b. $(-4)^0 = 1$ c. If $x = 0$, $x^0 \neq 1$. ■

Note that in Example c, if $x = 0$, then 0^0 is undefined. This follows from the fact that 0/0 is undefined.

Negative integer exponents

We would like the laws of exponents to hold also for negative exponents. We observe that for $a \neq 0$,

$$a^n \cdot a^{-n} = a^{n-n} = a^0 = 1.$$

Also, by the reciprocal property,

$$a^n \cdot \frac{1}{a^n} = 1 \qquad (a \neq 0).$$

Again, for consistency, we make the following definition.

$$\mathbf{a^{-n} = \frac{1}{a^n} \qquad (a \neq 0)} \tag{3}$$

Therefore, because a^{-n} is reciprocal of a^n, we have the following property.

$$\mathbf{\frac{1}{a^{-n}} = \frac{1}{\frac{1}{a^n}} = a^n \qquad (a \neq 0)} \tag{4}$$

EXAMPLE 2 a. $3^{-2} = \frac{1}{3^2}$
$= \frac{1}{9}$

b. $\frac{1}{2^{-3}} = 2^3$
$= 8$

c. $4^{-2} + 4^2 = \frac{1}{16} + 16$
$= \frac{1}{16} + \frac{256}{16}$
$= \frac{257}{16}$ ■

We can use Properties (3) and (4) above to rewrite algebraic expressions that contain variables with negative exponents as expressions in which all exponents are positive. Furthermore, we can now use the laws of exponents given in Section 6.1 to simplify expressions that have powers with negative exponents.

EXAMPLE 3

a. $x^{-3} \cdot x^5 = x^{-3+5} = x^2$

b. $(x^2y^{-3})^{-1} = x^{-2}y^3 = \dfrac{y^3}{x^2}$

c. $(3x^2 \cdot x^{-3})^2 = (3x^{-1})^2 = 9x^{-2} = \dfrac{9}{x^2}$ ■

Common Error *Note that in Example c,* $9x^{-2} = \dfrac{9}{x^2}$ *and not* $\dfrac{1}{9x^2}$. *The exponent only applies to the base x.*

$\left(\dfrac{a}{b}\right)^{-n} = \left(\dfrac{b}{a}\right)^n$

We can establish an additional property that will help us simplify quotients with negative exponents. From Law V on page 191,

$$\left(\frac{a}{b}\right)^{-n} = \frac{a^{-n}}{b^{-n}},$$

and then from (1) and (2) above,

$$\frac{a^{-n}}{b^{-n}} = \frac{b^n}{a^n} = \left(\frac{b}{a}\right)^n.$$

Hence, we have the following property.

> **VI.** $\left(\dfrac{a}{b}\right)^{-n} = \left(\dfrac{b}{a}\right)^n \quad (a, b \neq 0)$

The laws of exponents that we have now established can be applied in any order.

EXAMPLE 4 Write $\left(\dfrac{x^3}{x^2}\right)^{-3}$ as a quotient in which the variable occurs only once with a positive exponent.

Solution a. We can first apply Law IV to write

$$\left(\frac{x^3}{x^2}\right)^{-3} = (x)^{-3} = \frac{1}{x^3}.$$

b. Or we can first apply Law V and then Law II to write

$$\left(\frac{x^3}{x^2}\right)^{-3} = \frac{(x^3)^{-3}}{(x^2)^{-3}} = \frac{x^{-9}}{x^{-6}},$$

from which, by Law IV, we have

$$\frac{x^{-9}}{x^{-6}} = \frac{1}{x^{-6-(-9)}} = \frac{1}{x^3}.$$

c. Or we can first use Law VI to obtain

$$\left(\frac{x^3}{x^2}\right)^{-3} = \left(\frac{x^2}{x^3}\right)^{3} = \left(\frac{1}{x}\right)^{3} = \frac{1}{x^3}.$$ ■

The laws of exponents that we have established apply only to products and quotients. Special care must be taken when simplifying expressions that involve powers that contain sums and differences. In such cases it is best to first write any powers with positive exponents using Properties (1) and (2) above before using the other laws of exponents.

EXAMPLE 5 Write each expression as a single fraction involving positive exponents only.

a. $x^{-1} + y^{-2}$ b. $(x^{-1} + x^{-2})^{-1}$ c. $\dfrac{x^{-1} + y}{x^{-1}}$

Solutions a. $x^{-1} + y^{-2}$

$$= \frac{1}{x} + \frac{1}{y^2}$$

$$= \frac{(y^2)}{(y^2)}\frac{1}{x} + \frac{1(x)}{y^2(x)}$$

$$= \frac{y^2 + x}{xy^2}$$

b. $(x^{-1} + x^{-2})^{-1}$

$$= \frac{1}{\dfrac{1}{x} + \dfrac{1}{x^2}}$$

$$= \frac{1(x^2)}{\left(\dfrac{1}{x} + \dfrac{1}{x^2}\right)(x^2)}$$

$$= \frac{x^2}{x + 1}$$

c. $\dfrac{x^{-1} + y}{x^{-1}}$

$$= \frac{\dfrac{1}{x} + y}{\dfrac{1}{x}}$$

$$= \frac{\left(\dfrac{1}{x} + y\right)x}{\left(\dfrac{1}{x}\right)x}$$

$$= 1 + xy$$ ■

Common Error *Note in Example b that* $(x^{-1} + x^{-2})^{-1} \neq (x^{-1})^{-1} + (x^{-2})^{-1}$.

The laws of exponents can now be applied to simplify powers with variable exponents when the exponents are integers or the variables represent integers.

EXAMPLE 6 Write each expression as a product free of fractions in which each variable occurs only once.

a. $x^{-2n} \cdot x^n$ b. $\left(\dfrac{x^{1-n}}{x^{2-n}}\right)^{-2}$ c. $\left(\dfrac{x^n y^{2n-1}}{y^n}\right)^2$

Solutions

a. $x^{-2n} \cdot x^n$
$= x^{-2n+n}$
$= x^{-n}$

b. $\left(\dfrac{x^{1-n}}{x^{2-n}}\right)^{-2}$
$= (x^{(1-n)-(2-n)})^{-2}$
$= (x^{-1})^{-2}$
$= x^2$

c. $\left(\dfrac{x^n y^{2n-1}}{y^n}\right)^2$
$= (x^n y^{2n-1-n})^2$
$= (x^n y^{n-1})^2$
$= x^{2n} y^{2n-2}$ ■

EXERCISE 6.2

A

■ *Write each expression as a basic numeral or fraction in lowest terms. See Examples 1 and 2.*

1. 2^{-1}
2. 3^{-2}
3. $\dfrac{1}{3^{-1}}$
4. $\dfrac{3}{4^{-2}}$
5. $(-2)^{-3}$
6. $\dfrac{1}{(-3)^{-2}}$
7. $\dfrac{5^{-1}}{3^0}$
8. $\dfrac{2^0}{3^{-2}}$
9. $\left(\dfrac{3}{5}\right)^{-1}$
10. $\left(\dfrac{1}{3}\right)^{-2}$
11. $\dfrac{5^{-1}}{3^{-2}}$
12. $\dfrac{3^{-3}}{6^{-2}}$
13. $3^{-2} + 3^2$
14. $5^{-1} + 25^0$
15. $4^{-1} - 4^{-2}$
16. $8^{-2} - 2^0$

■ *Write each expression as a product or quotient of powers in which each variable occurs only once, and all exponents are positive. See Examples 3 and 4.*

17. x^2y^{-3}
18. $\dfrac{x^3}{y^{-2}}$
19. $(x^2 \cdot y)^{-3}$
20. $(xy^3)^{-2}$
21. $\left(\dfrac{x}{y^3}\right)^2$
22. $\left(\dfrac{2x}{y^2}\right)^3$
23. $\dfrac{(xy^2)^3}{(x^2y)^2}$
24. $\left(\dfrac{3x}{y^2}\right)^2\left(\dfrac{2y^3}{x}\right)^2$

25. $x^{-3} \cdot x^{7}$

26. $\dfrac{x^{3}}{x^{-2}}$

27. $(x^{-2}y^{0})^{3}$

28. $(x^{-2}y^{3})^{0}$

29. $\dfrac{x^{-1}}{y^{-1}}$

30. $\dfrac{x^{-3}}{y^{-2}}$

31. $\dfrac{8^{-1}x^{0}y^{-3}}{(2xy)^{-5}}$

32. $\left(\dfrac{x^{-1}y^{3}}{2x^{0}y^{-5}}\right)^{-2}$

33. $\left(\dfrac{x^{0}y^{2}}{z^{2}}\right)^{-1}$

34. $\left(\dfrac{x^{-1}yz^{0}}{xy^{-1}z}\right)^{-1}$

35. $\left(\dfrac{x^{2}y}{z^{3}}\right)\left(\dfrac{x}{z^{2}}\right)^{-1}$

36. $\left(\dfrac{2y^{-1}}{x^{2}}\right)^{-1} \cdot \dfrac{y^{2}}{x}$

37. $\left(\dfrac{x^{-2}y^{2}}{z^{-1}}\right)^{-1} \cdot \left(\dfrac{xy^{0}}{z}\right)^{-2}$

38. $\left(\dfrac{2y^{2}x}{3z}\right)^{2} \cdot \left(\dfrac{2x^{2}}{9z}\right)^{-2}$

■ *Write each expression as a single fraction involving positive exponents only. See Example 5.*

39. $x^{-2} + y^{-2}$

40. $x^{-1} - y^{-3}$

41. $\dfrac{x}{y^{-1}} + \dfrac{x^{-1}}{y}$

42. $\dfrac{x^{-1}}{y^{-1}} + \dfrac{y}{x}$

43. $(x - y)^{-2}$

44. $(x + y)^{-3}$

45. $x^{-1}y - xy^{-1}$

46. $xy^{-1} + x^{-1}y$

47. $\dfrac{x^{-1} - y}{x^{-1}}$

48. $\dfrac{x + y^{-1}}{y^{-1}}$

49. $\dfrac{x^{-1} + y^{-1}}{(xy)^{-1}}$

50. $\dfrac{x^{-2} - y^{-2}}{(xy)^{-1}}$

51. Use a counterexample to show that $(x + y)^{-2}$ is not equivalent to $\dfrac{1}{x^{2} + y^{2}}$.

52. Use a counterexample to show that $(x + y)^{-2}$ is not equivalent to $\dfrac{1}{x^{2}} + \dfrac{1}{y^{2}}$.

B

■ *Write each expression as a product free of fractions in which each variable occurs only once. See Example 6.*

53. $a^{3-n}a^{0}$

54. $x^{-n}x^{n+1}$

55. $\left(\dfrac{a^{2n}}{a^{n+1}}\right)^{-2}$

56. $\dfrac{x^{n}y^{n+1}}{x^{2n-1}y^{n}}$

57. $\dfrac{b^{n}c^{2n-1}}{b^{n+1}c^{2n}}$

58. $\left(\dfrac{x^{n-1}y^{n}}{x^{-2}y^{-n}}\right)^{2}$

59. $\left(\dfrac{x^{n}}{x^{n-1}}\right)^{-1}$

60. $\left(\dfrac{a^{2n}b^{n-1}}{a^{n-1}b}\right)^{2}$

6.3

SCIENTIFIC NOTATION

It is often very convenient to use an exponential form of notation in scientific applications of mathematics that involve very large or very small quantities.

For example, the mass of the earth is approximately

$$5{,}980{,}000{,}000{,}000{,}000{,}000{,}000{,}000{,}000 = 5.98 \times 10^{27} \text{ grams},$$

and the mass of a hydrogen atom is approximately

$$0.00000000000000000000000167 = 1.67 \times 10^{-24} \text{ gram}.$$

In each case, we have represented a number as the product of a number in the interval $[1, 10)$ and a power of 10; that is, we have factored a power of 10 from each number.

We first consider several simple cases to see how large and small numbers can be factored.

In the following factored forms of 38,400, one of the factors is a power of 10.

$$\begin{aligned} 38{,}400 &= 3840 \times 10 \\ &= 384 \times 10^2 \\ &= 38.4 \times 10^3 \\ &= 3.84 \times 10^4 \end{aligned}$$

Although any one of such factored forms may be more useful than the original form of the number, a special name is given to the last form. A number expressed as the product of a number between 1 and 10 (including 1) and a power of 10 is said to be in **scientific form** or **scientific notation.** For example,

$$4.18 \times 10^4, \quad 9.6 \times 10^2, \quad \text{and} \quad 4 \times 10^5$$

are in scientific form.

Now, let us consider some factored forms of 0.0057 in which one of the factors is a power of 10.

$$\begin{aligned} 0.0057 &= \frac{0.057}{10} = 0.057 \times \frac{1}{10} = 0.057 \times 10^{-1} \\ &= \frac{0.57}{100} = 0.57 \times \frac{1}{10^2} = 0.57 \times 10^{-2} \\ &= \frac{5.7}{1000} = 5.7 \times \frac{1}{10^3} = 5.7 \times 10^{-3} \end{aligned}$$

In this case, 5.7×10^{-3} is the scientific form for 0.0057.

These examples suggest the following procedure.

To Write a Number in Scientific Form:

1. Move the decimal point so that there is one nonzero digit to the left of the decimal point.
2. Multiply the result by a power of ten with an exponent equal to the number of places the decimal point was moved. The exponent is positive if the decimal point has been moved to the left and it is negative if the decimal point has been moved to the right.

EXAMPLE 1

a. $478{,}000 = 4.78000 \times 10^5$ (5 places)

$= 4.78 \times 10^5$

b. $0.00032 = 00003.2 \times 10^{-4}$ (4 places)

$= 3.2 \times 10^{-4}$ ■

A number written in scientific notation can be written in **standard form** by reversing the above procedure.

To Go from Scientific Notation to Decimal Notation:

Move the decimal point the number of places indicated by the exponent on 10—to the right if the exponent is positive and to the left if it is negative.

EXAMPLE 2

a. $3.75 \times 10^4 = 37500.$ (4 places)

$= 37{,}500$

b. $2.03 \times 10^{-3} = .00203$ (3 places)

$= 0.00203$

c. $\dfrac{1}{4 \times 10^3} = \dfrac{1}{4} \times \dfrac{1}{10^3}$

$= 0.25 \times 10^{-3}$

$= 0.00025$

d. $\dfrac{1}{5 \times 10^{-1}} = \dfrac{1}{5} \times \dfrac{1}{10^{-1}}$

$= 0.2 \times 10^1$

$= 2$ ■

Sometimes it is more convenient to express a number as a product of a power of 10 and a number that is not between 1 and 10. For example, under certain circumstances, any of the following forms may be a useful representation for 6280:

$$628 \times 10, \qquad 62.8 \times 10^2, \qquad 6.28 \times 10^3, \qquad 0.628 \times 10^4.$$

A factored form in which the factors do not contain decimals is generally easiest to use to simplify calculations.

EXAMPLE 3

a. $$\frac{10^2 \times 10^{-5} \times 10^4}{10^{-3} \times 10} = \frac{10^{2 + (-5) + 4}}{10^{-3 + 1}} = \frac{10}{10^{-2}} = 10^3$$

b. $$\frac{0.0024 \times 0.0007}{0.000021} = \frac{24 \times 10^{-4} \times 7 \times 10^{-4}}{21 \times 10^{-6}} = 8 \times 10^{-2} = 0.08$$ ■

Significant digits

When working with numbers, we sometimes use the term **significant digits** of a number to refer to the digits in the numbers that have meaning. In particular, the digits that are used to specify a number in scientific notation are significant digits.

EXAMPLE 4

a. 0.00321 has *three* significant digits because $0.00321 = 3.21 \times 10^{-3}$, and 3.21 has three digits.

b. 0.8005 has *four* significant digits because $0.8005 = 8.005 \times 10^{-1}$, and 8.005 has four digits. ■

Note that in Example a the zeros between the decimal point and the first nonzero digit 3 are not significant digits. However, in Example b the zeros between the nonzero digits 8 and 5 are significant.

Final zeros on the right of a decimal point are significant.

EXAMPLE 5

a. 4.300 has four significant digits.

b. 0.0530 has three significant digits because $0.0530 = 5.30 \times 10^{-2}$. ■

Final zeros on a natural number are assumed to be not significant unless further information is available. For example, in the statement

"The height of a building is 4300 feet,"

Without further information on how the measurement was made, we assume that 4300 has only two significant digits.

EXERCISE 6.3

A

■ *Express each number using scientific notation. See Example* 1.

1. 285
2. 3476
3. 21
4. 68,742
5. 8,372,000
6. 481,000
7. 0.024
8. 0.0063
9. 0.421
10. 0.000523
11. 0.000004
12. 0.0006

■ *Express each number using standard form. See Example* 2.

13. 2.4×10^2
14. 4.8×10^3
15. 6.87×10^5
16. 8.31×10^4
17. 5.0×10^{-3}
18. 8.0×10^{-1}
19. 2.02×10^{-2}
20. 4.31×10^{-3}
21. 12.27×10^3
22. 14.38×10^4
23. 23.5×10^{-4}
24. 621.0×10^{-2}
25. $\dfrac{1}{2 \times 10^3}$
26. $\dfrac{1}{4 \times 10^4}$
27. $\dfrac{1}{8 \times 10^{-2}}$
28. $\dfrac{1}{5 \times 10^{-3}}$
29. $\dfrac{3}{5 \times 10^4}$
30. $\dfrac{5}{8 \times 10^2}$

■ *Compute. See Example* 3.

31. $\dfrac{10^3 \times 10^{-6}}{10^2}$
32. $\dfrac{10^3 \times 10^{-7} \times 10^2}{10^{-2} \times 10^4}$
33. $\dfrac{10^2 \times 10^5 \times 10^{-3}}{10^2 \times 10^2}$
34. $\dfrac{(4 \times 10^3)(6 \times 10^{-2})}{3 \times 10^{-7}}$
35. $\dfrac{(2 \times 10^2)^2(3 \times 10^{-3})}{2 \times 10^4}$
36. $\dfrac{(3 \times 10)^3(2 \times 10^{-1})}{2 \times 10^{-2}}$
37. $\dfrac{(2 \times 10^{-3})(6 \times 10^2)^2}{(2 \times 10^{-2})^2}$
38. $\dfrac{(8 \times 10^4)^2(3 \times 10)^3}{(6 \times 10^{-2})^2}$

■ *Specify the number of significant digits in each number. See Examples* 4 *and* 5.

39. 783.2
40. 29.4
41. 0.023
42. 0.00472
43. 0.430
44. 0.600
45. 0.0503
46. 0.02004

■ *Compute. See Example 3.*

47. $\dfrac{0.6 \times 0.00084 \times 0.093}{0.00021 \times 0.00031}$

48. $\dfrac{0.065 \times 2.2 \times 50}{1.30 \times 0.011 \times 0.05}$

49. $\dfrac{28 \times 0.0006 \times 450}{1.5 \times 700 \times 0.018}$

50. $\dfrac{0.0054 \times 0.05 \times 300}{0.0015 \times 0.27 \times 80}$

51. $\dfrac{420 \times 0.0016 \times 800}{0.0028 \times 1200 \times 20}$

52. $\dfrac{0.0027 \times 0.004 \times 650}{260 \times 0.0001 \times 0.009}$

53. The speed of light is approximately 300,000,000 meters per second.
    a. Write this number in scientific notation.
    b. Express the speed of light in inches per second (1 inch equals 2.54 centimeters and 1 meter equals 100 centimeters).
54. One light-year is the number of miles traveled by light in 1 year (365 days), and the speed of light is approximately 186,000 miles per second. Express in scientific notation the number of miles in 1 light-year.
55. Light travels at a speed of 300,000,000 meters per second. Write this number in scientific form.
56. Visible blue light has a wavelength of 0.000 000 45 meters. Write this number in scientific form.
57. The average body cell of an animal has a diameter of 0.000 015 meters. Write this number in scientific form.
58. The diameter of the earth is approximately 6,450,000 meters. Write this number in scientific form.

6.4

RATIONAL EXPONENTS

If the laws of exponents developed in Section 6.1 are to hold for rational exponents, meanings consistent with these laws must be assigned to powers with rational exponents. Let us examine exponents that are the reciprocals of natural numbers, that is, exponents of the form $1/n$, where n is a natural number.

$a^{1/n}$, n a natural number

We shall first define powers of the form $a^{1/n}$ to be consistent with Law II of exponents (page 189). We shall make this definition in two parts: for n an even natural number; and for n an odd natural number.

> *If n is an even natural number and $a > 0$, then $a^{1/n}$ is the positive number such that*
>
> $$(a^{1/n})^n = a^{n/n} = a.$$

The number $a^{1/n}$ is called the **positive *n*th root of *a*.**

EXAMPLE 1

a. $16^{1/2} = 4$ because $4^2 = 16$

b. $-16^{1/2} = -(16)^{1/2} = -4$ because $-4^2 = -(4^2) = -16$

c. $(-16)^{1/2}$ is not defined in the set of real numbers because there is no real number a for which $a^2 = -16$ ■

The restriction that $a > 0$ in the above definition for $a^{1/n}$ is not necessary if n is an odd natural number.

> *If n is an odd natural number, then $a^{1/n}$ is the number such that*
>
> $$(a^{1/n})^n = a.$$

EXAMPLE 2

a. $(8)^{1/3} = 2$ because $2^3 = 8$

b. $(-8)^{1/3} = -2$ because $(-2)^3 = -8$

c. $(-64)^{1/3} = -4$ because $(-4)^3 = -64$ ■

> *If n is an even or odd natural number,*
>
> $$0^{1/n} = 0.$$

EXAMPLE 3

a. $0^{1/2} = 0$

b. $0^{1/3} = 0$ ■

$a^{m/n}$, *n* a natural number

Powers with positive bases and positive rational exponents can be defined in two ways.

> $$a^{m/n} = (a^{1/n})^m = (a^m)^{1/n} \qquad (a^{1/n} \text{ a real number})$$

Thus, we can look at $a^{m/n}$ either as the mth power of the nth root of a, or as the nth root of the mth power of a.

EXAMPLE 4 a. $8^{2/3} = (8^{1/3})^2 = (2)^2 = 4$ b. $8^{2/3} = (8^2)^{1/3} = (64)^{1/3} = 4$ ■

Hereafter, we shall use whichever form is most convenient for the purpose at hand. In Example 4a above, the form $(8^{1/3})^2$ is preferred because it is easier to extract the root first than it is to recognize the root after the number is squared.

To extend meaning to powers with negative rational exponents, we define $a^{-m/n}$ as follows.

> *For m and n positive integers,*
>
> $$a^{-m/n} = \frac{1}{a^{m/n}} \qquad \text{(} a^{1/n} \text{ a real number, } a \neq 0\text{).}$$

EXAMPLE 5 a. $27^{-2/3} = \dfrac{1}{27^{2/3}} = \dfrac{1}{(27^{1/3})^2} = \dfrac{1}{3^2} = \dfrac{1}{9}$ b. $(-8)^{-5/3} = \dfrac{1}{(-8)^{5/3}} = \dfrac{1}{[(-8)^{1/3}]^5} = \dfrac{1}{(-2)^5} = -\dfrac{1}{32}$ ■

With the definitions that we have made, it can be shown that powers with rational exponents—positive, negative, or 0—obey the laws of exponents set forth in previous sections.

EXAMPLE 6 a. $y^{3/4}y^{-1/2} = y^{3/4+(-1/2)} = y^{1/4}$ b. $\dfrac{x^{5/6}}{x^{2/3}} = \dfrac{x^{5/6}}{x^{4/6}} = x^{5/6-4/6} = x^{1/6}$ c. $\dfrac{(x^{1/2}y^2)^2}{(x^{2/3}y)^3} = \dfrac{xy^4}{x^2y^3} = \dfrac{y^{4-3}}{x^{2-1}} = \dfrac{y}{x}$ ■

EXAMPLE 7 a. $y^{1/3}(y + y^{2/3})$
$= y^{1/3+1} + y^{1/3+2/3}$
$= y^{4/3} + y$

b. $x^{-3/4}(x^{1/4} + x^{3/4})$
$= x^{-3/4+1/4} + x^{-3/4+3/4}$
$= x^{-2/4} + x^{0}$
$= x^{-1/2} + 1$ ■

EXAMPLE 8 a. $\dfrac{(x^n)^{3/2}}{x^{n/2}}$
$= x^{3n/2 - n/2}$
$= x^n$

b. $(y^{2n} \cdot y^{n/2})^4$
$= y^{8n} \cdot y^{2n}$
$= y^{10n}$

c. $\left(\dfrac{a^{n+2} \cdot b^{n/2}}{a^n}\right)^2$
$= (a^2 \cdot b^{n/2})^2$
$= a^4 b^n$ ■

Recall from Section 1.1 that any number that can be expressed as the quotient of two integers is called a rational number, and any real number that cannot be so expressed is called an irrational number. Some powers with rational-number exponents are rational numbers and some are irrational numbers. Any expression such as $a^{1/n}$ represents a rational number if and only if a is the nth power of a rational number. For example,

$4^{1/2}$, $(-27/8)^{1/3}$, and $(81)^{1/4}$ are rational numbers equal to 2, $-3/2$, and 3, respectively;

and

$2^{1/2}$, $5^{1/3}$, and $7^{1/4}$ are irrational numbers, such that $(2^{1/2})^2 = 2$, $(5^{1/3})^3 = 5$, and $(7^{1/4})^4 = 7$.

In Section 6.5, we shall consider how to obtain rational approximations for some irrational numbers.

EXERCISE 6.4

A

■ *Assume that all bases in this exercise are positive unless otherwise specified.*

■ *Write each expression using a basic numeral or fraction in lowest terms. See Examples* 1 *and* 2.

1. $9^{1/2}$
2. $25^{1/2}$
3. $32^{1/5}$
4. $27^{1/3}$
5. $(-8)^{1/3}$
6. $(-27)^{1/3}$
7. $64^{1/2}$
8. $81^{1/2}$

■ *See Examples 4 and 5.*

9. $81^{3/4}$ 10. $125^{2/3}$ 11. $(-8)^{4/3}$ 12. $(-64)^{2/3}$

13. $16^{-1/2}$ 14. $8^{-1/3}$ 15. $16^{-3/4}$ 16. $27^{-2/3}$

17. $27^{2/3}$ 18. $32^{3/5}$ 19. $(-27)^{4/3}$ 20. $8^{-2/3}$

■ *Write each expression as a product or quotient of powers in which each variable occurs only once and all exponents are positive. See Example 6.*

21. $x^{1/3}x^{1/3}$ 22. $y^{1/2}y^{3/2}$ 23. $\dfrac{x^{2/3}}{x^{1/3}}$ 24. $\dfrac{x^{3/4}}{x^{1/4}}$

25. $(a^{1/2})^3$ 26. $(b^6)^{2/3}$ 27. $x^{-3/4}x^{1/4}$ 28. $y^{-2/3}y^{5/3}$

29. $(a^{2/3}b)^{1/2}$ 30. $(a^{1/2}b^{1/3})^6$ 31. $\left(\dfrac{a^6}{b^3}\right)^{2/3}$ 32. $\left(\dfrac{a^{1/2}}{a^2}\right)^2$

33. $(r^{-2/3}t)^{-3}$ 34. $(x^{1/4}y^{1/2})^8$ 35. $\left(\dfrac{z^3}{t^6}\right)^{-1/3}$ 36. $\left(\dfrac{a^{-1/2}}{b^{1/3}}\right)^6$

37. $\left(\dfrac{x^{-2}y^{-1/3}z}{x^{-5/3}y^{-2/3}z^{2/3}}\right)^3$ 38. $\left(\dfrac{x^{1/4}y^{3/4}z^{-1}}{x^{-3/4}y^{1/4}z^0}\right)^2$

■ *Write each product so that each base of a power occurs at most once in each term. See Example 7.*

39. $x^{1/2}(x + x^{1/2})$ 40. $x^{1/5}(x^{2/5} + x^{4/5})$ 41. $x^{1/3}(x^{2/3} - x^{1/3})$

42. $x^{3/8}(x^{1/4} - x^{1/2})$ 43. $x^{-3/4}(x^{-1/4} + x^{3/4})$ 44. $y^{-1/4}(y^{3/4} + y^{5/4})$

45. $t^{3/5}(t^{2/5} + t^{-3/5})$ 46. $a^{-2/7}(a^{9/7} + a^{2/7})$ 47. $b^{3/4}(b^{1/4} + b^{-1/2})$

48. $x^{5/6}(x^{-5/6} + x^{1/6})$ 49. $x^{1/2}(x^{3/2} + 2x^{1/2} - x^{-1/2})$ 50. $x^{2/3}(x^{4/3} - x^{1/3} + x^{-1/3})$

51. $(2x^{1/2} - 1)(x^{1/2} + 1)$ 52. $(2x^{1/2} + 1)(x^{1/2} - 1)$ 53. $(x^{1/2} - 2)^2$

54. $(x^{1/2} + 3)^2$ 55. $(x^{1/2} - 2x)(x^{1/2} + x)$ 56. $(x^{1/2} - x)(x^{1/2} + 2x)$

57. Use a counterexample to show that $(a + b)^{1/2}$ is not equivalent to $a^{1/2} + b^{1/2}$.

58. Use a counterexample to show that $(a + b)^{-1/2}$ is not equivalent to $a^{-1/2} + b^{-1/2}$.

B

■ *Simplify. Assume that $m, n > 0$. See Example 8.*

59. $x^n \cdot x^{n/2}$ 60. $(a^2)^{n/2} \cdot (b^{2n})^{2/n}$ 61. $\dfrac{x^{2n}}{x^{n/2}}$

62. $\left(\dfrac{a^n}{b}\right)^{1/2}\left(\dfrac{b}{a^{2n}}\right)^{3/2}$ 63. $\dfrac{x^{3n}y^{2m+1}}{(x^ny^m)^{1/2}}$ 64. $\left(\dfrac{m^an^{2a}}{n^{4a}}\right)^{1/a}$

65. $\left(\dfrac{x^{2n} \cdot y^{3n}}{x^n}\right)^{1/3}$ 66. $\left(\dfrac{x^{n+1} \cdot y^{n+2}}{xy^2}\right)^{1/n}$

67. Which is greater, $16^{1/4}$ or $16^{1/2}$? Make a conjecture about the order of $a^{1/n}$ and $a^{1/m}$ when $n > m$ and $a > 1$.

68. Which is greater, $(1/16)^{1/4}$ or $(1/16)^{1/2}$? Make a conjecture about the order of $a^{1/n}$ and $a^{1/m}$ when $n > m$ and $0 < a < 1$.

6.5

RADICALS

Radical notation

In Section 6.4 we agreed to refer to $a^{1/n}$ (when it exists) as the nth root of a. An alternative symbol is often used for $a^{1/n}$.

> *For all natural numbers* $n \geqslant 2$,
>
> $$a^{1/n} = \sqrt[n]{a}.$$

In such a representation, the symbol $\sqrt{}$ is called a **radical,** a is called the **radicand,** n is called the **index,** and the expression is said to be a **radical of order n.** If no index is written, the index is understood to be 2.

EXAMPLE 1

a. $5^{1/2} = \sqrt{5}$

b. $x^{1/5} = \sqrt[5]{x}$

c. $2x^{1/3} = 2\sqrt[3]{x}$

d. $\sqrt{7} = 7^{1/2}$

e. $\sqrt[4]{2y} = (2y)^{1/4}$

f. $\sqrt[3]{2x} - 2\sqrt{y}$
$= (2x)^{1/3} - 2y^{1/2}$

■

We shall define $\sqrt[n]{a}$ to conform to our definition of $a^{1/n}$. It is convenient to do so in two parts: for n an even natural number; and for n an odd natural number.

> *For n an even natural number and $a \geqslant 0$, $\sqrt[n]{a}$ is the nonnegative number such that*
>
> $$(\sqrt[n]{a})^n = a.$$
>
> *In particular, for* $n = 2$,
>
> $$(\sqrt{a})^2 = \sqrt{a}\sqrt{a} = a.$$

EXAMPLE 2 a. $\sqrt[4]{81} = 3$ because $3^4 = 81$ b. $\sqrt{16} = 4$ because $4^2 = 16$

c. $(\sqrt{5})^2 = \sqrt{5}\sqrt{5} = 5$ d. $(\sqrt[4]{5})^4 = 5$ ■

Although each positive number a has two square roots, *the symbol* $\sqrt{a}$ *represents the positive value only*. Thus, the symbol $\sqrt{16}$ represents only the positive square root, 4, as in Example b above. The negative square root is denoted by $-\sqrt{16}$.

The restriction that $a > 0$ is not necessary if n is an odd natural number.

> *For n an odd natural number,* $\sqrt[n]{a}$ *is the number such that*
>
> $$(\sqrt[n]{a})^n = a.$$

EXAMPLE 3 a. $\sqrt[3]{8} = 2$ because $2^3 = 8$ b. $\sqrt[3]{-8} = -2$ because $(-2)^3 = -8$

c. $(\sqrt[3]{4})^3 = \sqrt[3]{4}\sqrt[3]{4}\sqrt[3]{4} = 4$ d. $(\sqrt[3]{-2})^3 = \sqrt[3]{-2}\sqrt[3]{-2}\sqrt[3]{-2} = -2$ ■

Since, from Section 6.4, for $a \geqslant 0$,

$$a^{m/n} = (a^m)^{1/n} = (a^{1/n})^m,$$

we may write a power with a rational exponent in radical form.

> $$a^{m/n} = \sqrt[n]{a^m} = (\sqrt[n]{a})^m \qquad (a \geqslant 0)$$

Note that the denominator of the exponent is the index of the radical, and the numerator of the exponent is either the exponent of the radicand or the exponent of the root.

EXAMPLE 4 a. $x^{2/3} = \sqrt[3]{x^2}$
$= (\sqrt[3]{x})^2$

b. $8^{2/3} = \sqrt[3]{8^2} = \sqrt[3]{64} = 4$ or
$8^{2/3} = (\sqrt[3]{8})^2 = 2^2 = 4$ ■

Since we have restricted the index of a radical to be a natural number, we must always express a fractional exponent in standard form (m/n or $-m/n$) before writing the power in radical form.

EXAMPLE 5 a. $x^{-(3/4)} = x^{(-3/4)}$
$= \sqrt[4]{x^{-3}}$

b. $8^{-(2/3)} = 8^{-2/3} = \sqrt[3]{8^{-2}}$
$= \sqrt[3]{\dfrac{1}{64}} = \dfrac{1}{4}$ ■

Sometimes you may find it helpful to simplify radical expressions by first writing such expressions in exponential form.

EXAMPLE 6 a. $\sqrt[3]{-8}$
$= -8^{1/3}$
$= -2$

b. $\sqrt[3]{x^6y^3}$
$= (x^6y^3)^{1/3}$
$= x^2y$

c. $-\sqrt[4]{81x^4}$
$= -(81x^4)^{1/4}$
$= -3x$ ■

We can also use fractional exponents to simplify a radical expression.

EXAMPLE 7 a. $\sqrt[4]{5^2} = 5^{2/4}$
$= 5^{1/2}$
$= \sqrt{5}$

b. $\sqrt[6]{9} = (3^2)^{1/6}$
$= 3^{1/3}$
$= \sqrt[3]{3}$

c. $\sqrt[8]{x^2} = x^{2/8}$
$= x^{1/4}$
$= \sqrt[4]{x}$ ■

$\sqrt[n]{a^n}$, *n* an even number

Because we have defined a radical with an even index to be a nonnegative number, we have the following special relationship for all values of a.

$$\sqrt[n]{a^n} = |a| \qquad (n \text{ even})$$

In particular,

$$\sqrt{a^2} = |a|.$$

In considering odd indices, there is no ambiguous interpretation possible.

For all a, $\sqrt[n]{a^n} = a \qquad$ **(*n* odd).**

EXAMPLE 8 a. $\sqrt{2^2} = |2| = 2$ b. $\sqrt{(-2)^2} = |-2| = 2$

c. $\sqrt[3]{4^3} = 4$ d. $\sqrt[3]{(-4)^3} = -4$ ■

If a radical has an even index and the variables in the radicand represent negative as well as positive real numbers, we must use absolute value notation when the expression is written without radical notation.

EXAMPLE 9 a. $\sqrt{16x^2} = 4|x|$ b. $\sqrt{x^2 - 2xy + y^2} = \sqrt{(x-y)^2}$

$= |x - y|$ ■

Because, as observed in Section 6.4, $a^{1/n}$ represents a rational number if and only if a is the nth power of a rational number, the same fact is true of $\sqrt[n]{a}$. Thus,

$$\sqrt{4}, \quad -\sqrt[3]{27}, \quad \sqrt[4]{81/16}, \quad \text{and} \quad \sqrt[5]{-32}$$

are rational numbers equal to

$$2, \quad -3, \quad 3/2, \quad \text{and} \quad -2.$$

Approximations for irrational numbers

Although irrational numbers such as $\sqrt{5}$, $\sqrt[3]{9}$, $\sqrt[4]{15}$, and $\sqrt[5]{61}$ do not have terminating or repeating decimal representations, we can obtain decimal approximations correct to any desired degree of accuracy. The Table of Squares, Square Roots, and Prime Factors on page 534, or some calculators, can be used to obtain approximations for some irrational numbers. For example,

$$\sqrt{2} \approx 1.414,$$

where the symbol $\approx$ is read "is approximately equal to."

EXAMPLE 10 Graph each set of numbers.

a. $\sqrt{4}$, $-\sqrt{3}$, $\sqrt{17}$, $-\sqrt{9}$ b. $\sqrt{2}$, $-\sqrt{6}$, 0, $\sqrt{16}$

Solutions

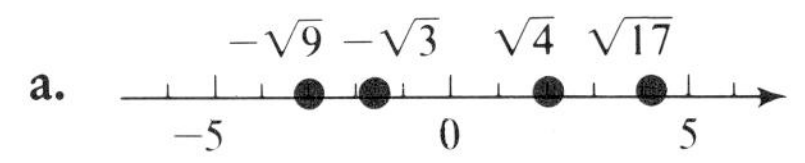

■

EXERCISE 6.5

A

■ *In Problems 1–76, assume each variable and each radicand represents a positive number.*

■ *Write each expression in exponential form. See Example 1.*

1. $\sqrt{7}$
2. $\sqrt{5}$
3. $\sqrt[3]{2x}$
4. $\sqrt[3]{4y}$
5. $\sqrt[3]{x} - 3\sqrt{y}$
6. $\sqrt{x} - 2\sqrt[3]{y}$
7. $\sqrt[3]{x}\sqrt{y}$
8. $\sqrt{x}\sqrt[3]{xy}$

■ *Write each expression in radical form. See Example 1.*

9. $3^{1/2}$
10. $7^{1/2}$
11. $4x^{1/3}$
12. $3x^{1/4}$
13. $(x - 2)^{1/4}$
14. $(y + 2)^{1/3}$
15. $3(xy)^{1/3}$
16. $2(x + y)^{1/5}$

■ *Find the root or power indicated. See Examples 2 and 3.*

17. $\sqrt{9}$
18. $\sqrt{36}$
19. $(\sqrt[4]{2})^4$
20. $(\sqrt{7})^2$
21. $\sqrt[3]{27}$
22. $\sqrt[3]{-27}$
23. $(\sqrt[3]{5})^3$
24. $(\sqrt[3]{-5})^3$

■ *Write each expression in radical notation. See Examples 4 and 5.*

25. $x^{2/3}$
26. $y^{3/4}$
27. $3x^{3/5}$
28. $5y^{2/3}$

29. $(x + 2y)^{3/2}$ 30. $(x - 2y)^{2/3}$ 31. $y^{-1/2}$ 32. $x^{-1/3}$

33. $x^{-2/3}$ 34. $y^{-2/7}$ 35. $3y^{-2/3}$ 36. $4x^{-3/2}$

■ *Write each expression with positive fractional exponents. See Example 6.*

37. $\sqrt[3]{x^2}$ 38. $\sqrt{y^3}$ 39. $\sqrt[3]{(xy)^2}$ 40. $\sqrt[3]{xy^2}$

41. $\sqrt{xy^3}$ 42. $\sqrt{(xy)^3}$ 43. $\dfrac{1}{\sqrt{x}}$ 44. $\dfrac{2}{\sqrt[3]{y}}$

■ *Find the indicated root. See Examples 2, 3, and 6.*

45. $\sqrt[3]{27}$ 46. $\sqrt[3]{125}$ 47. $\sqrt[3]{-64}$ 48. $\sqrt[5]{-32}$

49. $\sqrt[3]{x^3}$ 50. $\sqrt[5]{y^5}$ 51. $\sqrt{x^4}$ 52. $\sqrt{a^6}$

53. $\sqrt[3]{8y^6}$ 54. $\sqrt[3]{27y^9}$ 55. $-\sqrt{x^4y^6}$ 56. $-\sqrt{a^8b^{10}}$

57. $\sqrt{\dfrac{4}{9}x^2y^8}$ 58. $\sqrt{\dfrac{9}{16}a^2b^4}$ 59. $\sqrt[3]{\dfrac{-8}{125}x^3}$ 60. $\sqrt[3]{\dfrac{8}{27}a^3b^6}$

61. $\sqrt[4]{16x^4y^8}$ 62. $\sqrt[5]{-32x^5y^{10}}$ 63. $-\sqrt[3]{-8a^6b^9}$ 64. $-\sqrt[4]{81a^8b^{12}}$

■ *Reduce the order of each radical. See Example 7.*

65. $\sqrt[4]{3^2}$ 66. $\sqrt[6]{2^2}$ 67. $\sqrt[6]{3^3}$ 68. $\sqrt[8]{5^2}$

69. $\sqrt[6]{81}$ 70. $\sqrt[10]{32}$ 71. $\sqrt[6]{x^3}$ 72. $\sqrt[9]{y^3}$

■ *Graph each set of real numbers on a separate line graph. (Use a calculator or the Table of Squares, Square Roots, and Prime Factors in Appendix B to obtain rational-number approximations for irrational numbers.) See Example* 10.

73. $-\sqrt{7}, -\sqrt{1}, \sqrt{5}, \sqrt{9}$ 74. $-2/3, 0, \sqrt{3}, -\sqrt{11}$

75. $-\sqrt{20}, -\sqrt{6}, \sqrt{1}, 6$ 76. $\sqrt{41}, \sqrt{7}, -\sqrt{7}, 3/4$

B

■ *In the foregoing problems, variables and radicands were restricted to represent positive numbers. In Problems 77–82, consider variables and radicands to represent elements of the set of real numbers and use absolute-value notation as needed. See Example 9.*

77. $\sqrt{4x^2}$ 78. $\sqrt{9x^2y^4}$ 79. $\sqrt{x^2 + 2x + 1}$

80. $\sqrt{4x^2 - 4x + 1}$ 81. $\dfrac{2}{\sqrt{x^2 + 2xy + y^2}}$ 82. $\sqrt{x^4 + 2x^2y^2 + y^4}$

83. Use a counterexample to show that $\sqrt{a^2}$ is not equivalent to a.

84. Use a counterexample to show that $\sqrt{(a - 1)^2}$ is not equivalent to $a - 1$.

6.6

CHANGING FORMS OF RADICALS

From the definition of a radical and the laws of exponents, we can derive two important relationships.

Laws of radicals

> *For $a, b > 0$ and n a natural number,*
>
> $$\sqrt[n]{ab} = \sqrt[n]{a}\sqrt[n]{b}. \tag{1}$$

This property follows from the fact that

$$\sqrt[n]{ab} = (ab)^{1/n} = a^{1/n}b^{1/n} = \sqrt[n]{a}\sqrt[n]{b}.$$

Relationship (1) can be used to write a radical in a form in which the radicand contains no prime factor or polynomial factor raised to a power greater than or equal to the index of the radical.

EXAMPLE 1

a. $\sqrt{18} = \sqrt{3^2}\sqrt{2}$
$= 3\sqrt{2}$

b. $\sqrt[3]{x^7} = \sqrt[3]{x^3}\sqrt[3]{x^3}\sqrt[3]{x}$
$= x \cdot x\sqrt[3]{x}$
$= x^2\sqrt[3]{x}$

c. $\sqrt[3]{16x^3y^5} = \sqrt[3]{2^3}\sqrt[3]{x^3}\sqrt[3]{y^3}\sqrt[3]{2y^2}$
$= 2xy\sqrt[3]{2y^2}$

d. $\sqrt[4]{16x^5y} = \sqrt[4]{2^4}\sqrt[4]{x^4}\sqrt[4]{xy}$
$= 2x\sqrt[4]{xy}$

■

In each case the radicand was first factored into two or more factors, one or more of which consisted of factors raised to the same power as the index of the radical. This factor was then removed from the radicand.

We could have simplified the radical expressions in the examples above by first writing each radicand in factored form. Then, for each radical expression with index n, we could write a single factor of the radicand outside the radicand for each such n identical factors. Thus, Example 1b would appear as

$$\sqrt[3]{x^7} = \sqrt[3]{xxx\ \ xxx\ x} \quad \text{or} \quad \sqrt[3]{x^7} = \sqrt[3]{x^3\ \ x^3\ x}$$

$$= x \cdot x\sqrt[3]{x} \qquad\qquad = x \cdot x\sqrt[3]{x}$$

$$= x^2\sqrt[3]{x} \qquad\qquad = x^2\sqrt[3]{x}.$$

The second important relationship involves quotients.

For a, $b > 0$ and n a natural number,

$$\sqrt[n]{\frac{a}{b}} = \frac{\sqrt[n]{a}}{\sqrt[n]{b}}. \tag{2}$$

This property follows from the fact that

$$\sqrt[n]{\frac{a}{b}} = \left(\frac{a}{b}\right)^{1/n} = \frac{a^{1/n}}{b^{1/n}} = \frac{\sqrt[n]{a}}{\sqrt[n]{b}}.$$

We can use relationship (2) to write a radical in a form in which the radicand contains no fraction.

EXAMPLE 2 a. $\sqrt{\frac{3}{4}} = \frac{\sqrt{3}}{\sqrt{4}} = \frac{\sqrt{3}}{2}$ b. $\sqrt[3]{\frac{5}{8}} = \frac{\sqrt[3]{5}}{\sqrt[3]{8}} = \frac{\sqrt[3]{5}}{2}$ ■

If a radical expression in the denominator of a fraction cannot be written directly without radical notation, we can use the fundamental principle of fractions to obtain an equivalent form in which the denominator is free of radicals.

EXAMPLE 3 a. $\sqrt{\frac{1}{3}} = \frac{\sqrt{1}}{\sqrt{3}}$

$$= \frac{1\sqrt{3}}{\sqrt{3}\sqrt{3}} = \frac{\sqrt{3}}{3}$$

b. $\sqrt{\frac{2}{5x}} = \frac{\sqrt{2}}{\sqrt{5x}}$

$$= \frac{\sqrt{2}\sqrt{5x}}{\sqrt{5x}\sqrt{5x}} = \frac{\sqrt{10x}}{5x}$$ ■

The foregoing process is called *rationalizing the denominator* of a fraction.

EXAMPLE 4 Rationalize the denominator: $\dfrac{1}{\sqrt[3]{2x}}$.

Solution Because we need a third power, $(2x)^3$, beneath the radical sign in the denominator in order to write it without a radical sign, we must multiply $2x$ by two additional factors of $2x$. Thus, using the fundamental principle of fractions we obtain

$$\frac{1}{\sqrt[3]{2x}} = \frac{1\sqrt[3]{2x}\,\sqrt[3]{2x}}{\sqrt[3]{2x}\,\sqrt[3]{2x}\,\sqrt[3]{2x}}$$

$$= \frac{\sqrt[3]{(2x)^2}}{\sqrt[3]{(2x)^3}} = \frac{\sqrt[3]{4x^2}}{2x}. \qquad ■$$

EXAMPLE 5 Rationalize the denominator: $\sqrt[5]{\dfrac{6}{16x^3}}$.

Solution In this instance, we have

$$\sqrt[5]{\frac{6}{16x^3}} = \sqrt[5]{\frac{6}{2^4x^3}}.$$

Since we want a fifth power in the denominator, we need the factor $\sqrt[5]{2x^2}$. Thus,

$$\sqrt[5]{\frac{6}{16x^3}} = \frac{\sqrt[5]{6}\,\sqrt[5]{2x^2}}{\sqrt[5]{16x^3}\,\sqrt[5]{2x^2}}$$

$$= \frac{\sqrt[5]{12x^2}}{\sqrt[5]{32x^5}} = \frac{\sqrt[5]{12x^2}}{2x}. \qquad ■$$

Sometimes expressions that contain radical expressions in their numerators and denominators can be simplified by first applying Properties (1) and (2) above and then using the fundamental principle of fractions.

EXAMPLE 6 a. $\dfrac{\sqrt{a}\sqrt{ab^3}}{\sqrt{b}} = \sqrt{\dfrac{a^2b^3}{b}}$

$= \sqrt{a^2b^2}$

$= ab$

b. $\dfrac{\sqrt[3]{16y^4}}{\sqrt[3]{y}} = \sqrt[3]{\dfrac{16y^4}{y}}$

$= \sqrt[3]{2^4y^3}$

$= 2y\sqrt[3]{2}$ ■

Writing equivalent radical expressions

Application of Properties (1) and/or (2) on pages 214 and 215 can be used to rewrite radical expressions in various ways, and, in particular, to write them in what is called **simplest form**.

A Radical Expression Is in Simplest Form If:

1. The radicand contains no polynomial factor raised to a power equal to or greater than the index of the radical.
2. The radicand contains no fractions.
3. No radical expressions are contained in denominators of fractions.

Although we generally change the form of radicals to one of the forms described above, there are times when such forms are not preferred. For example, in certain situations, $\sqrt{\dfrac{1}{2}}$ or $\dfrac{1}{\sqrt{2}}$ may be more useful than the equivalent form $\dfrac{\sqrt{2}}{2}$. In such cases, we may want to rationalize the numerator of a fraction.

EXAMPLE 7 Rationalize each numerator.

a. $\dfrac{\sqrt{2}}{2}$

b. $\dfrac{\sqrt{2x}}{x}$

Solutions a. $\dfrac{\sqrt{2}}{2} = \dfrac{\sqrt{2}\cdot\sqrt{2}}{2\cdot\sqrt{2}}$

$= \dfrac{2}{2\sqrt{2}} = \dfrac{1}{\sqrt{2}}$

b. $\dfrac{\sqrt{2x}}{x} = \dfrac{\sqrt{2x}\cdot\sqrt{2x}}{x\cdot\sqrt{2x}}$

$= \dfrac{2x}{x\sqrt{2x}} = \dfrac{2}{\sqrt{2x}}$ ■

EXERCISE 6.6

A

■ *Assume that all variables in radicands in this exercise denote positive real numbers.*

■ *Change to simplest form. See Example 1.*

1. $\sqrt{18}$
2. $\sqrt{50}$
3. $\sqrt{20}$
4. $\sqrt{72}$
5. $\sqrt{75}$
6. $\sqrt{48}$
7. $\sqrt{x^4}$
8. $\sqrt{y^6}$
9. $\sqrt{x^3}$
10. $\sqrt{y^{11}}$
11. $\sqrt{9x^3}$
12. $\sqrt{4y^5}$
13. $\sqrt{8x^6}$
14. $\sqrt{18y^8}$
15. $\sqrt[4]{x^5}$
16. $\sqrt[3]{y^4}$
17. $\sqrt[5]{x^7y^9z^{11}}$
18. $\sqrt[5]{a^{12}b^{15}}$
19. $\sqrt[6]{a^7b^{12}c^{15}}$
20. $\sqrt[6]{m^8n^7}$
21. $\sqrt[7]{3^7a^8b^9c^{10}}$
22. $\sqrt[7]{4^8x^9y^{10}z^{14}}$
23. $\sqrt{18}\sqrt{2}$
24. $\sqrt{3}\sqrt{27}$
25. $\sqrt{xy}\sqrt{x^5y}$
26. $\sqrt{a}\sqrt{ab^2}$
27. $\sqrt[3]{2}\,\sqrt[3]{4}$
28. $\sqrt[4]{3}\,\sqrt[4]{27}$
29. $\sqrt{3 \times 10^2}$
30. $\sqrt{5 \times 10^3}$
31. $\sqrt{60{,}000}$
32. $\sqrt{800{,}000}$

■ *Rationalize each denominator. See Examples 2 and 3.*

33. $\sqrt{\dfrac{1}{5}}$
34. $\sqrt{\dfrac{2}{3}}$
35. $\dfrac{-1}{\sqrt{2}}$
36. $\dfrac{-\sqrt{3}}{\sqrt{7}}$
37. $\sqrt{\dfrac{x}{2}}$
38. $-\sqrt{\dfrac{y}{3}}$
39. $-\sqrt{\dfrac{y}{x}}$
40. $\sqrt{\dfrac{2a}{b}}$
41. $\dfrac{x}{\sqrt{x}}$
42. $\dfrac{-x}{\sqrt{2y}}$
43. $\sqrt{\dfrac{y}{2x}}$
44. $\sqrt{\dfrac{y}{6x}}$

■ *Rationalize each denominator. See Examples 4 and 5.*

45. $\dfrac{1}{\sqrt[3]{x^2}}$
46. $\dfrac{1}{\sqrt[4]{y^3}}$
47. $\sqrt[3]{\dfrac{2}{3y}}$
48. $\sqrt[4]{\dfrac{2}{3x}}$
49. $\sqrt[3]{\dfrac{x}{4y^2}}$
50. $\sqrt[4]{\dfrac{x}{8y^3}}$
51. $\sqrt[5]{\dfrac{3}{2x^3}}$
52. $\sqrt[5]{\dfrac{2}{9y^2}}$

■ *Change to simplest form. See Example 6.*

53. $\dfrac{\sqrt{a^5b^3}}{\sqrt{ab}}$
54. $\dfrac{\sqrt{x}\sqrt{xy^3}}{\sqrt{y}}$
55. $\dfrac{\sqrt{98x^2y^3}}{\sqrt{xy}}$
56. $\dfrac{\sqrt{45x^3}\sqrt{y^3}}{\sqrt{5y}}$
57. $\dfrac{\sqrt[3]{8b^4}}{\sqrt[3]{a^6}}$
58. $\dfrac{\sqrt[3]{16r^4}}{\sqrt[3]{4t^3}}$
59. $\dfrac{\sqrt[5]{a}\,\sqrt[5]{b^2}}{\sqrt[5]{ab}}$
60. $\dfrac{\sqrt[5]{x^2}\,\sqrt[5]{y^3}}{\sqrt[5]{xy^2}}$

■ *Rationalize each numerator. See Example 7.*

61. $\dfrac{\sqrt{3}}{3}$ 62. $\dfrac{\sqrt{2}}{3}$ 63. $\dfrac{\sqrt{x}}{\sqrt{y}}$ 64. $\dfrac{\sqrt{xy}}{x}$

65. Use a counterexample to show that $(\sqrt{a}+\sqrt{b})^2$ is not equivalent to $a+b$.

66. Use a counterexample to show that $\sqrt{a+b}$ is not equivalent to $\sqrt{a}+\sqrt{b}$.

B

■ *Simplify. Assume that all variables and radicands represent positive numbers.*

67. $\sqrt{8(x-1)^3}$ 68. $\sqrt{12(x+2)^3}$ 69. $\sqrt{x^3(y-2)^5}$

70. $\sqrt{y^5(x+1)^3}$ 71. $\sqrt{\dfrac{(y-3)^3}{xy^3}}$ 72. $\sqrt{\dfrac{(x+2)^5}{x^3y}}$

73. $\sqrt[3]{4x^5-x^3}$ 74. $\sqrt[3]{2y^4-y^3}$ 75. $\sqrt[3]{\dfrac{(x-1)^4}{xy^2}}$

76. $\sqrt[3]{\dfrac{(y+1)^4}{x^2y}}$ 77. $\dfrac{\sqrt{9(x-1)^2}}{\sqrt{3x}\sqrt{x^3-x^2}}$ 78. $\dfrac{\sqrt{4(y+2)^2}}{\sqrt{4y}\sqrt{y^3+2y^2}}$

6.7

EXPRESSIONS CONTAINING RADICALS

Sums and differences

The distributive property,

$$a(b+c)=ab+ac, \tag{1}$$

is assumed to hold for all real numbers. By the symmetric property of equality and the commutative property of multiplication, (1) can be written as

$$ba+ca=(b+c)a.$$

Since at this time all radical expressions have been defined so that they represent real numbers, the distributive property holds for radical expressions. Hence we may write sums or differences containing radicals of the same index and radicand as a single term.

EXAMPLE 1 a. $3\sqrt{3}+4\sqrt{3}=(3+4)\sqrt{3}$
$=7\sqrt{3}$

b. $7\sqrt{x}-2\sqrt{x}=(7-2)\sqrt{x}$
$=5\sqrt{x}$ ■

Sometimes it is necessary to simplify radical expressions in a sum or difference so that two or more terms contain radicals of the same index and radicand. They can then be written as a single term.

EXAMPLE 2 a. $3\sqrt{20} + \sqrt{45}$
$= 3 \cdot 2\sqrt{5} + 3\sqrt{5}$
$= 6\sqrt{5} + 3\sqrt{5}$
$= 9\sqrt{5}$

b. $\sqrt{32x} + \sqrt{2x} - \sqrt{18x}$
$= 4\sqrt{2x} + \sqrt{2x} - 3\sqrt{2x}$
$= 2\sqrt{2x}$ ■

Products and factors

A direct application of (1) permits us to write certain products that contain parentheses as expressions without parentheses.

EXAMPLE 3 a. $x(\sqrt{2} + \sqrt{3}) = x\sqrt{2} + x\sqrt{3}$

b. $\sqrt{3}(\sqrt{2x} + \sqrt{6}) = \sqrt{6x} + \sqrt{18}$
$= \sqrt{6x} + \sqrt{9}\sqrt{2}$
$= \sqrt{6x} + 3\sqrt{2}$

c. For $x \geqslant 0$,
$\sqrt{x}(\sqrt{2x} - \sqrt{x})$
$= \sqrt{2x^2} - x$
$= x\sqrt{2} - x$

d. For $x \geqslant 0$, $y \geqslant 0$,
$(\sqrt{x} - \sqrt{y})(\sqrt{x} + \sqrt{y})$
$= \sqrt{x^2} - \sqrt{y^2}$
$= x - y$ ■

In Chapter 4, we agreed to factor from each term of an expression only those common factors that are integers or positive integral powers of variables. However, we can (if we wish) consider other real numbers for factors. Thus, using the distributive property in the form

$$ab + ac = a(b + c),$$

radicals common to each term in an expression may be factored from the expression.

EXAMPLE 4 a. $3 + \sqrt{18}$
$= 3 + 3\sqrt{2}$
$= 3(1 + \sqrt{2})$

b. $\dfrac{\sqrt{a} + \sqrt{ab}}{\sqrt{a}} = \dfrac{\sqrt{a}(1 + \sqrt{b})}{\sqrt{a}}$
$= 1 + \sqrt{b}$ ■

Quotients

Recall from Section 6.6 that a monomial denominator of a fraction of the form $a/\sqrt{b}$ can be rationalized by multiplying the numerator and the denominator by $\sqrt{b}$. For example,

$$\frac{2}{\sqrt{3}} = \frac{2\sqrt{3}}{\sqrt{3}\sqrt{3}} = \frac{2\sqrt{3}}{3} \quad \text{and} \quad \frac{a}{\sqrt{b}} = \frac{a\sqrt{b}}{\sqrt{b}\sqrt{b}} = \frac{a\sqrt{b}}{b}.$$

The distributive property provides us with a means of rationalizing denominators of fractions in which radicals occur in one or both of the two terms of a binomial. To accomplish this, we first recall that

$$(a - b)(a + b) = a^2 - b^2,$$

where the product contains no first-degree term. Each of the two factors of a product exhibiting this property is said to be the **conjugate** of the other.

Now consider a fraction of the form

$$\frac{a}{b + \sqrt{c}} \qquad (b + \sqrt{c} \neq 0).$$

If we multiply the numerator and the denominator of this fraction by the conjugate of the denominator, the denominator of the resulting fraction will contain no term linear in $\sqrt{c}$; hence, it will be free of radicals. That is

$$\frac{a(b - \sqrt{c})}{(b + \sqrt{c})(b - \sqrt{c})} = \frac{ab - a\sqrt{c}}{b^2 - c} \qquad (b^2 - c \neq 0),$$

where the denominator has been rationalized.

This process is equally applicable to fractions of the form

$$\frac{a}{\sqrt{b} + \sqrt{c}},$$

since

$$\frac{a(\sqrt{b} - \sqrt{c})}{(\sqrt{b} + \sqrt{c})(\sqrt{b} - \sqrt{c})} = \frac{a\sqrt{b} - a\sqrt{c}}{b - c} \qquad (b - c \neq 0).$$

EXAMPLE 5 Rationalize each denominator.

a. $\dfrac{2}{\sqrt{3} - 1}$ b. $\dfrac{x}{2 + \sqrt{x}}$

Solutions a. $\dfrac{2}{\sqrt{3}-1} = \dfrac{2(\sqrt{3}+1)}{(\sqrt{3}-1)(\sqrt{3}+1)}$

$= \dfrac{2(\sqrt{3}+1)}{3-1}$

$= \sqrt{3}+1$

b. $\dfrac{x}{2+\sqrt{x}} = \dfrac{x(2-\sqrt{x})}{(2+\sqrt{x})(2-\sqrt{x})}$

$= \dfrac{x(2-\sqrt{x})}{4-x}$

$= \dfrac{2x - x\sqrt{x}}{4-x}$ ■

As noted in Section 6.6, we can rationalize the numerator of a fraction as well as the denominator.

EXAMPLE 6 Rationalize each numerator.

a. $\dfrac{\sqrt{2}-1}{2}$ b. $\dfrac{\sqrt{x}+1}{x}$

Solutions a. $\dfrac{\sqrt{2}-1}{2} = \dfrac{(\sqrt{2}-1)(\sqrt{2}+1)}{2(\sqrt{2}+1)}$

$= \dfrac{2-1}{2(\sqrt{2}+1)}$

$= \dfrac{1}{2\sqrt{2}+2}$

b. $\dfrac{\sqrt{x}+1}{x} = \dfrac{(\sqrt{x}+1)(\sqrt{x}-1)}{x(\sqrt{x}-1)}$

$= \dfrac{x-1}{x(\sqrt{x}-1)}$

$= \dfrac{x-1}{x\sqrt{x}-x}$ ■

EXERCISE 6.7

A

■ *Assume that all radicands and variables in this exercise are positive real numbers.*

■ *Write each expression as a single term. See Examples 1 and 2.*

1. $3\sqrt{7} + 2\sqrt{7}$
2. $5\sqrt{2} - 3\sqrt{2}$
3. $4\sqrt{3} - \sqrt{27}$
4. $\sqrt{75} + 2\sqrt{3}$
5. $\sqrt{50x} + \sqrt{32x}$
6. $\sqrt{8y} - \sqrt{18y}$
7. $3\sqrt{4xy^2} - 4\sqrt{9xy^2}$
8. $2\sqrt{8y^2z} + 3\sqrt{32y^2z}$

9. $3\sqrt{8a} + 2\sqrt{50a} - \sqrt{2a}$

10. $\sqrt{3b} - 2\sqrt{12b} + 3\sqrt{48b}$

11. $3\sqrt[3]{16} - \sqrt[3]{2}$

12. $\sqrt[3]{54} + 2\sqrt[3]{128}$

■ *Write each expression without parentheses and all radicals in simple form. See Example 3.*

13. $2(3 - \sqrt{5})$

14. $5(2 - \sqrt{7})$

15. $\sqrt{2}(\sqrt{6} + \sqrt{10})$

16. $\sqrt{3}(\sqrt{12} - \sqrt{15})$

17. $(3 + \sqrt{5})(2 - \sqrt{5})$

18. $(1 - \sqrt{2})(2 + \sqrt{2})$

19. $(\sqrt{x} - 3)(\sqrt{x} + 3)$

20. $(2 + \sqrt{x})(2 - \sqrt{x})$

21. $(\sqrt{2} - \sqrt{3})(\sqrt{2} + 2\sqrt{3})$

22. $(\sqrt{3} - \sqrt{5})(2\sqrt{3} + \sqrt{5})$

23. $(\sqrt{5} - \sqrt{2})^2$

24. $(\sqrt{2} - 2\sqrt{3})^2$

■ *Change each expression to the form indicated. See Example 4.*

25. $2 + 2\sqrt{3} = 2(\underline{?} + \underline{?})$

26. $5 + 10\sqrt{2} = 5(\underline{?} + \underline{?})$

27. $2\sqrt{27} + 6 = 6(\underline{?} + \underline{?})$

28. $5\sqrt{5} - \sqrt{25} = 5(\underline{?} - \underline{?})$

29. $4 + \sqrt{16y} = 4(\underline{?} + \underline{?})$

30. $3 + \sqrt{18x} = 3(\underline{?} + \underline{?})$

31. $\sqrt{2} - \sqrt{6} = \sqrt{2}(\underline{?} - \underline{?})$

32. $\sqrt{12} - 2\sqrt{6} = 2\sqrt{3}(\underline{?} - \underline{?})$

■ *Reduce each fraction to lowest terms after factoring the numerator. See Example 4.*

33. $\dfrac{2 + 2\sqrt{3}}{2}$

34. $\dfrac{6 + 2\sqrt{5}}{2}$

35. $\dfrac{6 + 2\sqrt{18}}{6}$

36. $\dfrac{8 - 2\sqrt{12}}{4}$

37. $\dfrac{x - \sqrt{x^3}}{x}$

38. $\dfrac{xy - x\sqrt{xy^2}}{xy}$

39. $\dfrac{x\sqrt{y} - \sqrt{y^3}}{\sqrt{y}}$

40. $\dfrac{\sqrt{x} - y\sqrt{x^3}}{\sqrt{x}}$

■ *Rationalize each denominator. See Example 5.*

41. $\dfrac{4}{1 + \sqrt{3}}$

42. $\dfrac{1}{2 - \sqrt{2}}$

43. $\dfrac{2}{\sqrt{7} - 2}$

44. $\dfrac{2}{4 - \sqrt{5}}$

45. $\dfrac{x}{\sqrt{x} - 3}$

46. $\dfrac{y}{\sqrt{3} - y}$

47. $\dfrac{\sqrt{6} - 3}{2 - \sqrt{6}}$

48. $\dfrac{\sqrt{x} + \sqrt{y}}{\sqrt{x} - \sqrt{y}}$

■ *Rationalize each numerator. See Example 6.*

49. $\dfrac{1 - \sqrt{2}}{2}$

50. $\dfrac{\sqrt{3} + \sqrt{2}}{\sqrt{3}}$

51. $\dfrac{\sqrt{x} - 1}{3}$

52. $\dfrac{4 - \sqrt{2y}}{2}$

53. $\dfrac{\sqrt{x} - \sqrt{y}}{x}$

54. $\dfrac{2\sqrt{x} + \sqrt{y}}{\sqrt{xy}}$

B

■ *Write each expression as a single fraction in which the denominator is rationalized.*

55. $\sqrt{x + 1} - \dfrac{x}{\sqrt{x + 1}}$

56. $\sqrt{x^2 - 2} - \dfrac{x^2 + 1}{\sqrt{x^2 - 2}}$

57. $\dfrac{x}{\sqrt{x^2+1}} - \dfrac{\sqrt{x^2+1}}{x}$

58. $\dfrac{x}{\sqrt{x^2-1}} + \dfrac{\sqrt{x^2-1}}{x}$

■ *Rationalize each numerator.*

59. $\dfrac{\sqrt{x-1}-1}{\sqrt{x}-1}$

60. $\dfrac{1-\sqrt{x+1}}{\sqrt{x+1}}$

61. $\dfrac{\sqrt{x+1}-\sqrt{x}}{\sqrt{x+1}+\sqrt{x}}$

62. $\dfrac{\sqrt{x-1}+\sqrt{x}}{\sqrt{x-1}-\sqrt{x}}$

6.8

COMPLEX NUMBERS

In Section 4.4 we considered only quadratic equations with solutions in the set of real numbers. Some quadratic equations do not have real solutions. For example, if $b > 0$, then $x^2 = -b$ has no real-number solution, because there is no real number whose square is negative. For this reason, the expression $\sqrt{-b}$, for $b \in R$, $b > 0$, is undefined in the set of real numbers. In this section, we wish to consider a set of numbers that contains members whose squares are negative real numbers and that also contains members that are real numbers. We shall see that this new set of numbers, called the set C of **complex numbers**, provides solutions for all quadratic equations in one variable with real coefficients.

Imaginary numbers

Let us first assume a set of numbers among whose members are square roots of negative real numbers. We define $\sqrt{-b}$ and $-\sqrt{-b}$, where b is a positive real number, to be numbers whose squares are equal to $-b$.

> *For* $b > 0$,
>
> $$(\sqrt{-b})^2 = -b \quad \textit{and} \quad (-\sqrt{-b})^2 = -b.$$
>
> *In particular,*
>
> $$(\sqrt{-1})^2 = -1 \quad \textit{and} \quad (-\sqrt{-1})^2 = -1.$$

It is customary to use the symbol i for $\sqrt{-1}$. With this convention, and assuming that $-i = -1 \cdot i$,

$$i = \sqrt{-1}, \qquad -i = -\sqrt{-1}, \qquad i^2 = -1, \quad \text{and} \quad (-i)^2 = -1.$$

Thus, in this new set of numbers, -1 has two square roots, i and $-i$.

Next, assuming that $-\sqrt{-b} = -1 \cdot \sqrt{-b}$ and that

$$(i\sqrt{b})(i\sqrt{b}) = i^2 b = -1 \cdot b = -b,$$

we have the alternate definition of $\sqrt{-b}$.

$$\boldsymbol{\sqrt{-b} = \sqrt{-1}\sqrt{b} = i\sqrt{b}} \quad \textit{and} \quad \boldsymbol{-\sqrt{-b} = -1 \cdot \sqrt{-1}\sqrt{b} = -i\sqrt{b}}.$$

Hence, a square root of any negative real number can be represented as the product of a real number and the number $\sqrt{-1}$ or i.

EXAMPLE 1 a. $\sqrt{-4} = \sqrt{-1}\sqrt{4}$
$= i\sqrt{4} = 2i$

b. $-\sqrt{-3} = -\sqrt{-1}\sqrt{3}$
$= -i\sqrt{3}$ ■

The numbers represented by the symbols $\sqrt{-b}$ and $-\sqrt{-b}$, where b is a real number greater than zero, are called **pure imaginary numbers**.

Now, consider all the possible expressions of the form $a + bi$, where $a, b \in R$ and $i = \sqrt{-1}$, which are the sums of all real numbers and all pure imaginary numbers. Such an expression names a complex number, that is, a number in the set C. If $b = 0$, then $a + bi = a$, and it is evident that the set of real numbers is contained in the set C of complex numbers. If $b \neq 0$, then $a + bi$ is called an **imaginary number** (see Figure 6.1), where a is the **real part** of the number and b is the **imaginary part.** For example, the numbers -7, $3 + 2i$, and $4i$ are all complex numbers. However, -7 is also a real number, and $3 + 2i$ and $4i$ are also imaginary numbers. Furthermore, $4i$ is a pure imaginary number.

Complex numbers: $C = \{a + bi \mid a, b \in R\}$

$(b = 0)$
Real numbers: $a + bi = a$*

$(b \neq 0)$
Imaginary numbers: $a + bi$

$(a = 0)$
Pure imaginary numbers: $a + bi = bi$

FIGURE 6.1

*See Figure 1.1, page 5, for sets of numbers that are contained in the set of real numbers.

EXAMPLE 2 Write each expression in the form $a + bi$ or $a + ib$.

a. $3\sqrt{-18}$

b. $2 - 3\sqrt{-16}$

Solutions

a.
$$\begin{aligned} 3\sqrt{-18} &= 3\sqrt{-1 \cdot 9 \cdot 2} \\ &= 3\sqrt{-1}\sqrt{9}\sqrt{2} \\ &= 3i(3)\sqrt{2} \\ &= 9i\sqrt{2} \end{aligned}$$

b.
$$\begin{aligned} 2 - 3\sqrt{-16} &= 2 - 3\sqrt{-1 \cdot 16} \\ &= 2 - 3\sqrt{-1}\sqrt{16} \\ &= 2 - 3i(4) \\ &= 2 - 12i \end{aligned}$$ ■

Sums and differences

To add or subtract complex numbers, we simply add or subtract their real parts and their imaginary parts.

EXAMPLE 3

a.
$$\begin{aligned} &(2 + 3i) + (5 - 4i) \\ &= (2 + 5) + (3 - 4)i \\ &= 7 - i \end{aligned}$$

b.
$$\begin{aligned} &(2 + 3i) - (5 - 4i) \\ &= (2 - 5) + [3 - (-4)]i \\ &= -3 + 7i \end{aligned}$$ ■

In general:

$$\boldsymbol{(a + bi) + (c + di) = (a + c) + (b + d)i}$$

and

$$\boldsymbol{(a + bi) - (c + di) = (a - c) + (b - d)i.}$$

Products

To multiply complex numbers, we treat them as though they were binomials and replace i^2 with -1.

EXAMPLE 4

a.
$$\begin{aligned} &(2 - i)(1 + 3i) \\ &= 2 + 6i - i - 3i^2 \\ &= 2 + 6i - i - 3(-1) \\ &= 2 + 6i - i + 3 \\ &= 5 + 5i \end{aligned}$$

b.
$$\begin{aligned} (3 - i)^2 &= (3 - i)(3 - i) \\ &= 9 - 3i - 3i + i^2 \\ &= 9 - 6i + (-1) \\ &= 8 - 6i \end{aligned}$$ ■

Quotients

Recall from Section 6.7 that, for $b > 0$, the *conjugate* of $a + \sqrt{b}$ is $a - \sqrt{b}$. Similarly, the conjugate of $a + \sqrt{-b}$ is $a - \sqrt{-b}$.

EXAMPLE 5

a. The conjugate of $2 + 3i$ is $2 - 3i$.

b. The conjugate of $-3 - i$ is $-3 + i$.

c. The conjugate of $2i$ is $-2i$.

d. The conjugate of $-4 + i$ is $-4 - i$. ■

The quotient

$$\frac{a + bi}{c + di}$$

of two complex numbers can be simplified by multiplying the numerator and the denominator by $c - di$, the conjugate of the denominator. That is,

$$\frac{a + bi}{c + di} = \frac{(a + bi)(c - di)}{(c + di)(c - di)}.$$

EXAMPLE 6

a.
$$\frac{4 - i}{-2i} = \frac{(4 - i)i}{-2i \cdot i} = \frac{4i - i^2}{-2i^2} = \frac{4i - (-1)}{-2(-1)} = \frac{4i + 1}{2} = \frac{1}{2} + 2i$$

b.
$$\frac{4 + i}{2 + 3i} = \frac{(4 + i)(2 - 3i)}{(2 + 3i)(2 - 3i)} = \frac{8 - 10i - 3i^2}{4 - 9i^2} = \frac{8 - 10i + 3}{4 + 9} = \frac{11}{13} - \frac{10}{13}i$$ ■

Radical notation

The symbol $\sqrt{-b}\ (b > 0)$ should be used with care, since certain relationships involving the square root symbol that are valid for real numbers are not valid when the symbol does not represent a real number. For instance,

$$\sqrt{-2}\sqrt{-3} = (i\sqrt{2})(i\sqrt{3}) = i^2\sqrt{6} = -\sqrt{6};$$

$$\sqrt{-2}\sqrt{-3} \neq \sqrt{(-2)(-3)} = \sqrt{6}.$$

To avoid difficulty with this point:

Rewrite all expressions of the form $\sqrt{-b}$ $(b > 0)$ *in the form* $i\sqrt{b}$ *before performing any computations.*

EXAMPLE 7

a. $\sqrt{-2}(3 - \sqrt{-5})$
$= i\sqrt{2}(3 - i\sqrt{5})$
$= 3i\sqrt{2} - i^2\sqrt{10}$
$= 3i\sqrt{2} - (-1)\sqrt{10}$
$= \sqrt{10} + 3i\sqrt{2}$

b. $(2 + \sqrt{-3})(2 - \sqrt{-3})$
$= (2 + i\sqrt{3})(2 - i\sqrt{3})$
$= 4 - 3i^2$
$= 4 - 3(-1)$
$= 7$

c. $\dfrac{2}{\sqrt{-3}}$
$= \dfrac{2 \cdot i}{3i \cdot i}$
$= \dfrac{2i}{3(-1)}$
$= \dfrac{-2}{3}i$

d. $\dfrac{1}{3 - \sqrt{-1}} = \dfrac{1}{3 - i}$
$= \dfrac{1 \cdot (3 + i)}{(3 - i)(3 + i)}$
$= \dfrac{3 + i}{9 - i^2}$
$= \dfrac{3 + i}{9 - (-1)} = \dfrac{3}{10} + \dfrac{1}{10}i$ ■

EXERCISE 6.8

A

■ *Write each expression in the form* $a + bi$ *or* $a + ib$. *See Examples* 1 *and* 2.

1. $\sqrt{-4}$
2. $\sqrt{-9}$
3. $\sqrt{-32}$
4. $\sqrt{-50}$
5. $3\sqrt{-8}$
6. $4\sqrt{-18}$
7. $3\sqrt{-24}$
8. $2\sqrt{-40}$
9. $5\sqrt{-64}$
10. $7\sqrt{-81}$
11. $-2\sqrt{-12}$
12. $-3\sqrt{-75}$
13. $4 + 2\sqrt{-1}$
14. $5 - 3\sqrt{-1}$
15. $3\sqrt{-50} + 2$

16. $5\sqrt{-12} - 1$
17. $\sqrt{4} + \sqrt{-4}$
18. $\sqrt{20} - \sqrt{-20}$

■ *Write each expression in the form* $a + bi + a + ib$. *See Example* 3.

19. $(2 + 4i) + (3 + i)$
20. $(2 - i) + (3 - 2i)$
21. $(4 - i) - (6 - 2i)$
22. $(2 + i) - (4 - 2i)$
23. $3 - (4 + 2i)$
24. $(2 - 6i) - 3$

■ *See Example* 4.

25. $(2 - i)(3 + 2i)$
26. $(1 - 3i)(4 - 5i)$
27. $(3 + 2i)(5 + i)$
28. $(-3 - i)(2 - 3i)$
29. $(6 - 3i)(4 - i)$
30. $(7 + 3i)(-2 - 3i)$
31. $(2 - i)^2$
32. $(2 + 3i)^2$
33. $(2 - i)(2 + i)$
34. $(1 - 2i)(1 + 2i)$

■ *See Examples* 5 *and* 6.

35. $\frac{1}{3i}$
36. $\frac{-2}{5i}$
37. $\frac{3 - i}{5i}$
38. $\frac{4 + 2i}{3i}$
39. $\frac{2}{1 - i}$
40. $\frac{-3}{2 + i}$
41. $\frac{2 + i}{1 + 3i}$
42. $\frac{3 - i}{1 + i}$
43. $\frac{2 - 3i}{3 - 2i}$
44. $\frac{6 + i}{2 - 5i}$
45. $\frac{3 + 2i}{5 - 3i}$
46. $\frac{-4 - 3i}{2 + 7i}$

■ *See Example* 7.

47. $\sqrt{-4}(1 - \sqrt{-4})$
48. $\sqrt{-9}(3 + \sqrt{-16})$
49. $(2 + \sqrt{-9})(3 - \sqrt{-9})$
50. $(4 - \sqrt{-2})(3 + \sqrt{-2})$
51. $\frac{3}{\sqrt{-4}}$
52. $\frac{-1}{\sqrt{-25}}$
53. $\frac{2 - \sqrt{-1}}{2 + \sqrt{-1}}$
54. $\frac{1 + \sqrt{-2}}{3 - \sqrt{-2}}$
55. For what values of x will $\sqrt{x - 5}$ be real? Imaginary?
56. For what values of x will $\sqrt{x + 3}$ be real? Imaginary?

B

57. Simplify. [*Hint:* $i^2 = -1$ and $i^4 = 1$.]
    a. i^6 b. i^{12} c. i^{15} d. i^{102}
58. Express with a positive exponent and simplify.
    a. i^{-1} b. i^{-2} c. i^{-3} d. i^{-6}
59. Evaluate $x^2 + 2x + 3$ for $x = 1 + i$.
60. Evaluate $2y^2 - y + 2$ for $y = 2 - i$.

CHAPTER SUMMARY

[6.1–6.2] For n a natural number,

$$a^n = a \cdot a \cdot a \cdot \cdots \cdot a \qquad (n \text{ factors});$$

for n an integer,

$$a^{-n} = \frac{1}{a^n} \qquad (a \neq 0);$$

also,

$$a^0 = 1 \qquad (a \neq 0).$$

The laws of exponents are determined by the definitions adopted for powers:

I. $a^m \cdot a^n = a^{m+n}$

II. $(a^m)^n = a^{mn}$

III. $(ab)^n = a^n b^n$

IV. $\dfrac{a^m}{a^n} = a^{m-n} \qquad (a \neq 0)$

V. $\left(\dfrac{a}{b}\right)^n = \dfrac{a^n}{b^n} \qquad (b \neq 0)$

VI. $\left(\dfrac{a}{b}\right)^{-n} = \left(\dfrac{b}{a}\right)^n \qquad (a, b \neq 0)$

[6.3] A number is expressed in **scientific notation** when it is expressed as a product of a number between 1 and 10 and a power of 10.

[6.4] For each natural number n, the number $a^{1/n}$, when it exists, is called an ***n*th root of *a***.

If n is an even natural number and $a \geqslant 0$, then $a^{1/n}$ is the nonnegative number such that

$$(a^{1/n})^n = a.$$

If n is an odd natural number, the $a^{1/n}$ is the number such that

$$(a^{1/n})^n = a.$$

For $a^{1/n}$ a real number,

$$a^{m/n} = (a^{1/n})^m = (a^m)^{1/n}.$$

[6.5] An nth root of a is also designated by the **radical expression** $\sqrt[n]{a}$, where n is the **index** and a is the **radicand.** For all natural numbers $n \geq 2$,

$$\sqrt[n]{a} = a^{1/n}.$$

For n an even natural number:

If $a > 0$, $\sqrt[n]{a}$ is the positive number such that

$$(\sqrt[n]{a})^n = a.$$

In particular, for $n = 2$,

$$(\sqrt{a})^2 = \sqrt{a}\sqrt{a} = a.$$

For $a < 0$,

$$\sqrt[n]{a^n} = |a|.$$

For n an odd natural number,

$$(\sqrt[n]{a})^n = a.$$

For $\sqrt[n]{a}$ a real number,

$$a^{m/n} = \sqrt[n]{a^m} = (\sqrt[n]{a})^m.$$

[6.6] Two laws of radicals follow from corresponding laws for exponents:

$$\sqrt[n]{ab} = \sqrt[n]{a}\sqrt[n]{b} \quad \text{and} \quad \sqrt[n]{\frac{a}{b}} = \frac{\sqrt[n]{a}}{\sqrt[n]{b}}.$$

A radical expression is in simplest form if:

1. The radicand contains no polynomial factor raised to a power equal to or greater than the index of the radical.
2. The radicand contains no fractions.
3. No radical expressions are contained in denominators of fractions.

[6.7] Expressions containing radicals can be rewritten using properties of real numbers. In particular, the fundamental principle of fractions can be used to rationalize the denominator (or the numerator) of a fraction containing radicals.

[6.8] A number of the form $a + bi$, where $a, b \in R$, is called a **complex number.**

It is a real number if $b = 0$, it is an **imaginary number** if $b \neq 0$, and it is a **pure imaginary number** if $a = 0$ and $b \neq 0$.

The sum and the product of two complex numbers are defined so that they conform with the sum and the product of two binomials, respectively.

REVIEW EXERCISES

A

■ *Assume that all variables denote positive numbers.*

[6.1–6.2]

■ *Simplify. In Problems 1–18, assume that no denominator equals 0.*

1. a. $\dfrac{x^4y^3}{xy}$ b. $(3x^2y)^3$

2. a. $\dfrac{x^{-2} \cdot y^{-1}}{x}$ b. $\left(\dfrac{x^2y^{-2}}{x^{-2}y}\right)^{-2}$

3. a. $3^{-2} + 2^{-1}$ b. $xy^{-1} + x^{-1}y$

4. a. $\dfrac{(2x + 1)^{-1}}{(2x)^{-1} + 1^{-1}}$ b. $\dfrac{x^{-2} - y^{-2}}{(x - y)^{-2}}$

[6.3]

■ *Write each number in scientific notation.*

5. a. 0.000000000023 b. 307,000,000,000

■ *Express each number using standard form.*

6. a. 3.49×10^{-5} b. $\dfrac{3}{4 \times 10^2}$

■ *Simplify.*

7. a. $\dfrac{(2 \times 10^{-3})(3 \times 10^4)}{6 \times 10^{-1}}$ b. $\dfrac{(4 \times 10^3)(6 \times 10^{-4})}{3 \times 10^{-2}}$

■ *Specify the number of significant digits in each number.*

8. a. 0.0702 b. 0.4030

[6.4]

■ *Simplify.*

9. a. $(-27)^{2/3}$ b. $4^{-1/2}$

10. a. $x^{3/2} \cdot x^{1/2}$ b. $\left(\dfrac{x^{2/3}y^{1/2}}{x^{1/3}}\right)^6$

■ *Multiply.*

11. a. $x^{2/3}(x^{1/3} - x)$ b. $y^{-1/4}(y^{5/4} - y^{1/4})$

■ *Factor as indicated.*

12. a. $x^{4/5} = x^{1/5}(\underline{?})$ b. $y^{-3/4} = y^{-1/2}(\underline{?})$

[6.5]

■ *Express in radical notation.*

13. a. $(1 - x^2)^{2/3}$ b. $(1 - x^2)^{-2/3}$

■ *Express in exponential notation.*

14. a. $\sqrt[3]{x^2y}$ b. $\dfrac{1}{\sqrt[3]{(a + b)^2}}$

■ *Find each root indicated.*

15. a. $\sqrt{4y^2}$ b. $\sqrt[3]{-8x^3y^6}$

16. a. $\sqrt{\dfrac{1}{4}x^4y^6}$ b. $\sqrt[4]{\dfrac{1}{16}y^4}$

[6.6]

■ *Simplify.*

17. a. $\sqrt{180}$ b. $\sqrt[4]{32x^4y^5}$

18. a. $\dfrac{1}{\sqrt{2}}$ b. $\dfrac{1}{\sqrt{3}}$

19. a. $\dfrac{x}{\sqrt{xy}}$ b. $\dfrac{\sqrt{xy}\sqrt{6x^3y}}{\sqrt{2xy}}$

20. a. $\sqrt[3]{\dfrac{1}{2}}$ b. $\sqrt[4]{\dfrac{2}{3x}}$

21. a. $\sqrt[3]{\dfrac{x^7y^5}{x^2y}}$ b. $\dfrac{\sqrt[3]{x^5y^4}}{\sqrt[3]{x^2y}}$

22. a. $\dfrac{\sqrt[3]{12x^4}\,\sqrt[3]{4y^4}}{\sqrt[3]{6x}}$ b. $\dfrac{\sqrt[5]{x^3}\,\sqrt[5]{(xy)^3}}{\sqrt[5]{xy}}$

■ *Rationalize each numerator.*

23. a. $\dfrac{\sqrt{3x}}{2}$ b. $\dfrac{\sqrt{2x}}{y}$

24. a. $\dfrac{3\sqrt{2xy}}{\sqrt{y}}$ b. $\dfrac{2\sqrt{3xy}}{\sqrt{3y}}$

[6.7]

■ *Simplify.*

25. a. $4\sqrt{12} + 2\sqrt{75}$ b. $3\sqrt{2x} - \sqrt{32x} + 4\sqrt{50x}$
26. a. $9\sqrt{2xy^2} - 3y\sqrt{8x}$ b. $\sqrt{75x^3} - x\sqrt{3x}$

■ *Write each expression without parentheses and then write all radicals in simple form.*

27. a. $\sqrt{3}(\sqrt{6} - 2)$ b. $\sqrt{5}(\sqrt{10} - \sqrt{5})$
28. a. $(2 - \sqrt{3})(3 - 2\sqrt{3})$ b. $(\sqrt{x} + 2)(\sqrt{x} - 2)$

■ *Simplify each expression.*

29. a. $\dfrac{3 + 2\sqrt{18}}{3}$ b. $\dfrac{4 - 2\sqrt{8}}{4}$
30. a. $\dfrac{x + \sqrt{x^3y}}{x}$ b. $\dfrac{xy - \sqrt{x^3y^2}}{xy}$
31. a. $\dfrac{3}{2\sqrt{x}} - \dfrac{1}{\sqrt{x}}$ b. $\dfrac{4}{\sqrt{y}} - \dfrac{2}{3\sqrt{y}}$

■ *Rationalize each denominator.*

32. a. $\dfrac{4}{2 - \sqrt{3}}$ b. $\dfrac{2}{\sqrt{5} + 2}$
33. a. $\dfrac{y}{\sqrt{y} - 3}$ b. $\dfrac{x - y}{\sqrt{x} + \sqrt{y}}$

■ *Rationalize each numerator.*

34. a. $\dfrac{4 + \sqrt{3}}{2}$ b. $\dfrac{\sqrt{x} + 4}{x}$

[6.8]

■ *Write each expression in the form* $a + bi$ *or* $a + ib$.

35. a. $4 + 2\sqrt{-9}$ b. $5 - 3\sqrt{-12}$
36. a. $(2 - 3i) + (4 + 2i)$ b. $(6 + 2i) - (1 - i)$
37. a. $(5 + i)(2 + 3i)$ b. $i(3 - 4i)$
38. a. $\dfrac{4}{3i}$ b. $\dfrac{1 - 3i}{2 - i}$

B

■ *Simplify.*

39. a. $\left(\dfrac{x^{2n}y^{2n-1}}{y^n}\right)^2$ b. $\left(\dfrac{x^{3n}y^{1-n}}{x^{4n-1}}\right)^{-1}$

40. a. $(x^3)^{n/3}(y^{2n})^{1/n}$ b. $\left(\dfrac{x^n \cdot y^{n+3}}{y^3}\right)^{1/n}$

41. Compute: $\dfrac{0.0016 \times 0.00012 \times 270}{0.08 \times 0.00004 \times 81}$.

42. Write $\sqrt{(x-5)^2}$ $(x \in R)$ without radical notation.

43. Write $\sqrt{4x^2 - 8x + 4}$ $(x \in R)$ without radical notation.

44. Simplify $\sqrt[6]{16}$.

45. Simplify i^{-5} and express with positive exponents.

7 Nonlinear Equations and Inequalities

In Section 4.4 we solved quadratic equations of the form

$$ax^2 + bx + c = 0,$$

where in each case the left-hand member could be factored and the solutions were rational numbers. In this chapter we will solve quadratic equations by other methods. Furthermore, these methods will enable us to obtain solutions that may be irrational numbers or imaginary numbers.

7.1

SOLUTION OF EQUATIONS OF THE FORM $x^2 = b$; COMPLETING THE SQUARE

Quadratic equations of the form

$$x^2 = b$$

may be solved by a method often termed the **extraction of roots**. If the equation has a solution, then from the definition of a square root, x must be a square root of b.

Since each nonzero real number b has two square roots (either real or imaginary), we have two solutions. These are given by

$$x = \sqrt{b} \quad \text{and} \quad x = -\sqrt{b},$$

and the solution set is $\{\sqrt{b}, -\sqrt{b}\}$, where elements are real if $b > 0$ and imaginary if $b < 0$. If $b = 0$, we have one number, 0, satisfying the equation; the solution set of $x^2 = 0$ is $\{0\}$. The same conclusion can be reached by noting that, if we factor

$$x^2 - b = 0$$

over the complex numbers, we have

$$(x - \sqrt{b})(x + \sqrt{b}) = 0,$$

which when solved by the methods of Section 4.4, also leads to the solution set $\{\sqrt{b}, -\sqrt{b}\}$.

EXAMPLE 1 Solve.

a. $2x^2 - 6 = 0$ b. $5x^2 = 0$ c. $4x^2 + 3 = 0$

Solutions a.
$$\begin{aligned} 2x^2 - 6 &= 0 \\ 2x^2 &= 6 \\ x^2 &= 3 \\ x = \sqrt{3} \quad \text{or} \quad x &= -\sqrt{3} \end{aligned}$$
The solution set is $\{\sqrt{3}, -\sqrt{3}\}$.

b.
$$\begin{aligned} 5x^2 &= 0 \\ x^2 &= 0 \\ x &= 0 \end{aligned}$$
The solution set is $\{0\}$.

c. $4x^2 + 3 = 0$

$$x^2 = -\frac{3}{4}; \quad x = \sqrt{-\frac{3}{4}} \quad \text{or} \quad x = -\sqrt{\frac{-3}{4}}$$

The solution set is $\left\{\frac{1}{2}i\sqrt{3}, -\frac{1}{2}i\sqrt{3}\right\}$. ■

Equations of the form

$$(x - p)^2 = q \qquad (q \geqslant 0)$$

may also be solved by the method of extraction of roots.

EXAMPLE 2 Solve $(x - 2)^2 = 16$.

Solution The equation

$$(x - 2)^2 = 16$$

implies that $(x - 2)$ is a number whose square is 16. Hence,

$$x - 2 = 4 \quad \text{or} \quad x - 2 = -4,$$

from which we have

$$x = 6 \quad \text{or} \quad x = -2.$$

The solution set is $\{6, -2\}$. ■

EXAMPLE 3 Solve $(x + 4)^2 = -9$.

Solution The equation

$$(x + 4)^2 = -9$$

implies that $(x + 4)$ is a number whose square is -9. Hence,

$$x + 4 = \sqrt{-9} \quad \text{or} \quad x + 4 = -\sqrt{-9},$$

from which we have

$$x = -4 + 3i \quad \text{or} \quad x = -4 - 3i.$$

The solution set is $\{-4 + 3i, -4 - 3i\}$. ■

Solution by completing the square

The method of extraction of roots can be used to find the solution set of any quadratic equation. Let us first consider a specific example,

$$x^2 - 4x - 12 = 0,$$

which can be written

$$x^2 - 4x \qquad = 12.$$

If the square of one-half of the coefficient of the first-degree term,

$$\left[\frac{1}{2}(-4)\right]^2,$$

equal to 4 is added to each member, we obtain

$$x^2 - 4x + 4 = 12 + 4,$$

in which the left-hand member is the square of $(x - 2)$. Therefore, the equation can be written

$$(x - 2)^2 = 16,$$

and the solution set is obtained as in Example 2.

Now consider the general quadratic equation in standard form,

$$ax^2 + bx + c = 0,$$

for the special case where $a = 1$; that is,

$$x^2 + bx + c = 0. \tag{1}$$

We begin by adding $-c$ to each member of (1), which yields

$$x^2 + bx \qquad = -c. \tag{2}$$

If we then add the square of one-half of the coefficient of x, $(^b/_2)^2$, to each member of (2), the result is

$$x^2 + bx + \left(\frac{b}{2}\right)^2 = -c + \left(\frac{b}{2}\right)^2, \tag{3}$$

where the left-hand member is equivalent to $(x + {}^b/_2)^2$, and we have

$$\left(x + \frac{b}{2}\right)^2 = -c + \frac{b^2}{4}. \tag{4}$$

Since we have performed only elementary transformations, (4) is equivalent to (1) and we can solve (4) by the method used in the examples above.

The technique used to obtain Equations (3) and (4) is called **completing the square.** Note that we completed the square in (2) by dividing the coefficient b of the linear term by 2 and squaring the result. The expression obtained,

$$x^2 + bx + \left(\frac{b}{2}\right)^2,$$

was then written in the equivalent form $(x + {}^b/_2)^2$. When the left-hand member of an equation is in this form and the right-hand member does not contain the variable x, the equation can be solved by the method of extraction of roots. The following example shows the step by step procedure that is used.

EXAMPLE 4 Solve $x^2 - 3x - 1 = 0$.

Solution First rewrite the equation with the constant term as the right-hand member.

$$x^2 - 3x \qquad = 1$$

Add $(-3/2)^2$, the square of one-half of the coefficient of the first-degree term, to each member.

$$x^2 - 3x + \left(-\frac{3}{2}\right)^2 = 1 + \left(-\frac{3}{2}\right)^2$$

Rewrite the left-hand member as the square of a binomial and simplify the right-hand member.

$$\left(x - \frac{3}{2}\right)^2 = \frac{13}{4}$$

Set $(x - 3/2)$ equal to each square root of $13/4$.

$$x - \frac{3}{2} = \sqrt{\frac{13}{4}} \quad \text{or} \quad x - \frac{3}{2} = -\sqrt{\frac{13}{4}}$$

$$x = \frac{3}{2} + \frac{\sqrt{13}}{2} \qquad\qquad x = \frac{3}{2} - \frac{\sqrt{13}}{2}$$

The solution set is $\left\{\dfrac{3 + \sqrt{13}}{2}, \dfrac{3 - \sqrt{13}}{2}\right\}$. ■

We began with the special case,

$$x^2 + bx + c = 0, \tag{5}$$

rather than the general form,

$$ax^2 + bx + c = 0,$$

because the term necessary to complete the square is obvious when $a = 1$. However, a quadratic equation in standard form can always be written in the form (5) by multiplying each member of

$$ax^2 + bx + c = 0$$

by $^1/_a$ $(a \neq 0)$ and obtaining

$$x^2 + \frac{b}{a}x + \frac{c}{a} = 0.$$

EXAMPLE 5 Solve $2x^2 + x - 1 = 0$.

Solution Rewrite the equation with the constant term as the right-hand member and the coefficient of x^2 equal to 1.

$$x^2 + \frac{1}{2}x \quad = \frac{1}{2}$$

Add $[\frac{1}{2}(\frac{1}{2})]^2$, the square of one-half of the coefficient of the first-degree term, to each member.

$$x^2 + \frac{1}{2}x + \frac{1}{16} = \frac{1}{2} + \frac{1}{16}$$

Rewrite the left-hand member as the square of a binomial.

$$\left(x + \frac{1}{4}\right)^2 = \frac{9}{16}$$

Set $(x + \frac{1}{4})$ equal to each square root of $\frac{9}{16}$.

$$x + \frac{1}{4} = \frac{3}{4} \quad \text{or} \quad x + \frac{1}{4} = -\frac{3}{4}$$

$$x = \frac{1}{2} \qquad\qquad x = -1$$

The solution set is $\{\frac{1}{2}, -1\}$. ■

The methods used to solve equations in this section can also be used to solve equations that involve more than one variable.

EXAMPLE 6 Solve $x^2 - 4y^2 = 16$ for y in terms of x.

Solution First write the equation as

$$4y^2 = x^2 - 16,$$

and then as

$$y^2 = \frac{x^2 - 16}{4}.$$

From which

$$y = \pm\sqrt{\frac{x^2 - 16}{4}}$$
$$= \pm\frac{1}{2}\sqrt{x^2 - 16}. \quad ■$$

EXERCISE 7.1

A

■ *Solve by the extraction of roots. See Example* 1.

1. $x^2 = 100$
2. $x^2 = 16$
3. $9x^2 = 25$
4. $4x^2 = 9$
5. $2x^2 = 14$
6. $3x^2 = 15$
7. $4x^2 + 24 = 0$
8. $3x^2 + 9 = 0$
9. $\dfrac{2x^2}{3} = 4$
10. $\dfrac{3x^2}{5} = 3$
11. $\dfrac{4x^2}{3} = -27$
12. $\dfrac{9x^2}{2} = -50$

■ *Solve. See Examples* 2 *and* 3.

13. $(x - 2)^2 = 9$
14. $(x + 3)^2 = 4$
15. $(2x - 1)^2 = 16$
16. $(3x + 1)^2 = 25$
17. $(x + 2)^2 = -3$
18. $(x - 5)^2 = -7$
19. $\left(x - \dfrac{1}{2}\right)^2 = \dfrac{3}{4}$
20. $\left(x - \dfrac{2}{3}\right)^2 = \dfrac{5}{9}$
21. $\left(x + \dfrac{1}{3}\right)^2 = \dfrac{1}{81}$
22. $\left(x + \dfrac{1}{2}\right)^2 = \dfrac{1}{16}$
23. $(2x - 5)^2 = 9$
24. $(3x + 4)^2 = 16$
25. $(3x + 5)^2 = 9$
26. $(2x + 3)^2 = 36$
27. $(7x - 1)^2 = -15$
28. $(5x + 3)^2 = -7$
29. $(8x - 7)^2 = -8$
30. $(5x - 12)^2 = -24$

■ *Solve by completing the square. See Examples 4 and 5.*

31. $x^2 + 4x - 12 = 0$
32. $x^2 - x - 6 = 0$
33. $x^2 - 2x + 1 = 0$
34. $x^2 + 4x + 4 = 0$
35. $x^2 + 9x + 20 = 0$
36. $x^2 - x - 20 = 0$
37. $x^2 - 2x - 1 = 0$
38. $x^2 + 3x - 1 = 0$
39. $x^2 = 3 - 3x$
40. $x^2 = 5 - 5x$
41. $2x^2 + 4x - 3 = 0$
42. $3x^2 + x - 4 = 0$
43. $2x^2 - 5 = 3x$
44. $4x^2 - 3 = 2x$
45. $x^2 - x + 3 = 0$
46. $x^2 - 5x + 8 = 0$
47. $2x^2 + 4x = -3$
48. $3x^2 + x = -4$

B

■ *Solve for x in terms of a, b, and c. See Example 6.*

49. $x^2 - a = 0$
50. $x^2 - 2a = 0$
51. $\frac{ax^2}{b} = c$
52. $\frac{bx^2}{c} - a = 0$
53. $(x - a)^2 = 16$
54. $(x + a)^2 = 36$
55. $(ax + b)^2 = 9$
56. $(ax - b)^2 = 25$
57. Solve $a^2 + b^2 = c^2$ for b in terms of a and c.
58. Solve $S = h - r^2$ for r in terms of S and h.
59. Solve $A = P(1 + r)^2$ for r in terms of A and P.
60. Solve $s = \frac{1}{2}gt^2 + c$ for t in terms of s, g, and c.
61. Solve $x^2 + 9y^2 = 9$ for y in terms of x.
62. Solve $9x^2 + 4y^2 = 36$ for y in terms of x.
63. Solve $4x^2 - 9y^2 = 36$ for y in terms of x.
64. Solve $9x^2 - 25y^2 = 0$ for y in terms of x.
65. Solve $ax^2 + bx + c = 0$ for x in terms of a, b, and c by using the method of completing the square.

7.2

THE QUADRATIC FORMULA

In Section 4.4 we considered quadratic equations that we were able to solve by factoring over the integers. Let us now develop a formula that can be used to solve all quadratic equations, including those having solutions in the set of complex numbers.

When a quadratic equation is written in its standard form,

$$ax^2 + bx + c = 0, \tag{1}$$

we can solve it by completing the square as follows:

$$\begin{aligned}
x^2 + \frac{b}{a}x + \frac{c}{a} &= 0 \\
x^2 + \frac{b}{a}x + \left(\frac{b}{2a}\right)^2 &= -\frac{c}{a} + \left(\frac{b}{2a}\right)^2 \\
\left(x + \frac{b}{2a}\right)^2 &= \frac{b^2}{4a^2} - \frac{c}{a} \\
\left(x + \frac{b}{2a}\right)^2 &= \frac{b^2 - 4ac}{4a^2} \\
x + \frac{b}{2a} &= \pm\sqrt{\frac{b^2 - 4ac}{4a^2}} \\
x &= -\frac{b}{2a} \pm \frac{\sqrt{b^2 - 4ac}}{2a}.
\end{aligned}$$

When the right-hand member is written as a single fraction, the resulting equation is called the **quadratic formula.**

$$\boldsymbol{x = \frac{-b \pm \sqrt{b^2 - 4ac}}{2a}}$$

This is a formula for the solutions of a quadratic equation expressed in terms of the coefficients. The symbol $\pm$ is used to condense the two equations

$$\boldsymbol{x = \frac{-b + \sqrt{b^2 - 4ac}}{2a}} \quad \text{or} \quad \boldsymbol{x = \frac{-b - \sqrt{b^2 - 4ac}}{2a}}$$

into a single equation. We need only substitute the coefficients a, b, and c of a given quadratic equation in the formula to find the solution set for the equation.

Although the quadratic formula can be used to solve any quadratic equation in standard form, it is particularly useful when the left-hand member of a quadratic equation cannot be factored easily.

EXAMPLE 1 Solve $2x^2 - x - 2 = 0$.

Solution Substitute 2 for a, -1 for b, and -2 for c in the quadratic formula.

$$x = \frac{-(-1) \pm \sqrt{(-1)^2 - 4(2)(-2)}}{2(2)}$$
$$= \frac{1 \pm \sqrt{1 + 16}}{4}$$

The solution set is $\left\{\frac{1 + \sqrt{17}}{4}, \frac{1 - \sqrt{17}}{4}\right\}$. ■

Notice in the above example that we could have expressed the solutions as $\frac{1}{4} + \frac{\sqrt{17}}{4}$ and $\frac{1}{4} - \frac{\sqrt{17}}{4}$. In fact, it is clear from the form

$$x = \frac{-b}{2a} \pm \frac{\sqrt{b^2 - 4ac}}{2a}$$

that, if the solutions of a quadratic equation are irrational or imaginary, they are conjugates:

$$\frac{-b}{2a} + \frac{\sqrt{b^2 - 4ac}}{2a} \quad \text{and} \quad \frac{-b}{2a} - \frac{\sqrt{b^2 - 4ac}}{2a}.$$

A quadratic equation must be written in standard form before the values of a, b, and c can be determined.

EXAMPLE 2 Solve $\frac{x^2}{4} + x = \frac{5}{4}$.

Solution Write in standard form.

$$4\left(\frac{x^2}{4} + x\right) = \left(\frac{5}{4}\right)(4)$$
$$x^2 + 4x = 5$$
$$x^2 + 4x - 5 = 0$$

Substitute 1 for a, 4 for b, and -5 for c in the quadratic formula.*

*Note that the equation could be solved by the factoring method, as could many of the following exercises. However, the intent in this section is to practice using the quadratic formula.

$$x = \frac{-(4) \pm \sqrt{(4)^2 - 4(1)(-5)}}{2(1)}$$

$$= \frac{-4 \pm \sqrt{16 + 20}}{2} = \frac{-4 \pm 6}{2}$$

The solution set is $\{-5, 1\}$. ■

EXAMPLE 3 Solve $x = \dfrac{-2}{2x - 1}$.

Solution First write the equation in standard form to obtain

$$(2x - 1)x = \frac{-2}{2x - 1}(2x - 1),$$

$$2x^2 - x = -2,$$

$$2x^2 - x + 2 = 0.$$

Substitute 2 for a, -1 for b, and 2 for c in the quadratic equation to obtain

$$x = \frac{-(-1) \pm \sqrt{(-1)^2 - 4(2)(2)}}{2(2)}$$

$$= \frac{1 \pm \sqrt{-15}}{4} = \frac{1 \pm i\sqrt{15}}{4}$$

The solution set is $\left\{\dfrac{1 - i\sqrt{15}}{4}, \dfrac{1 + i\sqrt{15}}{4}\right\}$. ■

Discriminant

In the quadratic formula, the number represented by $b^2 - 4ac$ is called the **discriminant** of the equation. If a, b, and c are real numbers, the discriminant affects the solution set in the following ways:

1. If $b^2 - 4ac = 0$, there is one real solution (of multiplicity two).
2. If $b^2 - 4ac > 0$, there are two unequal real solutions.
3. If $b^2 - 4ac < 0$, there are two unequal imaginary solutions.

EXAMPLE 4 Find the discriminant for each equation and determine whether the solution(s) are real or imaginary.

a. $x^2 - x - 3 = 0$ **b.** $2x^2 + x + 1 = 0$

Solution a. Substitute 1 for a, -1 for b, and -3 for c in the discriminant.

$$b^2 - 4ac = (-1)^2 - 4(1)(-3)$$
$$= 1 + 12 = 13$$

Because $13 > 0$, the solutions are real and unequal.

b. Substitute 2 for a, 1 for b, and 1 for c in the discriminant.

$$b^2 - 4ac = 1^2 - 4(2)(1)$$
$$= 1 - 8 = -7.$$

Because $-7 < 0$, the solutions are imaginary. ■

We can use the quadratic formula to solve a quadratic equation in more than one variable for a specified variable in terms of the other variables.

EXAMPLE 5 Solve $x^2 - xy + y = 2$ for x in terms of y.

Solution Write the equation in standard form.

$$x^2 - yx + (y - 2) = 0$$

Substitute 1 for a, $-y$ for b, and $y - 2$ for c in the quadratic formula.

$$x = \frac{-(-y) \pm \sqrt{(-y)^2 - 4(1)(y - 2)}}{2(1)}$$
$$= \frac{y \pm \sqrt{y^2 - 4y + 8}}{2}$$ ■

EXERCISE 7.2

A

■ *In Problems 1–24, solve for x, y, or z using the quadratic formula. See Examples* 1, 2, *and* 3.

1. $x^2 - 5x + 4 = 0$
2. $x^2 - 4x + 4 = 0$
3. $y^2 + 3y = 4$
4. $y^2 - 5y = 6$
5. $z^2 = 3z - 1$
6. $2z^2 = 7z - 6$

7. $0 = x^2 - \frac{5}{3}x + \frac{1}{3}$
8. $0 = x^2 - \frac{1}{2}x + \frac{1}{2}$
9. $5z + 6 = 6z^2$
10. $13z + 5 = 6z^2$
11. $x^2 - 5x = 0$
12. $y^2 + 3y = 0$
13. $4y^2 + 8 = 0$
14. $2z^2 + 1 = 0$
15. $2y^2 = y - 1$
16. $x^2 + 2x = -5$
17. $y = \frac{1}{y - 3}$
18. $2z = \frac{3}{z - 2}$
19. $3y = \frac{1 + y}{y - 1}$
20. $2x = \frac{x + 1}{x - 1}$
21. $2y^2 = y - 2 - y^2$
22. $3z^2 + 2z + 2 = z$
23. $2x = \frac{-1}{2x - 1}$
24. $y = \frac{-1}{y - 1}$

■ *In Problems 25–30, find only the discriminant and determine whether the solution(s) are:*
a. *one real;* b. *real and unequal;* c. *imaginary and unequal.*
See Example 4.

25. $x^2 - 7x + 12 = 0$
26. $y^2 - 2y - 3 = 0$
27. $5x^2 + 2x + 1 = 0$
28. $2y^2 + 3y + 7 = 0$
29. $9x^2 - 6x + 1 = 0$
30. $4y^2 - 12y + 9 = 0$

B

■ *Solve for x in terms of the other variables or constants. See Example 5.*

31. $x^2 - kx - 2k^2 = 0 \quad k \geq 0$
32. $2x^2 - kx + 3 = 0$
33. $ax^2 - x + c = 0$
34. $x^2 + 2x + k + 3 = 0$
35. $x^2 + 2x - y = 0$
36. $2x^2 - 3x + 2y = 0$
37. $3x^2 + xy + y^2 = 2$
38. $x^2 - 3xy + y^2 = 3$
39. In Problem 37 solve for y in terms of x.
40. In Problem 38 solve for y in terms of x.
41. If r_1 and r_2 are solutions of $ax^2 + bx + c = 0$, show that

$$r_1 + r_2 = -\frac{b}{a} \quad \text{and} \quad r_1 \cdot r_2 = \frac{c}{a}.$$

7.3

EQUATIONS INVOLVING RADICALS

To solve equations containing radicals, we shall make use of the following property.

If each member of an equation is raised to the same natural-number power, the solutions of the original equation are contained in the solution set of the resulting equation.

This property can be expressed in symbols as follows.

For n a natural number, the solution set of

$$[P(x)]^n = [Q(x)]^n \tag{1}$$

contains all the solutions of

$$P(x) = Q(x). \tag{2}$$

Extraneous solutions

Note that an application of this property does not always result in an equivalent equation. Equation (1) actually may have additional solutions that are not solutions of (2). With respect to (2), these are called **extraneous solutions**. For example, the solution set of the equation

$$x^2 = 9,$$

obtained from

$$x = 3$$

by squaring each member, is $\{3, -3\}$. This set contains -3 as an extraneous solution, since -3 does not satisfy the equation $x = 3$.

Because the result of applying the foregoing principle is not always an equivalent equation, each solution obtained through its use *must* be checked in the original equation to verify its validity. The check is part of the solution process.

EXAMPLE 1 Solve $\sqrt{x - 3} = 2$.

Solution First square each member to remove the radical, and then solve for x.

$$(\sqrt{x - 3})^2 = 2^2$$
$$x - 3 = 4$$
$$x = 7$$

Check Does $\sqrt{7 - 3} = 2$? Yes. The solution set is $\{7\}$. ■

EXAMPLE 2 Solve $\sqrt{x + 2} + 4 = x$.

Solution Obtain $\sqrt{x + 2}$ as the only term in one member.

$$\sqrt{x + 2} = x - 4$$

Square each member.

$$(\sqrt{x + 2})^2 = (x - 4)^2$$
$$x + 2 = x^2 - 8x + 16$$

Solve the quadratic equation.

$$x^2 - 9x + 14 = 0$$
$$(x - 7)(x - 2) = 0$$
$$x = 7 \qquad x = 2$$

Check Does $\sqrt{7 + 2} + 4 = 7$? Yes. Does $\sqrt{2 + 2} + 4 = 2$? No. 2 is not a solution. The solution set is $\{7\}$. ■

Sometimes it is necessary to square each member of an equation more than once in order to obtain an equation free of radicals.

EXAMPLE 3 Solve $\sqrt{x - 7} + \sqrt{x} = 7$.

Solution First add $-\sqrt{x}$ to each member to obtain

$$\sqrt{x - 7} = 7 - \sqrt{x},$$

which contains only one term with a radical in the left-hand member. Now square each member to remove one radical.

$$(\sqrt{x - 7})^2 = (7 - \sqrt{x})^2$$
$$x - 7 = 49 - 14\sqrt{x} + x$$
$$-56 = -14\sqrt{x}$$
$$4 = \sqrt{x}$$

Now square each member again to obtain

$$(4)^2 = (\sqrt{x})^2,$$
$$16 = x.$$

Check Does $\sqrt{16 - 7} + \sqrt{16} = 7$? Yes. The solution set is $\{16\}$. ■

Using the property introduced on page 250, we can solve formulas that involve radicals for specified variables in terms of the other variables in the formulas.

EXAMPLE 4 Solve $t = \sqrt{\dfrac{1 + s^2}{g}}$, for s.

Solution Squaring each member yields

$$t^2 = \frac{1 + s^2}{g},$$

from which

$$gt^2 = 1 + s^2,$$
$$gt^2 - 1 = s^2;$$

hence,

$$s = \pm\sqrt{gt^2 - 1}.$$ ■

EXERCISE 7.3

A

■ *Solve and check. If there is no solution, so state. Assume that all variables represent real numbers. See Examples 1 and 2.*

1. $\sqrt{x} - 5 = 3$
2. $\sqrt{x} - 4 = 1$
3. $\sqrt{y + 6} = 2$
4. $\sqrt{y - 3} = 5$
5. $3z + 4 = \sqrt{3z + 10}$
6. $2z - 3 = \sqrt{7z - 3}$
7. $2x + 1 = \sqrt{10x + 5}$
8. $4x + 5 = \sqrt{3x + 4}$
9. $\sqrt{y + 4} = y - 8$
10. $4\sqrt{x - 4} = x$
11. $\sqrt{2y - 1} = \sqrt{3y - 6}$
12. $\sqrt{4y + 1} = \sqrt{6y - 3}$

13. $\sqrt{x - 3}\sqrt{x} = 2$ 14. $\sqrt{x}\sqrt{x - 5} = 6$ 15. $\sqrt[3]{x} = -3$

16. $\sqrt[3]{x} = -4$ 17. $\sqrt[4]{x - 1} = 2$ 18. $\sqrt[4]{x - 1} = 3$

■ *See Example* 3.

19. $\sqrt{y + 4} = \sqrt{y + 20} - 2$ 20. $4\sqrt{y} + \sqrt{1 + 16y} = 5$

21. $\sqrt{x} + \sqrt{2} = \sqrt{x + 2}$ 22. $\sqrt{4x + 17} = 4 - \sqrt{x + 1}$

23. $(5 + x)^{1/2} + x^{1/2} = 5$ 24. $(y + 7)^{1/2} + (y + 4)^{1/2} = 3$

25. $(y^2 - 3y + 5)^{1/2} - (y + 2)^{1/2} = 0$ 26. $(z - 3)^{1/2} + (z + 5)^{1/2} = 4$

■ *Solve. Leave the results in the form of an equation. Assume that no variable takes a value for which any denominator is* 0. *See Example* 4.

27. $r = \sqrt{\dfrac{A}{\pi}}$, for A 28. $t = \sqrt{\dfrac{2v}{g}}$, for g

29. $R\sqrt{RS} = 1$, for S 30. $P = \sqrt{\dfrac{l}{g}}$, for g

31. $r = \sqrt{t^2 - s^2}$, for t 32. $q - 1 = 2\sqrt{\dfrac{r^2 - 1}{3}}$, for r

33. $A = B + C\sqrt{D + E^2}$, for E 34. $A = B - C\sqrt{D - E^2}$, for E

B

35. In Problem 27, specify the restrictions on each variable so that the solutions are real numbers.

36. In Problem 28, specify the restrictions on each variable so that the solutions are real numbers.

37. The base of an isosceles triangle is 6 centimeters. Find the altitude if the perimeter is 20 centimeters.

38. The longer leg of a right triangle is 1 centimeter shorter than the hypotenuse. Find the hypotenuse if the perimeter is 30 centimeters.

39. Using the information given in the figure, find the length of the segment BC, if $AC + CD$ equals 11 feet.

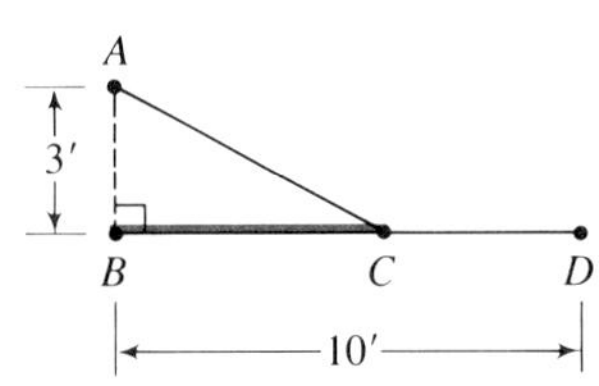

40. Using the information given in the figure, find the length of the segment AB, if $AB + BD$ equals 14 centimeters.

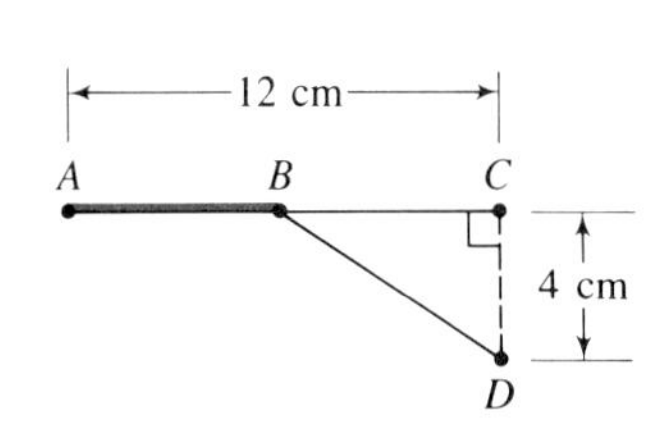

7.4

EQUATIONS THAT ARE QUADRATIC IN FORM

Some equations that are not quadratic equations are nevertheless quadratic in form—that is, they are of the form

$$au^2 + bu + c = 0, \tag{1}$$

where u represents some expression in terms of another variable. For example,

$$x^4 - 10x^2 + 9 = 0 \quad \text{or} \quad (x^2)^2 - 10(x^2) + 9 = 0 \tag{2}$$

is quadratic in x^2,

$$y - 2\sqrt{y} - 8 = 0 \quad \text{or} \quad (\sqrt{y})^2 - 2(\sqrt{y}) - 8 = 0 \tag{3}$$

is quadratic in $\sqrt{y}$,

$$y^{2/3} - 5y^{1/3} + 4 = 0 \quad \text{or} \quad (y^{1/3})^2 - 5(y^{1/3}) + 4 = 0 \tag{4}$$

is quadratic in $y^{1/3}$.

We can solve an equation of the form (1) by using a procedure shown in the following examples.

EXAMPLE 1 Solve $x^4 - 10x^2 + 9 = 0$. (2)

Solution If we let $x^2 = u$, the equation reduces to

$$u^2 - 10u + 9 = 0.$$

Solving this equation, we obtain

$$(u - 9)(u - 1) = 0$$
$$u = 9 \quad \text{or} \quad u = 1.$$

Since $u = x^2$, we have

$$x^2 = 9 \quad \text{or} \quad x^2 = 1,$$

from which we obtain the solution set $\{3, -3, 1, -1\}$. ■

The method illustrated in the above example is commonly called *substitution of variables*.

Note that the equation in the above example could be solved by factoring the left-hand member to obtain

$$(x^2 - 9)(x^2 - 1) = 0,$$

from which

$$x^2 - 9 = 0 \quad \text{or} \quad x^2 - 1 = 0$$
$$x = \pm 3 \qquad\qquad x = \pm 1,$$

and the solution set is $\{3, -3, 1, -1\}$.

Because the substitution of variables method is a useful tool (there are many cases where the factoring method shown above is not possible), we will use it in all the exercises in this section.

EXAMPLE 2 Solve $y - 2\sqrt{y} - 8 = 0.$ (3)

Solution Let $\sqrt{y} = u$; therefore, $y = u^2$. Substitute for y and $\sqrt{y}$.

$$u^2 - 2u - 8 = 0$$

Solve for u.

$$(u - 4)(u + 2) = 0$$
$$u = 4 \quad \text{or} \quad u = -2$$

Replace u with $\sqrt{y}$ and solve for y. Since $\sqrt{y}$ cannot be negative, only 4 need be considered.

$$\sqrt{y} = 4$$
$$y = 16$$

Check Does $16 - 2\sqrt{16} - 8 = 0$? Yes.

Hence, $\{16\}$ is the solution set. ■

EXAMPLE 3 Solve $y^{2/3} - 5y^{1/3} + 4 = 0.$ (4)

Solution Let $y^{1/3} = u$; therefore $y^{2/3} = u^2$. Substitute for $y^{2/3}$ and $y^{1/3}$.

$$u^2 - 5u + 4 = 0$$

Solve for u.

$$(u - 4)(u - 1) = 0$$
$$u = 4 \quad \text{or} \quad u = 1$$

Replace u with $y^{1/3}$ and solve for y.

$$y^{1/3} = 4 \quad \text{or} \quad y^{1/3} = 1$$
$$y = 64 \quad \text{or} \quad y = 1$$

Check Does $64^{2/3} - 5(64)^{1/3} + 4 = 0$? Does $(1)^{2/3} - 5(1)^{1/3} + 4 = 0$?
Does $16 - 20 + 4 = 0$? Yes Does $1 - 5 + 4 = 0$? Yes

Hence, $\{1, 64\}$ is the solution set. ■

EXERCISE 7.4

A

■ *Solve. See Example* 1.

1. $x^4 - 5x^2 + 4 = 0$
2. $x^4 - 13x^2 + 36 = 0$
3. $y^4 - 4y^2 + 3 = 0$
4. $y^4 - 6y^2 + 5 = 0$
5. $2x^4 + 17x^2 - 9 = 0$
6. $x^4 - 2x^2 - 24 = 0$
7. $z^4 + 4z^2 + 3 = 0$
8. $z^4 + 7z^2 + 10 = 0$

■ *Solve. See Example* 2.

9. $x - 2\sqrt{x} - 15 = 0$
10. $x + 3\sqrt{x} - 10 = 0$
11. $y + 3\sqrt{y} + 2 = 0$
12. $y - 2\sqrt{y} + 1 = 0$
13. $y^2 + 7 - \sqrt{y^2 + 7} - 12 = 0$
14. $y^2 - 5 - 5\sqrt{y^2 - 5} + 6 = 0$
15. $x^2 - 1 - \sqrt{x^2 - 1} - 6 = 0$
16. $x^2 - 4 + 4\sqrt{x^2 - 4} + 3 = 0$

■ *Solve. See Example* 3.

17. $y^{2/3} - 2y^{1/3} - 8 = 0$
18. $z^{2/3} - 2z^{1/3} = 35$
19. $x^{2/3} - 3x^{1/3} = 4$
20. $2y^{2/3} + 5y^{1/3} = 3$
21. $x - 9x^{1/2} + 18 = 0$
22. $z + z^{1/2} = 72$
23. $2x - 9x^{1/2} = -4$
24. $8x^{1/2} + 7x^{1/4} = 1$

B

25. $y^{-2} - y^{-1} - 12 = 0$

26. $z^{-2} + 9z^{-1} - 10 = 0$

27. $(x - 1)^{1/2} - 2(x - 1)^{1/4} - 15 = 0$

28. $(x - 2)^{1/2} - 11(x - 2)^{1/4} + 18 = 0$

■ *Solve Problems 29–32 in two ways:*

a. *By the method of Section 7.3* b. *By the method of this section.*

29. $y - 7\sqrt{y} + 12 = 0$

30. $x + \sqrt{x} - 6 = 0$

31. $x + 18 = 11\sqrt{x}$

32. $y + 2\sqrt{y} = 15$

7.5

QUADRATIC INEQUALITIES

As in the case with first-degree inequalities, we can generate equivalent second-degree inequalities by applying Properties (1)–(3) and Equations (3)–(5) of Section 4.5. Additional procedures are necessary, however, to obtain the solution sets of such inequalities. For example, consider the inequality

$$x^2 + 4x < 5.$$

To determine values of x for which this condition holds, we might first rewrite the inequality equivalently as

$$x^2 + 4x - 5 < 0,$$

and then as

$$(x + 5)(x - 1) < 0.$$

In this case, the left-hand member will be negative (less than 0) only for those values of x for which the factors $x + 5$ and $x - 1$ are opposite in sign. These can be determined analytically by noting that $(x + 5)(x - 1) < 0$ implies either

$$x + 5 < 0 \quad \text{and} \quad x - 1 > 0 \tag{1}$$

or

$$x + 5 > 0 \quad \text{and} \quad x - 1 < 0. \tag{2}$$

Each of these two cases can be considered separately.

The inequalities for (1),

$$x + 5 < 0 \quad \text{and} \quad x - 1 > 0,$$

imply

$$x < -5 \quad \textit{and} \quad x > 1,$$

a condition which is not satisfied by any values of x. Therefore, the solution set of (1) is $\varnothing$.

The inequalities for (2),

$$x + 5 > 0 \quad \text{and} \quad x - 1 < 0,$$

imply

$$x > -5 \quad \textit{and} \quad x < 1,$$

which leads to the solution set

$$\{x | -5 < x < 1\}.$$

Sign arrays

The solution of the inequality $(x + 5)(x - 1) < 0$ in the above example can be interpreted through the use of an array of signs which are determined by the factors in the left-hand member. Note that in Figure 7.1 the product is negative (less than zero) wherever the factors have opposite signs.

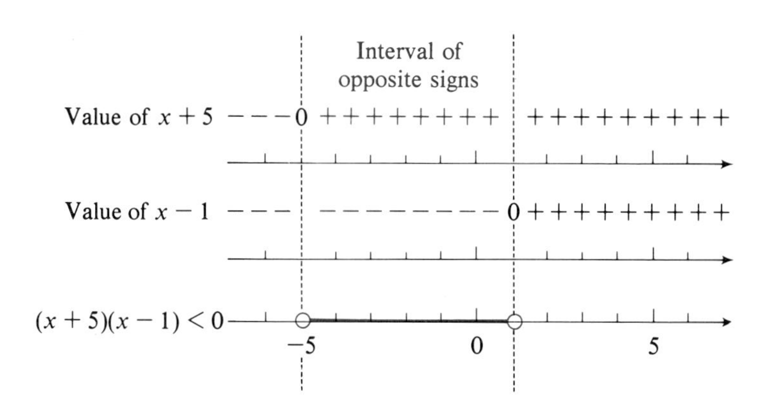

FIGURE 7.1

Critical numbers

Sign arrays such as the one shown in Figure 7.1 can be used to solve quadratic inequalities. However, solutions of such inequalities can also be obtained by using the notion of a *critical number*.

Notice in the foregoing solution process that the numbers -5 and 1, which can be obtained by solving the equation $(x + 5)(x - 1) = 0$, separate the set of real numbers into the three intervals,

$$(-\infty, -5), \qquad (-5, 1), \quad \text{and} \quad (1, +\infty),$$

which are shown in Figure 7.2. Each of these intervals either is or is not a part of the solution set of the inequality $x^2 + 4x < 5$.

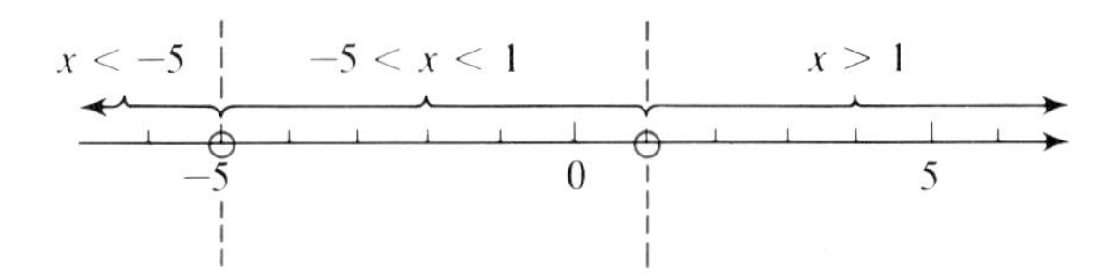

FIGURE 7.2

To determine which, we need only substitute an arbitrarily selected number from each interval and test it in the inequality. Let us use -6, 0, and 2.

$(-6)^2 + 4(-6) \overset{?}{<} 5$	$0^2 + 4(0) \overset{?}{<} 5$	$2^2 + 4(2) \overset{?}{<} 5$
$36 - 24 \overset{?}{<} 5$	$0 + \overset{?}{<} 5$	$4 + 8 \overset{?}{<} 5$
$12 \overset{?}{<} 5$	$0 \overset{?}{<} 5$	$12 \overset{?}{<} 5$
no	yes	no

The only interval involved that is in the solution set is $(-5, 1)$, and hence this interval is the solution set. The graph is shown in Figure 7.3. If the inequality in this example were $x^2 + 4x - 5 \leq 0$, the endpoints on the graph would be shown as closed dots.

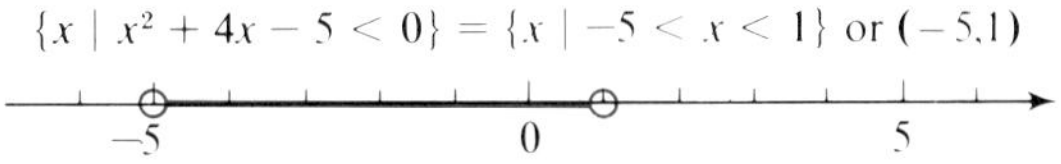

$\{x \mid x^2 + 4x - 5 < 0\} = \{x \mid -5 < x < 1\}$ or $(-5, 1)$

−5 0 5

FIGURE 7.3

Real numbers such as -5 and 1 in the preceding example are called *critical numbers*. In general,

> *The values of x for which* $P(x) = 0$ *or* $P(x)$ *is undefined are called* **critical numbers** *of* $P(x) < 0$ *or* $P(x) > 0$.

EXAMPLE 1 a. Critical numbers for $2x^2 - x - 1 > 0$ are $-\frac{1}{2}$ and 1, because

$$2x^2 - x - 1 = (2x + 1)(x - 1) = 0$$

for these values.

b. Critical numbers for $\dfrac{1}{x^2 - 4} < 0$ are 2 and -2, because

$$\frac{1}{x^2 - 4} = \frac{1}{(x - 2)(x + 2)}$$

is not defined for either number. ■

EXAMPLE 2 Solve and graph the solution set of

$$x^2 - 3x - 4 \geqslant 0. \tag{3}$$

Solution Factoring the left-hand member yields

$$(x + 1)(x - 4) \geqslant 0,$$

and we note that -1 and 4 are critical numbers because $(x + 1)(x - 4) = 0$ for these values. Hence, we check the intervals on the number line in Figure a. We substitute selected arbitrary values in each interval, say, -2, 0, and 5, for the variable in (3).

$(-2)^2 - 3(-2) - 4 \overset{?}{\geqslant} 0$	$(0)^2 - 3(0) - 4 \overset{?}{\geqslant} 0$	$(5)^2 - 3(5) - 4 \overset{?}{\geqslant} 0$
yes	no	yes

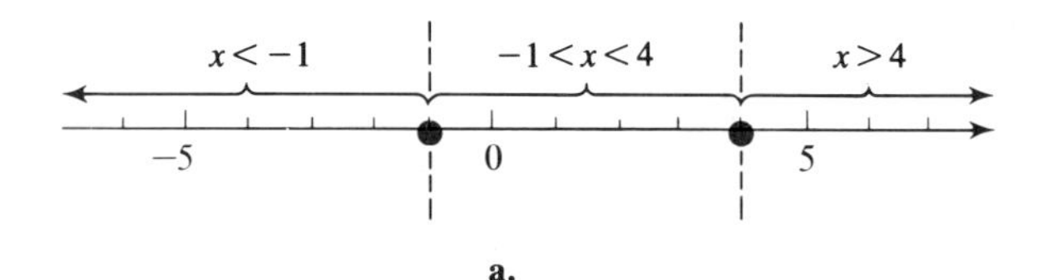

a.

Hence, the solution set is $(-\infty, -1] \cup [4, +\infty)$. The graph is shown in Figure b.

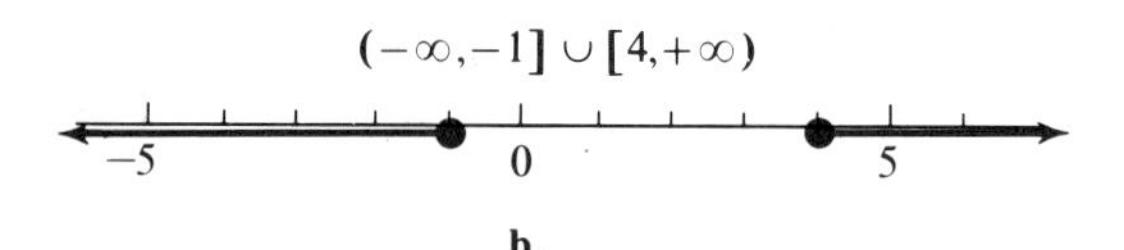

b.

■

Variable in a denominator

Inequalities involving fractions must be approached with care if any fraction contains a variable in the denominator. If each member of such an inequality is multiplied by an expression containing the variable, we must be careful either to distinguish between those values of the variable for which the expression denotes a positive and a negative number, or to make certain that the expression by which we multiply is always positive. However, we can use the notion of critical numbers to avoid these complications.

EXAMPLE 3 Solve and graph the solution set of

$$\frac{x}{x-2} \geqslant 5. \tag{4}$$

Solution We first write (4) equivalently as

$$\frac{x}{x-2} - 5 \geqslant 0,$$

from which

$$\frac{x - 5(x-2)}{x-2} \geqslant 0,$$

$$\frac{-4x + 10}{x-2} \geqslant 0. \tag{4a}$$

In this case the critical numbers are $5/2$ and 2, because

$$\frac{-4x + 10}{x-2}$$

equals 0 for $x = 5/2$ and is undefined for $x = 2$. Thus, we want to check the intervals shown on the number line in Figure a on page 262. Substituting arbitrary values for the variable in (4) or (4a) in each of the three intervals, say, 0, $9/4$, and 3, we can identify the solution set

$$(2, 5/2],$$

as we did in the example on page 260. The graph is shown in Figure b. Note that the left-hand endpoint is an open dot (2 is not a member of the solution set) because the left-hand member of (4) is undefined for $x = 2$.

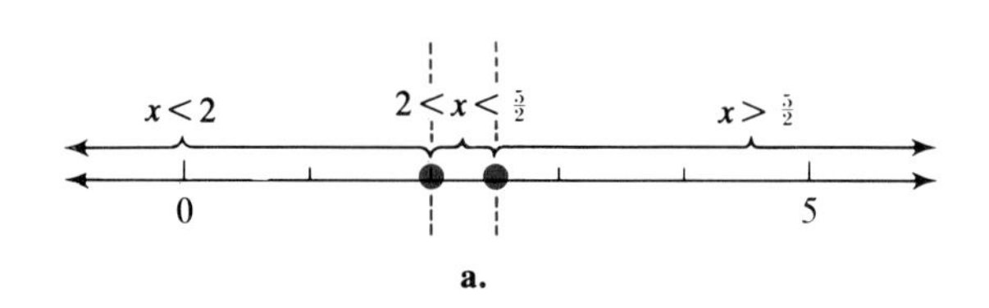

a.

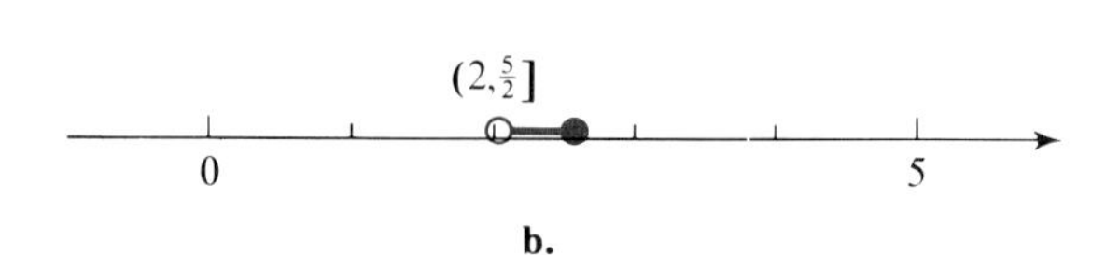

b.

■

The procedures shown in the examples above are also applicable if more than two critical numbers are associated with an inequality.

It is sometimes possible to determine the solution set of a quadratic inequality simply by inspection. For example, the solution set of

$$4x^2 + 6 > 0$$

is $\{x \mid x \text{ a real number}\}$ because the left-hand member is greater than zero for all real-number replacements of x. For the same reason, the solution set of

$$4x^2 + 6 < 0$$

is $\varnothing$.

EXERCISE 7.5

A

■ *Solve and represent each solution set on a line graph. See Examples 1 and 2.*

1. $(x + 1)(x - 2) > 0$
2. $(x - 3)(x + 2) > 0$
3. $(x + 3)(x - 4) < 0$
4. $(x + 2)(x + 5) \leq 0$
5. $x(x - 2) \leq 0$
6. $x(x + 4) < 0$
7. $x(x - 5) > 0$
8. $x(x + 3) \geq 0$
9. $x^2 - 3x - 4 > 0$
10. $x^2 - 5x - 6 \geq 0$
11. $x^2 - x - 6 \leq 0$
12. $x^2 + x - 12 < 0$
13. $y(y + 2) \leq 15$
14. $y(y - 3) < 10$
15. $y(y - 1) > 12$
16. $y(y + 3) \geq 18$
17. $x^2 < 5$
18. $x^2 \leq 7$
19. $x^2 + 5 < 0$
20. $4x^2 + 1 < 0$

■ *See Example 3.*

21. $\dfrac{x + 3}{x - 2} < 0$
22. $\dfrac{x - 1}{x + 4} > 0$
23. $\dfrac{x}{x + 2} - 4 > 0$
24. $\dfrac{x + 2}{x - 2} - 6 \geq 0$

25. $\dfrac{x-1}{x} \leqslant 3$

26. $\dfrac{x}{x-1} \geqslant 5$

27. $\dfrac{1}{x-1} \leqslant 1$

28. $\dfrac{x+1}{x-1} < 1$

B

■ *Solve and represent each solution set on a line graph.*

29. $\dfrac{4}{x+1} < \dfrac{3}{x}$

30. $\dfrac{3}{x-1} \geqslant \dfrac{1}{x}$

31. $\dfrac{2}{x-2} \geqslant \dfrac{4}{x}$

32. $\dfrac{3}{4x+1} > \dfrac{2}{x-5}$

33. $5 < x^2 + 1 < 10$

34. $-2 < y^2 - 3 < 13$

35. A projectile fired from ground level is at a height of $320t - 16t^2$ feet after t seconds. During what period of time is it higher than 1024 feet?

36. A ball thrown vertically reaches a height h in feet given by $h = 56t - 16t^2$, where t is time measured in seconds. During what period(s) of time is the ball between 40 feet and 48 feet high?

CHAPTER SUMMARY

[7.1] To solve an equation by the **extraction of roots** method, we use the fact that:

$$\text{If} \quad x^2 = b, \quad \text{then} \quad x = \sqrt{b} \quad \text{or} \quad x = -\sqrt{b}.$$

To solve an equation of the form $x^2 + bx = c$ by completing the square, we first add $({}^{b}/{}_{2})^2$ to both members.

[7.2] The solutions of $ax^2 + bx + c = 0$ $(a \neq 0)$ are given by

$$x = \frac{-b + \sqrt{b^2 - 4ac}}{2a} \quad \text{and} \quad x = \frac{-b - \sqrt{b^2 - 4ac}}{2a}.$$

The number represented by $b^2 - 4ac$ is the **discriminant** of the quadratic equation $ax^2 + bx + c = 0$.

[7.3] To solve equations containing radicals, we use the fact that if each member of an equation is raised to the same power, the solutions of the original equation are contained in the solution set of the resulting equation. That is:

For n a natural number, the solution set of $[P(x)]^n = [Q(x)]^n$ contains all the solutions of $P(x) = Q(x)$.

[7.4] An equation of the form

$$au^2 + bu + c = 0,$$

where u represents some expression in another variable, is said to be **quadratic in form**. Such equations may be solved by a method commonly called *substitution of variables*.

[7.5] Quadratic inequalities may be solved by using a sign array or by first identifying **critical numbers**.

REVIEW EXERCISES

A

[7.1]

■ *Solve for x by the extraction of roots.*

1. a. $2x^2 = 50$ b. $3x^2 + 7 = 0$
2. a. $(x + 3)^2 = 25$ b. $(x - 4)^2 = 15$
3. a. $\left(x - \frac{1}{3}\right)^2 = \frac{2}{9}$ b. $\left(x + \frac{2}{3}\right) = \frac{5}{9}$
4. a. $(2x - 3)^2 = -5$ b. $(3x + 2) = -7$

■ *Solve by completing the square.*

5. a. $x^2 - 4x - 6 = 0$ b. $2x^2 + 3x - 3 = 0$
6. a. $x^2 + 3x = 3$ b. $3x^2 = 2x + 4$
7. a. $x^2 - x + 2 = 0$ b. $x^2 + 2x + 3 = 0$
8. a. $2x^2 = 2x - 3$ b. $3x^2 = 3x - 1$

[7.2]

■ *Solve for x using the quadratic formula.*

9. a. $\frac{1}{2}x^2 + 1 = \frac{3}{2}x$ b. $x^2 - 3x + 7 = 0$
10. a. $x^2 - 3x + 1 = 0$ b. $2x^2 + x - 3 = 0$
11. a. $x^2 - x + 2 = 0$ b. $x^2 - 2x + 4 = 0$
12. a. $2x^2 + 3x + 2 = 0$ b. $2x^2 - x + 3 = 0$

[7.3]

■ *Solve.*

13. $x - 3\sqrt{x} + 2 = 0$

14. $\sqrt{x + 1} + \sqrt{x + 8} = 7$

15. $p = \sqrt{\dfrac{1 - 2t^2}{s}}$ for t

16. $R = \dfrac{1 + \sqrt{p^2 + 1}}{2}$ for p

[7.4]

■ *Solve.*

17. $x^4 - 3x^2 - 4 = 0$

18. $x - x^{1/2} = 12$

19. $y - 2\sqrt{y} - 8 = 0$

20. $y + 2 + 4\sqrt{y + 2} - 12 = 0$

[7.5]

■ *Solve each inequality and graph the solution set on a line graph.*

21. a. $x^2 - 9x > 0$
    b. $x^2 + 5x + 6 < 0$

22. a. $x(x + 2) \leqslant 8$
    b. $x(x - 3) > 4$

23. a. $\dfrac{x + 1}{x - 3} \geqslant 0$
    b. $\dfrac{x}{x + 2} \leqslant 2$

B

24. Solve $(ax - b)^2 = 4$ for x in terms of a and b.

25. Solve $2x^2 - kx + 1 = 0$ for x in terms of k.

26. Solve $2x^2 + xy + y^2 = 0$ for y in terms of x.

27. Determine k so that the solutions of $x^2 - 2x - k + 1 = 0$ are real numbers.

28. Solve $(x - 1)^{1/2} - (x - 1)^{1/4} - 6 = 0$.

■ *Solve and represent each solution on a line graph.*

29. $\dfrac{1}{x - 2} \leqslant \dfrac{2}{x}$

30. $\dfrac{2}{x + 1} > \dfrac{3}{x}$

Cumulative Review Chapters 1-7

The numbers in brackets refer to the sections where such problems are first considered.

1. Subtract the sum of $2t^2 - 7t + 3$ and $-5t^2 - 5t - 2$ from $2t^3 - 3t - 1$. [2.2]
2. Multiply $(3x^2 - 4x + 2)(2x^2 + x - 3)$. [2.4]
3. Graph $\{x \mid 3x - 1 > 8\} \cap \{x \mid -2x + 4 > -4\}$. [3.3]
4. Reduce $\dfrac{4 - 4y + y^2}{4 - y^2}$. [5.1]
5. Use synthetic division to divide $\dfrac{x^6 + 2}{x + 2}$. [5.3]
6. Express $\dfrac{3}{-2x + 4}$ as an equivalent fraction with denominator $2x^3 - 8x$. [5.4]

■ *Express as a single fraction in lowest terms.*

7. $1 + \dfrac{1}{x} - \dfrac{x + 1}{x^2 - x}$ [5.4]
8. $\left(\dfrac{1}{x - 1} - \dfrac{1}{x + 2}\right)\left(\dfrac{1}{2} + \dfrac{1}{x}\right)$ [5.4]
9. $\dfrac{\dfrac{x}{y} - \dfrac{y}{x}}{\dfrac{1}{x^2} - \dfrac{1}{y^2}}$ [5.5]
10. Write as a complex fraction, and then as a single fraction in lowest terms:

$$\left(2x - 1 - \frac{14}{x - 2}\right) \div \left(x - \frac{8}{x - 2}\right). \quad [5.5]$$

■ *Factor completely.*

11. $2(3x - 1)^2 + 6(2x + 1)(3x - 1)$ [4.1]
12. $4x^2y^2 - x^2y^4$ [4.2]
13. $9x^2 - 12x + 4 - 4y^2$ [4.3]
14. $8x^3 - 27y^3$ [4.3]

■ *Simplify.*

15. $5x - 2x[x - 3(2 - 4x)(1 + x) - x^2]$ [2.4]
16. $(2x^2)^2(xy^2) - 3xy^2(xy)^2 + (4xy)^2xy^2$ [2.3]

■ *Write each expression as a product or quotient in which each variable occurs only once, and all exponents are positive.*

17. $\left(\dfrac{a^3b}{c^2x^2}\right)^3 \left[\dfrac{bxy^4}{a^2z^3}\left(\dfrac{c^3xz}{y^2}\right)^3\right]^2$ [6.1]
18. $\left(\dfrac{3x^{-2}y^3}{z^{-1}}\right)^{-2}\left(\dfrac{2x^0z^4}{xy^{-2}}\right)^3$ [6.2]
19. $\dfrac{xy^{-1}}{x^{-1} - y^{-1}}$ [6.2]
20. $\dfrac{(4x10^3)^3(3x10^{-5})}{(2x10^{-2})^2}$ [6.3]
21. $\left(\dfrac{a^{-1/2}b^{1/3}}{c^{-5/6}}\right)^{-3}$ [6.4]
22. Simplify $(-64)^{-4/3}$. [6.4]
23. Multiply $-2x^{1/3}(x^{-2/3} - 1 + 2x^{-1/3})$. [6.4]
24. Write $2x^{-1/3}(x + y)^{2/3}$ in radical form. [6.5]
25. Find the indicated root: $\sqrt[3]{-27x^3y^{30}}$. [6.5]
26. Reduce the order of $\sqrt[9]{64}$. [6.5]
27. Factor: $\sqrt{18} - 2\sqrt{6} = \sqrt{3}\,(\underline{?} - \underline{?})$. [6.7]
28. Reduce $\dfrac{6\sqrt{xy} - \sqrt{12x^3}}{2\sqrt{x}}$. [6.7]
29. Rationalize the denominator: $\sqrt[4]{\dfrac{3}{8x^2}}$. [6.6]
30. Rationalize the denominator: $\dfrac{\sqrt{x} + 2}{1 - 3\sqrt{x}}$. [6.7]

■ *Write in the form $a + bi$:*

31. $\dfrac{-2 - i}{4 - 3i}$ [6.8]
32. $(2 - \sqrt{-3})(-1 + \sqrt{-3})$ [6.8]

■ *Solve.*

33. $(2x - 1)(x - 3) - 4x^2 = -2[(x + 4)^2 - x]$ [3.1]
34. $|5x - 7| \geq 6$ [3.4]
35. $(x - 4)(x + 3) = -1 - (x + 1)(x - 3)$ [4.4]
36. $4 - \dfrac{3}{x - 2} = \dfrac{15}{x}$ [5.6]
37. $x^2 + 3 = 2x$ [6.8]

38. Solve by extracting roots: $(3x - 5)^2 = -18$. [7.1]

39. Solve by completing the square: $3x^2 = 2x - 1$. [7.1]

40. Solve by using the quadratic formula: $x = \dfrac{-2}{3x - 4}$. [7.2]

41. Solve for d: $C = \dfrac{s}{1 - d} + \dfrac{t}{1 - e}$. [5.6]

■ *Solve.*

42. $\sqrt[3]{x - 2} = -2$ [7.3]

43. $(x - 5)^{1/2} + (x + 7)^{1/2} = 6$ [7.3]

44. $x^{2/3} - 7x^{1/3} = 8$ [7.4]

45. $x(x + 3) > 4$ [7.5]

46. Solve for Q: $A = \sqrt[3]{\dfrac{-b + Q}{2}}$. [7.3]

47. Use the discriminant to classify the solutions of $8x^2 - 13x + 4$. [7.2]

48. Tulip bulbs cost \$0.69 apiece, and daffodils cost \$0.89. How many of each should be included in a package of 50 bulbs which sells for \$40.50? [3.2]

49. To be eligible for the second prize in his company's sales competition, a salesperson must average between \$80 and \$120 a week in commissions for five weeks. If Willie has made \$92, \$65, \$103, and \$79 in commissions over the past four weeks, how much must he make next week to be eligible for second prize? [3.3]

50. One leg of a right triangle is 3 feet shorter than the hypotenuse. If the other leg is 9 feet long, find the length of the hypotenuse. [4.5]

51. If a car travels 504 miles on 18 gallons of gas, how much gas will be needed for a 602-mile trip? [5.9]

52. A balloonist travels 120 miles against a head wind of 8 miles per hour and returns with a tail wind of 8 mph. The round trip takes 8 hours. What is the speed of the balloon in still air? [5.9]

B

53. If $P(x) = x - x^3$ and $Q(x) = 1 + x + x^2$, find $P[Q(-3)]$. [2.1]

54. Multiply $(a^{3n} - b^n)(2a^{2n} + 2b^{2n})$. [2.4]

55. Divide $\dfrac{y^5 - 1}{y^2 - 2y + 1}$. [5.2]

56. If $P(x) = \dfrac{x^2 - 1}{x}$ and $Q(x) = \dfrac{(x - 1)^2}{x^2}$, find $P(x) \div Q(x)$. [5.4]

■ *Factor.*

57. $x^n + 2x^{n+2} + x^{2n}$ [4.1]

58. $4x^4 - 4x^2y^2 - 35y^4$ [4.2]

59. $(x + 1)^3 - 8$ [4.3]

60. Write as a product or quotient in which each variable occurs only once: $\frac{(x^{n+4}y^{2-n})^3}{x^{2n}(y^n)^{n+1}}$. [6.1]

61. Compute $\frac{48{,}000 \times 0.000072}{270 \times 0.064 \times 0.5}$. [6.3]

62. For what values of x does $\sqrt{1 - 2x + x^2}$ equal $1 - x$? [6.5]

63. Simplify $\sqrt[3]{\frac{a^6 - 3a^4}{(a + 2)^2}}$. [6.6]

64. Rationalize the denominator: $\frac{1 - \sqrt{x - 1}}{1 + \sqrt{x - 1}}$. [6.7]

65. Write i^{-5} with a positive exponent, and simplify. [6.8]

66. Solve for x in terms of y: $2x^2 - 2xy + y^2 = 4$. [7.2]

67. Solve for v in terms of E, m, and P: $E = \frac{1}{2}mv^2 + P$. [7.1]

68. Solve $(x - 1)^6 + 4(x - 1)^3 - 32 = 0$. [7.4]

69. Solve $\frac{3}{x - 1} \leqslant \frac{2}{x + 3}$. [7.5]

70. The altitude of an isosceles triangle is 12 inches. Find the base if the perimeter is 36 inches. [7.3]

351

8 Equations and Inequalities in Two Variables

8.1

SOLUTION OF AN EQUATION IN TWO VARIABLES

Ordered pairs

An equation in two variables, such as $y = 2x + 3$, is said to be *satisfied* if the variables are replaced with a pair of numbers—one from the replacement set of x and one from the replacement set of y—that make the resulting statement true. The pair of numbers, usually written in the form (x, y), is a **solution** of the equation or inequality.

The pair (x, y) is called an **ordered pair**, because it is understood that the numbers are considered in a particular order, x first and y second. These numbers are called the **first** and **second components** of the ordered pair, respectively. Although any letters may be used as variables, in this book we shall almost always use x and y. Note that parentheses have previously been used to denote an open interval. The intended meaning should be clear from the context in which it is used.

To find ordered pairs that are solutions of a given equation, we can assign any real-number value to one of the variables and then determine the related value, if any, of the other. For example, for

$$y - x = 1,$$

we can obtain solutions (ordered pairs) by assigning to x any real number as a value and then determining the corresponding value of y. For example, substituting 2, 3, and 4 for x, we have

$$y - (2) = 1, \quad \text{from which} \quad y = 3;$$
$$y - (3) = 1, \quad \text{from which} \quad y = 4;$$
$$y - (4) = 1, \quad \text{from which} \quad y = 5.$$

Thus, (2, 3), (3, 4), and (4, 5) are three solutions of $y - x = 1$. There is, of course, an infinite number of solutions to this equation.

EXAMPLE 1 Find each missing component so that each ordered pair is a solution of the equation $y - 2x = 4$.

a. $(0, \underline{?})$ b. $(\underline{?}, 0)$ c. $(3, \underline{?})$

Solutions

a. $y - 2x = 4$
$y - 2(0) = 4$
$y = 4$
$(0, 4)$

b. $y - 2x = 4$
$(0) - 2x = 4$
$x = -2$
$(-2, 0)$

c. $y - 2x = 4$
$y - 2(3) = 4$
$y = 10$
$(3, 10)$ ■

EXAMPLE 2 List the ordered pairs that satisfy the equation $y = 3x + 1$ and have the x-components 1, 2, and 3.

Solution

For $x = 1$,
$y = 3(1) + 1$
$= 4.$

For $x = 2$,
$y = 3(2) + 1$
$= 7$

For $x = 3$,
$y = 3(3) + 1$
$= 10$

Thus, the desired solutions are (1, 4), (2, 7), and (3, 10). ■

Equivalent equations

In finding ordered pairs that satisfy a given equation, it is sometimes helpful if one variable is first expressed in terms of the other. The properties in Section 2.2 that are used to generate equivalent equations in one variable apply also in transforming equations in two or more variables. For example, we can transform the equation

$$y - 3x = 4 \qquad (1)$$

to the equivalent equation

$$y = 3x + 4. \qquad (2)$$

Both equations have the same solution set. In (1) the variables x and y are said to be **implicitly** related, while in (2), y is said to be expressed **explicitly** in terms of x.

EXAMPLE 3 Transform each equation into an equation in which y is expressed explicitly in terms of x, and then find values of y for the given values of x.

a. $3y - xy = 4;\quad 2, 5$

b. $x^2 + 4y^2 = 5;\quad 0, 1$

Solutions

a.
$$3y - xy = 4$$
$$y(3 - x) = 4$$
$$y = \frac{4}{3 - x}$$

When $x = 2$, $y = \dfrac{4}{3 - (2)} = 4$;

when $x = 5$, $y = \dfrac{4}{3 - (5)} = -2$.

b.
$$x^2 + 4y^2 = 5$$
$$4y^2 = 5 - x^2$$
$$y^2 = \frac{1}{4}(5 - x^2)$$
$$y = \pm\frac{1}{2}\sqrt{5 - x^2}$$

When $x = 0$, $y = \pm\frac{1}{2}\sqrt{5}$;

when $x = 1$, $y = \pm 1$. ■

EXERCISE 8.1

A

■ *Find each missing component so that each ordered pair is a solution of the equation. See Example 1.*

1. $y = x + 7$ a. $(0, \underline{?})$ b. $(2, \underline{?})$ c. $(-2, \underline{?})$
2. $y = 6 - 2x$ a. $(0, \underline{?})$ b. $(\underline{?}, 0)$ c. $(-1, \underline{?})$
3. $3x - 4y = 6$ a. $(0, \underline{?})$ b. $(\underline{?}, 0)$ c. $(-5, \underline{?})$
4. $x + 2y = 5$ a. $(0, \underline{?})$ b. $(5, \underline{?})$ c. $(-3, \underline{?})$

■ *List the ordered pairs that satisfy each equation and have the given x-components. See Example 2.*

5. $y = x - 4;\quad -3, 0, 3$
6. $y = 2x + 6;\quad -2, 0, 2$
7. $y = \dfrac{3}{x + 2};\quad 1, 2, 3$
8. $y = \dfrac{4x}{x^2 - 1};\quad 0, 2, 4$
9. $y = \sqrt{x^2 - 1};\quad 1, 2, 3$
10. $y = \dfrac{1}{2}\sqrt{4 - x^2};\quad 0, 1, 2$

■ *Transform each equation into one in which y is expressed explicitly in terms of x, and then find values for y for the given values of x. See Example 3.*

11. $2x + y = 6$; 2, 4
12. $4x - y = 2$; $-2, -4$
13. $xy - x = 2$; $-2, 2$
14. $3x - xy = 6$; 1, 3
15. $xy - y = 4$; 4, 8
16. $x^2y - xy = -5$; 2, 4
17. $x^2y - 4y = xy + 2$; $-1, 1$
18. $x^2y - xy + 3 = 5y$; $-2, 2$
19. $4 = \dfrac{x}{y^2 - 2}$; $-1, 3$
20. $3 = \dfrac{x}{y^2 + 1}$; $-2, 1$
21. $3x^2 - 4y^2 = 4$; 2, 3
22. $5x^2 - 4y^2 = 2$; 2, 4

B

■ *Find the solution of each equation for the specified component.*

23. $y = |x| - 3$, for $x = -4$
24. $y = |x| + 5$, for $x = -2$
25. $y = |x - 1| + |x|$, for $x = -1$
26. $y = |2x + 1| - |x|$, for $x = -3$
27. $y = 2^x$, for $x = 3$
28. $y = 2^x$, for $x = -3$
29. $y = 2^{-x}$, for $x = 3$
30. $y = 2^{-x}$, for $x = -3$
31. Find k if (1, 2) is a solution of $x + ky = 8$.
32. Find k if $(-6, 1)$ is a solution of $kx - 2y = 6$.

8.2

GRAPHS OF LINEAR EQUATIONS

Cartesian coordinate system

The graph of all ordered pairs that can be formed with real number components, an infinite set of points, is the entire geometric plane. We shall assume that the ordered pairs (x, y) can be placed in a one-to-one correspondence with the geometric points in the plane by any arbitrary scaling of the axes of the familiar **Cartesian** (or **rectangular**) **coordinate system** shown in Figure 8.1. That is, for each point in the plane, there corresponds a unique ordered pair and vice versa.

This correspondence is established by using the first component, x, of the ordered pair to denote the directed (perpendicular) distance of the point from the vertical axis (the **abscissa** of the point); x is positive if it falls to the right of the vertical axis and negative if it falls to the left. The second component, y, of the ordered pair is used to denote the directed (perpendicular) distance of the point from the horizontal axis (the **ordinate** of the point); y is positive if it falls above the horizontal axis and negative if it falls below.

Each axis is labeled with the variable it represents. Most commonly, the horizontal axis is called the **x-axis** and the vertical axis is called the **y-axis**, and their point of intersection is called the **origin**. The components of the ordered

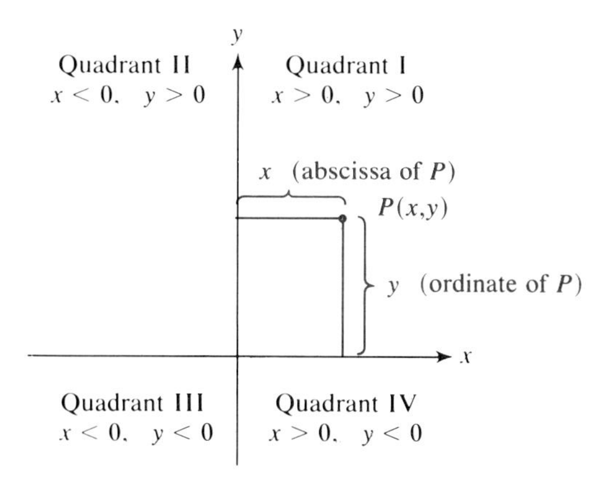

FIGURE 8.1

pair that corresponds to a given point are called the **coordinates** of the point, and the point is called the **graph** of the ordered pair. Each of the four regions into which the axes divide the plane is called a **quadrant**, and they are referred to by Roman numerals, as illustrated in Figure 8.1.

Graphs of linear equations

First-degree equations in two variables have infinitely many solutions that are ordered pairs of real numbers. Hence, these solutions can be displayed on a Cartesian coordinate system. For example, two solutions of

$$y = x + 2$$

are (1, 3) and (2, 4). The graphs of these two points are shown in Figure 8.2a. Figure 8.2b shows the location of the graphs of three additional solutions, namely, $(-2, 0)$, $(-1, 1)$, and (0, 2).

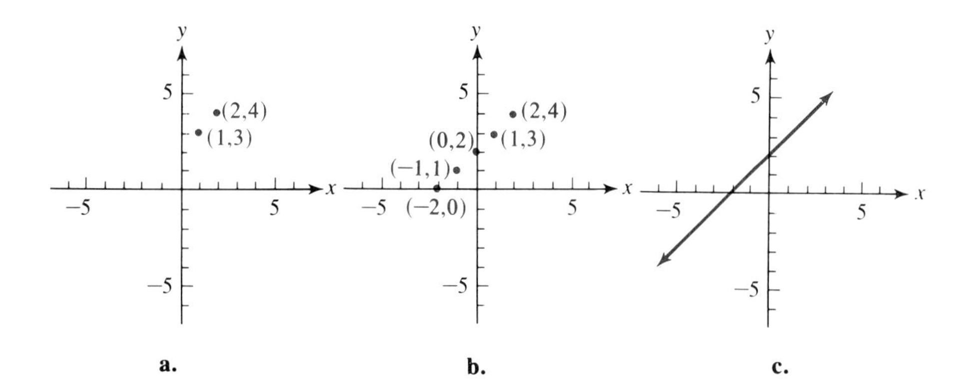

FIGURE 8.2

In general, we are interested in the graphs of equations in two variables over the set of all real numbers x for which y is a real number. The graph of

$y = x + 2$ for all real numbers x (an infinite set of points) is shown in Figure 8.2c. It can be shown that the coordinates of each point on the line in part c satisfy $y = x + 2$ and, conversely, that every solution of the equation corresponds to a point on the line. The line is referred to as the **graph of the solution set of the equation** or as the **graph of the equation.**

Although we will not prove it here, any first-degree equation in two variables—that is, any equation that can be written equivalently in the form

$$ax + by + c = 0 \qquad (a \text{ and } b \text{ not both } 0),$$

where a, b, and c are real numbers—has a graph that is a straight line. For this reason, such equations are often called **linear equations.** Since any two distinct points determine a straight line, it is evident that two solutions of such an equation determine its graph.

Intercept method of graphing

In practice, the two solutions that are easiest to find are those with second and first components, respectively, equal to 0—that is, the solutions $(x_1, 0)$ and $(0, y_1)$. Since these two points are the points where the graph intersects the x- and y-axes, they are easy to locate. The numbers x_1 and y_1 are, respectively, called the ***x*- and *y*-intercepts** of the graph. To find the x-intercept, substitute 0 for y in the equation and solve for x; to find the y-intercept, substitute 0 for x and solve for y.

EXAMPLE 1 Graph $3x + 4y = 12$.

Solution If $y = 0$, then

$$\begin{aligned} 3x + 4(0) &= 12, \\ 3x &= 12, \\ x &= 4. \end{aligned}$$

The x-intercept is 4 with coordinates (4, 0).

If $x = 0$, then

$$\begin{aligned} 3(0) + 4y &= 12, \\ 4y &= 12, \\ y &= 3. \end{aligned}$$

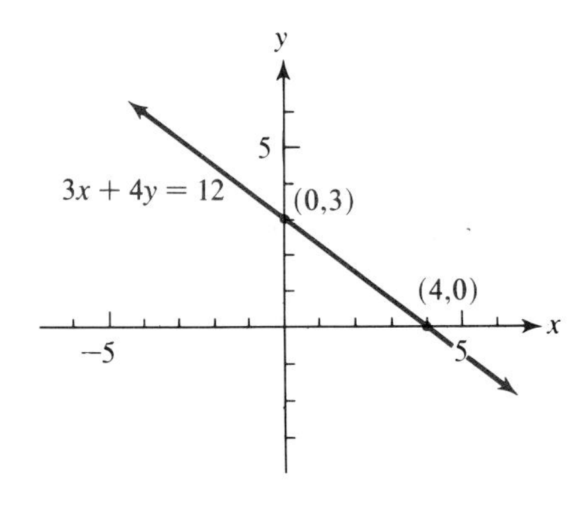

The y-intercept is 3 with coordinates (0, 3). Thus, the graph of the equation appears as shown. ■

EXAMPLE 2 Graph $3x - 4y = 24$.

Solution If $x = 0$, then

$$3(0) - 4y = 24$$
$$y = -6.$$

If $y = 0$, then

$$3x - 4(0) = 24$$
$$x = 8.$$

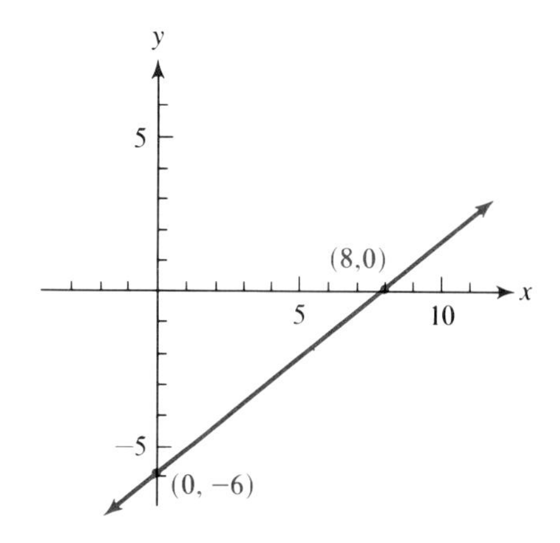

Hence, -6 is the y-intercept and 8 is the x-intercept. Graph these intercepts and draw a line through the points. ■

If the graph intersects the axes at or near the origin, the intercepts either do not represent two separate points, or the points are too close together to be of much use in drawing the graph. It is then necessary to plot *at least* one other point at a distance far enough removed from the origin to establish the line with accuracy.

EXAMPLE 3 Graph $y = 3x$.

Solution If $x = 0$, then $y = 0$, and both intercepts of the graph are at the point (0, 0). Assigning any other replacement for x, say, 2, we obtain a second solution (2, 6). We first graph the ordered pairs (0, 0) and (2, 6) and then complete the graph, as shown.

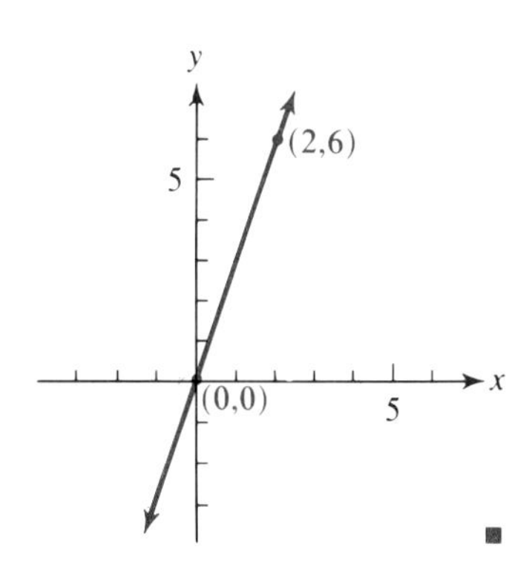

■

Special cases of linear equations

There are two special cases of linear equations worth noting. First, an equation such as

$$y = 4$$

can be considered an equation in two variables,

$$0x + y = 4.$$

For each x, this equation assigns $y = 4$. That is, any ordered pair of the form $(x, 4)$ is a solution of the equation. For example,

$$(-1, 4), \quad (2, 4), \quad \text{and} \quad (4, 4)$$

are all solutions of the equation. If we draw a straight line through the graphs of these points, we obtain the graph shown in Figure 8.3.

The other special case of a linear equation is of the type

$$x = 3,$$

which may be looked upon as an equation in two variables,

$$x + 0y = 3.$$

Here, only one value is permissible for x, namely, 3, while any value may be assigned to y. That is, any ordered pair of the form $(3, y)$ is a solution of this equation. If we choose two solutions, say, $(3, 1)$ and $(3, 3)$, and draw a straight line through the graphs of these points, we have the graph of the equation shown in Figure 8.4.

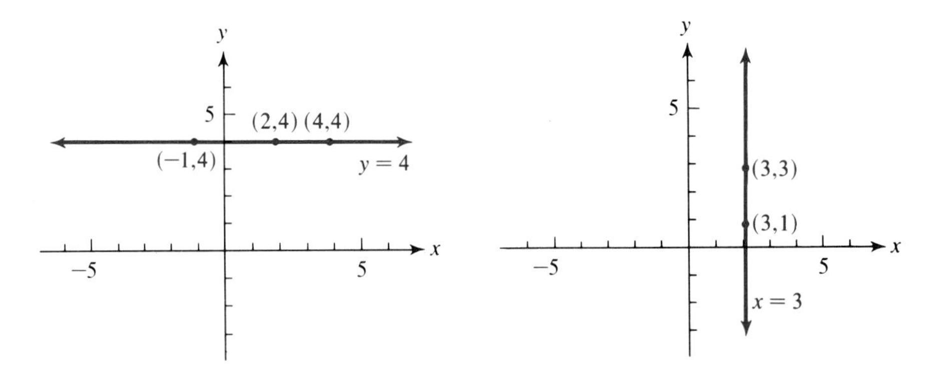

FIGURE 8.3

FIGURE 8.4

If k represents a constant (real number), then, in general, the graph of $y = k$ is a horizontal line and the graph of $x = k$ is a vertical line.

EXAMPLE 4 a. Graph $y = 2$. b. Graph $x = 4$.

Solutions a.

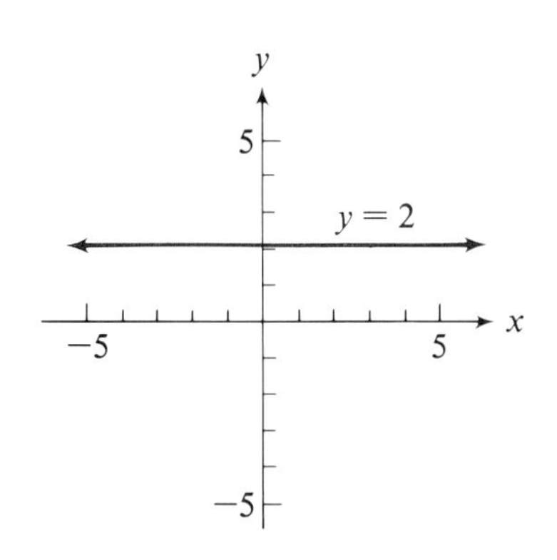

b.

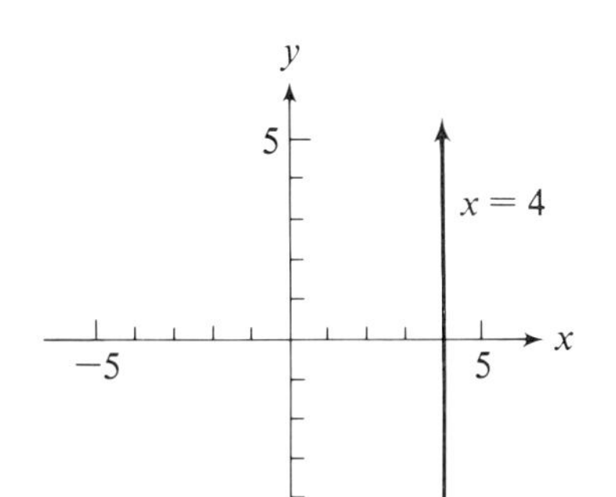

■

EXERCISE 8.2

A

■ *Graph each equation. See Examples 1 and 2.*

1. $y = x - 5$
2. $y = x + 3$
3. $y = 3x + 6$
4. $y = 4x - 8$
5. $x + 2y = 8$
6. $2x - y = 6$
7. $3x - 4y = 12$
8. $2x + 6y = 6$
9. $6x = y + 5$
10. $4x = y - 6$
11. $2x = 3y - 4$
12. $3x = 4y + 6$

■ *Graph each equation. See Example 3.*

13. $2x - y = 0$
14. $x + 3y = 0$
15. $y = -3x$
16. $x = 2y$
17. $4x - y = 0$
18. $4x + y = 0$
19. $x + y = 0$
20. $x - y = 0$

■ *Graph each equation. See Example 4.*

21. $y = -3$
22. $x = -2$
23. $2x = 8$
24. $3y = 15$
25. $x = 0$
26. $y = 0$

B

■ *Graph each pair of equations on the same coordinate system. Estimate the coordinates of each point of intersection.*

27. $y = 2x + 4$ and $y = x - 2$
28. $x - 3y = 6$ and $x + 2y = 4$
29. $y = x - 5$ and $x = 2$
30. $y = 2x - 4$ and $y = -3$

8.3

DISTANCE AND SLOPE FORMULAS

Distance formula

Any two distinct points in a plane can be looked upon as the endpoints of a line segment. We shall discuss two fundamental properties of a line segment—its length and its slope with respect to the x-axis. We first observe that any two distinct points P_1 and P_2 either lie on the same vertical line or one is to the right of the other; the point on the right has a greater x-coordinate than the point on the left. If we construct a line parallel to the y-axis through P_2 and a line parallel to the x-axis through P_1, the lines will meet at a point P_3, as shown in either part a or part b of Figure 8.5.

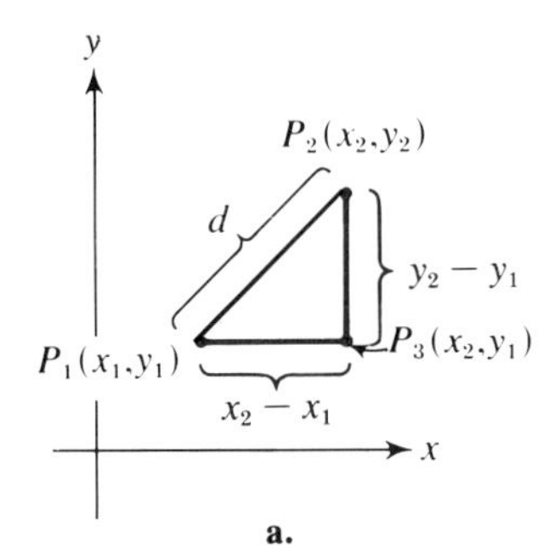

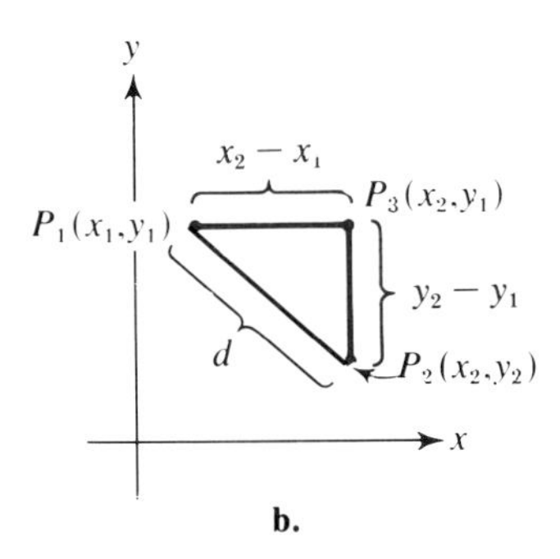

FIGURE 8.5

The x-coordinate of P_3 is evidently the same as the x-coordinate of P_2, while the y-coordinate of P_3 is the same as that of P_1; hence, the coordinates of P_3 are (x_2, y_1). By inspection, we observe that the distance between P_2 and P_3 is $(y_2 - y_1)$ and the distance between P_1 and P_3 is $(x_2 - x_1)$.

The **Pythagorean theorem** may be used to find the length of the line segment joining P_1 and P_2. This theorem asserts that the square of the length of the hypotenuse of any right triangle is equal to the sum of the squares of the lengths of the legs. Thus, referring to Figure 8.5, we note that

$$d^2 = (x_2 - x_1)^2 + (y_2 - y_1)^2.$$

If we consider only the positive square root of the right-hand member, we have the **distance formula.**

$$d = \sqrt{(x_2 - x_1)^2 + (y_2 - y_1)^2}. \quad (1)$$

EXAMPLE 1 Find the distance between (2, −1) and (4, 3).

Solution Substituting (2, −1) for $P_1(x_1, y_1)$ and (4, 3) for $P_2(x_2, y_2)$ in the distance formula, we obtain

$$\begin{aligned} d &= \sqrt{(x_2 - x_1)^2 + (y_2 - y_1)^2} \\ &= \sqrt{[4 - 2]^2 + [3 - (-1)]^2} \\ &= \sqrt{4 + 16} \\ &= \sqrt{20} = 2\sqrt{5}. \quad ■ \end{aligned}$$

Notice that in the above example, we would obtain the same answer if we used (4, 3) for P_1 and (2, −1) for P_2:

$$\begin{aligned} d &= \sqrt{[2 - 4]^2 + [(-1) - 3]^2} \\ &= \sqrt{4 + 16} = 2\sqrt{5}. \end{aligned}$$

If the points P_1 and P_2 lie on the same horizontal line ($y_2 = y_1$) and if we are concerned only with distance and not direction, then

$$d = \sqrt{(x_2 - x_1)^2 + 0^2} = |x_2 - x_1|.$$

If they lie on the same vertical line ($x_2 = x_1$), then

$$d = \sqrt{0^2 + (y_2 - y_1)^2} = |y_2 - y_1|.$$

EXAMPLE 2 a. The points (6, 2) and (−3, 2) lie on the same horizontal line, namely, $y = 2$. The distance between the points is

$$|x_2 - x_1| = |6 - (-3)| = 9.$$

b. The points $(4, -5)$ and $(4, -2)$ lie on the same vertical line, namely, $x = 4$. The distance between the points is

$$|y_2 - y_1| = |(-5) - (-2)| = 3. \quad \blacksquare$$

Slope formula

A second useful property of a line segment that joins two points is its orientation in the plane. This property can be measured by comparing the change in y values with the change in x values, as shown in Figure 8.6.

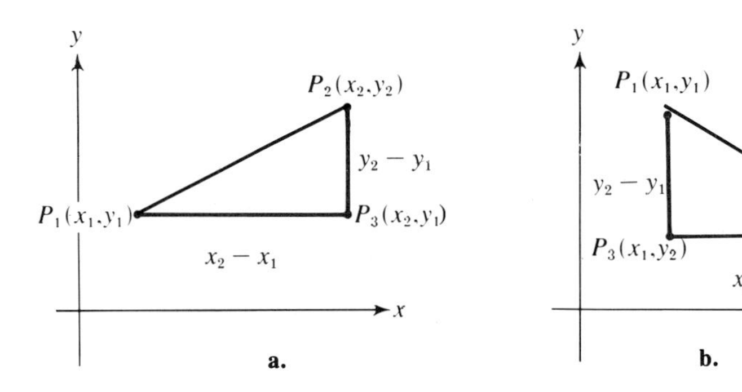

FIGURE 8.6

The ratio of the change in y values to the change in x values is the **slope** of the segment and is designated by the letter m. Thus, since the change in y values is $(y_2 - y_1)$ and the change in x values is $(x_2 - x_1)$, the slope of the segment joining P_1 and P_2 is given by the following formula.

$$\textbf{slope:} \quad m = \frac{y_2 - y_1}{x_2 - x_1} \quad (x_2 \neq x_1). \tag{2}$$

Taking P_2 to the right of P_1, the "run" $(x_2 - x_1)$ is positive and the slope is positive or negative as the "rise" $(y_2 - y_1)$ is positive or negative—that is, according to whether the line slopes upward or downward from P_1 to P_2. A positive slope indicates that a line is rising to the right; a negative slope indicates that it is falling to the right. Since

$$\frac{y_2 - y_1}{x_2 - x_1} = \frac{-(y_1 - y_2)}{-(x_1 - x_2)} = \frac{y_1 - y_2}{x_1 - x_2},$$

the restriction that P_2 be to the right of P_1 is not necessary, and the order in which the points are considered is immaterial.

EXAMPLE 3 Find the slope of the line segment joining the points $(2, -1)$ and $(4, 3)$.

Solution Let $(2, -1)$ be (x_1, y_1) and $(4, 3)$ be (x_2, y_2). Then

$$m = \frac{y_2 - y_1}{x_2 - x_1} = \frac{3 - (-1)}{4 - 2}$$
$$= \frac{4}{2} = 2. \quad \blacksquare$$

Several lines with different slopes are shown in Figure 8.7.

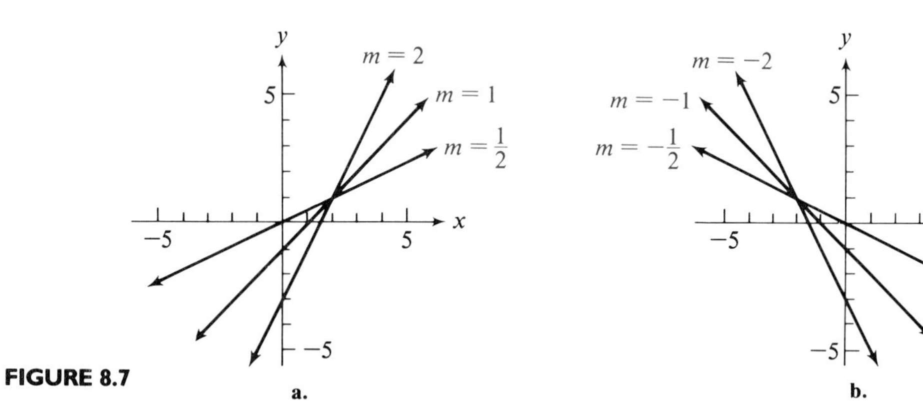

FIGURE 8.7

If a segment is parallel to the x-axis, as shown in Figure 8.8a, then $y_2 - y_1 = 0$ and it will have a slope of 0. If a segment is parallel to the y-axis, as in Figure 8.8b, $x_2 - x_1 = 0$ and its slope is not defined.

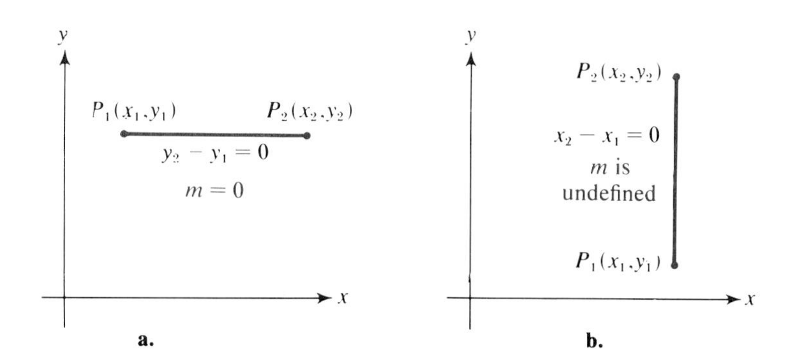

FIGURE 8.8

Parallel and perpendicular lines

The following two properties of line segments are important in mathematics and applications of mathematics.

> *Two line segments with slopes m_1 and m_2 are:*
>
> *Parallel if* $\mathbf{m_1 = m_2}$;
>
> *Perpendicular if* $\mathbf{m_1 m_2 = -1}$

EXAMPLE 4 a. In Figure a, line segment AB with slope

$$m_1 = \frac{5-3}{5-2} = \frac{2}{3}$$

is parallel to line segment CD with slope

$$m_2 = \frac{1-(-1)}{4-1} = \frac{2}{3}.$$

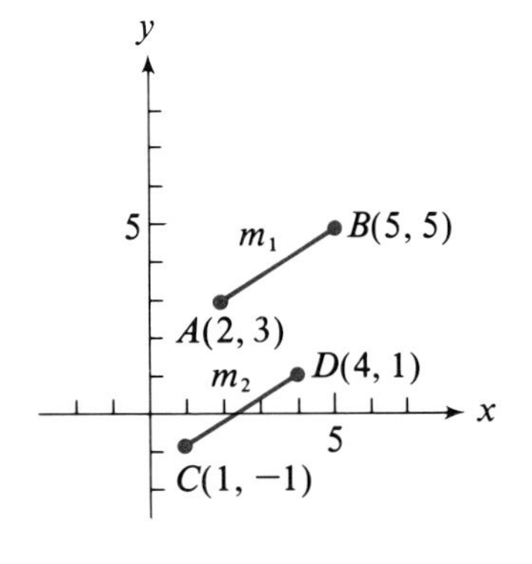

a.

b. In Figure b, line segment DE with slope

$$m_2 = \frac{1-7}{4-0} = \frac{-6}{4} = \frac{-3}{2}$$

is perpendicular to line segment AB with slope

$$m_1 = \frac{5-3}{5-2} = \frac{2}{3},$$

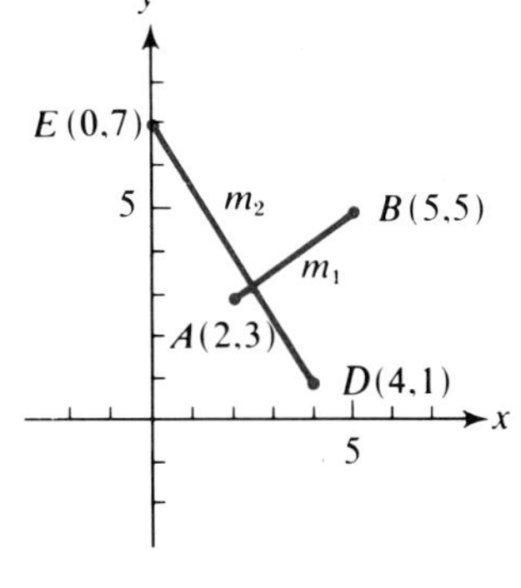

b.

because

$$m_1 m_2 = \frac{2}{3}\left(\frac{-3}{2}\right) = -1. \quad ■$$

EXERCISE 8.3

A

■ *Find the distance between each of the given pairs of points, and find the slope of the line segment joining them. Sketch each line segment in the coordinate plane. See Examples 1–3.*

1. (1, 1), (4, 5)
2. (−1, 1), (5, 9)
3. (−3, 2), (2, 14)
4. (−4, −3), (1, 9)
5. (2, 1), (4, 0)
6. (−3, 2), (0, 0)
7. (5, −4), (−1, 1)
8. (2, −3), (−2, −1)
9. (3, 5), (−2, 5)
10. (2, 0), (−2, 0)
11. (0, 5), (0, −5)
12. (−2, −5), (−2, 3)

■ *Use the distance formula to find the perimeter of the triangle whose vertices are given. Sketch each triangle in the coordinate plane.*

13. (0, 6), (9, −6), (−3, 0)
14. (10, 1), (3, 1), (5, 9)
15. (5, 6), (11, −2), (−10, −2)
16. (−1, 5), (8, −7), (4, 1)

■ *See Example 4 for Problems 17–20.*

17. Show that the two line segments whose endpoints are (5, 4), (3, 0) and (−1, 8), (−4, 2) are parallel.
18. Show that the two line segments whose endpoints are (−4, 2), (2, −2) and (3, 0), (−3, 4) are parallel.
19. Show that the two line segments whose endpoints are (0, −7), (8, −5) and (5, 7), (8, −5) are perpendicular.
20. Show that the two line segments whose endpoints are (8,0), (6, 6) and (−3, 3), (6, 6) are perpendicular.

■ *Specify the slope of each line.*

21. a. l_1 b. l_2

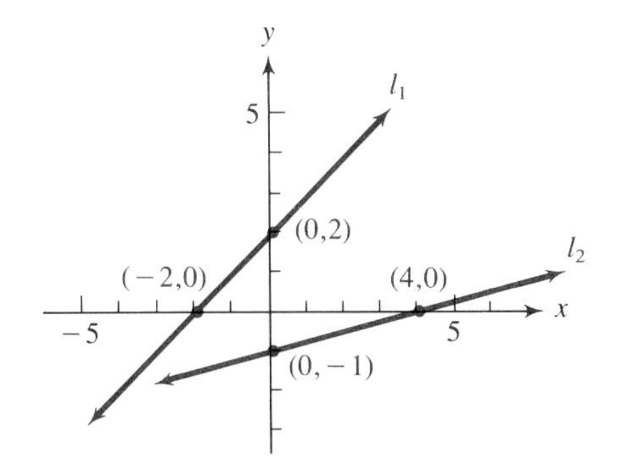

22. a. l_1 b. l_2

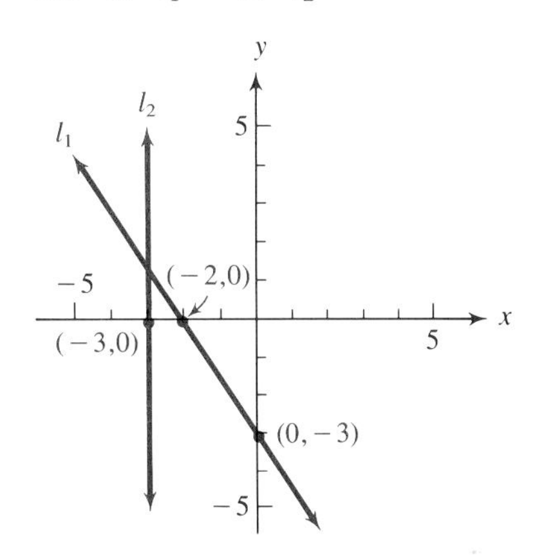

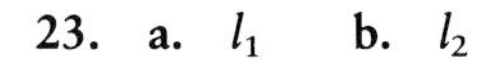

23. a. l_1 b. l_2

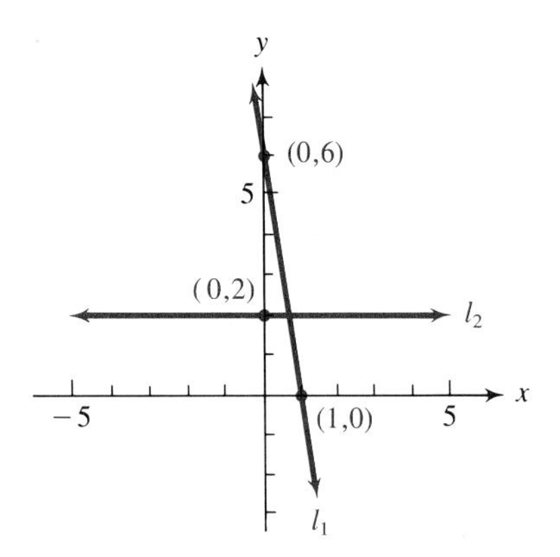

24. a. l_1 b. l_2

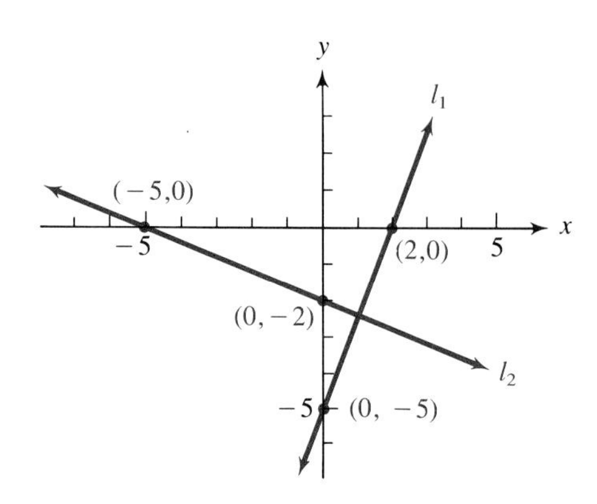

B

25. Show that the triangle described in Problem 13 is a right triangle. [*Hint:* Use the converse of the Pythagorean theorem—that is, if $c^2 = a^2 + b^2$, the triangle is a right triangle. Alternatively, show that two sides are perpendicular.]
26. Show that the triangle with vertices at (0, 0), (6, 0), and (3, 3) is a right isosceles triangle—that is, a right triangle with two sides that have the same length.
27. Show that the points (2, 4), (3, 8), (5, 1), and (4, −3) are the vertices of a parallelogram. [*Hint:* A four-sided figure is a parallelogram if the opposite sides are parallel.]
28. Show that the points (−5, 4), (7, −11), (12, 25), and (0, 40) are the vertices of a parallelogram.
29. Given the points $P_1(4, -1)$, $P_2(2, 7)$, and $P_3(-3, 4)$, find the value of k in the ordered pair $P_4(5, k)$ that makes P_1P_2 parallel to P_3P_4.
30. Using the points in Problem 29, find the k that makes P_1P_2 perpendicular to P_3P_4.

8.4

FORMS OF LINEAR EQUATIONS

Let us designate

$$\mathbf{ax + by + c = 0} \quad \text{or} \quad \mathbf{ax + by = c} \tag{1}$$

as **standard form** for a linear equation. We shall now consider two alternative forms that display useful aspects.

Point-slope form

Assuming that the slope of the line segment joining any two points on a line does not depend upon the points, consider a line on the plane with given slope m and passing through a given point (x_1, y_1), as shown in Figure 8.9. If we choose any other point on the line and assign to it the coordinates (x, y), the slope of the line is given by

$$\frac{y - y_1}{x - x_1} = m \qquad (x \neq x_1), \tag{2}$$

from which

$$y - y_1 = m(x - x_1) \qquad (x \neq x_1). \tag{3}$$

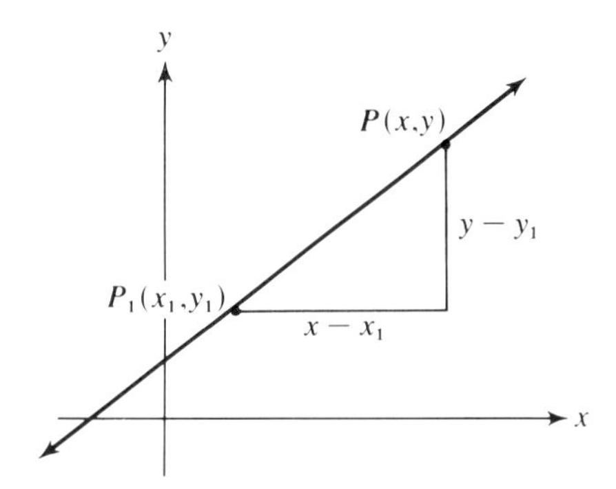

FIGURE 8.9

The ordered pair (x_1, y_1) corresponds to a point on the line, and the coordinates of (x_1, y_1) also satisfy (3). Hence, the graph of (3) contains all points in the graph of (2) and, in addition, the point (x_1, y_1) and Equation (4) holds for all values of x.

$$\mathbf{y - y_1 = m(x - x_1)} \tag{4}$$

Equation (4) is called the **point-slope form** for a linear equation.

EXAMPLE 1 Find an equation of the line that goes through the point $(1, -4)$ with slope $-3/4$.

Solution Substituting -4 for y_1, 1 for x_1, and $-3/4$ for m in the point-slope form for a linear equation, we obtain

$$y - (-4) = -\frac{3}{4}(x - 1).$$

To change the equation to standard form we multiply each member by 4 to obtain

$$4y + 16 = -3(x - 1),$$
$$4y + 16 = -3x + 3,$$
$$3x + 4y + 13 = 0. \quad ■$$

We can now find an equation of the line whose graph includes two given points. We first use the slope formula developed in Section 8.3 and then use the point-slope formula with either of the two given points.

EXAMPLE 2 Find an equation of the line whose graph includes the points (2, 2) and (−4, 1).

Solution First find the slope, selecting either point as (x_1, y_1) and the other point as (x_2, y_2).

$$m = \frac{y_2 - y_1}{x_2 - x_1} = \frac{1 - 2}{-4 - 2}$$
$$= \frac{-1}{-6} = \frac{1}{6}.$$

Then use the point-slope formula with either ordered pair. Using (2, 2) for (x_1, y_1) in the formula $y - y_1 = m(x - x_1)$ yields

$$y - 2 = \frac{1}{6}(x - 2), \quad (5)$$

from which

$$6y - 12 = x - 2,$$
$$x - 6y = -10. \quad ■$$

Two-point formula

Note that we can substitute $\dfrac{y_2 - y_1}{x_2 - x_1}$ for m in the point-slope formula to obtain Equation (6).

$$y - y_1 = \frac{y_2 - y_1}{x_2 - x_1}(x - x_1) \tag{6}$$

This equation is called the **two-point formula.** Given two points, we can make substitutions directly in this formula to obtain an equation of the line containing the two points. For example, substituting the coordinates of the points (2, 2) and (−4, 1) from Example 2 for the coordinates (x_1, y_1) and (x_2, y_2), respectively, in the two-point formula we obtain

$$y - 2 = \frac{1 - 2}{-4 - 2}(x - 2),$$

which simplifies to Equation (5) in Example 2.

Slope-intercept form

Now consider the equation of the line passing through a given point on the y-axis with coordinates $(0, b)$ and slope m, as shown in Figure 8.10. Substituting $(0, b)$ in the point-slope form of a linear equation,

$$y - y_1 = m(x - x_1),$$

we obtain

$$y - b = m(x - 0),$$

from which we get Equation (7).

$$\boldsymbol{y = mx + b.} \tag{7}$$

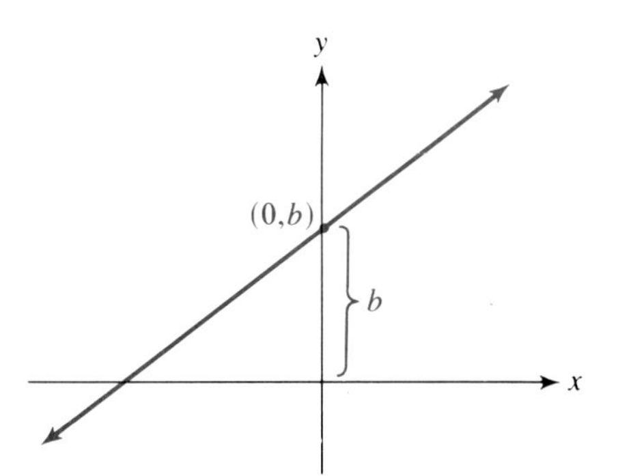

FIGURE 8.10

Equation (7) is called the **slope-intercept form** for a linear equation. Note that b is the y-intercept of the graph of the equation.

EXAMPLE 3 Write $3x + 4y = 6$ in slope-intercept form and specify the slope of the line and the y-intercept.

Solution We first solve the equation explicitly for y:

$$4y = -3x + 6$$
$$y = \frac{-3}{4}x + \frac{3}{2}.$$

Hence, the slope is $-3/4$, the coefficient of x, and the y-intercept is $3/2$. ■

If the x and y intercepts are a and b ($a, b \neq 0$), respectively, as shown in Figure 8.11, then $m = -b/a$. Replacing m in the point-slope formula we have

$$y = -\frac{b}{a}x + b,$$
$$ay = -bx + ab,$$
$$bx + ay = ab.$$

Then multiplying each member by $1/ab$, we obtain Equation (8), called the **intercept form** for a linear equation.

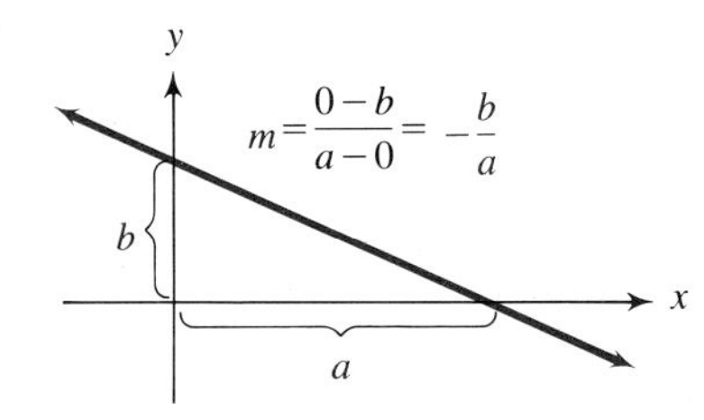

FIGURE 8.11

$$\frac{x}{a} + \frac{y}{b} = 1 \qquad (8)$$

EXAMPLE 4 Given that the x and y intercepts of a line are 2 and -3, respectively, write an equation of the line in standard form.

Solution Substituting 2 and -3 for a and b, respectively, in the intercept form we obtain

$$\frac{x}{2} + \frac{y}{-3} = 1,$$

from which by multiplying each member by the LCD, -6, we obtain

$$-3x + 2y = -6,$$

and

$$3x - 2y = 6. \qquad ■$$

The next example shows how we can write an equation of a line that is parallel to or perpendicular to a given line.

EXAMPLE 5 Write the equations of the line that is parallel to, and the line that is perpendicular to, the graph of $2x + 3y = 6$ and passes through $(3, 3)$.

Solution First find the slope of the graph of $2x + 3y = 6$:

$$3y = -2x + 6$$

$$y = \frac{-2}{3}x + 2.$$

Hence, the slope is $-2/3$ and the slope m_1 of the line parallel to this line is also $-2/3$.

The slope m_2 of the line perpendicular to the given line is

$$m_2 = \frac{-1}{m_1} = -\frac{1}{\frac{-2}{3}} = \frac{3}{2}.$$

Using the point-slope formula, the two equations of the lines passing through the point $(3, 3)$ with slopes $m_1 = -2/3$ and $m_2 = 3/2$ are:

a. $y - 3 = \frac{-2}{3}(x - 3)$

$3y - 9 = -2x + 6,$

$2x + 3y = 15$

b. $y - 3 = \frac{3}{2}(x - 3)$

$2y - 6 = 3x - 9,$

$3x - 2y = 3.$

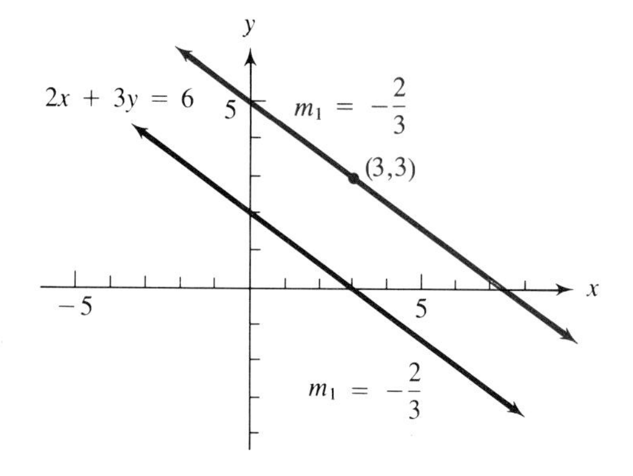

a.

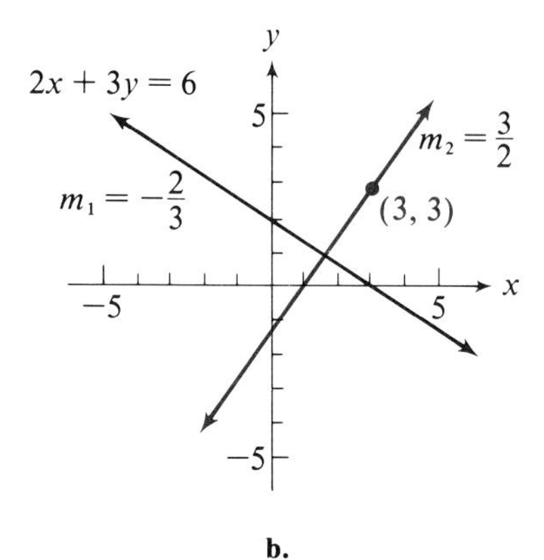

b. ■

EXERCISE 8.4

A

■ *Find an equation of the line that goes through each of the given points and has the given slope. Write the equation in standard form. See Example 1.*

1. $(-1, 3)$; $m = 2$
2. $(2, -5)$; $m = -3$
3. $(-2, 6)$; $m = -1$
4. $(-6, -1)$; $m = 4$
5. $(0, 3)$; $m = \frac{1}{2}$
6. $(2, 0)$; $m = -\frac{1}{3}$
7. $(-1, 2)$; $m = -\frac{3}{2}$
8. $(2, -1)$; $m = \frac{5}{3}$
9. $(-3, -5)$; $m = 0$
10. $(0, -6)$; $m = 0$
11. $(-3, 2)$; parallel to y-axis
12. $(0, 0)$; $m = 1$

■ *Find an equation of the line whose graph includes the two given points. Write the equation in the form* $ax + by = c$ *or* $ax + by + c = 0$. *See Example 2.*

13. $(-4, 2), (3, 3)$
14. $(5, -1), (2, -3)$
15. $(-1, -3), (2, 0)$
16. $(0, 5), (3, -4)$
17. $(-3, -4), (2, -2)$
18. $(-1, 4), (-3, -3)$
19. $(0, -4), (3, 0)$
20. $(-1, 0), (0, -4)$

■ **a.** *Write each equation in slope-intercept form.* **b.** *Specify the slope of the line and the y-intercept. See Example 3.*

21. $x + y = 3$
22. $2x + y = -1$
23. $3x + 2y = 1$
24. $3x - y = 7$
25. $x - 3y = 2$
26. $2x - 3y = 0$
27. $8x - 3y = 0$
28. $-x = 2y - 5$
29. $y + 2 = 0$
30. $y - 3 = 0$

■ *Write an equation in standard form of the line with x-intercept a and y-intercept b. See Example 4.*

31. $a = 3$ and $b = -2$
32. $a = -4$ and $b = 1$
33. $a = -2$ and $b = 4$
34. $a = 5$ and $b = -3$
35. $a = -3$ and $b = -6$
36. $a = -4$ and $b = -2$
37. $a = -\frac{1}{2}$ and $b = \frac{3}{4}$
38. $a = \frac{2}{3}$ and $b = -\frac{1}{4}$

B

■ *See Example 5 for Problems 39–42.*

39. Write an equation of the line that is parallel to the graph of $x - 2y = 5$ and passes through the origin. Draw the graphs of both equations.
40. Write an equation of the line that passes through (0, 5) parallel to $2y - 3x = 5$. Draw the graphs of both equations.
41. Write an equation of the line that is perpendicular to the graph of $x - 2y = 5$ and passes through the origin. Draw the graphs of both equations.
42. Write an equation of the line that is perpendicular to the graph of $2y - 3x = 5$ and passes through (0, 5). Draw the graphs of both equations.

8.5

GRAPHS OF LINEAR INEQUALITIES

The solutions of inequalities of the form

$$\mathbf{ax + by + c > 0} \quad \text{or} \quad \mathbf{ax + by + c < 0},$$

where a, b, and c are real numbers, are ordered pairs of real numbers. The solution sets can be graphed on the plane, but the graph will be a region of the plane rather than a straight line. As an example, consider the inequality

$$2x + y - 3 < 0. \tag{1}$$

Rewritten in the form

$$y < -2x + 3, \tag{2}$$

we have that y is less than $-2x + 3$ for each x. The graph of the equation

$$y = -2x + 3 \tag{3}$$

is simply a straight line, as illustrated in Figure 8.12. Therefore, to graph (2), we need only observe that any point below this line has x- and y-coordinates that satisfy (2). For example, if $x = 2$, then all ordered pairs of the form (2, y) such that

$$y < -2(2) + 3$$

or

$$y < -1$$

are in the solution set and their graphs are below the line, as shown in Figure 8.12.

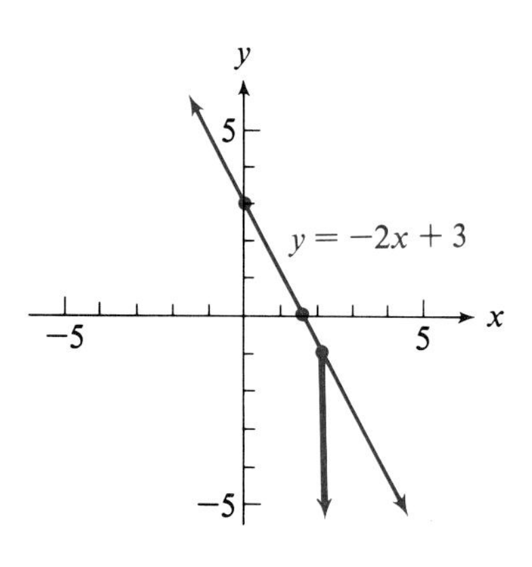

FIGURE 8.12

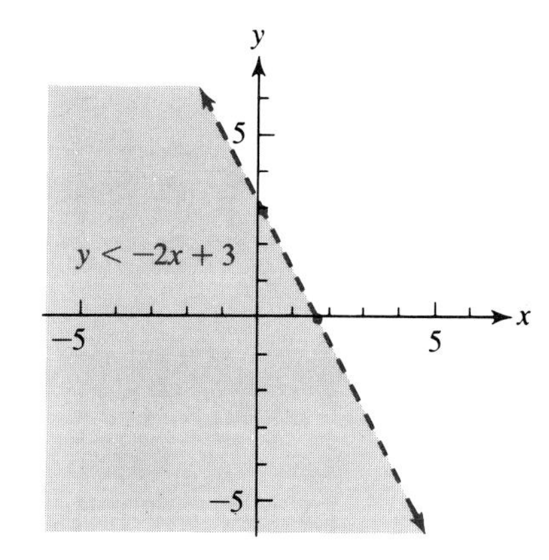

FIGURE 8.13

The solution set of (2) corresponds to the entire region below the line. The region is indicated on the graph (Figure 8.13) by shading. The broken line in Figure 8.13 indicates that the points on the line do not correspond to elements in the solution set of the inequality. If the original inequality were

$$2x + y - 3 \leq 0,$$

the line would be a part of the graph of the solution set and would be shown as a solid line.

In general, if $b \neq 0$,

$$ax + by + c < 0 \qquad \text{or} \qquad ax + by + c > 0$$

will have as a solution set all ordered pairs associated with the points in a **half-plane** either above or below the line with equation

$$ax + by + c = 0,$$

depending upon the inequality symbols involved.

To determine which of the half-planes should be shaded, substitute the coordinates of any point not on the line into the original inequality and note whether or not they satisfy the inequality. If they do, then the half-plane containing the point is shaded; if they do not, then the other half-plane is shaded. A good point to use in this process is the origin, with coordinates (0, 0), if the origin does not lie in the line.

EXAMPLE 1 Graph $3x - 2y < 6$.

Solution We first graph $3x - 2y = 6$ using the intercept method.

Then, substituting 0 for x and 0 for y in the inequality, we obtain

$$3(0) - 2(0) < 6.$$

Since this is a true statement, we shade the half-plane that contains the origin.

In this case, the edge of the half-plane is a dashed line because the original inequality does not contain the "equal to" symbol.

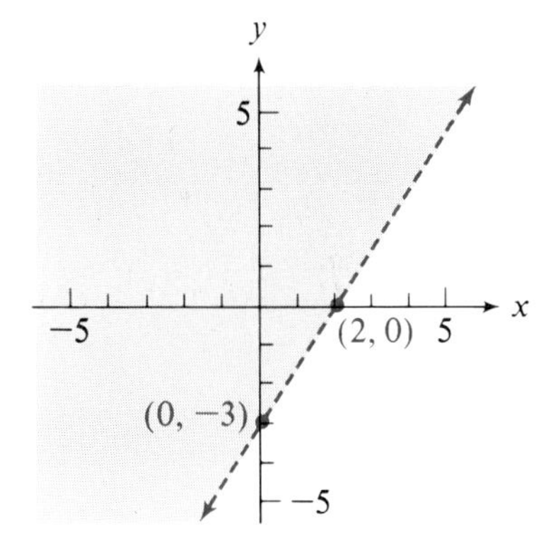

■

In Section 2.3 we graphed solution sets of inequalities on a number line, where it was understood that we were graphing an equation in one variable. Now, just as we view an equation such as $x = 2$ to represent the equation $x + 0y = 2$, we also view an inequality such as $x > 2$ to represent the inequality $x + 0y > 2$.

EXAMPLE 2 Graph $x \geq 2$.

Solution Graph the equality $x = 2$. The solution set consists of all ordered pairs where x is greater than 2. Hence, shade the region to the right of the line representing $x = 2$. The line is part of the graph.

In this case, the edge of the half-plane is a solid line because the original inequality contains the "equal to" symbol. ■

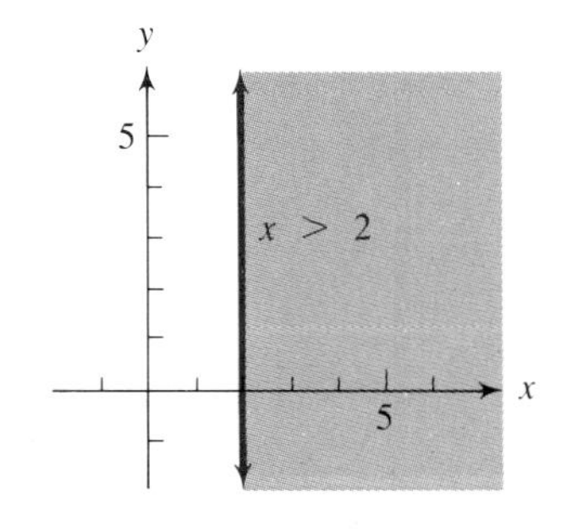

EXERCISE 8.5

A

■ *Graph each inequality. See Examples 1 and 2.*

1. $y > x + 3$
2. $y < x + 4$
3. $y \leq x + 2$
4. $y \geq x - 2$
5. $x + y < 5$
6. $x - y < 3$
7. $2x + y < 2$
8. $x - 2y < 5$
9. $x \leq 2y + 4$
10. $2x \leq y + 1$
11. $0 \geq x - y$
12. $0 \geq x + 3y$
13. $x > 0$
14. $y < 0$
15. $y \geq 3$
16. $x < -2$
17. $-1 < x < 5$
18. $-2 \leq x < 0$
19. $-4 < y \leq 0$
20. $0 \leq y \leq 1$

B

■ *Graph each pair of inequalities on the same coordinate system and use double shading to indicate the region common to both.*

21. $x > 0$ and $y > 0$
22. $x \leq 0$ and $y \geq 0$
23. $x > 0$ and $y < 0$
24. $x \leq 0$ and $y \leq 0$
25. $x > 4$ and $y > 2$
26. $x < 2$ and $y < 2$
27. $x + y \leq 6$ and $x + y \geq 4$
28. $2x - y \leq 4$ and $x + 2y > 6$
29. $y < x$ and $y \geq -3$
30. $y \geq -x$ and $y < 2$

CHAPTER SUMMARY

[8.1] The solution of an equation or inequality in two variables is an **ordered pair** of numbers. The replacement set for possible solutions is the infinite set of ordered pairs $\{(x, y) \mid x \text{ and } y \text{ are real numbers}\}$.

[8.2] We can use a **Cartesian** (or **rectangular**) **coordinate system** to graph ordered pairs of numbers on a plane. The components of a given ordered pair are called the **coordinates** of its graph.

The graph of a first-degree (linear) equation in two variables is a straight line. The x-coordinate of a point of intersection of a graph with the x-axis is an **x-intercept** of the graph, and the y-coordinate of a point of intersection of a graph with the y-axis is a **y-intercept** of the graph. These intercepts are obtained by setting $y = 0$ and $x = 0$, respectively, in the equation.

[8.3] For any two points in a geometric plane corresponding to (x_1, y_1) and (x_2, y_2), the distance d between the points is given by the **distance formula**,

$$d = \sqrt{(x_2 - x_1)^2 + (y_2 - y_1)^2}.$$

The **slope** m of the line containing the points (x_1, y_1) and (x_2, y_2) is given by

$$m = \frac{y_2 - y_1}{x_2 - x_1} \qquad (x_2 \neq x_1).$$

The slopes of parallel lines are equal ($m_1 = m_2$), and the nonzero slopes of perpendicular lines are the negative reciprocals of each other ($m_1 = -1/m_2$).

[8.4] Any of the forms of linear equations discussed in this section may be used in working with their graphs.

Form	Data Required	Equation
Point-slope	Slope, m Point, (x_1, y_1)	$y - y_1 = m(x - x_1)$
Two-point	Two points, (x_1, y_1) and (x_2, y_2)	$y - y_1 = \frac{y_2 - y_1}{x_2 - x_1}(x - x_1)$
Slope-intercept	Slope, m y-intercept, b	$y = mx + b$
Intercept	x-intercept, a y-intercept, b	$\frac{x}{a} + \frac{y}{b} = 1$

[8.5] The graph of a first-degree (linear) inequality in two variables,

$$ax + by < c \quad \text{or} \quad ax + by > c,$$

is a **half-plane.** The edge of the half-plane is not included; it is shown as a dashed line. The graph of

$$ax + by \leqslant c \quad \text{or} \quad ax + by \geqslant c$$

is also a half-plane; it includes the edge, which is shown as a solid line.

REVIEW EXERCISES

A

[8.1]

1. Find the missing component in each solution of $2x - 6y = 12$.
   a. $(0, ?)$ b. $(?, 0)$ c. $(3, ?)$
2. List the solutions of $y = 2x - 3$, where x is 2, 4, and 6.
3. In the equation $xy = 2x^2y + 3$, express y explicitly in terms of x.
4. In the equation $x^2 - 9y^2 = 4$, express y explicitly in terms of x.

[8.2]

■ *Graph each equation.*

5. $y = 3x + 1$
6. $y = 2x - 5$
7. $3x - y = 6$
8. $3x - y = 0$
9. $x + y = 0$
10. $x - 2y = 6$
11. $2x = y - 8$
12. $3x = y - 6$
13. $2x = -4$
14. $3x - 6 = 0$

[8.3]

15. a. Find the distance between the points $(3, -5)$ and $(6, 8)$.
    b. Find the slope of the line segment joining the points in part a.
16. a. Find the distance between the points $(4, 2)$ and $(7, -4)$.
    b. Find the slope of the line segment joining the points in part a.
17. a. Given points $P_1(2, 2)$, $P_2(-2, -2)$, $P_3(0, 4)$, and $P_4(-3, 1)$, show that the line segments P_1P_2 and P_3P_4 are parallel.
    b. Show that the line segments P_1P_2 and P_1P_3 are perpendicular.

[8.4]

18. Find the equation in standard form of the line through $(3, 5)$ with slope 2.
19. Find the equation in standard form of the line through $(-4, 2)$ with slope $-\frac{1}{2}$.

20. a. Write $3x - y = 4$ in slope-intercept form.
    b. Specify the slope and the y-intercept of its graph.
21. a. Write $2x + 3y = 6$ in slope-intercept form.
    b. Specify the slope and the y-intercept of its graph.
22. Write an equation in standard form of the line through $(3, -1)$ and $(4, 5)$.
23. Write an equation of the line that passes through $(-4, 2)$ and $(0, 3)$.
24. Write an equation of the line with x-intercept -5 and y-intercept 2.

[8.5]

■ *Graph each inequality.*

25. $y > x + 2$
26. $y \leq 4x + 4$
27. $x - 2y < 6$
28. $2x - y \geq 6$
29. $y > -3$
30. $-2 < x \leq 3$

B

31. Given the points $P_1(2, -3)$, $P_2(3, 4)$, and $P_3(4, -1)$, find the value of k in the ordered pair $P_4(6, k)$ that makes P_3P_4 perpendicular to P_1P_2.
32. Graph $y = 2x$ and $x + y = 4$ on the same coordinate system and estimate the coordinates of the point of intersection.
33. Write an equation in standard form of the line that is perpendicular to the graph of $2y + x = 6$ and passes through $(2, -3)$.
34. Graph $y > -3x$ and $x - 2y \leq 6$ on the same coordinate system and use double shading to indicate the region common to both.

9 Systems of Linear Equations and Inequalities

9.1

SYSTEMS IN TWO VARIABLES

In Chapter 8 we considered the solution set of equations or inequalities in two variables which contain infinitely many ordered pairs of numbers. It is often necessary to consider *pairs* of equations or inequalities and to inquire whether or not their solution sets contain ordered pairs in common. That is, we are interested in determining the members of the *intersection* of their solution sets.

Systems of equations

We shall begin by considering the system (in standard form),

$$a_1x + b_1y = c_1 \qquad (a_1, b_1 \text{ not both } 0)$$
$$a_2x + b_2y = c_2 \qquad (a_2, b_2 \text{ not both } 0).$$

In a geometric sense, because the graphs of both of these equations are straight lines, we are confronted with three possibilities, as illustrated in Figure 9.1:

a. The graphs are the same line.
b. The graphs are parallel but distinct lines.
c. The graphs intersect at one and only one point.

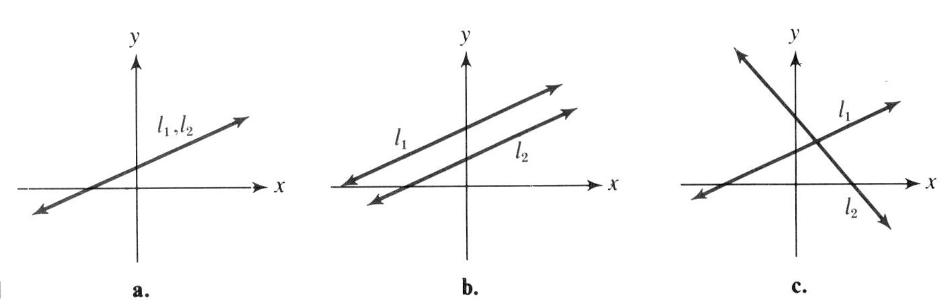

FIGURE 9.1

These possibilities lead, correspondingly, to the conclusion that one and only one of the following is true for any given system of two such linear equations in x and y:

- **a.** The solution sets of the equations are equal, and their intersection contains all (an infinite number of) ordered pairs found in either one of the solution sets.
- **b.** The intersection of the two solution sets is the null set.
- **c.** The intersection of the two solution sets contains one and only one ordered pair.

In case a, the linear equations in x and y are said to be **dependent**, and, in case b, the equations are said to be **inconsistent**. In case c, the equations are **consistent** and **independent** and the system has one and only one solution.

Consider the system

$$\begin{aligned} x + y &= 5 \\ x - y &= 1. \end{aligned}$$

From the graph of the system in Figure 9.2, it is evident that the ordered pair (3, 2) is the only ordered pair common to the solution sets of both equations. We can verify that (3, 2) is indeed the ordered pair in question by substituting (3, 2) into each equation and observing that a true statement results in each case.

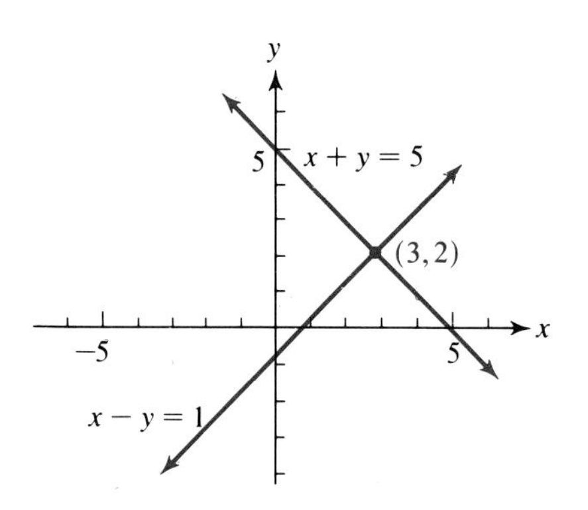

FIGURE 9.2

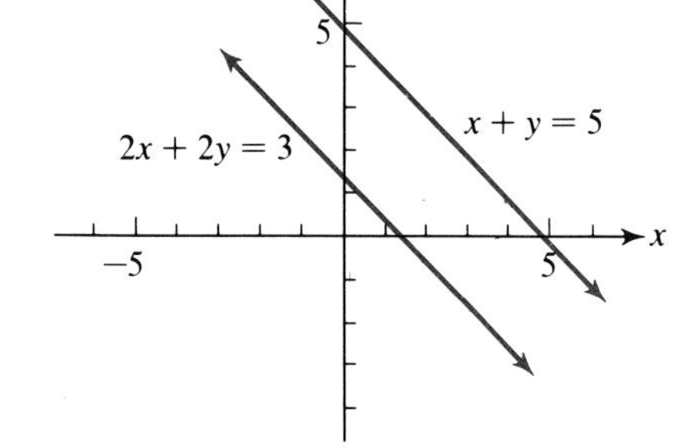

FIGURE 9.3

As another example, consider the system

$$\begin{aligned} x + y &= 5 \\ 2x + 2y &= 3, \end{aligned}$$

and the graphs of these equations in Figure 9.3, where the lines appear to be parallel. We conclude from this that the solution set of this system is $\emptyset$.

Linear combinations

Because graphing equations is a time-consuming process, and, more important, because graphic results are not always precise, solutions to systems of linear equations are usually sought by algebraic methods. One such method depends on the following property.

Any ordered pair (x, y) *that satisfies the equations*

$$a_1x + b_1y = c_1 \quad (1)$$

$$a_2x + b_2y = c_2 \quad (2)$$

will also satisfy the equation

$$A(a_1x + b_1y) + B(a_2x + b_2y) = Ac_1 + Bc_2 \quad (3)$$

for all real numbers A and B.

The left-hand member of (3) is called a **linear combination** of the left-hand members of (1) and (2).

The property stated above asserts that any ordered pair satisfying both (1) and (2) must also satisfy the sum of any real-number multiples of (1) and (2). This fact can be used to identify any such ordered pairs.

The concept of a linear combination can be used to solve a system by choosing multipliers A and B so that the coefficients of *one* of the variables, x or y, are additive inverses of each other, resulting in an equation free of *one* of the variables.

EXAMPLE 1 Solve the system

$$2x + 3y = 8 \quad (4)$$

$$3x - 4y = -5. \quad (5)$$

Solution We first want to rewrite one or both of the equations in equivalent forms so that the coefficients of the same variable (either x or y) will be additive inverses of each other. In this example, we shall obtain the coefficients of x as additive inverses by multiplying each member of (4) by 3 and each member of (5) by -2 to obtain

$$6x + 9y = 24 \quad (4a)$$

$$-6x + 8y = 10. \quad (5a)$$

Adding the corresponding members of (4a) and (5a), we obtain

$$17y = 34 \quad (6)$$

$$y = 2, \quad (6a)$$

which must contain in its solution set any solution common to the solution sets of the two original equations. But any solution of (6) or (6a) is of the form $(x, 2)$—that is, it has y-component 2 for any value of x. Now, substituting 2 for y in either (4) or (5), we can determine the x-component for the ordered pair $(x, 2)$ that satisfies both (4) and (5). If (4) is used, we have

$$2x + 3(2) = 8$$

$$x = 1;$$

and if (5) is used, we have

$$3x - 4(2) = -5$$

$$x = 1.$$

Since the ordered pair $(1, 2)$ satisfies both (4) and (5), the required solution set is $\{(1, 2)\}$. ■

It is helpful to rewrite any equation in a system that has fractional coefficients as an equivalent equation without fractions before attempting to use the procedure followed in Example 1.

EXAMPLE 2 Solve the system by linear combinations.

$$\frac{2}{3}x - y = 2 \quad (1)$$

$$x + \frac{1}{2}y = 7 \quad (2)$$

Solution Multiply each member of Equation (1) by 3 and each member of Equation (2) by 2.

$$2x - 3y = 6 \qquad (1a)$$
$$2x + y = 14 \qquad (2a)$$

Add -1 times Equation (1a) to 1 times Equation (2a) and solve for y.

$$4y = 8$$
$$y = 2$$

Substitute 2 for y in (1), (2), (1a), or (2a) and solve for x. In this example, Equation (2) is used.

$$x + \frac{1}{2}(2) = 7$$
$$x = 6$$

The solution set is $\{(6, 2)\}$.

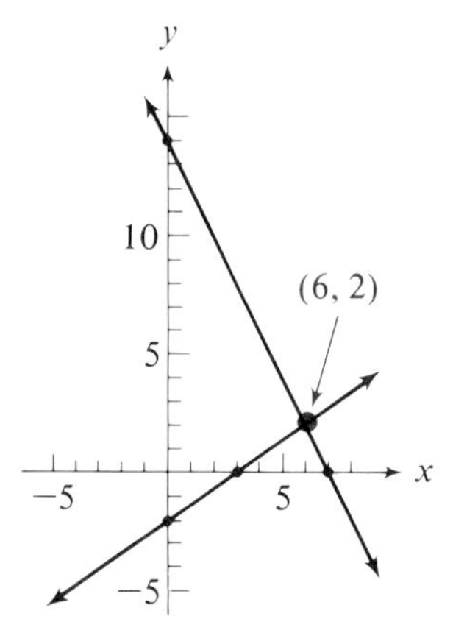

■

The sketch in the above example provides a check on the algebraic solution.

Test for a unique solution

We can readily determine whether a system has a unique solution or whether the equations are inconsistent or dependent by first writing each equation in slope-intercept form. In general, the system in standard form

$$a_1x + b_1y = c_1$$
$$a_2x + b_2y = c_2$$

can be written equivalently as

$$y = \frac{-a_1}{b_1}x + \frac{c_1}{b_1}$$
$$y = \frac{-a_2}{b_2}x + \frac{c_2}{b_2}.$$

Now, if the slopes $-a_1/b_1$ and $-a_2/b_2$ are unequal, the graphs will intersect at one point and there will be one and only one solution. If $a_1/b_1 = a_2/b_2$, then the graphs are either the same line (if $c_1/b_1 = c_2/b_2$) or parallel lines (if $c_1/b_1 \neq c_2/b_2$). These conclusions are summarized in the following property in a modified form.

Any system of the standard form

$$a_1x + b_1y = c_1$$
$$a_2x + b_2y = c_2$$

has one and only one solution if

$$\frac{a_1}{a_2} \neq \frac{b_1}{b_2}, \tag{I}$$

has no solution if

$$\frac{a_1}{a_2} = \frac{b_1}{b_2} \neq \frac{c_1}{c_2}, \tag{II}$$

and has an infinite number of solutions if

$$\frac{a_1}{a_2} = \frac{b_1}{b_2} = \frac{c_1}{c_2}. \tag{III}$$

EXAMPLE 3 a. For the system

$$2x = 2 - 3y$$
$$6y = 7 - 4x,$$

we first rewrite it in standard form as

$$2x + 3y = 2$$
$$4x + 6y = 7,$$

and then note that

$$\frac{2}{4} = \frac{3}{6} \neq \frac{2}{7}.$$

Hence, from Property (II) above, the system does not have a solution.

b. For the system

$$\begin{aligned} 3y &= 5 - 2x \\ -4x &= 7y - 8, \end{aligned}$$

we first rewrite it in standard form as

$$\begin{aligned} 2x + 3y &= 5 \\ 4x + 7y &= 8, \end{aligned}$$

and then note that

$$\frac{2}{4} \neq \frac{3}{7}.$$

Hence, from Property (I) above, the system has one and only one solution. ■

Systems of inequalities

We sometime want to find the solutions that are common to the solution sets of two or more inequalities; that is, we want to find the *intersection* of their solution sets. In general, we can obtain good approximations to such solutions by graphical methods.

In Section 8.5 we observed that the graph of the solution set of an inequality in two variables is a region in the plane. The intersection of the graphs of two inequalities might also be a region in the plane.

Perhaps you have already graphed the systems of inequalities that were given in the "B" section of Exercise 8.5. We will now look at several more examples.

EXAMPLE 4 Graph the system

$$y > x \quad \text{and} \quad y > 2$$

using double shading.

Solution 1. Graph each equation

$$y = x \quad \text{and} \quad y = 2.$$

2. Shade the appropriate half-plane for each inequality. Use dashed lines for the edge of each half-plane.
3. Use a heavy shading (or color) to note the region common to both solution sets.

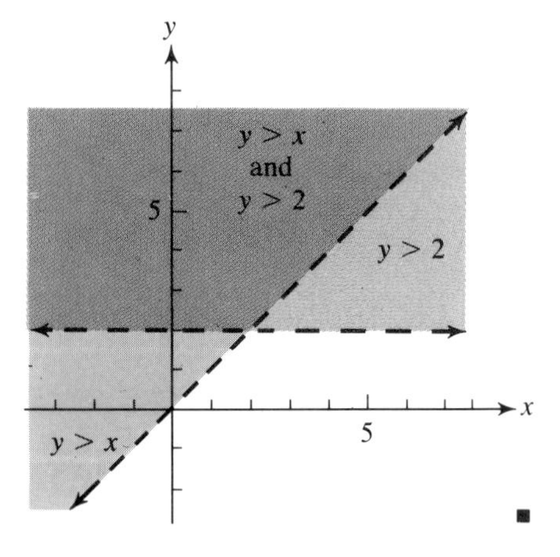

■

Applications of systems of three or more inequalities in two variables often involve locating vertices of the boundary of the region that is the graph of the solution of the system.

EXAMPLE 5 Graph the region that represents the solution set of the system, and find the coordinates of the vertices (approximations as necessary) of the boundary.

$$x \geqslant 0$$
$$y \geqslant 0$$
$$x - y - 2 \leqslant 0$$
$$x + 2y - 6 \leqslant 0$$

Solution Graph the associated equations

$$x = 0$$
$$y = 0$$
$$x - y - 2 = 0$$
$$x + 2y - 6 = 0$$

as shown. Note which half-plane is determined by each of the four inequalities and shade (or color) the intersections of the half-planes.

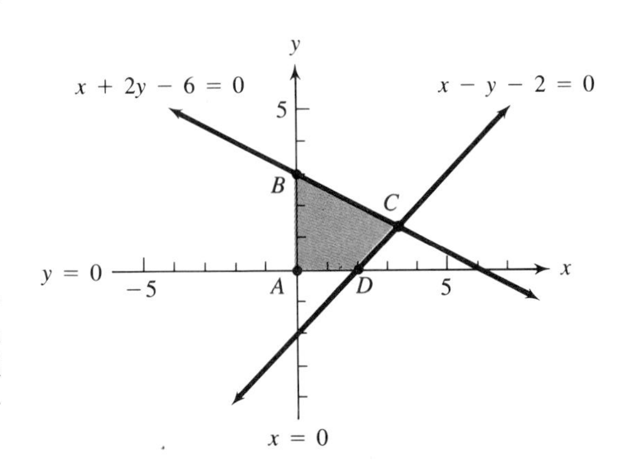

Determine the coordinates of each vertex A, B, C, and D.

A: (0, 0) B: (0, 3) C: (3½, 1½) (estimate) D: (2, 0) ■

EXERCISE 9.1

A

■ *Solve each system by linear combinations. Sketch the graphs of the equations. See Examples* 1 *and* 2.

1. $x - y = 1$, $x + y = 5$
2. $2x - 3y = 6$, $x + 3y = 3$
3. $3x + y = 7$, $2x - 5y = -1$
4. $2x - y = 7$, $3x + 2y = 14$
5. $5x - y = -29$, $2x + 3y = 2$
6. $x + 4y = -14$, $3x + 2y = -2$
7. $3x + 2y = 7$, $x + y = 3$
8. $2x - 3y = 8$, $x + y = -1$
9. $3y = x - 1$, $y = 6x - 6$
10. $3x = 3y - 3$, $2y = 6x + 14$
11. $2x - 3y = -4$, $5x + 2y = 9$
12. $3x + 5y = 1$, $2x - 3y = 7$
13. $5x + 2y = 3$, $x = 0$
14. $2x - y = 0$, $x = -3$
15. $3x - 2y = 4$, $y = -1$
16. $x + 2y = 6$, $x = 2$
17. $\frac{1}{4}x - \frac{1}{3}y = -\frac{5}{12}$, $\frac{1}{10}x + \frac{1}{5}y = \frac{1}{2}$
18. $\frac{2}{3}x - y = 4$, $x - \frac{3}{4}y = 6$
19. $\frac{1}{7}x = \frac{3}{7}y + 1$, $y = 2x + 4$
20. $\frac{1}{3}x = \frac{2}{3}y + 2$, $2y = x - 6$

■ *Use the properties on page* 308 *to determine whether each of the following systems has: one solution, no solution, or an infinite number of solutions. DO NOT SOLVE. See Example* 3.

21. $x + 3y = 6$, $2x + 6y = 12$
22. $3x - 2y = 6$, $6x - 4y = 8$
23. $2x - y = 1$, $8x - 4y = 3$
24. $6x + 2y = 1$, $12x + 4y = 2$
25. $x - 3y = 4$, $2x + y = 6$
26. $2x + y = 4$, $x - 3y = 2$

27. $x = 2y - 4$
$2x - 4y = 6$

28. $y = 3x + 4$
$6x - 2y = 4$

29. $y - 4 = x$
$2x + 8 = 4y$

30. $x + 2 = y$
$3y - 6 = 3x$

31. $x = 2y + 3$
$y = 2x - 3$

32. $2x - 3 = y$
$2y + 3 = x$

■ *Graph each system of inequalities. See Example 4.*

33. $x + y < 4$
$x < 2$

34. $x + y > 6$
$y < 3$

35. $2x - y \geqslant 6$
$x + 2y \leqslant 4$

36. $x + 2y \geqslant 6$
$2x - y \leqslant 4$

37. $2x + 3y - 6 < 0$
$y \geqslant 0$

38. $3x - 2y + 6 \geqslant 0$
$x < 0$

B

■ *Graph the region that represents the solution set of the system, and find the coordinates of the vertices (approximations as necessary) of the boundary. See Example 5.*

39. $x \geqslant 0$
$y \geqslant 0$
$x + y - 4 \leqslant 0$

40. $x \geqslant 0$
$y \leqslant 0$
$x - y - 3 \geqslant 0$

41. $y \leqslant 0$
$x + y \geqslant 0$
$x - y - 4 \leqslant 0$

42. $x \geqslant 0$
$x - y \geqslant 0$
$x + y - 4 \leqslant 0$

43. $x \geqslant 0$
$y \geqslant 0$
$y - 3 \leqslant 0$
$x - 4 \leqslant 0$

44. $x \leqslant 0$
$y \geqslant 0$
$x + 3 \geqslant 0$
$x + y \leqslant 0$

■ *Solve each system. Hint: Set $u = 1/x$ and $v = 1/y$, solve for u and v, then solve for x and y.*

45. $\dfrac{1}{x} + \dfrac{1}{y} = 7$
$\dfrac{2}{x} + \dfrac{3}{y} = 16$

Hint: Set $u = 1/x$ and $v = 1/y$ yields $u + v = 7$, $2u + 3v = 16$. Solve for u and v. Then use the fact that $u = 1/x$ and $v = 1/y$ to solve for x and y.

46. $\dfrac{1}{x} + \dfrac{2}{y} = -\dfrac{11}{12}$
$\dfrac{1}{x} + \dfrac{1}{y} = -\dfrac{7}{12}$

47. $\dfrac{5}{x} - \dfrac{6}{y} = -3$
$\dfrac{10}{x} + \dfrac{9}{y} = 1$

48. $\dfrac{1}{x} + \dfrac{2}{y} = 11$
$\dfrac{1}{x} + \dfrac{2}{y} = -1$

49. $\dfrac{1}{x} - \dfrac{1}{y} = 4$
$\dfrac{2}{x} - \dfrac{1}{2y} = 11$

50. $\dfrac{2}{3x} + \dfrac{3}{4y} = \dfrac{7}{12}$
$\dfrac{4}{x} - \dfrac{3}{4y} = \dfrac{7}{4}$

9.2

SYSTEMS IN THREE VARIABLES

A solution of an equation in three variables, such as

$$x + 2y - 3z = -4,$$

is an ordered triple of numbers (x, y, z), because all three of the variables must be replaced by numerals before we can decide whether the result is an equality. Thus, $(0, -2, 0)$ and $(-1, 0, 1)$ are solutions of this equation, while $(1, 1, 1)$ is not. There are, of course, infinitely many members in the solution set.

The solution set of a system of three linear equations in three variables, such as

$$x + 2y - 3z = -4 \qquad (1)$$

$$2x - y + z = 3 \qquad (2)$$

$$3x + 2y + z = 10, \qquad (3)$$

is the intersection of the solution sets of all three equations in the system. We seek solution sets of systems such as these by methods similar to those used in solving linear systems in two variables.

In the system presented above, we might begin by multiplying Equation (1) by -2 and adding the result to 1 times Equation (2), to produce

$$\begin{aligned} (-2)(x + 2y - 3z) &= (-2)(-4) \\ \underline{(1)(2x) - y + z)} &= \underline{(1)(3)} \\ -5y + 7z &= 11 \qquad (4) \end{aligned}$$

which is satisfied by any ordered triple (x, y, z) that satisfies (1) and (2). Similarly, we can add -3 times Equation (1) to 1 times Equation (3) to obtain

$$\begin{aligned} (-3)(x + 2y - 3z) &= (-3)(-4) \\ \underline{(1)(3x + 2y + z)} &= \underline{(1)(10)} \\ -4y + 10z &= 22 \qquad (5) \end{aligned}$$

which is satisfied by any ordered triple (x, y, z) that satisfies both (1) and (3).

We can now argue that any ordered triple satisfying the system (1), (2), and (3) will also satisfy the system

$$-5y + 7z = 11 \qquad (4)$$

$$-4y + 10z = 22. \qquad (5)$$

Since the system (4) and (5) does not depend on x, the problem has been reduced to one of finding only the y- and z-components of the solution. The system (4) and (5) can be solved by the method of Section 9.1, which leads to the values $y = 2$ and $z = 3$. Now, since any solution of (1), (2), and (3) must be of the form $(x, 2, 3)$, we can substitute 2 for y and 3 for z in (1) to obtain $x = 1$. The desired solution set is

$$\{(1, 2, 3)\}.$$

The process of solving a system of equations can be reduced to a series of mechanical procedures, as illustrated by the following example.

EXAMPLE 1 Solve the system:

$$x + 2y - z = -3 \tag{1}$$

$$x - 3y + z = 6 \tag{2}$$

$$2x + y + 2z = 5. \tag{3}$$

Solution First obtain a system of two equations in two variables. Multiply Equation (1) by -1 and add the result to 1 times Equation (2) to get (4):

$$\begin{aligned}(-1)(x + 2y - z) &= (-1)(-3)\\ \underline{(1)(x - 3y + z)} &= \underline{(1)(6)}\\ -5y + 2z &= 9\end{aligned} \tag{4}$$

Multiply Equation (1) by -2 and add the result to 1 times Equation (3) to get (5):

$$\begin{aligned}(-2)(x + 2y - z) &= (-2)(-3)\\ \underline{(1)(2x + y + 2z)} &= \underline{(1)5}\\ -3y + 4z &= 11\end{aligned} \tag{5}$$

Now solve the system in two variables formed by Equations (4) and (5). Multiply Equation (4) by -2 and add the result to 1 times Equation (5) to obtain

$$(-2)(-5y + 2z) = (-2)(9) \tag{4'}$$

$$\underline{(1)(-3y + 4z)} = \underline{(1)(11)} \tag{5'}$$

$$7y = -7$$

$$y = -1$$

Substitute -1 for y in either (4) or (5)—we shall use (4)—and solve for z.

$$\begin{aligned} -5(-1) + 2z &= 9 \\ z &= 2 \end{aligned}$$

Substitute -1 for y and 2 for z in (1), (2), or (3)—we shall use (1)—and solve for x.

$$\begin{aligned} x + 2(-1) - 2 &= -3 \\ x &= 1 \end{aligned}$$

The solution set is $\{(1, -1, 2)\}$. ■

As suggested in Section 9.1, it is usually helpful to first rewrite any equation that has fractional coefficients as an equivalent equation without fractions.

EXAMPLE 2 Solve the system:

$$x + 2y + z = -3 \tag{6}$$

$$\frac{1}{3}x - y + \frac{1}{3}z = 2 \tag{7}$$

$$x + \frac{1}{2}y + z = \frac{5}{2} \tag{8}$$

Solution Multiply each member of Equation (7) by 3 and each member of Equation (8) by 2, to obtain the equivalent system

$$\begin{aligned} x + 2y - z &= -3 \\ 3\left(\frac{1}{3}x\right) - 3(y) + 3\left(\frac{1}{3}z\right) &= 3(2) \\ 2(x) + 2\left(\frac{1}{2}y\right) + 2(z) &= 2\left(\frac{5}{2}\right). \end{aligned}$$

When the equations are simplified, the system (and its solution) is identical to the system in Example 1. ■

Graphs in three dimensions By establishing a three-dimensional Cartesian coordinate system, a one-to-one correspondence can be established between the points in a three-dimensional space and ordered triples of real numbers. If this is done, it can be shown that the graph of a first-degree equation in three variables is a plane. Hence, the solution set of a system of three such equations consists of the coordinates of the common intersection of three planes.

Although it is not practical to solve a system of three first-degree equations in three variables by graphing, it is interesting to see the possibilities for the relative positions of the plane graphs (Figure 9.4).

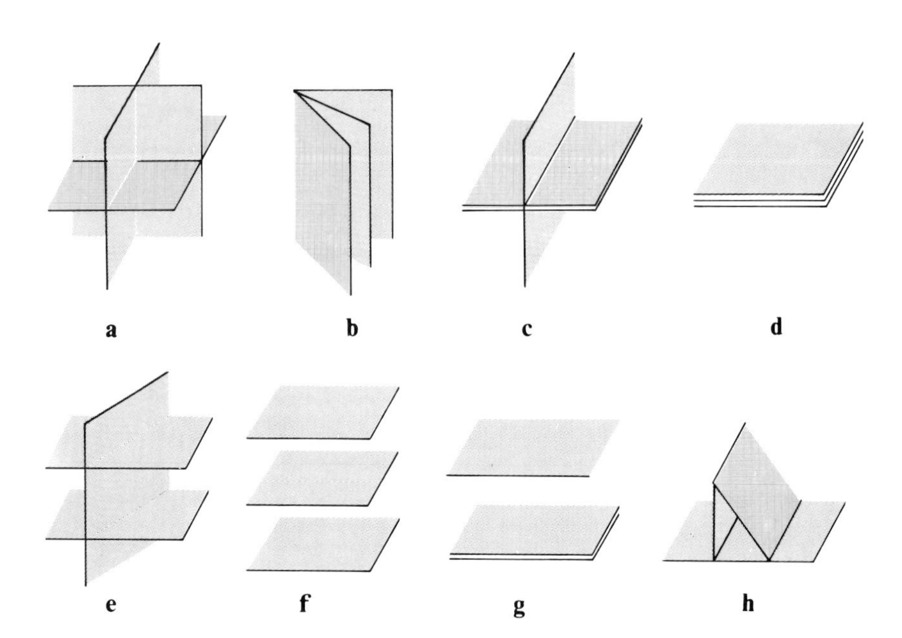

FIGURE 9.4

In case (a), the common intersection consists of a single point, and hence the solution set of the corresponding system of three equations contains a single member. In cases (b), (c), and (d), the intersection is a line or a plane, and the solution of the corresponding system has infinitely many members. In cases (e), (f), (g), and (h), the three planes have no common intersection, and the solution set of the corresponding system is the null set.

If at any step in forming linear combinations the resulting linear combination vanishes or yields a contradiction, then the system does not have a unique solution. It either has an infinite number of members or no members in its solution set.

EXAMPLE 3 Solve the system:

$$3x + y - 2z = 1 \tag{9}$$

$$6x + 2y - 4z = 5 \tag{10}$$

$$-2x - y + 3z = -1 \tag{11}$$

Solution To eliminate y in Equations (9) and (10) we multiply Equation (9) by -2 and add the result to 1 times Equation (10) to obtain

$$\begin{array}{rl} 6x + 2y - 4z = 5 & (10) \\ -6x - 2y + 4z = -2 & -2 \text{ times } (9) \\ \hline 0 = 3 & \end{array}$$

Since the resulting linear combination yields a contradictory result, the system does not have a unique solution. ■

EXAMPLE 4 Solve the system:

$$-x + 3y - z = -2 \qquad (12)$$

$$2x + y - 4z = 6 \qquad (13)$$

$$2x - 6y + 2z = 4 \qquad (14)$$

Solution To eliminate x in Equations (12) and (14) we multiply Equation (12) by 2 and add the result to 1 times Equation (14) to obtain

$$\begin{array}{rl} 2x - 6y + 2z = 4 & (14) \\ -2x + 6y - 2z = -4 & 2 \text{ times } (12) \\ \hline 0 = 0 & \end{array}$$

Since the resulting linear combination vanishes, the system does not have a unique solution. ■

EXERCISE 9.2

A

■ *Solve. See Example* 1.

1. $x + y + z = 2$
   $2x - y + z = -1$
   $x - y - z = 0$

2. $x + y + z = 1$
   $2x - y + 3z = 2$
   $2x - y - z = 2$

3. $x + y + 2z = 0$
   $2x - 2y + z = 8$
   $3x + 2y + z = 2$

4. $x - 2y + 4z = -3$
   $3x + y - 2z = 12$
   $2x + y - 3z = 11$

5. $x - 2y + z = -1$
   $2x + y - 3z = 3$
   $3x + 3y - 2z = 10$

6. $x + 5y - z = 2$
   $3x - 9y + 3z = 6$
   $x - 3y - z = -6$

7.
$$\begin{aligned} 2x - y + z &= 6 \\ x - 3y - z &= 7 \\ 3x + 2y + z &= 6 \end{aligned}$$

8.
$$\begin{aligned} x - 2y + 4z &= 10 \\ 2x + 3y - z &= -7 \\ x - y + 2z &= 4 \end{aligned}$$

9.
$$\begin{aligned} 3x - y &= 6 \\ x - 2z &= -7 \\ 2y + z &= -2 \end{aligned}$$

10.
$$\begin{aligned} 2x + z &= 5 \\ 3y + 2z &= 6 \\ x - 2z &= 10 \end{aligned}$$

11.
$$\begin{aligned} 4x + z &= 3 \\ 2x - y &= 2 \\ 3y + 2z &= 0 \end{aligned}$$

12.
$$\begin{aligned} 3y + z &= 3 \\ -2x + 3y &= 7 \\ 3x - 2z &= -6 \end{aligned}$$

■ *Solve. See Example 2.*

13.
$$\begin{aligned} x - \frac{1}{2}y - \frac{1}{2}z &= 4 \\ x - \frac{3}{2}y - 2z &= 3 \\ \frac{1}{4}x + \frac{1}{4}y - \frac{1}{4}z &= 0 \end{aligned}$$

14.
$$\begin{aligned} x + 2y + \frac{1}{2}z &= 0 \\ x + \frac{3}{5}y - \frac{2}{5}z &= \frac{1}{5} \\ 4x - 7y - 7z &= 6 \end{aligned}$$

15.
$$\begin{aligned} x + y - z &= 2 \\ \frac{1}{2}x - y + \frac{1}{2}z &= -\frac{1}{2} \\ x + \frac{1}{3}y - \frac{2}{3}z &= \frac{4}{3} \end{aligned}$$

16.
$$\begin{aligned} x + y - 2z &= 3 \\ x - \frac{1}{3}y + \frac{1}{3}z &= \frac{5}{3} \\ \frac{1}{2}x - \frac{1}{2}y - z &= \frac{3}{2} \end{aligned}$$

17.
$$\begin{aligned} x &= -y \\ x + z &= \frac{5}{6} \\ y - 2z &= -\frac{7}{6} \end{aligned}$$

18.
$$\begin{aligned} x &= y + \frac{1}{2} \\ y &= z + \frac{5}{4} \\ 2z &= x - \frac{7}{4} \end{aligned}$$

■ *Solve. If there is no unique solution, so state. See Examples* 1–4.

19.
$$\begin{aligned} 3x - 2y + z &= 6 \\ 2x + y - z &= 2 \\ 4x + 2y - 2z &= 3 \end{aligned}$$

20.
$$\begin{aligned} x + 3y - z &= 4 \\ -2x - 6y + 2z &= 1 \\ x + 2y - z &= 3 \end{aligned}$$

21.
$$\begin{aligned} 2x + 3y - z &= -2 \\ x - y + \frac{1}{2}z &= 2 \\ 4x - \frac{1}{3}y + 2z &= 8 \end{aligned}$$

22.
$$\begin{aligned} 3x + 6y + 2z &= -2 \\ \frac{1}{2}x - 3y - z &= 1 \\ 4x + y + \frac{1}{3}z &= -\frac{1}{3} \end{aligned}$$

23.
$$\begin{aligned} 2x + y &= 6 \\ x - z &= 4 \\ 3x + y - z &= 10 \end{aligned}$$

24.
$$\begin{aligned} x - 2y + z &= 5 \\ -x + y &= -2 \\ y - z &= -3 \end{aligned}$$

25.
$$\begin{aligned} x &= 2y - 7 \\ y &= 4z + 3 \\ z &= 3x + y \end{aligned}$$

26.
$$\begin{aligned} x &= y + z \\ y &= 2x - z \\ z &= 3x - y \end{aligned}$$

27.
$$\begin{aligned} \frac{1}{2}x + y &= \frac{1}{2}z \\ x - y &= -z - 2 \\ -x - 2y &= -z + \frac{4}{3} \end{aligned}$$

28.
$$\begin{aligned} x &= \frac{1}{2}y - \frac{1}{2}z + 1 \\ x &= 2y + z - 1 \\ x &= \frac{1}{2}y - \frac{1}{2}z + \frac{1}{4} \end{aligned}$$

29.
$$\begin{aligned} x - y &= 0 \\ 2x + 2y + z &= 5 \\ 2x + y - \frac{1}{2}z &= 0 \end{aligned}$$

30.
$$\begin{aligned} 3x + y &= 1 \\ 2x - y + z &= -1 \\ x - 3y - z &= -\frac{2}{3} \end{aligned}$$

B

■ *Solve each system. Hint: Use substitutions* $u = 1/x$, $v = 1/y$ *and* $w = 1/z$; *solve for* u, v *and* w, *then solve for* x, y *and* z.

31. $\dfrac{1}{x} + \dfrac{1}{y} - \dfrac{1}{z} = 1$

$\dfrac{2}{x} - \dfrac{2}{y} + \dfrac{1}{z} = 1$

$\dfrac{-3}{x} + \dfrac{1}{y} - \dfrac{1}{z} = -3$

32. $\dfrac{4}{x} - \dfrac{2}{y} + \dfrac{1}{z} = 4$

$\dfrac{3}{x} - \dfrac{1}{y} + \dfrac{2}{z} = 0$

$\dfrac{-1}{x} + \dfrac{3}{y} - \dfrac{2}{z} = 0$

33. $\dfrac{1}{x} + \dfrac{2}{y} - \dfrac{2}{z} = 3$

$\dfrac{2}{x} - \dfrac{4}{y} + \dfrac{2}{z} = -2$

$\dfrac{4}{x} - \dfrac{2}{y} - \dfrac{4}{z} = 5$

34. $\dfrac{2}{x} - \dfrac{1}{y} - \dfrac{1}{z} = -1$

$\dfrac{4}{x} - \dfrac{2}{y} + \dfrac{1}{z} = -5$

$\dfrac{2}{x} + \dfrac{1}{y} - \dfrac{4}{z} = 4$

9.3

SOLUTION OF SYSTEMS USING SECOND-ORDER DETERMINANTS

We now introduce a mathematical tool called a *determinant* that can be used to solve linear systems. In this section we limit our study to the solution of systems of two equations in two variables.

2×2 determinants

An expression of the form

$$\begin{vmatrix} a_1 & b_1 \\ a_2 & b_2 \end{vmatrix}$$

is called a **determinant**. The numbers a_1, b_1, a_2, and b_2 are called elements of the determinant. Because this determinant has two rows and two columns of elements, it is called a two-by-two (2×2) determinant or a determinant of order two. We define this determinant as follows.

$$\begin{vmatrix} \mathbf{a_1} & \mathbf{b_1} \\ \mathbf{a_2} & \mathbf{b_2} \end{vmatrix} = \mathbf{a_1 b_2 - a_2 b_1}.$$

This value is obtained by multiplying the elements on the diagonals and subtracting the second product from the first product. The process can be shown schematically as

$$\begin{vmatrix} a_1 & b_1 \\ a_2 & b_2 \end{vmatrix} = a_1b_2 - a_2b_1.$$

A determinant is simply one way to represent a single number.

EXAMPLE 1 a. $\begin{vmatrix} 1 & 2 \\ -1 & 3 \end{vmatrix} \begin{aligned} &= (1)(3) - (-1)(2) \\ &= 3 - (-2) = 5 \end{aligned}$

b. $\begin{vmatrix} 0 & -1 \\ -1 & 7 \end{vmatrix} \begin{aligned} &= (0)(7) - (-1)(-1) \\ &= 0 - 1 = -1. \end{aligned}$

c. $\begin{vmatrix} -2 & 3 \\ 4 & -1 \end{vmatrix} \begin{aligned} &= (-2)(-1) - (4)(3) \\ &= 2 - 12 = -10 \end{aligned}$

d. $\begin{vmatrix} 4 & -2 \\ 1 & 0 \end{vmatrix} \begin{aligned} &= (4)(0) - (1)(-2) \\ &= 0 - (-2) = 2 \end{aligned}$ ■

Solutions of linear systems

Determinants can be used to solve linear systems. In this section, we shall confine our attention to linear systems of two equations in two variables of the form

$$a_1x + b_1y = c_1 \tag{1}$$

$$a_2x + b_2y = c_2. \tag{2}$$

If this system is solved by means of a linear combination, we have, upon multiplication of Equation (1) by $-a_2$ and Equation (2) by a_1, the equations

$$-a_1a_2x - a_2b_1y = -a_2c_1 \tag{1a}$$

$$a_1a_2x + a_1b_2y = a_1c_2. \tag{2a}$$

The sum of the members of (1a) and (2a) is

$$a_1b_2y - a_2b_1y = a_1c_2 - a_2c_1.$$

Now, factoring y from each term in the left-hand member, we have

$$(a_1b_2 - a_2b_1)y = a_1c_2 - a_2c_1,$$

from which

$$y = \frac{a_1c_2 - a_2c_1}{a_1b_2 - a_2b_1} \qquad (a_1b_2 - a_2b_1 \neq 0). \tag{3}$$

Now, the numerator of (3) is just the value of the determinant

$$\begin{vmatrix} a_1 & c_1 \\ a_2 & c_2 \end{vmatrix},$$

which we designate as D_y, and the denominator is the value of the determinant

$$\begin{vmatrix} a_1 & b_1 \\ a_2 & b_2 \end{vmatrix},$$

which we designate as D; so Equation (3) can be written as follows.

$$\boldsymbol{y = \frac{D_y}{D} = \frac{\begin{vmatrix} a_1 & c_1 \\ a_2 & c_2 \end{vmatrix}}{\begin{vmatrix} a_1 & b_1 \\ a_2 & b_2 \end{vmatrix}}}. \tag{4}$$

The elements of the determinant in the denominator of (4) are the coefficients of the variables in (1) and (2). The elements of the determinant in the numerator of (4) are identical to those in the denominator, except that *the elements in the column containing the coefficients of y have been replaced by c_1 and c_2*, the constant terms of (1) and (2).

By exactly the same procedure, we can show the following.

$$\boldsymbol{x = \frac{D_x}{D} = \frac{\begin{vmatrix} c_1 & b_1 \\ c_2 & b_2 \end{vmatrix}}{\begin{vmatrix} a_1 & b_1 \\ a_2 & b_2 \end{vmatrix}}}. \tag{5}$$

Equations (4) and (5) together yield the components of the solution of the system. The use of determinants in this way is known as **Cramer's rule** for the solution of a system of linear equations.

If $D = 0$ when using Cramer's rule, the equations in the system are either dependent or inconsistent, depending upon whether or not D_y and D_x are both 0. This follows from the discussion on page 308, where these conditions are considered in terms of the coefficients a_1, b_1, a_2, b_2, and the constant terms c_1 and c_2.

EXAMPLE 2 Solve the following system using Cramer's rule.

$$\begin{aligned} 2x - 3y &= 6 \\ 2x + y &= 14 \end{aligned}$$

Solution

$$D = \begin{vmatrix} 2 & -3 \\ 2 & 1 \end{vmatrix} = (2)(1) - (2)(-3) = 8$$

The elements in D_x are obtained from the elements in D by replacing the elements in the column containing the coefficients of x with the corresponding constants 6 and 14.

$$D_x = \begin{vmatrix} 6 & -3 \\ 14 & 1 \end{vmatrix} = (6)(1) - (14)(-3) = 48$$

The elements in D_y are obtained from the elements in D by replacing the elements in the column containing the coefficients of y with the corresponding constants 6 and 14.

$$D_y = \begin{vmatrix} 2 & 6 \\ 2 & 14 \end{vmatrix} = (2)(14) - (2)(6) = 16$$

Values for x and y can now be determined by Cramer's rule.

$$x = \frac{D_x}{D} = \frac{48}{8} = 6 \qquad y = \frac{D_y}{D} = \frac{16}{8} = 2$$

The solution set is $\{(6, 2)\}$. ■

EXERCISE 9.3

A

■ *Evaluate. See Example 1.*

1. $\begin{vmatrix} 1 & 0 \\ 2 & 1 \end{vmatrix}$

2. $\begin{vmatrix} 3 & -2 \\ 4 & 1 \end{vmatrix}$

3. $\begin{vmatrix} -5 & -1 \\ 3 & 3 \end{vmatrix}$

4. $\begin{vmatrix} 1 & -2 \\ -1 & 2 \end{vmatrix}$

5. $\begin{vmatrix} -1 & 6 \\ 0 & -2 \end{vmatrix}$

6. $\begin{vmatrix} 20 & 3 \\ -20 & -2 \end{vmatrix}$

7. $\begin{vmatrix} -2 & -1 \\ -3 & -4 \end{vmatrix}$

8. $\begin{vmatrix} -1 & -5 \\ -2 & -6 \end{vmatrix}$

■ *Solve each system using Cramer's rule. See Example 2.*

9. $2x - 3y = -1$
   $x + 4y = 5$

10. $3x - 4y = -2$
    $x - 2y = 0$

11. $3x - 4y = -2$
    $6x + 12y = 36$

12. $2x - 4y = 7$
    $x - 2y = 1$

13. $\frac{1}{3}x - \frac{1}{2}y = 0$
    $\frac{1}{2}x + \frac{1}{4}y = 4$

14. $\frac{2}{3}x + y = 1$
    $x - \frac{4}{3}y = 0$

15. $x - 2y = 5$
    $\frac{2}{3}x - \frac{4}{3}y = 6$

16. $\frac{1}{2}x + y = 3$
    $-\frac{1}{4}x - y = -3$

17. $x - 3y = 1$
    $y = 1$

18. $2x - 3y = 12$
    $x = 4$

19. $ax + by = 1$
    $bx + ay = 1$

20. $x + y = a$
    $x - y = b$

B

■ *Show that each statement is true for every real value of each variable.*

21. $\begin{vmatrix} a & a \\ b & b \end{vmatrix} = 0$

22. $\begin{vmatrix} a_1 & b_1 \\ a_2 & b_2 \end{vmatrix} = -\begin{vmatrix} a_2 & b_2 \\ a_1 & b_1 \end{vmatrix}$

23. $\begin{vmatrix} a_1 & b_1 \\ a_2 & b_2 \end{vmatrix} = -\begin{vmatrix} b_1 & a_1 \\ b_2 & a_2 \end{vmatrix}$

24. $\begin{vmatrix} ka_1 & b_1 \\ ka_2 & b_2 \end{vmatrix} = k\begin{vmatrix} a_1 & b_1 \\ a_2 & b_2 \end{vmatrix}$

25. $\begin{vmatrix} ka & a \\ kb & b \end{vmatrix} = 0$

26. $\begin{vmatrix} a_1 & b_1 \\ ka_2 & kb_2 \end{vmatrix} = k\begin{vmatrix} a_1 & b_1 \\ a_2 & b_2 \end{vmatrix}$

27. Show that if both $D_y = 0$ and $D_x = 0$, it follows that $D = 0$ when c_1 and c_2 are not both 0, and the equations in the system

$$a_1x + b_1y = c_1$$
$$a_2x + b_2y = c_2$$

are dependent. [*Hint:* Show that the first two determinant equations imply that $a_1c_2 = a_2c_1$ and $b_1c_2 = b_2c_1$ and that the rest follows from the formation of a proportion with these equations.]

28. Show that for the system given in Problem 27, if $D = 0$ and $D_x = 0$, then $D_y = 0$.

9.4

SOLUTION OF SYSTEMS USING THIRD-ORDER DETERMINANTS

A third-order determinant is defined as follows.

$$\begin{vmatrix} a_1 & b_1 & c_1 \\ a_2 & b_2 & c_2 \\ a_3 & b_3 & c_3 \end{vmatrix} = a_1b_2c_3 - a_1b_3c_2 + a_3b_1c_2 - a_2b_1c_3 + a_2b_3c_1 - a_3b_2c_1. \quad (1)$$

Again, we note that a 3×3 determinant is simply a number, namely, that number represented by the expression in the right-hand member of (1). We can rewrite (1) in a simpler form in terms of 2×2 determinants formed from the elements of the 3×3 determinant.

Minor of a determinant

The **minor** of an element in a determinant is defined as the determinant that remains after deleting the row and column in which the element appears. In the determinant (1), for example,

$$\text{the minor of the element } a_1 \text{ is } \begin{vmatrix} b_2 & c_2 \\ b_3 & c_3 \end{vmatrix},$$

$$\text{the minor of the element } b_1 \text{ is } \begin{vmatrix} a_2 & c_2 \\ a_3 & c_3 \end{vmatrix},$$

$$\text{the minor of the element } c_1 \text{ is } \begin{vmatrix} a_2 & b_2 \\ a_3 & b_3 \end{vmatrix}.$$

Expansion by minors

If, by suitably factoring pairs of terms in the right-hand member, (1) is rewritten in the form

$$\begin{vmatrix} a_1 & b_1 & c_1 \\ a_2 & b_2 & c_2 \\ a_3 & b_3 & c_3 \end{vmatrix} = a_1(b_2c_3 - b_3c_2) - b_1(a_2c_3 - a_3c_2) + c_1(a_2b_3 - a_3b_2), \quad (2)$$

we observe that the sums enclosed in parentheses in the right-hand member of (2) are the respective minors (second-order determinants) of the elements a_1, b_1, and c_1. Therefore, (2) can be written

$$\begin{vmatrix} a_1 & b_1 & c_1 \\ a_2 & b_2 & c_2 \\ a_3 & b_3 & c_3 \end{vmatrix} = a_1\begin{vmatrix} b_2 & c_2 \\ b_3 & c_3 \end{vmatrix} - b_1\begin{vmatrix} a_2 & c_2 \\ a_3 & c_3 \end{vmatrix} + c_1\begin{vmatrix} a_2 & b_2 \\ a_3 & b_3 \end{vmatrix}. \quad (3)$$

The right-hand member of (3) is called the **expansion** of the determinant by minors *about the first row.*

Suppose, instead of factoring the right-hand member of (1) into the right-hand member of (2), we factor it as

$$\begin{vmatrix} a_1 & b_1 & c_1 \\ a_2 & b_2 & c_2 \\ a_3 & b_3 & c_3 \end{vmatrix} = a_1(b_2c_3 - b_3c_2) - a_2(b_1c_3 - b_3c_1) + a_3(b_1c_2 - b_2c_1). \quad (4)$$

Then we have the expansion of the determinant by minors *about the first column,*

$$\begin{vmatrix} a_1 & b_1 & c_1 \\ a_2 & b_2 & c_2 \\ a_3 & b_3 & c_3 \end{vmatrix} = a_1\begin{vmatrix} b_2 & c_2 \\ b_3 & c_3 \end{vmatrix} - a_2\begin{vmatrix} b_1 & c_1 \\ b_3 & c_3 \end{vmatrix} + a_3\begin{vmatrix} b_1 & c_1 \\ b_2 & c_2 \end{vmatrix}. \quad (5)$$

With the proper use of signs, it is possible to expand a determinant by minors about *any* row or *any* column and obtain an expression equivalent to a factored form of the right-hand member of (1). A helpful device for determining the signs of the terms in an expansion of a third-order determinant by minors is the array of alternating signs

$$\begin{matrix} + & - & + \\ - & + & - \\ + & - & + \end{matrix}$$

which we call the **sign array** for the determinant. To obtain an expansion of (1) about a given row or column, the appropriate sign from the sign array is prefixed to each term in the expansion.

EXAMPLE 1 a. If we first expand the determinant

$$\begin{vmatrix} 1 & 2 & -3 \\ 0 & 2 & -1 \\ 1 & 1 & 0 \end{vmatrix} \tag{6}$$

about the second row, we have

$$\begin{aligned} \begin{vmatrix} 1 & 2 & -3 \\ 0 & 2 & -1 \\ 1 & 1 & 0 \end{vmatrix} &= -0\begin{vmatrix} 2 & -3 \\ 1 & 0 \end{vmatrix} + 2\begin{vmatrix} 1 & -3 \\ 1 & 0 \end{vmatrix} - (-1)\begin{vmatrix} 1 & 2 \\ 1 & 1 \end{vmatrix} \\ &= 0 + 2(0 + 3) + 1(1 - 2) \\ &= 6 - 1 = 5. \end{aligned}$$

b. If we expand (6) about the third row, we have

$$\begin{aligned} \begin{vmatrix} 1 & 2 & -3 \\ 0 & 2 & -1 \\ 1 & 1 & 0 \end{vmatrix} &= 1\begin{vmatrix} 2 & -3 \\ 2 & -1 \end{vmatrix} - 1\begin{vmatrix} 1 & -3 \\ 0 & -1 \end{vmatrix} + 0\begin{vmatrix} 1 & 2 \\ 0 & 2 \end{vmatrix} \\ &= 1(-2 + 6) - 1(-1 - 0) + 0 \\ &= 4 + 1 = 5. \end{aligned}$$

You should expand this determinant about the first row and about each column to verify that the result is the same in each expansion. ■

The expansion of a higher-order determinant by minors can be accomplished in the same way. By continuing the pattern of alternating signs used for third-order determinants, the sign array extends to higher-order determinants. The determinants in each term in the expansion will be of order one less than the order of the original determinant.

Cramer's rule Consider the linear system in three variables

$$a_1x + b_1y + c_1z = d_1 \tag{1}$$

$$a_2x + b_2y + c_2z = d_2 \tag{2}$$

$$a_3x + b_3y + c_3z = d_3. \tag{3}$$

By solving this system using the methods of Section 9.5, it can be shown that Cramer's rule is applicable to such systems and, in fact, to all similar systems as well as to linear systems in two variables.

$$x = \frac{D_x}{D}, \qquad y = \frac{D_y}{D}, \qquad z = \frac{D_z}{D},$$

where

$$D = \begin{vmatrix} a_1 & b_1 & c_1 \\ a_2 & b_2 & c_2 \\ a_3 & b_3 & c_3 \end{vmatrix}, \qquad D_x = \begin{vmatrix} d_1 & b_1 & c_1 \\ d_2 & b_2 & c_2 \\ d_3 & b_3 & c_3 \end{vmatrix},$$

$$D_y = \begin{vmatrix} a_1 & d_1 & c_1 \\ a_2 & d_2 & c_2 \\ a_3 & d_3 & c_3 \end{vmatrix}, \qquad D_z = \begin{vmatrix} a_1 & b_1 & d_1 \\ a_2 & b_2 & d_2 \\ a_3 & b_3 & d_3 \end{vmatrix}.$$

Note that the elements of the determinant D in each denominator are the coefficients of the variables in (1), (2), and (3). Note also that the numerators are formed from D by replacing the elements in the x, y, or z column, respectively, by d_1, d_2, and d_3.

EXAMPLE 2 Solve the following system using Cramer's rule.

$$\begin{aligned} x + 2y - 3z &= -4 \\ 2x - y + z &= 3 \\ 3x + 2y + z &= 10 \end{aligned}$$

Solution The determinant D, with elements that are the coefficients of the variables, is given by

$$D = \begin{vmatrix} 1 & 2 & -3 \\ 2 & -1 & 1 \\ 3 & 2 & 1 \end{vmatrix}.$$

We can expand the determinant about the first column, obtaining

$$D = \begin{vmatrix} 1 & 2 & -3 \\ 2 & -1 & 1 \\ 3 & 2 & 1 \end{vmatrix} = 1\begin{vmatrix} -1 & 1 \\ 2 & 1 \end{vmatrix} - 2\begin{vmatrix} 2 & -3 \\ 2 & 1 \end{vmatrix} + 3\begin{vmatrix} 2 & -3 \\ -1 & 1 \end{vmatrix}$$

$$= -3 - 16 - 3 = -22.$$

Replacing the first column in D with -4, 3, and 10, we obtain

$$D_x = \begin{vmatrix} -4 & 2 & -3 \\ 3 & -1 & 1 \\ 10 & 2 & 1 \end{vmatrix}.$$

Expanding D_x about the third column, we have

$$D_x = \begin{vmatrix} -4 & 2 & -3 \\ 3 & -1 & 1 \\ 10 & 2 & 1 \end{vmatrix}$$

$$= -3\begin{vmatrix} 3 & -1 \\ 10 & 2 \end{vmatrix} - 1\begin{vmatrix} -4 & 2 \\ 10 & 2 \end{vmatrix} + 1\begin{vmatrix} -4 & 2 \\ 3 & -1 \end{vmatrix}$$

$$= -48 + 28 - 2 = -22.$$

In a similar fashion, we can compute D_y and D_z:

$$D_y = \begin{vmatrix} 1 & -4 & -3 \\ 2 & 3 & 1 \\ 3 & 10 & 1 \end{vmatrix} = -44, \qquad D_z = \begin{vmatrix} 1 & 2 & -4 \\ 2 & -1 & 3 \\ 3 & 2 & 10 \end{vmatrix} = -66.$$

We then have

$$x = \frac{D_x}{D} = \frac{-22}{-22} = 1, \qquad y = \frac{D_y}{D} = \frac{-44}{-22} = 2, \qquad \text{and} \qquad z = \frac{D_z}{D} = \frac{-66}{-22} = 3.$$

The solution set of the system is therefore $\{(1, 2, 3)\}$. ■

As noted on page 322 for a linear system in two variables, if $D = 0$ for a linear system in three variables, the system does not have a unique solution.

EXERCISE 9.4

A

■ *Evaluate each determinant. See Example* 1.

1. $\begin{vmatrix} 2 & 0 & 1 \\ 1 & 1 & 2 \\ -1 & 0 & 1 \end{vmatrix}$

2. $\begin{vmatrix} 1 & 3 & 1 \\ -1 & 2 & 1 \\ 0 & 2 & 0 \end{vmatrix}$

3. $\begin{vmatrix} 2 & -1 & 0 \\ -3 & 1 & 2 \\ 1 & -3 & 1 \end{vmatrix}$

4. $\begin{vmatrix} 2 & 4 & -1 \\ -1 & 3 & 2 \\ 4 & 0 & 2 \end{vmatrix}$

5. $\begin{vmatrix} 1 & 2 & 3 \\ 3 & -1 & 2 \\ 2 & 0 & 2 \end{vmatrix}$

6. $\begin{vmatrix} 1 & 0 & 0 \\ 0 & 1 & 2 \\ 0 & 3 & 4 \end{vmatrix}$

7. $\begin{vmatrix} -1 & 0 & 2 \\ -2 & 1 & 0 \\ 0 & 1 & -3 \end{vmatrix}$

8. $\begin{vmatrix} 2 & 1 & 4 \\ 3 & 2 & 6 \\ 5 & -3 & 10 \end{vmatrix}$

9. $\begin{vmatrix} 2 & 5 & -1 \\ 1 & 0 & 2 \\ 0 & 0 & 1 \end{vmatrix}$

10. $\begin{vmatrix} 2 & 3 & 1 \\ 0 & 1 & 0 \\ -4 & 2 & 1 \end{vmatrix}$

11. $\begin{vmatrix} a & b & 1 \\ a & b & 1 \\ 1 & 1 & 1 \end{vmatrix}$

12. $\begin{vmatrix} a & a & a \\ 1 & 2 & 3 \\ 4 & 5 & 6 \end{vmatrix}$

13. $\begin{vmatrix} x & 0 & 0 \\ 0 & x & 0 \\ 0 & 0 & x \end{vmatrix}$

14. $\begin{vmatrix} 0 & 0 & x \\ 0 & x & 0 \\ x & 0 & 0 \end{vmatrix}$

15. $\begin{vmatrix} x & y & 0 \\ x & y & 0 \\ 0 & 0 & 1 \end{vmatrix}$

16. $\begin{vmatrix} 0 & a & b \\ a & 0 & a \\ b & a & 0 \end{vmatrix}$

17. $\begin{vmatrix} a & b & 0 \\ b & 0 & b \\ 0 & b & a \end{vmatrix}$

18. $\begin{vmatrix} 0 & b & 0 \\ b & a & b \\ 0 & b & 0 \end{vmatrix}$

■ *Solve each system using Cramer's rule. If a unique solution does not exist ($D = 0$), so state. See Example 2.*

19. $\begin{aligned} x + y &= 2 \\ 2x - z &= 1 \\ 2y - 3z &= -1 \end{aligned}$

20. $\begin{aligned} 2x - 6y + 3z &= -12 \\ 3x - 2y + 5z &= -4 \\ 4x + 5y - 2z &= 10 \end{aligned}$

21. $\begin{aligned} x - 2y + z &= -1 \\ 3x + y - 2z &= 4 \\ y - z &= 1 \end{aligned}$

22. $\begin{aligned} 2x + 5z &= 9 \\ 4x + 3y &= -1 \\ 3y - 4z &= -13 \end{aligned}$

23. $\begin{aligned} 2x + 2y + z &= 1 \\ x - y + 6z &= 21 \\ 3x + 2y - z &= -4 \end{aligned}$

24. $\begin{aligned} 4x + 8y + z &= -6 \\ 2x - 3y + 2z &= 0 \\ x + 7y - 3z &= -8 \end{aligned}$

25. $\begin{aligned} x + y + z &= 0 \\ 2x - y - 4z &= 15 \\ x - 2y - z &= 7 \end{aligned}$

26. $\begin{aligned} x + y - 2z &= 3 \\ 3x - y + z &= 5 \\ 3x + 3y - 6z &= 9 \end{aligned}$

27. $\begin{aligned} x - 2y + 2z &= 3 \\ 2x - 4y + 4z &= 1 \\ 3x - 3y - 3z &= 4 \end{aligned}$

28. $\begin{aligned} 3x - 2y + 5z &= 6 \\ 4x - 4y + 3z &= 0 \\ 5x - 4y + z &= -5 \end{aligned}$

29. $\begin{aligned} \frac{1}{4}x - z &= -\frac{1}{4} \\ x + y &= \frac{2}{3} \\ 3x + 4z &= 5 \end{aligned}$

30. $\begin{aligned} 2x - \frac{2}{3}y + z &= 2 \\ \frac{1}{2}x - \frac{1}{3}y - \frac{1}{4}z &= 0 \\ 4x + 5y - 3z &= -1 \end{aligned}$

31. $\begin{aligned} x + 4z &= 3 \\ y + 3z &= 9 \\ 2x + 5y - 5z &= -5 \end{aligned}$

32. $\begin{aligned} 2x + y &= 18 \\ y + z &= -1 \\ 3x - 2y - 5z &= 38 \end{aligned}$

B

33. Show that for all values of x, y, and z,

$$\begin{vmatrix} x & x & a \\ y & y & b \\ z & z & c \end{vmatrix} = 0.$$

[*Hint:* Expand about the elements of the third column.] Make a conjecture about determinants that contain two identical columns.

34. Show that

$$\begin{vmatrix} 0 & 0 & 0 \\ a & b & c \\ d & e & f \end{vmatrix} = 0$$

for all values of a, b, c, d, e, and f. Make a conjecture about determinants that contain a row of 0 elements.

35. Show that

$$\begin{vmatrix} 1 & 2 & 3 \\ 4 & 5 & 6 \\ 0 & 0 & 1 \end{vmatrix} = -\begin{vmatrix} 4 & 5 & 6 \\ 1 & 2 & 3 \\ 0 & 0 & 1 \end{vmatrix}.$$

Make a conjecture about the result of interchanging any two rows of a determinant.

36. Show that

$$\begin{vmatrix} 2 & 0 & 1 \\ 4 & 1 & -2 \\ 6 & 1 & 1 \end{vmatrix} = 2\begin{vmatrix} 1 & 0 & 1 \\ 2 & 1 & -2 \\ 3 & 1 & 1 \end{vmatrix}.$$

Make a conjecture about the result of factoring a common factor from each element of a column in a determinant.

37. Show that the graph of

$$\begin{vmatrix} x & y & 1 \\ 4 & -1 & 1 \\ 2 & 3 & 1 \end{vmatrix} = 0$$

is a line containing the points $(4, -1)$ and $(2, 3)$.

38. Show that the slope-intercept form of the equation of a line can be written

$$\begin{vmatrix} x & y & 1 \\ 0 & b & 1 \\ 1 & m & 0 \end{vmatrix} = 0.$$

9.5

SOLUTION OF SYSTEMS USING MATRICES

In previous sections we solved linear systems by using linear combinations and determinants. In this section we consider another mathematical tool called a *matrix* (plural: matrices) that has wide applications in mathematics, and in particular, we will see how it can also be used to solve linear systems.

A **matrix** is a rectangular array of elements or **entries** (in this book, real numbers). These entries are ordinarily displayed using brackets or parentheses (we shall use brackets). Thus,

$$\begin{bmatrix} 1 & 2 & 3 \\ 4 & 5 & 6 \\ 7 & 8 & 9 \end{bmatrix}, \quad \begin{bmatrix} 2 & -1 & 3 \\ 4 & 0 & 2 \end{bmatrix}, \quad \begin{bmatrix} 4 \\ 5 \\ 6 \end{bmatrix}$$

are matrices with real-number elements. The **order**, or **dimension**, of a matrix is the ordered pair having as first component the number of (horizontal) rows and as second component the number of (vertical) columns in the matrix. Thus, the matrices above are 3×3 (read "three-by-three"), 2×3 (read "two-by-three"), and 3×1 (read "three-by-one"), respectively. Note that the first matrix—where the number of rows is equal to the number of columns—is an example of a **square** matrix.

Elementary transformations

We can transform one matrix into another in a variety of ways. However, here we are only concerned with the following kinds of transformations:

1. Multiplying the entries of any row by a nonzero real number.
2. Interchanging two rows.
3. Multiplying the entries of any row by a real number and adding the results to the corresponding elements of another row.

Such transformations are called **elementary transformations** or **row operations**, and, if a matrix A is transformed into a matrix B by a finite succession of such transformations, then we say that A and B are **row-equivalent**. We represent this by writing $A \sim B$ (read "A is row-equivalent to B"). For example,

$$A = \begin{bmatrix} 1 & 3 & -1 \\ 2 & 1 & 4 \\ 6 & 2 & -1 \end{bmatrix} \quad \text{and} \quad B = \begin{bmatrix} 1 & 3 & -1 \\ 6 & 3 & 12 \\ 6 & 2 & -1 \end{bmatrix}$$

are row-equivalent, because we can multiply each entry in row 2 of A by 3 to obtain B;

$$A = \begin{bmatrix} 3 & -1 & 2 \\ 2 & 1 & 4 \\ 3 & 1 & 9 \end{bmatrix} \quad \text{and} \quad B = \begin{bmatrix} 3 & 1 & 9 \\ 2 & 1 & 4 \\ 3 & -1 & 2 \end{bmatrix}$$

are row-equivalent, because we can interchange rows 1 and 3 of A to obtain B; and

$$A = \begin{bmatrix} 1 & 2 & 1 \\ 2 & 0 & -1 \\ 3 & 1 & 2 \end{bmatrix} \quad \text{and} \quad B = \begin{bmatrix} 1 & 2 & 1 \\ 0 & -4 & -3 \\ 3 & 1 & 2 \end{bmatrix}$$

are row-equivalent, because we can multiply each entry in row 1 of A by -2 and add the results to the corresponding entries in row 2 of A to obtain B.

It is often convenient to perform more than one elementary transformation on a given matrix. For example, if in the matrix

$$\begin{bmatrix} 1 & -2 & 1 \\ 2 & 1 & 3 \\ -3 & 0 & 0 \end{bmatrix}$$

we add $-2 \cdot$ (row 1) to row 2, and $3 \cdot$ (row 1) to row 3, we obtain the row-equivalent matrix

$$\begin{bmatrix} 1 & -2 & 1 \\ 0 & 5 & 1 \\ 0 & -6 & 3 \end{bmatrix}.$$

EXAMPLE 1 Use row operations on the first matrix to form an equivalent matrix with the given elements.

$$\begin{bmatrix} 1 & -4 \\ 3 & 6 \end{bmatrix} \sim \begin{bmatrix} 1 & -4 \\ 0 & ? \end{bmatrix}$$

Solution

$$\begin{matrix} \\ -3(\text{row }1) + \text{row }2 \end{matrix} \begin{bmatrix} 1 & -4 \\ 3 & 6 \end{bmatrix} \rightarrow \begin{bmatrix} 1 & -4 \\ 0 & 18 \end{bmatrix} \quad \blacksquare$$

EXAMPLE 2 Use row operations on the first matrix to form an equivalent matrix with the given elements.

$$\begin{bmatrix} 1 & -3 & 1 \\ 3 & 1 & -1 \\ 2 & -2 & 3 \end{bmatrix} \sim \begin{bmatrix} 1 & -3 & 1 \\ 0 & ? & ? \\ 0 & ? & ? \end{bmatrix}$$

Solution

$$\begin{matrix} \\ -3(\text{row }1) + \text{row }2 \\ -2(\text{row }1) + \text{row }3 \end{matrix} \begin{bmatrix} 1 & -3 & 1 \\ 3 & 1 & -1 \\ 2 & -2 & 3 \end{bmatrix} \rightarrow \begin{bmatrix} 1 & -3 & 1 \\ 0 & 10 & -4 \\ 0 & 4 & 1 \end{bmatrix} \quad \blacksquare$$

EXAMPLE 3 Use row operations on the first matrix to form an equivalent matrix with the given elements.

$$\begin{bmatrix} 1 & -3 & 1 \\ 0 & 10 & -4 \\ 0 & 4 & 1 \end{bmatrix} \sim \begin{bmatrix} 1 & -3 & 1 \\ 0 & 5 & -2 \\ 0 & 0 & 13/5 \end{bmatrix}$$

Solution First obtain a row-equivalent matrix with second row $[0 \quad 5 \quad -2]$.

$$\begin{matrix} \\ \tfrac{1}{2}(\text{row } 2) \\ \\ \end{matrix} \begin{bmatrix} 1 & -3 & 1 \\ 0 & 10 & -4 \\ 0 & 4 & 1 \end{bmatrix} \to \begin{bmatrix} 1 & -3 & 1 \\ 0 & 5 & -2 \\ 0 & 4 & 1 \end{bmatrix}$$

Now obtain a row-equivalent matrix with third row $[0 \quad 0 \quad 13/5]$.

$$\begin{matrix} \\ \\ -\tfrac{4}{5}(\text{row } 2) + \text{row } 3 \end{matrix} \begin{bmatrix} 1 & -3 & 1 \\ 0 & 5 & -2 \\ 0 & 4 & 1 \end{bmatrix} \to \begin{bmatrix} 1 & -3 & 1 \\ 0 & 5 & -2 \\ 0 & 0 & 13/5 \end{bmatrix}$$ ■

Solution of linear systems

In a system of linear equations of the form

$$\begin{aligned} a_1x + b_1y + c_1z &= d_1 \\ a_2x + b_2y + c_2z &= d_2 \\ a_3x + b_3y + c_3z &= d_3, \end{aligned}$$

the matrices

$$\begin{bmatrix} a_1 & b_1 & c_1 \\ a_2 & b_2 & c_2 \\ a_3 & b_3 & c_3 \end{bmatrix} \quad \text{and} \quad \left[\begin{array}{ccc|c} a_1 & b_1 & c_1 & d_1 \\ a_2 & b_2 & c_2 & d_2 \\ a_3 & b_3 & c_3 & d_3 \end{array}\right]$$

are called the **coefficient matrix** and the **augmented matrix**, respectively.

By performing elementary transformations on the augmented matrix of a system of equations, we can obtain a matrix from which the solution set of the system is readily determined. The validity of the method, as illustrated in the examples below, stems from the fact that performing elementary row operations on the augmented matrix of a system corresponds to forming equivalent systems of equations.

For example, the augmented matrix of

$$\begin{aligned} x + 2y &= 7 \\ 2x - y &= 4 \end{aligned} \quad \text{is} \quad \left[\begin{array}{cc|c} 1 & 2 & 7 \\ 2 & -1 & 4 \end{array}\right].$$

Performing elementary transformations, we have:

$$-2(\text{row } 1) + \text{row } 2 \left[\begin{array}{cc|c} 1 & 2 & 7 \\ 0 & -5 & -10 \end{array}\right].$$

This matrix corresponds to the system

$$\begin{aligned} x + 2y &= 7 \\ -5y &= -10. \end{aligned}$$

From the last equation, $y = 2$. Substituting 2 for y in the first equation, we obtain $x = 3$. Hence, the solution set is $\{(3, 2)\}$.

Note that in the above example the elements $[0 \quad -5 \quad -10]$ in the second row of the final augmented matrix correspond to an equation in one variable $-5y = -10$. In general, to solve a system using matrices, we perform elementary transformations to obtain a coefficient matrix with only one nonzero element in the last row.

EXAMPLE 4 Use matrices to solve the system

$$\begin{aligned} x - 2y &= -5 \\ 2x + 3y &= 11. \end{aligned}$$

Solution The augmented matrix is

$$\left[\begin{array}{cc|c} 1 & -2 & -5 \\ 2 & 3 & 11 \end{array}\right].$$

Perform elementary operations to obtain a coefficient matrix with only one nonzero element in the second row.

$$-2(\text{row } 1) + \text{row } 2 \left[\begin{array}{cc|c} 1 & -2 & -5 \\ 2 & 3 & 11 \end{array}\right] \rightarrow \left[\begin{array}{cc|c} 1 & -2 & -5 \\ 0 & 7 & 21 \end{array}\right]$$

The last matrix corresponds to the system

$$x - 2y = -5 \qquad (1)$$

$$7y = 21. \qquad (2)$$

From Equation (2), $y = 3$. Substitute 3 for y in (1).

$$x - 2(3) = 5; \qquad x = 1.$$

The solution set is $\{(1, 3)\}$. ■

The next example involves a system of three equations in three variables.

EXAMPLE 5 Solve the system

$$\begin{aligned} x - 3y + z &= -2 \\ 3x + y - z &= 8 \\ 2x - 2y + 3z &= -1. \end{aligned}$$

Solution The augmented matrix of the system is

$$\left[\begin{array}{ccc|c} 1 & -3 & 1 & -2 \\ 3 & 1 & -1 & 8 \\ 2 & -2 & 3 & -1 \end{array}\right].$$

Perform elementary row operations to obtain a coefficient matrix with only one nonzero element in the third row.

$$\begin{array}{r} \\ -3(\text{row } 1) + \text{row } 2 \\ -2(\text{row } 1) + \text{row } 3 \end{array} \left[\begin{array}{ccc|c} 1 & -3 & 1 & -2 \\ 3 & 1 & -1 & 8 \\ 2 & -2 & 3 & -1 \end{array}\right]$$

$$\downarrow$$

$$\begin{array}{r} \tfrac{1}{2}(\text{row } 2) \\ \\ \\ \end{array} \left[\begin{array}{ccc|c} 1 & -3 & 1 & -2 \\ 0 & 10 & -4 & 14 \\ 0 & 4 & 1 & 3 \end{array}\right] \qquad \begin{aligned} x - 3y + z &= -2 \\ 0x + 10y - 4z &= 14 \\ 0x + 4y + z &= 3 \end{aligned}$$

$$\downarrow$$

$$\begin{array}{r} \\ \\ -\tfrac{4}{5}(\text{row } 2) + \text{row } 3 \end{array} \left[\begin{array}{ccc|c} 1 & -3 & 1 & -2 \\ 0 & 5 & -2 & 7 \\ 0 & 4 & 1 & 3 \end{array}\right] \qquad \begin{aligned} x - 3y + z &= -2 \\ 0x + 5y - 2z &= 7 \\ 0x + 4y + z &= 3 \end{aligned}$$

$$\downarrow$$

$$\begin{array}{r} \\ \\ 5(\text{row } 3) \end{array} \left[\begin{array}{ccc|c} 1 & -3 & 1 & -2 \\ 0 & 5 & -2 & 7 \\ 0 & 0 & \tfrac{13}{5} & -\tfrac{13}{5} \end{array}\right] \qquad \begin{aligned} x - 3y + z &= -2 \\ 0x + 5y - 2z &= 7 \\ 0x + 0y + (\tfrac{13}{5})z &= -\tfrac{13}{5} \end{aligned}$$

$$\downarrow$$

$$\left[\begin{array}{ccc|c} 1 & -3 & 1 & -2 \\ 0 & 5 & -2 & 7 \\ 0 & 0 & 13 & -13 \end{array}\right] \qquad \begin{aligned} x - 3y + z &= -2 \\ 0x + 5y - 2z &= 7 \\ 0x + 0y + 13z &= -13 \end{aligned}$$

The last matrix corresponds to the system

$$\begin{aligned} x - 3y + z &= -2 \\ 5y - 2z &= 7 \\ 13z &= -13. \end{aligned}$$

Since, from the last equation, $z = -1$, we can substitute -1 for z in the second equation to obtain $y = 1$. Finally, substituting -1 for z and 1 for y in the first equation, we have $x = 2$; so the solution set is $\{(2, 1, -1)\}$. ■

If any step in this procedure results in the equation $0x + 0y + 0z = 0$ or a contradiction such as $0x + 0y + 0z = 2$, then the system contains dependent or inconsistent equations, and it has either an infinite number of members or no members in its solution set. In either case we say the system has no unique solution.

EXERCISE 9.5

A

■ *Use row operations on the first matrix to form an equivalent matrix with the given elements. See Examples* 1, 2, *and* 3.

1. $\begin{bmatrix} 1 & -3 \\ 2 & 1 \end{bmatrix} \sim \begin{bmatrix} 1 & -3 \\ 0 & ? \end{bmatrix}$

2. $\begin{bmatrix} -2 & 3 \\ 4 & 1 \end{bmatrix} \sim \begin{bmatrix} -2 & 3 \\ 0 & ? \end{bmatrix}$

3. $\begin{bmatrix} 2 & 6 \\ 5 & 3 \end{bmatrix} \sim \begin{bmatrix} 2 & 6 \\ ? & 0 \end{bmatrix}$

4. $\begin{bmatrix} 6 & 4 \\ -1 & -2 \end{bmatrix} \sim \begin{bmatrix} 6 & 4 \\ ? & 0 \end{bmatrix}$

5. $\begin{bmatrix} 1 & -2 & 2 \\ 2 & 3 & -1 \\ 4 & 1 & -3 \end{bmatrix} \sim \begin{bmatrix} 1 & -2 & 2 \\ 0 & ? & ? \\ 0 & ? & ? \end{bmatrix}$

6. $\begin{bmatrix} 2 & -1 & 3 \\ -4 & 0 & 4 \\ 6 & 2 & -1 \end{bmatrix} \sim \begin{bmatrix} 2 & -1 & 3 \\ 0 & ? & ? \\ 0 & ? & ? \end{bmatrix}$

7. $\begin{bmatrix} -1 & 4 & 3 \\ 2 & -2 & -4 \\ 1 & 2 & 3 \end{bmatrix} \sim \begin{bmatrix} -1 & 4 & 3 \\ ? & 0 & ? \\ ? & 0 & ? \end{bmatrix}$

8. $\begin{bmatrix} 3 & -2 & 4 \\ 2 & 2 & 1 \\ -1 & 1 & 5 \end{bmatrix} \sim \begin{bmatrix} 3 & -2 & 4 \\ ? & ? & 0 \\ ? & ? & 0 \end{bmatrix}$

9. $\begin{bmatrix} -2 & 1 & -3 \\ 4 & 2 & 0 \\ -6 & -1 & 2 \end{bmatrix} \sim \begin{bmatrix} -2 & 1 & -3 \\ 0 & ? & ? \\ 0 & 0 & ? \end{bmatrix}$

10. $\begin{bmatrix} -1 & 2 & 3 \\ 4 & 0 & 1 \\ -2 & 2 & -3 \end{bmatrix} \sim \begin{bmatrix} -1 & 2 & 3 \\ 0 & ? & ? \\ 0 & 0 & ? \end{bmatrix}$

■ *Use row operations on the augmented matrix to solve each system. See Example 4 for Problems 11–18.*

11. $x + 3y = 11$
$2x - y = 1$

12. $x - 5y = 11$
$2x + 3y = -4$

13. $x - 4y = -6$
$3x + y = -5$

14. $x + 6y = -14$
$5x - 3y = -4$

15. $2x + y = 5$
$3x - 5y = 14$

16. $3x - 2y = 16$
$4x + 2y = 12$

17. $x - y = -8$
$x + 2y = 9$

18. $4x - 3y = 16$
$2x + y = 8$

■ *See Example 5 for Problems 19–26.*

19. $x + 3y - z = 5$
$3x - y + 2z = 5$
$x + y + 2z = 7$

20. $x - 2y + 3z = -11$
$2x + 3y - z = 6$
$3x - y - z = 3$

21. $2x - y + z = 5$
$x - 2y - 2z = 2$
$3x + 3y - z = 4$

22. $x - 2y - 2z = 4$
$2x + y - 3z = 7$
$x - y - z = 3$

23. $2x - y - z = -4$
$x + y + z = -5$
$x + 3y - 4z = 12$

24. $x - 2y - 5z = 2$
$2x + 3y + z = 11$
$3x - y - z = 11$

25. $2x - y = 0$
$3y + z = 7$
$2x + 3z = 1$

26. $3x - z = 7$
$2x + y = 6$
$3y - z = 7$

9.6

APPLICATIONS

In previous sections we solved a variety of applied problems by constructing first-degree equations in one variable. This required us to first express each of the quantities we wished to find in terms of one variable.

It is generally easier to assign different variables to represent different quantities. However, we then must construct two or more equations from the stated conditions in a problem.

In writing systems of equations, we must be careful that the conditions giving rise to one equation are independent of the conditions giving rise to any other equation. Note that the approach we use to solve equations in the following examples is similar to the five-step procedure we used to solve word problems using one variable, even though we do not list the steps explicitly.

In Examples 1 and 2 below only two numbers are to be found. Hence it is only necessary to use two variables and a system of two independent equations.

EXAMPLE 1 The sum of two numbers is 17 and one of the numbers is 4 less than 2 times the other. Find the numbers.

Solution Represent each number by a separate variable.

One number: x
Other number: y

Represent the two independent conditions stated in the problem by two equations.

$$x + y = 17$$
$$x = 2y - 4$$

Rewrite the equations in the form

$$x + y = 17$$
$$x - 2y = -4.$$

Any one of the procedures that we considered in the previous sections can be used to solve this system. The numbers are 10 and 7. ■

EXAMPLE 2 A company offers split-rail fencing for sale in two options. One option consists of four posts and six rails for \$31; the other consists of three posts and four rails for \$22. What are the costs of one post and one rail?

Solution Represent each quantity by a separate variable.

Cost of one post: p
Cost of one rail: r

Represent the two independent conditions stated in the problem by two equations.

$$4p + 6r = 31 \qquad (1)$$
$$3p + 4r = 22 \qquad (2)$$

Again, any one of the procedures that we have used to solve systems can be used here. One post costs \$4.00 and one rail costs \$2.50. ■

The following examples illustrate how a system in three variables can be constructed when we want to find three quantities. Again in writing such systems, we must be careful that the conditions giving rise to one equation are independent of the conditions giving rise to any other equation.

EXAMPLE 3 The sum of three numbers is 12. Twice the first number is equal to the second, and the third is equal to the sum of the other two. Find the numbers.

Solution Represent each number by a separate variable.

First number: x
Second number: y
Third number: z

Write the three conditions stated in the problem as three equations.

$$x + y + z = 12$$
$$2x = y$$
$$x + y = z$$

Rewrite the equations in the form

$$x + y + z = 12 \quad (1)$$
$$2x - y = 0 \quad (2)$$
$$x + y - z = 0. \quad (3)$$

Solving the system by any of the methods in the preceding sections gives $x = 2$, $y = 4$, and $z = 6$. Hence, the first number is 2, the second number is 4, and the third number is 6. ■

EXAMPLE 4 Three solutions of $ax + by + cz = 1$ are $(0, 3, -2)$, $(3, -4, 0)$, and $(1, 0, 0)$. Find the coefficients a, b, and c.

Solution In this case we want to solve for a, b, and c. Substituting the coordinates of the given points for x, y, and z in

$$ax + by + cz = 1$$

we obtain

$$a(0) + b(3) + c(-2) = 1$$
$$a(3) + b(-4) + c(0) = 1$$
$$a(1) + b(0) + c(0) = 1.$$

The system can now be written as

$$\begin{aligned} 3b - 2c &= 1 \\ 3a - 4b &= 1 \\ a &= 1. \end{aligned}$$

Solving the system gives $a = 1$, $b = 1/2$, and $c = 1/4$. ■

EXERCISE 9.6

A

■ *Solve each problem using a system of equations. See Examples* 1 *and* 2.

1. The sum of two numbers is 24 and one of the numbers is 6 less than the other. Find the numbers.
2. The difference of two numbers is 14 and one of the numbers is 1 more than 2 times the other. Find the numbers.
3. If 1/3 of an integer is added to 1/2 the next consecutive integer, the sum is 33. Find the integers.
4. If 1/2 of an integer is added to 1/5 the next consecutive integer, the sum is 17. Find the integers.
5. The admission at a baseball game was \$1.50 for adults and \$0.85 for children. The receipts were \$93.10 for 82 paid admissions. How many adults and how many children attended the game?
6. In an election, 7179 votes were cast for two candidates. If 6 votes had switched from the winner to the loser, the loser would have won by 1 vote. How many votes were cast for each candidate?
7. A sum of \$2000 is invested, part at 10% and the remainder at 8%. Find the amount invested at each rate if the yearly income from the two investments is \$184.
8. A woman has \$1200 invested in two stocks, one of which returns 8% per year and the other 12% per year. How much has she invested in each stock if the income from the 8% stock is \$3 more than the income from the 12% stock?
9. On an airplane flight for which first-class fare is \$80 and tourist fare is \$64 there were 42 passengers. If receipts for the flight totaled \$2880, how many first-class and how many tourist passengers were on the flight?
10. A hardware retailer bought 50 machine screws—1-inch screws at 18¢ each and 3/4-inch screws at 15¢ each. If the total cost was \$8.10, how many screws of each length did the retailer purchase?

11. An airplane travels 1260 miles in the same time that an automobile travels 420 miles. If the rate of the airplane is 120 miles per hour greater than the rate of the automobile, find the rate of each.

12. Two cars start together and travel in the same direction, one going twice as fast as the other. At the end of 3 hours, they are 96 miles apart. How fast is each traveling?

■ *Solve each problem using a system of equations. See Examples* 3 *and* 4.

13. The sum of three numbers is 15. The second equals 2 times the first and the third equals the second. Find the numbers.

14. The sum of three numbers is 2. The first number is equal to the sum of the other two, and the third number is the result of subtracting the first from the second. Find the numbers.

15. A box contains \$6.25 in nickels, dimes, and quarters. There are 85 coins in all with 3 times as many nickels as dimes. How many coins of each kind are there?

16. A man has \$446 in ten-dollar, five-dollar, and one-dollar bills. There were 94 bills in all and 10 more five-dollar bills than ten-dollar bills. How many bills of each kind did he have?

17. The perimeter of a triangle is 155 inches. Side x is 20 inches shorter than side y, and side y is 5 inches longer than side z. Find the lengths of the sides of the triangle.

18. One angle of a triangle measures 10° more than a second angle, and the third angle is 10° more than six times the measure of the smallest angle. Find the measure of each angle.

B

19. In 1986 a vintner has a white wine that is 4 years older than a certain red wine. In 1976, the white wine was 2 times as old as the red wine. In what years were the wines produced?

20. In 1984, Mr. Evans died leaving a will saying that when his son was 2 times as old as his daughter, both would receive the funds from a trust. If the boy was 3 times as old as his sister in 1984 and if they receive their funds in 1989, how old was each when their father died?

21. A record company determines that each production run to manufacture a record involves an initial set-up cost of \$20 and \$0.40 for each record produced. The records sell for \$1.20 each.

    a. Express the cost C of production in terms of the number x of records produced.

    b. Express the revenue R in terms of the number of records sold.

    c. How many records must be sold for the record company to break even (no profit and no loss) on a particular production?

22. How many records must the record company in Problem 21 sell in order to break even if the price of each record is lowered to \$0.60?

23. Find a and b so that the graph of $ax + by + 3 = 0$ passes through the points $(-1, 2)$ and $(-3, 0)$. [*Hint:* If the graph of the equation passes through the points, the components of each ordered pair must be valid replacements for x and y. Substitute -1 for x and 2 for y and -3 for x and 0 for y to obtain a system in a and b.]

24. Find a and b so that the solution set of the system below is $\{(1, 2)\}$.

$$\begin{aligned} ax + by &= 4 \\ bx - ay &= -3 \end{aligned}$$

25. Three solutions of the equation $ax + by + cz = 1$ are $(-2, 0, 4)$, $(6, -1, 0)$, and $(0, 3, 0)$. Find the coefficients a, b, and c.

26. Three solutions of the equation $ax + by + bz = 1$ are $(0, 4, 2)$, $(-1, 3, 0)$, and $(-1, 0, 2)$. Find the coefficients a, b, and c.

CHAPTER SUMMARY

[9.1] The system of linear equations (in standard form)

$$\begin{aligned} a_1x + b_1y &= c_1 \\ a_2x + b_2y &= c_2 \end{aligned}$$

has no solution if the equations are **inconsistent**, infinitely many solutions if the equations are **dependent**, and exactly one solution when the equations are **consistent** and **independent.**

The system has exactly one solution if $\frac{a_1}{a_2} \neq \frac{b_1}{b_2}$.

The equations are inconsistent if $\frac{a_1}{a_2} = \frac{b_1}{b_2} \neq \frac{c_1}{c_2}$.

The equations are dependent if $\frac{a_1}{a_2} = \frac{b_1}{b_2} = \frac{c_1}{c_2}$.

A unique solution of a system can be found by using a linear combination of two independent equations to obtain one equation in one variable.

Solutions of systems of inequalities in two variables can be approximated by graphical means.

[9.2] The solution of a system of three linear equations in three variables (if a solution exists) can be obtained by first using linear combinations to form a system of two equations in two variables. The solution to this latter system contains components that are the respective components of the solution of the original system in three variables. The third component can be obtained by substituting these two values into any one of the equations of the original system.

[9.3] The order of a **determinant** is the number of rows or columns in the determinant. A second-order determinant is defined by

$$\begin{vmatrix} a_1 & b_1 \\ a_2 & b_2 \end{vmatrix} = a_1b_2 - a_2b_1.$$

Cramer's rule can be used to solve systems of two linear equations in two variables. If

$$a_1x + b_1y = c_1$$
$$a_2x + b_2y = c_2,$$

then, for $D \neq 0$,

$$x = \frac{D_x}{D} = \frac{\begin{vmatrix} c_1 & b_1 \\ c_2 & b_2 \end{vmatrix}}{\begin{vmatrix} a_1 & b_1 \\ a_2 & b_2 \end{vmatrix}} \quad \text{and} \quad y = \frac{D_y}{D} = \frac{\begin{vmatrix} a_1 & c_1 \\ a_2 & c_2 \end{vmatrix}}{\begin{vmatrix} a_1 & b_1 \\ a_2 & b_2 \end{vmatrix}}.$$

[9.4] A third-order determinant is defined by

$$\begin{vmatrix} a_1 & b_1 & c_1 \\ a_2 & b_2 & c_2 \\ a_3 & b_3 & c_3 \end{vmatrix} = a_1b_2c_3 - a_1b_3c_2 + a_3b_1c_2 - a_2b_1c_3 + a_2b_3c_1 - a_3b_2c_1.$$

The **minor** of an element in a determinant D is the determinant that results when the row and column containing the element are deleted. A determinant of an order greater than or equal to three can be evaluated by expansion by minors about any row or column.

Cramer's rule can also be used to solve systems of three linear equations in three variables. If

$$a_1x + b_1y + c_1z = d_1$$
$$a_2x + b_2y + c_2z = d_2$$
$$a_3x + b_3y + c_3z = d_3,$$

then, for $D \neq 0$,

$$x = \frac{D_x}{D}, \quad y = \frac{D_y}{D}, \quad \text{and} \quad z = \frac{D_z}{D}.$$

[9.5] A **matrix** is a rectangular array of elements or **entries.** Two matrices are said to be **row-equivalent** if one matrix can be transformed to the other by one or more of the following transformations:

1. Multiplying the entries of any row by a nonzero real number.
2. Interchanging two rows.
3. Multiplying the entries of any row by a real number and adding the results to the corresponding elements of another row.

We can use these transformations to solve linear systems of equations.

[9.6] Systems of equations in several variables can be used to solve a variety of applied problems. Different variables are used for different unknown quantities and two or more independent equations involving these variables are constructed.

REVIEW EXERCISES

A

[9.1]

■ *Solve each system by linear combinations.*

1. $x + 5y = 18$
   $x - y = -3$

2. $x + 5y = 11$
   $2x + 3y = 8$

3. $\frac{2}{3}x - 3y = 8$
   $x + \frac{3}{4}y = 12$

4. $3x = 5y - 6$
   $3y = 10 - 11x$

■ *State whether the equations in each system have a unique solution, are dependent, or inconsistent.*

5. a. $2x - 3y = 4$
   $x + 2y = 7$

   b. $2x - 3y = 4$
   $6x - 9y = 12$

6. a. $2x - 3y = 4$
   $6x - 9y = 4$

   b. $x - y = 6$
   $x - y = 6$

■ *Graph the solution set of each system of inequalities.*

7. a. $x < 3$
   $2x - y > 0$

   b. $y \leq 2$
   $x + 3y > 0$

8. a. $\begin{aligned} y - x &> 0 \\ 2x + 3y &> 6 \end{aligned}$

b. $\begin{aligned} y - x &\leq 0 \\ 3x - 2y &\geq 6 \end{aligned}$

[9.2]

■ *Solve each system using linear combinations.*

9. $\begin{aligned} x + 3y - z &= 3 \\ 2x - y + 3z &= 1 \\ 3x + 2y + z &= 5 \end{aligned}$

10. $\begin{aligned} x + y + z &= 2 \\ 3x - y + z &= 4 \\ 2x + y + 2z &= 3 \end{aligned}$

11. $\begin{aligned} x + z &= 5 \\ y - z &= -8 \\ 2x + z &= 7 \end{aligned}$

12. $\begin{aligned} x + 4y + 4z &= -20 \\ 3x - 2y + z &= -4 \\ 2x - 4y + z &= -4 \end{aligned}$

13. $\begin{aligned} \frac{1}{2}x + y + z &= 3 \\ x - 2y - \frac{1}{3}z &= -5 \\ \frac{1}{2}x - 3y - \frac{2}{3}z &= -6 \end{aligned}$

14. $\begin{aligned} \frac{3}{4}x - \frac{1}{2}y + 6z &= 2 \\ \frac{1}{2}x + y - \frac{3}{4}z &= 0 \\ \frac{1}{4}x + \frac{1}{2}y - \frac{1}{2}z &= 0 \end{aligned}$

[9.3]

15. Evaluate $\begin{vmatrix} 3 & -2 \\ 1 & -5 \end{vmatrix}$.

16. Evaluate $\begin{vmatrix} -4 & 0 \\ 2 & -6 \end{vmatrix}$.

■ *Solve each system using Cramer's rule.*

17. $\begin{aligned} x - 2y &= 6 \\ 3x + y &= 25 \end{aligned}$

18. $\begin{aligned} 2x + 3y &= -2 \\ x - 8y &= -39 \end{aligned}$

19. $\begin{aligned} \frac{1}{4}x - \frac{1}{3}y &= -\frac{5}{12} \\ \frac{1}{10}x + \frac{1}{5}y &= \frac{1}{2} \end{aligned}$

20. $\begin{aligned} \frac{2}{3}x - y &= 4 \\ x - \frac{3}{4}y &= 6 \end{aligned}$

[9.4]

■ *Evaluate each determinant.*

21. $\begin{vmatrix} 2 & 1 & 3 \\ 0 & 4 & -1 \\ 2 & 0 & 3 \end{vmatrix}$

22. $\begin{vmatrix} 3 & -1 & 2 \\ -2 & 1 & 0 \\ 2 & 4 & 1 \end{vmatrix}$

■ *Solve each system using Cramer's rule.*

23. $\begin{aligned} x + y &= 3 \\ y + z &= 5 \\ x - y + 2z &= 5 \end{aligned}$

24. $\begin{aligned} 2x + 3y - z &= -2 \\ x - y + z &= 6 \\ 3x - y + z &= 10 \end{aligned}$

25. $$\begin{aligned} x + y + z &= 2 \\ 2x - y + z &= -1 \\ x - y - z &= 0 \end{aligned}$$

26. $$\begin{aligned} x - 2y &= -3 \\ y + 3z &= -1 \\ x - z &= 2 \end{aligned}$$

[9.5]

■ *Use row operations on a matrix to solve each system.*

27. $$\begin{aligned} x - 2y &= 5 \\ 2x + y &= 5 \end{aligned}$$

28. $$\begin{aligned} 4x - 3y &= 16 \\ 2x + y &= 8 \end{aligned}$$

29. $$\begin{aligned} 2x - y &= 7 \\ 3x + 2y &= 14 \end{aligned}$$

30. $$\begin{aligned} 2x - y + 3z &= -6 \\ x + 2y - z &= 7 \\ 3x + y + z &= 2 \end{aligned}$$

31. $$\begin{aligned} x + 2y - z &= -3 \\ 2x - 3y + 2z &= 2 \\ x - y + 4z &= 7 \end{aligned}$$

32. $$\begin{aligned} x + y + z &= 1 \\ 2x - y - z &= 2 \\ 2x - y + 3z &= 2 \end{aligned}$$

[9.6]

■ *Solve each problem using two or three variables.*

33. A collection of coins consisting of dimes and quarters has a value of \$4.95. How many dimes are in the collection if there are 25 more dimes than quarters?
34. The first-class fare on an airplane flight is \$280 and the tourist fare is \$160. If 64 passengers paid a total of \$12,160 for the flight, how many of each ticket were sold?
35. A woman has invested \$8000, part in a bank at 10% and part in a savings and loan association at 12%. If her annual return is \$844, how much has she invested at each rate?
36. A sum of \$2400 is split between an investment in a mutual fund paying 14% and one in corporate bonds paying 11%. If the return on the 14% investment exceeds that on the 11% investment by \$111 per year, how much is invested at each rate?
37. An airplane travels 840 miles in the same time that an automobile travels 210 miles. If the rate of the airplane is 180 miles per hour greater than the rate of the automobile, find the rate of each.
38. One woman drives 180 miles in the same time that a second woman drives 200 miles. If the second woman drives 5 miles per hour faster than the first woman, find the speed of each woman.
39. The perimeter of a triangle is 30 centimeters. The length of one side is 7 centimeters shorter than a second side, and the third side is 1 centimeter longer than the second side. Find the length of each side.
40. One angle of a triangle measures 20° more than a second angle and the third angle is three times the measure of the first angle. Find the measure of each angle.

B

41. Graph the region that represents the solution set of the system.

$$\begin{aligned} x &\leq 0 \\ y + 3 &\geq 0 \\ y - x &\leq 0 \end{aligned}$$

Locate the coordinates of the vertices (approximation as necessary) of the boundary.

42. Use a substitution of variables to solve:

$$\begin{aligned} \frac{3}{x} - \frac{1}{y} &= \frac{7}{2} \\ \frac{2}{x} + \frac{3}{y} &= 7 \end{aligned}$$

43. Find a and b so that the graph of $y = ax + b$ passes through the points $(0, 3)$ and $(-2, 2)$.

44. Find a and b so that the graph of $ax + by - 4 = 0$ passes through the points $(2, -2)$ and $(1, 3)$.

45. Three solutions of the equation $ax + by + cz = 1$ are $(0, 2, 1)$, $(6, -1, 2)$, and $(0, 2, 0)$. Find the coefficients a, b, and c.

46. Three solutions of the equation $ax + by + cz = 1$ are $(2, 1, 0)$, $(-1, 3, 2)$, and $(3, 0, 0)$. Find the coefficients a, b, and c.

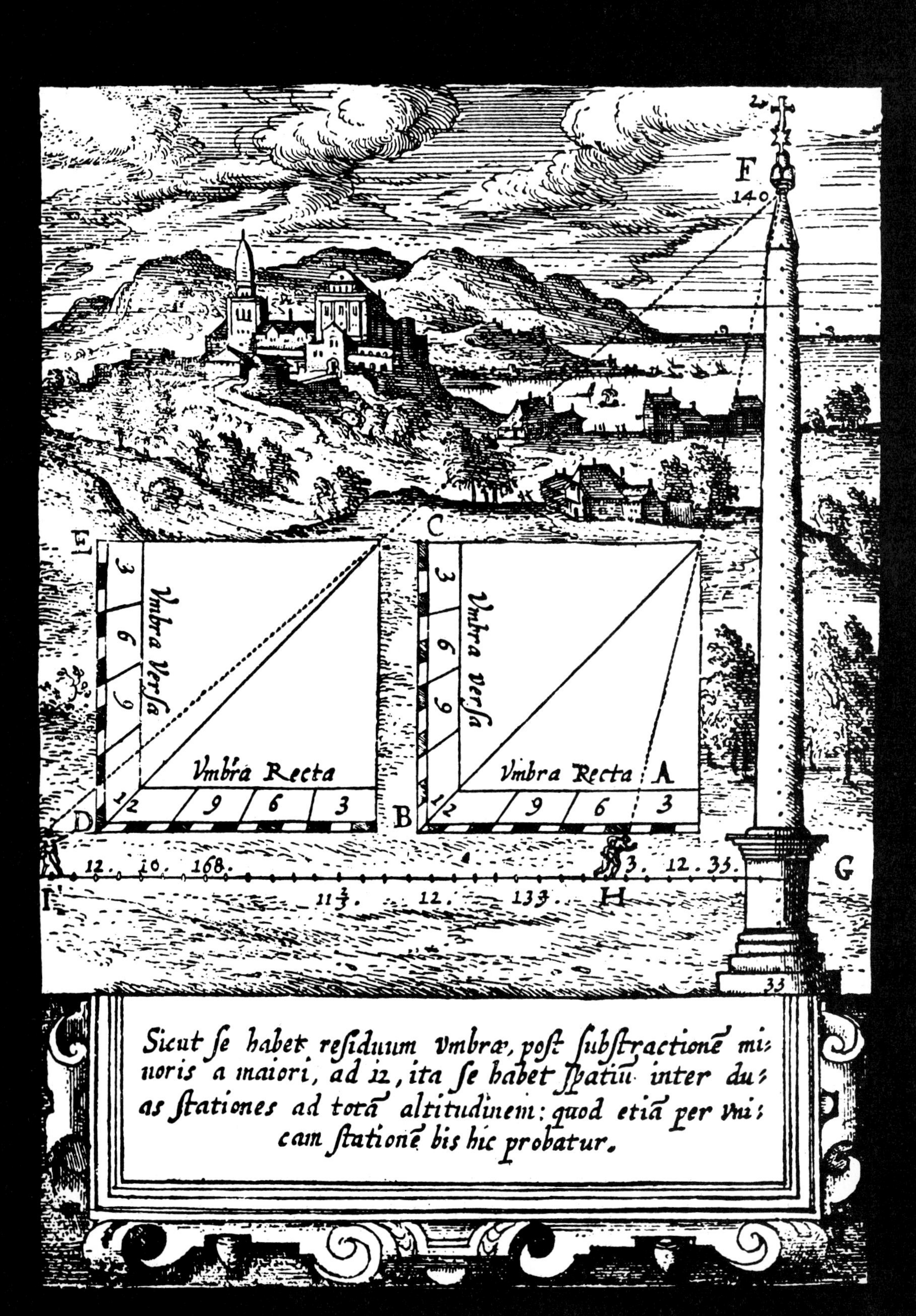
F
140
E
C
Vmbra Versa
Vmbra Versa
Vmbra Recta
Vmbra Recta
A
D
B
G
H
I
12. 10. 168.
3. 12. 33.
33
Sicut se habet residuum umbræ, post substractione minoris a maiori, ad 12, ita se habet spatiu inter duas stationes ad tota altitudinem: quod etia per vnicam statione bis hic probatur.

10 Quadratic Equations and Inequalities in Two Variables

In this chapter we consider second-degree equations in two variables whose graphs are called *circles, ellipses, parabolas,* and *hyperbolas.* Because these curves can be formed by the intersections of a plane and a cone as shown in Figure 10.1, they are called **conic sections**, or simply **conics**.

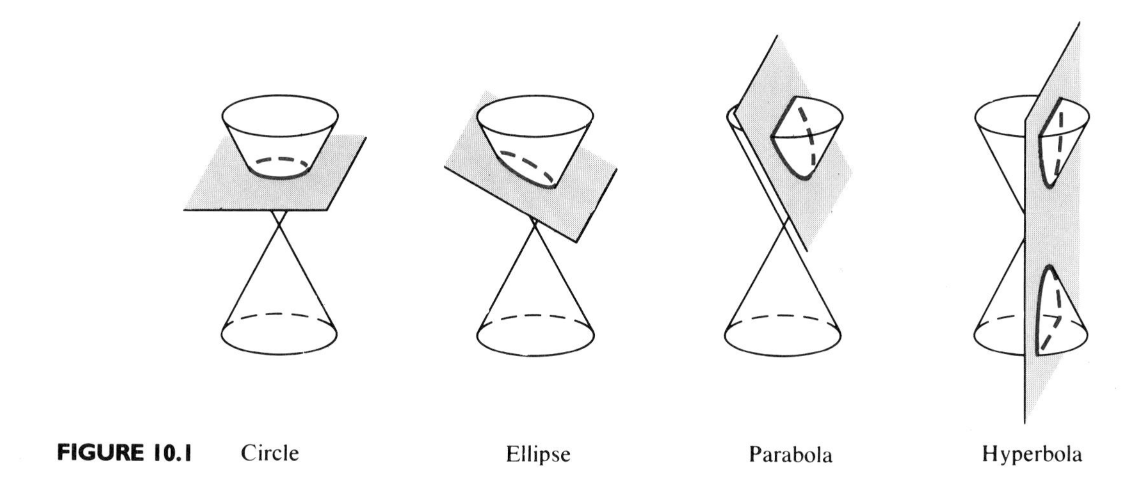

FIGURE 10.1

10.1

CIRCLES AND ELLIPSES

Circles

A circle is defined as follows:

> *A **circle** is a set of points in a plane that are at a given distance **r** (the **radius**) from a fixed point (the **center**).*

All points $P(x, y)$ on the circle in Figure 10.2a are equidistant from the center at the *origin*. Using the distance formula and the definition above, we have that the distance r of any point $P(x, y)$ from the origin is given by

$$(x - 0)^2 + (y - 0)^2 = r^2$$
$$x^2 + y^2 = r^2. \tag{1}$$

By the distance formula and the definition above, the distance r of any point $P(x, y)$ on a circle from a center (h, k) (see Figure 10.2b) is given by

$$(x - h)^2 + (y - k)^2 = r^2. \tag{1'}$$

The axes x' and y' in Figure 10.2b are called the **translated axes** of the circle.

EXAMPLE 1 a. The graph of $(x - 2)^2 + (y + 3)^2 = 16$ is a circle with radius 4 and center $(2, -3)$.

b. The graph of $x^2 + (y - 4)^2 = 7$ is a circle with radius $\sqrt{7}$ and center $(0, 4)$.

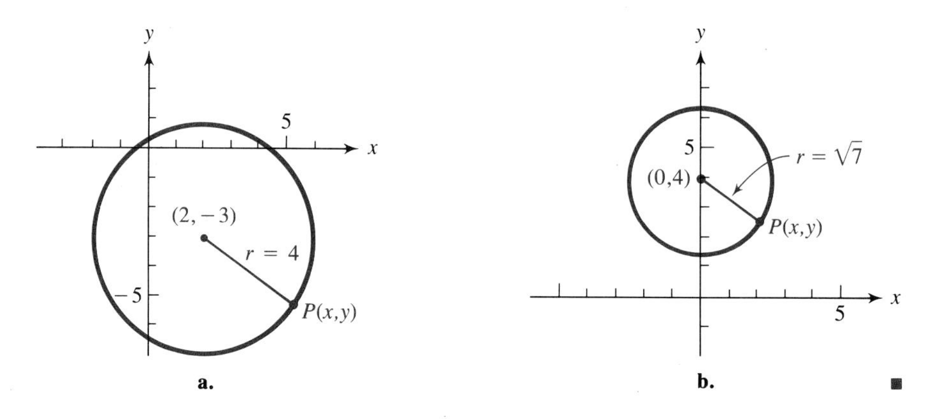

■

Equations of the form (1) and (1′) are called **standard forms** for the equation of a circle. We can rewrite an equation of form (1′) as an equivalent equation without parentheses. For example, the equation in Example 1a,

$$(x - 2)^2 + (y + 3)^2 = 16,$$

is equivalent to

$$x^2 - 4x + 4 + y^2 + 6y + 9 = 16,$$

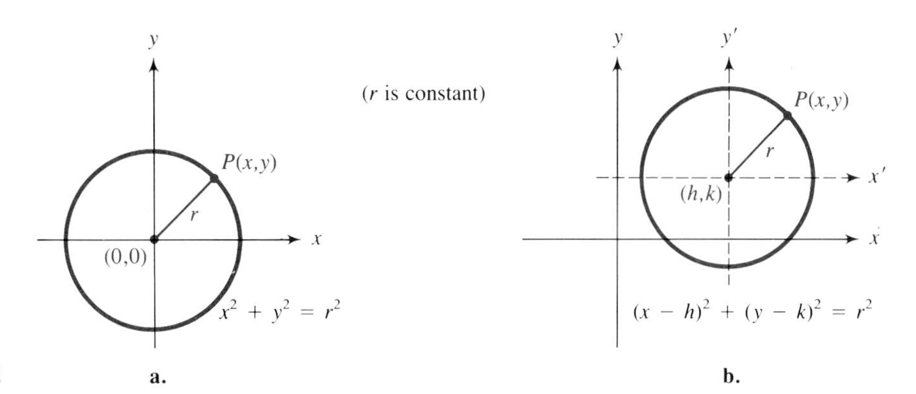

FIGURE 10.2 a. b.

from which we obtain

$$x^2 + y^2 - 4x + 6y - 3 = 0.$$

Note that the coefficients of both x^2 and y^2 are 1.

If an equation that is in the form $x^2 + y^2 + ax + by + c = 0$ is written in standard form we can determine the center directly. We can do this by completing the square in both variables.

EXAMPLE 2 a. Write the equation

$$x^2 + y^2 + 8x - 2y + 6 = 0$$

in standard form.

b. Graph the equation.

Solution a. We first prepare to complete the square in both x and y and write

$$(x^2 + 8x \qquad) + (y^2 - 2y \qquad) = -6.$$

Completing the square in x by adding 16 to each member and completing the square in y by adding 1 to each member, we have

$$(x^2 + 8x + 16) + (y^2 - 2y + 1) = -6 + 16 + 1,$$

from which we obtain the standard form

$$(x + 4)^2 + (y - 1)^2 = 11.$$

b. The center of the circle is $(-4, 1)$ and the radius is $\sqrt{11}$.

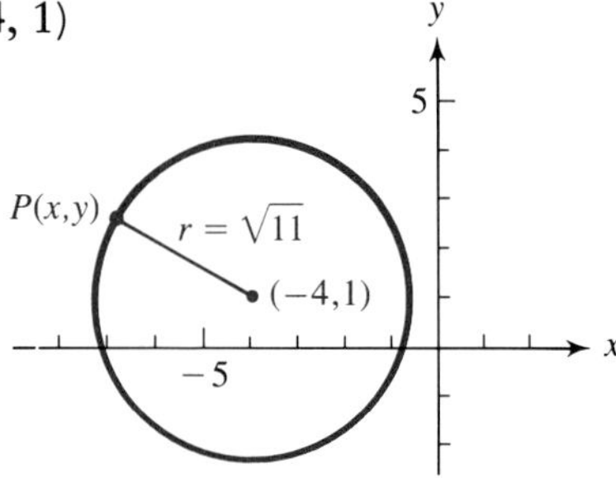

■

Ellipses

An ellipse is defined as follows:

*An **ellipse** is a set of points in a plane such that for each point P(x,y) the sum of the distances from two fixed points (**foci**) is a constant (see Figure 10.3).*

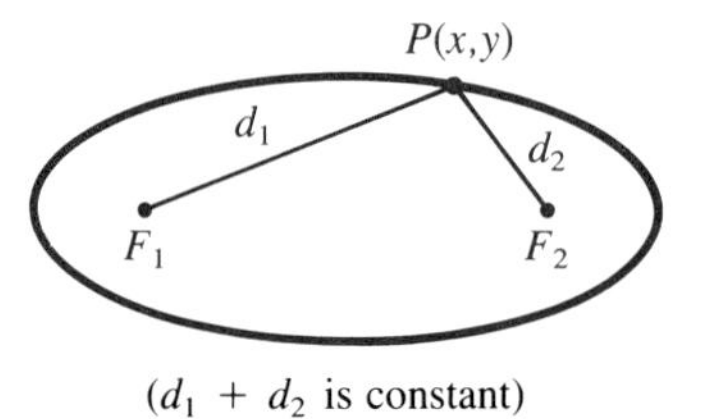

FIGURE 10.3 $(d_1 + d_2$ is constant$)$

Using this definition and the distance formula it can be shown that an equation of an ellipse with center at the origin can take the form

$$\frac{x^2}{a^2} + \frac{y^2}{b^2} = 1 \qquad (a > b) \tag{2}$$

when the foci are in the x-axis (see Figure 10.4a); the x-intercepts are a and $-a$ and the y-intercepts are b and $-b$. An equation of an ellipse will take the form

$$\frac{x^2}{b^2} + \frac{y^2}{a^2} = 1 \qquad (a > b) \tag{3}$$

when the foci are in the y-axis (see Figure 10.4b). In this case the x-intercepts are b and $-b$ and the y-intercepts are a and $-a$. In both figures the segment of length $2a$ is called the **major axis** and the segment of length $2b$ is called the **minor axis.** The endpoints of the major axis are called the **vertices.**

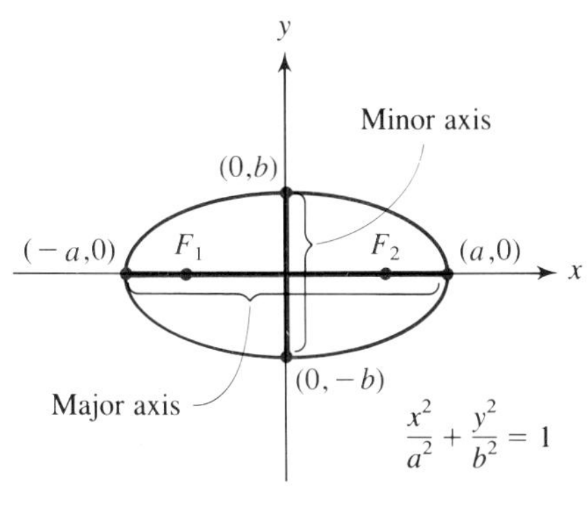

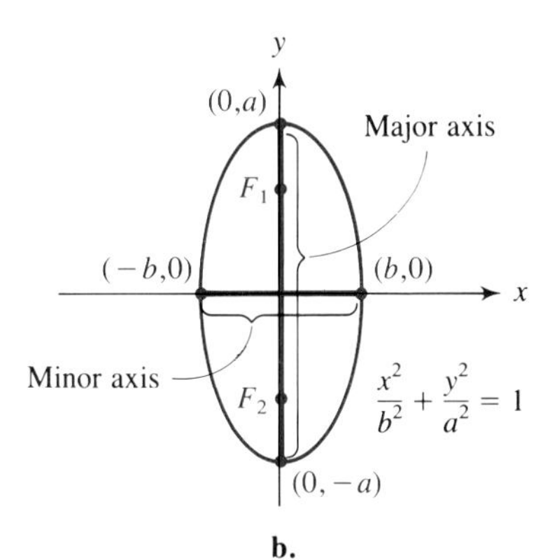

FIGURE 10.4

EXAMPLE 3 Graph $\frac{x^2}{9} + \frac{y^2}{4} = 1$.

Solution The graph of the equation is an ellipse with center at the origin. Since $a^2 = 9$ and $b^2 = 4$, the x-intercepts are 3 and -3 and the y-intercepts are 2 and -2. The major axis is $2a = 6$ and the vertices are the endpoints 3 and -3 of this axis.

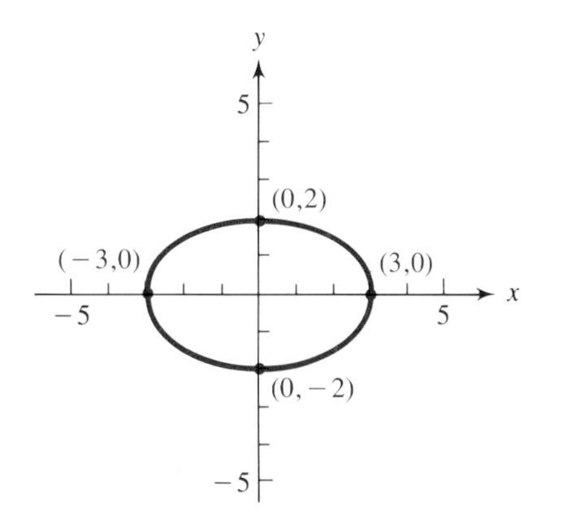

■

It can be shown that Equation (2) can take the form

$$\frac{(x-h)^2}{a^2} + \frac{(y-k)^2}{b^2} = 1 \qquad (a > b) \tag{2'}$$

for an ellipse with center (h, k) whose foci are on the translated x'-axis, and that Equation (3) can take the form

$$\frac{(x - h)^2}{b^2} + \frac{(y - k)^2}{a^2} = 1 \qquad (a > b) \tag{3'}$$

for an ellipse with center (h, k) whose foci are in the translated y'-axis. For both cases $2a$ is the length of the major axis and $2b$ is the length of the minor axis. See Figures 10.5a and b. Equations (2), (2′), (3), and (3′) are the **standard forms** for an ellipse.

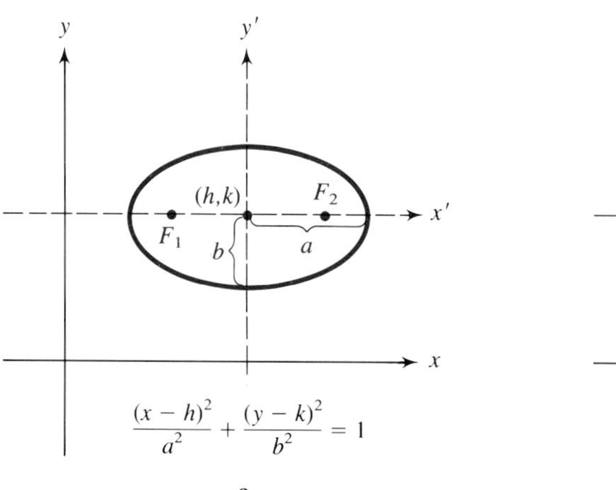

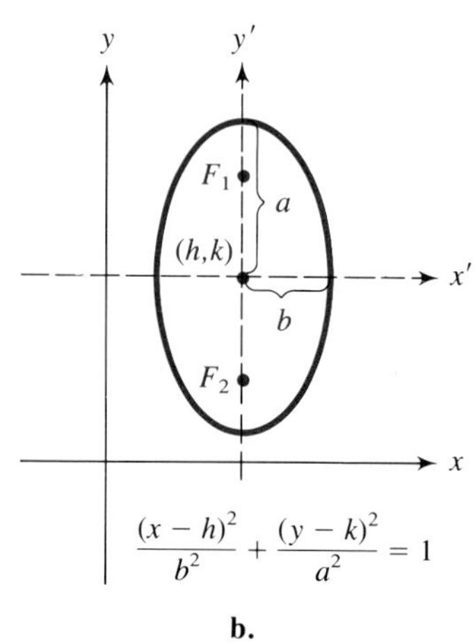

FIGURE 10.5 a. b.

EXAMPLE 4 Graph $\dfrac{(x - 3)^2}{8} + \dfrac{(y + 2)^2}{25} = 1$.

Solution The graph is an ellipse with center $(3, -2)$, $a^2 = 25$, and $b^2 = 8$. Hence the intercepts on the y'-axis are 5 and -5, and the intercepts on the x'-axis are $\sqrt{8}$ and $-\sqrt{8}$.

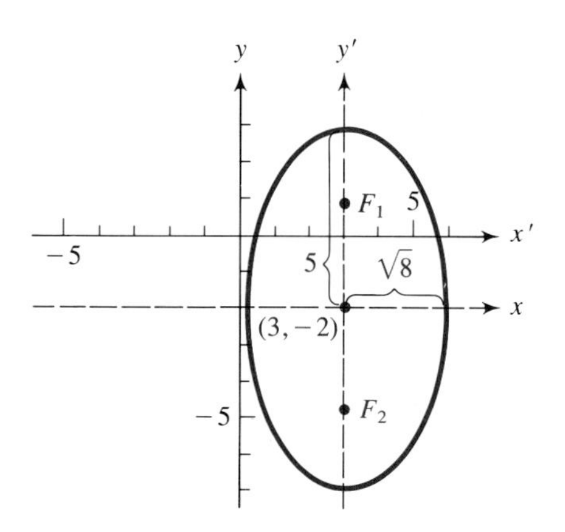

■

If an equation of an ellipse is not in standard form, we can use the method of completing the square that we used for a circle to write the equation in standard form.

EXAMPLE 5 a. Write the equation

$$4x^2 + 9y^2 - 16x - 18y - 11 = 0.$$

in standard form.

b. Graph the equation.

Solution a. We first prepare to complete the square in both x and y and write

$$4(x^2 - 4x \qquad) + 9(y^2 - 2y \qquad) = 11.$$

Completing the square in x by adding 16 $(4 \cdot 4)$ to each member and completing the square in y by adding 9 $(9 \cdot 1)$ to each member, we obtain

$$4(x^2 - 4x + 4) + 9(y^2 - 2y + 1) = 11 + 16 + 9$$

and then

$$4(x - 2)^2 + 9(y - 1)^2 = 36,$$

from which we obtain the standard form by dividing each member by 36.

$$\frac{(x - 2)^2}{9} + \frac{(y - 1)^2}{4} = 1$$

b. The graph is an ellipse with center at (2, 1), $a^2 = 9$, and $b^2 = 4$. Hence, the intercepts on the x'-axis are 3 and -3, and the intercepts on the y'-axis are 2 and -2.

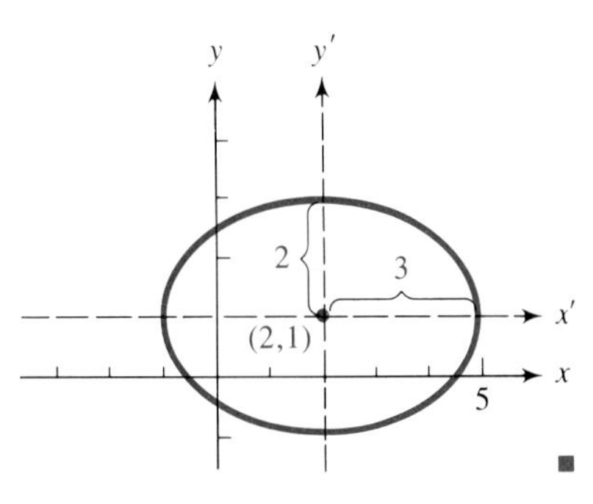

■

Note that the parts of circles and ellipses on either side of their axes are *mirror images*. Hence we say that they are **symmetric** with respect to their axes.

EXERCISE 10.1

A

■ *Graph each equation. See Examples 1 and 2.*

1. $x^2 + y^2 = 25$
2. $x^2 + y^2 = 16$
3. $4x^2 + 4y^2 = 16$
4. $2x^2 + 2y^2 = 18$
5. $(x - 4)^2 + (y - 2)^2 = 9$
6. $(x - 1)^2 + (y - 3)^2 = 16$
7. $(x + 3)^2 + y^2 = 10$
8. $x^2 + (y + 4)^2 = 12$
9. $x^2 + (y + 3)^2 = 15$
10. $(x + 2)^2 + y^2 = 18$
11. $(x - 6)^2 + (y + 2)^2 = 12$
12. $(x + 6)^2 + (y - 3)^2 = 8$
13. $x^2 + y^2 + 2x - 4y - 6 = 0$
14. $x^2 + y^2 - 6x + 2y - 4 = 0$
15. $x^2 + y^2 + 8x - 4 = 0$
16. $x^2 + y^2 - 10y - 2 = 0$
17. $x^2 + y^2 + 6y = 0$
18. $x^2 + y^2 - 6x = 0$

■ *Graph each equation. See Examples 3 and 4.*

19. $\dfrac{x^2}{16} + \dfrac{y^2}{4} = 1$
20. $\dfrac{x^2}{9} + \dfrac{y^2}{16} = 1$
21. $\dfrac{x^2}{10} + \dfrac{y^2}{25} = 1$
22. $\dfrac{x^2}{16} + \dfrac{y^2}{12} = 1$
23. $x^2 + \dfrac{y^2}{14} = 1$
24. $\dfrac{x^2}{8} + y^2 = 1$
25. $\dfrac{(x - 3)^2}{16} + \dfrac{(y - 4)^2}{9} = 1$
26. $\dfrac{(x - 2)^2}{4} + \dfrac{(y - 5)^2}{25} = 1$
27. $\dfrac{(x + 2)^2}{6} + \dfrac{(y - 5)^2}{12} = 1$
28. $\dfrac{(x - 5)^2}{15} + \dfrac{(y + 3)^2}{8} = 1$
29. $\dfrac{x^2}{16} + \dfrac{(y + 4)^2}{6} = 1$
30. $\dfrac{(x - 5)^2}{15} + \dfrac{y^2}{25} = 1$

■ *Graph each equation. See Example 5.*

31. $9x^2 + 4y^2 - 16y = 20$
32. $x^2 + 16y^2 + 6x = 7$
33. $9x^2 + 16y^2 - 18x + 96y + 9 = 0$
34. $16x^2 + 9y^2 + 64x - 18y - 71 = 0$
35. $x^2 + 4y^2 + 4x - 16y + 4 = 0$
36. $2x^2 + y^2 - 16x + 6y + 11 = 0$
37. $6x^2 + 5y^2 - 12x + 20y - 4 = 0$
38. $5x^2 + 8y^2 - 20x + 16y - 12 = 0$
39. $8x^2 + y^2 - 48x + 4y + 68 = 0$
40. $x^2 + 10y^2 + 4x + 20y + 4 = 0$

10.2

PARABOLAS

A parabola is defined as follows:

*A **parabola** is a set of points $P(x,y)$ in a plane whose distances from a fixed line l (the **directrix**) and a fixed point F (the **focus**) are equal. (See Figure 10.6.)*

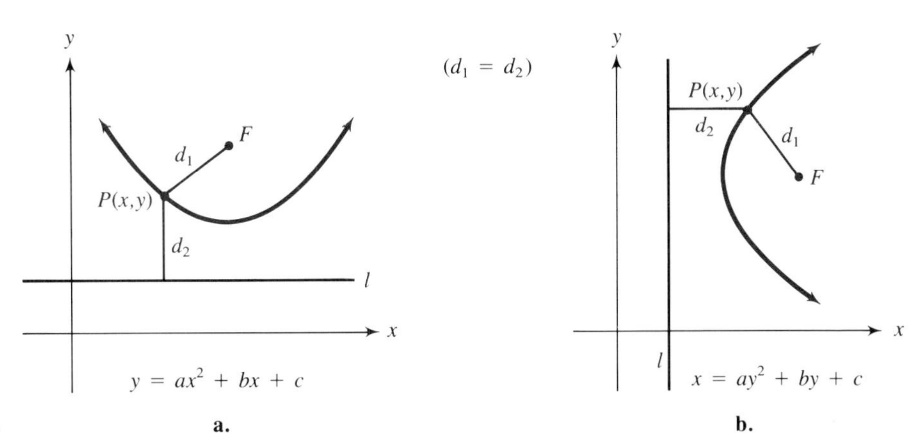

FIGURE 10.6

Using the above definition and the distance formula it can be shown that equations of parabolas that "open" upward or downward are of the form

$$y = ax^2 + bx + c \qquad (a \neq 0),$$

and that equations of parabolas that "open" to the right or left are of the form

$$x = ay^2 + by + c \qquad (a \neq 0).$$

As a simple example, we first consider the quadratic equation in two variables

$$y = x^2 - 4. \tag{1}$$

As with linear equations in two variables, solutions of this equation are ordered pairs. We need replacements for both x and y to obtain a statement we can judge to be true. Such ordered pairs can be found by arbitrarily assigning values to x and computing related values for y. For instance, assigning the value -3 to x in (1), we have

$$\begin{aligned} y &= (-3)^2 - 4 \\ &= 5, \end{aligned}$$

and so $(-3, 5)$ is a solution of Equation (1). Similarly, we find that

$$(-2, 0), \quad (-1, -3), \quad (0, -4), \quad (1, -3), \quad (2, 0), \quad \text{and} \quad (3, 5)$$

are also solutions of (1).

Plotting the corresponding points in the plane, we have the graph shown in Figure 10.7a. Clearly, these points do not lie in a straight line, and we might reasonably inquire whether the graph of

$$y = x^2 - 4$$

forms any kind of meaningful pattern in the plane. By plotting additional solutions with x-components between those already found, we may be able to obtain a clearer picture. Accordingly, we find the solutions

$$\left(\frac{-5}{2}, \frac{9}{4}\right), \quad \left(\frac{-3}{2}, \frac{-7}{4}\right), \quad \left(\frac{-1}{2}, \frac{-15}{4}\right), \quad \left(\frac{1}{2}, \frac{-15}{4}\right), \quad \left(\frac{3}{2}, \frac{-7}{4}\right), \quad \left(\frac{5}{2}, \frac{9}{4}\right),$$

and, by plotting these points in addition to those found earlier, we have the graph shown in Figure 10.7b. It now appears reasonable to connect these points in sequence by a smooth curve, as in Figure 10.7c, and to assume that the curve, a parabola, is a good approximation to the graph of (1).

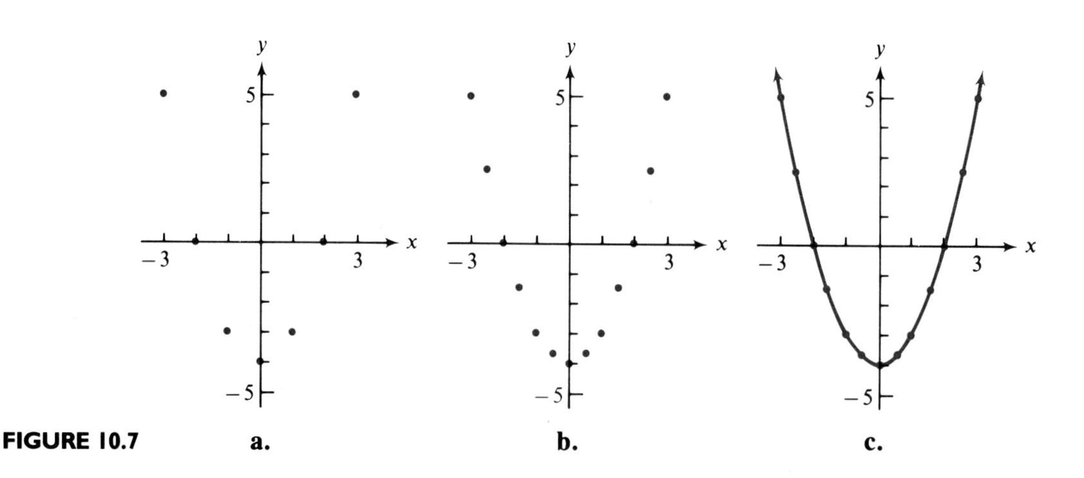

FIGURE 10.7

Orientation of parabolas

Observe that the coefficient, 1, of the second-degree term x^2 in $y = x^2 - 4$ is positive and that its graph, the parabola in Figure 10.7c, opens upward. In general, the graphs of all equations of the form

$$y = ax^2 + bx + c \tag{2}$$

open upward if $a > 0$ and downward if $a < 0$.

The graph of a quadratic equation of the form

$$x = ay^2 + by + c \qquad (a \neq 0) \tag{3}$$

is also a parabola. Such parabolas open to the right if $a > 0$ and to the left if $a < 0$. See the following chart.

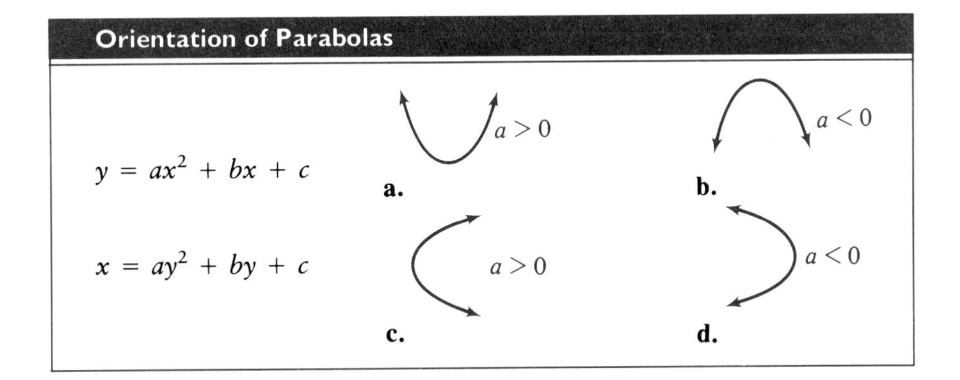

EXAMPLE 1

a. The graph of $y = 2x^2 - x + 3$ opens upward because 2, the coefficient of x^2, is positive. (Figure a above.)

b. The graph of $y = -2x^2 + x - 3$ opens downward because -2, the coefficient of x^2, is negative. (Figure b above.)

c. The graph of $x = y^2 + 2y$ opens to the right because 1, the coefficient of y^2, is positive. (Figure c above.)

d. The graph of $x = -y^2 + 6$ opens to the left because -1, the coefficient of y^2, is negative. (Figure d above.) ■

In this section we will be primarily interested in sketching parabolas of the form $y = ax^2 + bx + c$ that either open upward or downward.

Sketching parabolas

When graphing a quadratic equation in two variables, it is desirable to select first components for the ordered pairs that ensure that the most significant parts of the graph are displayed. For a parabola, these parts include the intercepts, if they exist, and the **maximum** or **minimum** (highest or lowest) point on the curve.

Intercepts

In determining the x- and y-intercepts of the graph of

$$y = ax^2 + bx + c,$$

we let $x = 0$ to find the y-intercept and let $y = 0$ to find the x-intercepts.

EXAMPLE 2 The y-intercept of

$$y = -2x^2 - 5x + 3 \tag{4}$$

can be found by assigning the value of 0 to x:

$$y = -2(0)^2 - 5(0) + 3 = 3.$$

Letting $y = 0$ in (1) yields

$$0 = -2x^2 - 5x + 3.$$

Thus, the x-intercepts of (1) are the solutions of

$$-2x^2 - 5x + 3 = 0.$$

These can be found by writing the equation equivalently as

$$-1(x + 3)(2x - 1) = 0.$$

By inspection, the solutions—and therefore the x-intercepts of the graph of (4) are -3 and ½. ■

From the y-intercept 3 and the x-intercepts -3 and ½ that we obtained in Example 2, and the fact that the graph of (4) *opens downward* because the coefficient of x^2 is -2, we can sketch part of the graph shown in Figure 10.8. Our next step is to find the high point.

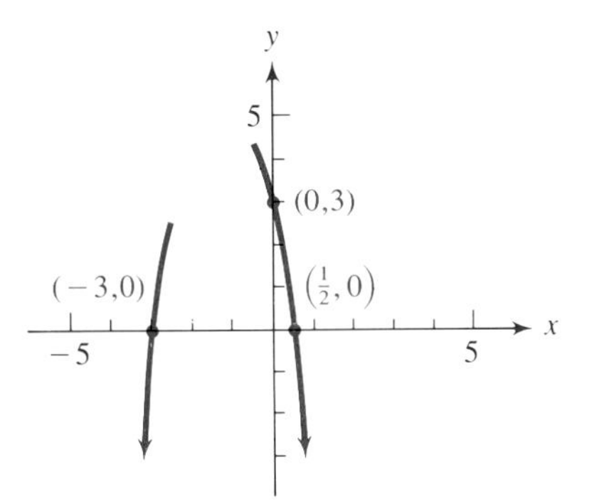

FIGURE 10.8

Low or high point

We can obtain the coordinates of the lowest (or the highest) point on a parabola by the procedure of completing the square that we considered in Section 7.2. For example, Equation (4) in Example 2 can be rewritten as

$$y = -2\left(x^2 + \frac{5}{2}x \qquad \right) + 3 \tag{4'}$$

and then as

$$y = -2\left(x^2 + \frac{5}{2}x + \frac{25}{16}\right) + 3 + 2\left(\frac{25}{16}\right)$$

where $2(25/16)$ was both subtracted from and added to the right-hand member. Since

$$x^2 + \frac{5}{2}x + \frac{25}{16} = \left(x + \frac{5}{4}\right)^2$$

and

$$3 + 2\left(\frac{25}{16}\right) = \frac{49}{8},$$

Equation (4′) is equivalent to

$$y = -2\left(x + \frac{5}{4}\right)^2 + \frac{49}{8}.$$

For $x = -5/4$, the expression $(x + 5/4)$ equals 0; hence y has a maximum value $49/8$. Therefore, the high point corresponds to $(-5/4, 49/8)$. We can now complete the graph in Figure 10.8 as shown in Figure 10.9.

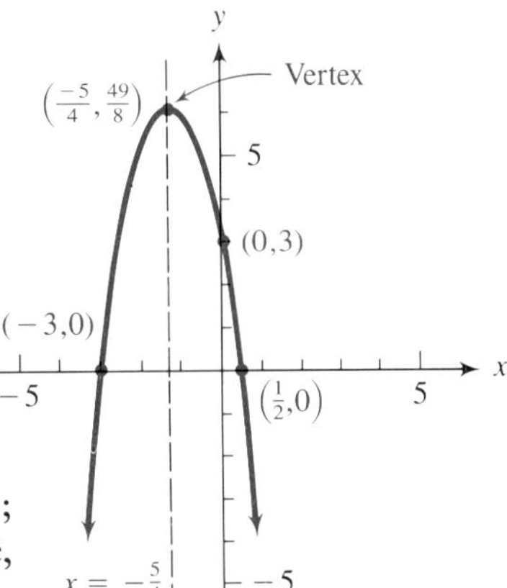

FIGURE 10.9

The maximum or minimum point of a parabola is called the **vertex** and, as the example above suggests, can readily be obtained from the equation of a parabola in the form

$$y = a(x - h)^2 + k,$$

where (h, k) is the vertex.

EXAMPLE 3 Specify the coordinates of the vertex of the graph of each equation.

a. $y = 2(x - 3)^2 + 4$ b. $y = 4\left(x + \frac{1}{2}\right)^2 - \frac{7}{4}$

c. $y = 2x^2 - 4x + 6$

Solution a. $y = 2(x - 3)^2 + 4$; $(3, 4)$ b. $y = 4\left(x - \left(-\frac{1}{2}\right)\right)^2 - \frac{7}{4}$; $\left(-\frac{1}{2}, -\frac{7}{4}\right)$

c. Completing the square in x yields

$$\begin{aligned} y &= 2(x^2 - 2x \quad\;) + 6 \\ &= 2(x^2 - 2x + 1) + 6 - 2 \\ &= 2(x - 1)^2 + 4. \end{aligned}$$

Hence the coordinates of the vertex are $(1, 4)$. ■

Symmetry of a parabola

Notice in Figure 10.9 that a line through the vertex of a parabola and parallel to the y-axis divides the parabola into two parts, each the mirror image of the other. This line is called the **axis** of parabola; the curve is **symmetric** with respect to this axis. For example, the axis of the graph of Equation (4) is given by $x = -5/4$, the equation of the line through the vertex.

Number of x-intercepts

Recall from Section 7.4 that a quadratic equation in one variable

$$ax^2 + bx + c = 0$$

may have no real solution, one real solution, or two real solutions. If the equation has no real solution, the graph of the related quadratic equation in two variables

$$y = ax^2 + bx + c$$

does not touch the x-axis; if there is one solution, the graph is tangent to the x-axis at a point; if there are two real solutions, the graph crosses the x-axis at two distinct points. This is shown in Figure 10.10 on page 364.

For example, the graph of

$$y = x^2 + 1$$

has no x-intercepts because $x^2 + 1 = 0$ has no real solution. The graph of

$$y = x^2 - 4x + 4$$

has one x-intercept because

$$x^2 - 4x + 4 = (x - 2)(x - 2) = 0$$

has only one solution. The graph of

$$y = x^2 - 5x + 4$$

has two x-intercepts because

$$x^2 - 5x + 4 = (x - 1)(x - 4) = 0$$

has two solutions.

The properties of parabolas that we have considered to assist us in sketching their graphs are summarized in the following table.

To Sketch Parabolas with Equation $y = ax^2 + bx + c$

1. Determine whether the curve opens upward (if $a > 0$) or downward (if $a < 0$).
2. Find the coordinates (h, k) of the vertex by writing the equation as

$$y = a(x - h)^2 + k.$$

3. Identify and graph intercepts, if any.
   a. Let $x = 0$ to obtain y-intercept.
   b. Let $y = 0$ to obtain x-intercepts, if any.
4. a. If the x-intercepts are real numbers, you will have enough points to sketch the graph.
   b. If the x-values are imaginary when $y = 0$, locate one or two points on the curve and use symmetry to complete the sketch.

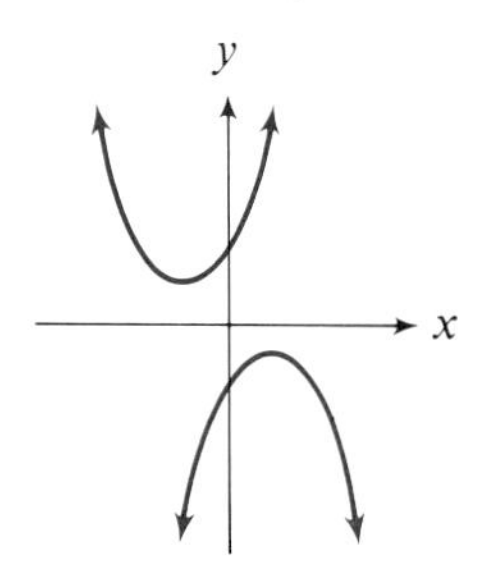

No intersection
No real solutions

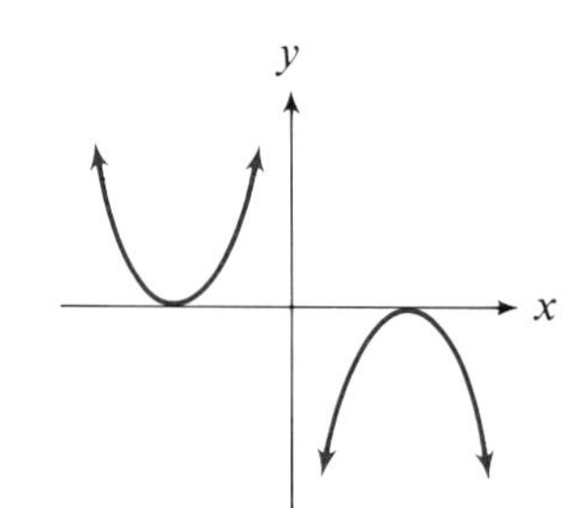

One intersection
One real solution

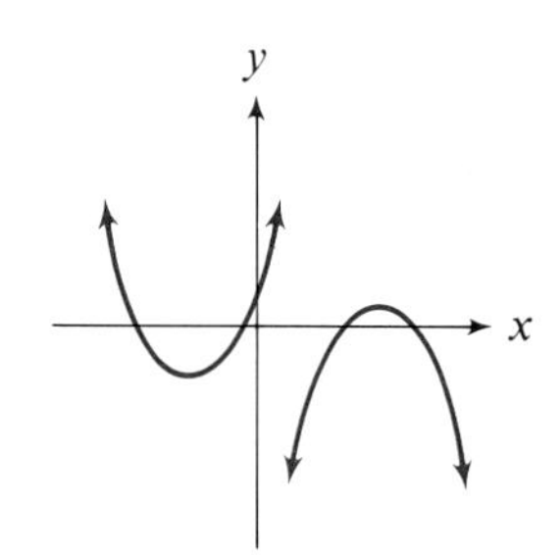

Two intersections
Two real solutions

FIGURE 10.10 a. b. c.

EXAMPLE 4 Graph $y = x^2 - 3x - 4$.

Solution We proceed as follows:

1. Since the coefficient of x^2 is positive, the curve opens upward.
2. To find the coordinates of the vertex we complete the square in the variable x to obtain

$$y = \left(x^2 - 3x + \frac{9}{4}\right) - 4 - \frac{9}{4},$$
$$y = \left(x - \frac{3}{2}\right)^2 - \frac{25}{4}.$$

The coordinates are $(3/2, -25/4)$.

3. If $x = 0$,

$$y = 0^2 - 3(0) - 4 = -4,$$

and the y-intercept is -4. If $y = 0$, then

$$x^2 - 3x - 4 = 0$$
$$(x - 4)(x + 1) = 0$$
$$x = 4 \quad \text{or} \quad x = -1;$$

so the x-intercepts are 4 and -1.

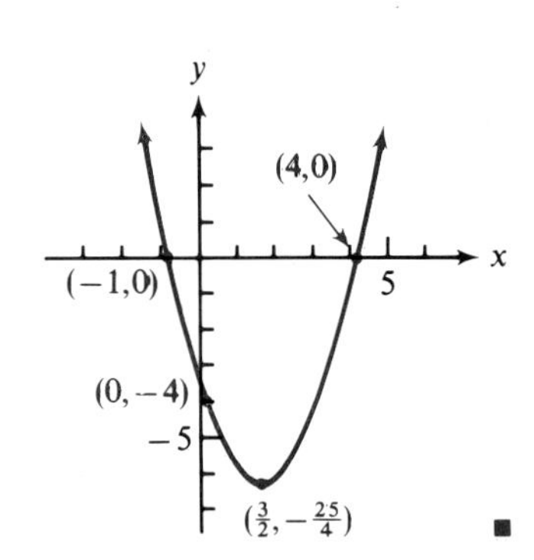

4. The curve is sketched as shown. ■

In Example 4 above, the two intercepts on the x-axis helped us to sketch the parabola. In the following example, the parabola does not have x-intercepts and we have to obtain several additional points to complete the sketch as suggested by 4.b. in the table on page 363.

EXAMPLE 5 Graph $y = -2x^2 + x - 1$.

Solution

1. Since the coefficient of x^2 is negative, the curve opens downward.
2. Complete the square in x as follows.

$$\begin{aligned} y &= -2\left(x^2 - \frac{1}{2}x\right) - 1, \\ &= -2\left(x^2 - \frac{1}{2}x + \frac{1}{16}\right) - 1 + \frac{1}{8} \\ &= -2\left(x^2 - \frac{1}{4}\right)^2 - \frac{7}{8} \end{aligned}$$

The coordinates of the vertex are $(1/4, -7/8)$.

3. If $x = 0$, then $y = -1$; so the y-intercept is -1.
If $y = 0$, then $-2x^2 + x - 1 = 0$. The solutions are imaginary; so there are no x-intercepts.
4. To find two additional points, let $x = -1$ and $x = 1$.
If $x = -1$,

$$y = -2(-1)^2 + (-1) - 1 = -4;$$

so $(-1, -4)$ is on the curve.
If $x = 1$,

$$y = -2(1)^2 + 1 - 1 = -2;$$

so $(1, -2)$ is on the curve.

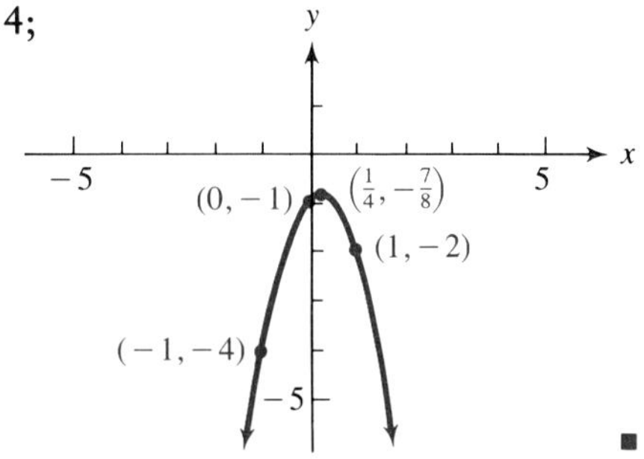

y
x
−5
5
(0, −1)
$\left(\frac{1}{4}, -\frac{7}{8}\right)$
(1, −2)
(−1, −4)
−5

Use symmetry to sketch the graph as shown. ■

Many parabolas can be readily sketched by using only one or two of the steps suggested in the table on page 363.

EXAMPLE 6 To graph $y = x^2 + 2$, we first note that the parabola opens upward with a vertex at (0, 2) because 2 is the least value for y and this is the value when $x = 0$. Hence, the y-axis is the axis of symmetry. Plotting a point, say (2, 6), and using symmetry, we can quickly graph the parabola shown.

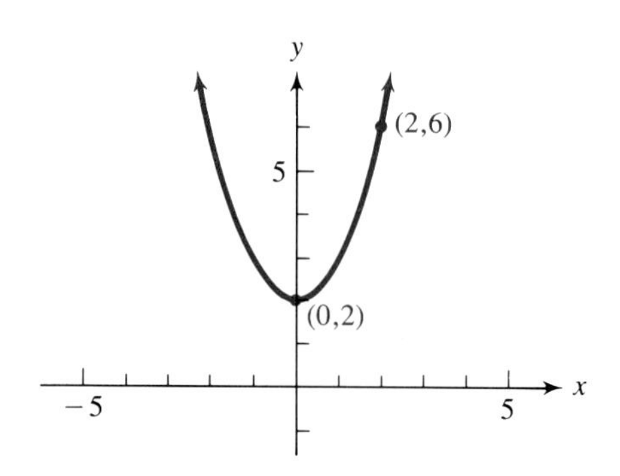

■

Finding maximum or minimum values

Sometimes in a relation between two variables we want to find a value of one variable which will give us the maximum or minimum value of the second variable. A wide variety of applications are concerned with such problems, and their solutions are considered in depth in advanced courses. However, we can at this time solve some of them.

EXAMPLE 7 Find two numbers whose sum is 18 and whose product is as large as possible.

Solution

First number: x
Second number: $18 - x$

Their product:

$$P = x(18 - x)$$
$$= -x^2 + 18x.$$

Completing the square in x yields

$$P = -(x^2 - 18x + 81) + 81$$
$$= -(x - 9)^2 + 81.$$

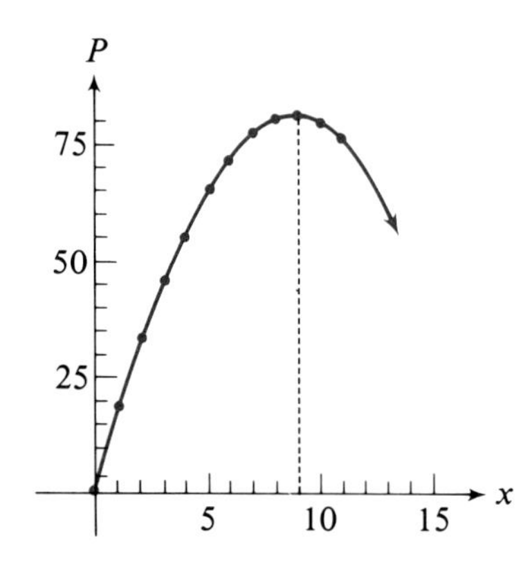

Hence 9 is the x-coordinate of the vertex and yields the maximum value for P. Since $18 - x = 9$, the numbers are 9 and 9. ■

EXERCISE 10.2

A

■ a. *Specify whether the graph of each equation opens upward or downward.* b. *Find the intercepts. See Examples 1 and 2.*

1. $y = x^2 - 5x + 4$
2. $y = x^2 + 5x + 6$
3. $y = -x^2 + x + 6$
4. $y = -x^2 - x + 12$
5. $y = 2x^2 + 5x - 3$
6. $y = 3x^2 - x - 2$
7. $y = -3x^2 - 5x + 2$
8. $y = -2x^2 + x + 6$
9. $y = 2x^2 - 4x$
10. $y = 3x^2 - 12x$
11. $y = -4x^2 - 8$
12. $y = -2x^2 - 6$

■ *Specify the coordinates of the vertex of the graph of each equation. See Example 3.*

13. $y = 3(x - 2)^2 - 5$
14. $y = 4(x + 3)^2 + 4$
15. $y = -4x^2 + 8$
16. $y = 2x^2 - 12$
17. $y = 2(x + 3)^2$
18. $y = -3(x - 2)^2$
19. $y = x^2 + 6x + 5$
20. $y = x^2 - 8x + 2$
21. $y = 2x^2 + 12x + 12$
22. $y = 3x^2 - 6x + 4$
23. $y = -3x^2 + 9x - 2$
24. $y = -2x^2 + 5x - 1$

■ *Graph each equation. See Examples 4, 5, and 6.*

25. $y = x^2 - 5x + 4$
26. $y = x^2 + x - 6$
27. $y = x^2 - 4x$
28. $y = x^2 + 6x$
29. $y = x^2 + 4x - 5$
30. $y = x^2 + 6x + 8$
31. $y = x^2 + 1$
32. $y = x^2 + 4$
33. $y = x^2 - 3$
34. $y = x^2 - 5$
35. $y = -x^2 + 4$
36. $y = -x^2 + 5$
37. $y = 3x^2 + x$
38. $y = 5x^2 - x$
39. $y = x^2 + 2x + 1$
40. $y = x^2 - 2x + 1$
41. $y = -2x^2 + x - 3$
42. $y = -x^2 + 2x - 1$
43. $y = -x^2 + x + 1$
44. $y = -3x^2 + x - 2$

■ *Use graphical methods in Problems 45–48. See Example 7.*

45. Find two numbers whose sum is 12 and whose product is a maximum.
46. Find the maximum area of a rectangle whose perimeter is 100 inches. [*Hint:* Let x represent the length; then $50 - x$ represents the width and $A = x(50 - x)$.]
47. The equation $d = 32t - 8t^2$ relates the distance d (feet) above the ground reached in time t (seconds) by an object thrown vertically upward. Sketch the graph of the equation for $0 \leq t \leq 4$ and estimate the time it will take the object to reach its greatest height.
48. In Problem 47, how many seconds is the object in the air?

B

49. Sketch the family of four curves $y = kx^2$ $(k = 1, 2, 3, 4)$ on a single set of axes.
50. Sketch the family of four curves $y = kx^2$ $(k = -1, -2, -3, -4)$ on a single set of axes.

51. Sketch the family of four curves $y = x^2 + k$ $(k = -2, 0, 2, 4)$ on a single set of axes.
52. Sketch the family of four curves $y = x^2 + kx$ $(k = -2, 0, 2, 4)$ on a single set of axes.
53. Find values for a, b, and c such that the graph of $y = ax^2 + bx + c$ will contain the points $(-1, 2)$, $(1, 6)$, and $(2, 11)$. *Hint:* Use a system of equations.
54. Find values for a, b, and c such that the graph of $y = ax^2 + bx + c$ will contain the points $(1, 0)$, $(3, -2)$, and $(5, 4)$.

10.3

HYPERBOLAS

A hyperbola is defined as follows:

> *A **hyperbola** is the set of points $P(x,y)$ in a plane such that for each point the difference of its distances from two fixed points (**foci**) is a constant. (See Figure* 10.11.)

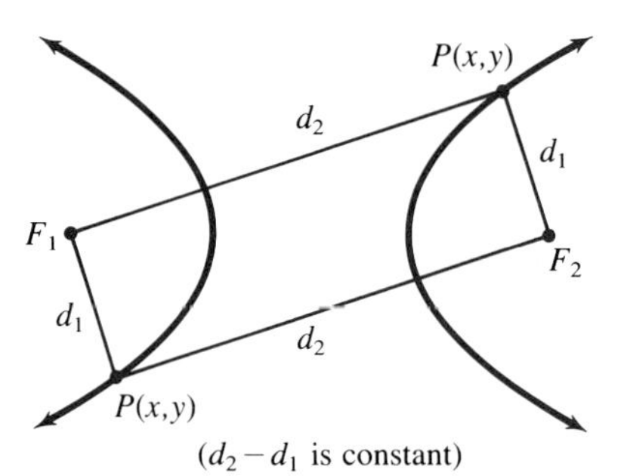

FIGURE 10.11 ($d_2 - d_1$ is constant)

Using this definition and the distance formula it can be shown that an equation of a hyperbola with center at the origin and foci in the x-axis can take the form

$$\frac{x^2}{a^2} - \frac{y^2}{b^2} = 1; \qquad (1)$$

the x-intercepts are a and $-a$ (see Figure 10.12a). An equation of a hyperbola can take the form

$$\frac{y^2}{a^2} - \frac{x^2}{b^2} = 1 \qquad (2)$$

if the foci are in the y-axis; in this case the y-intercepts are a and $-a$ (see Figure 10.12b).

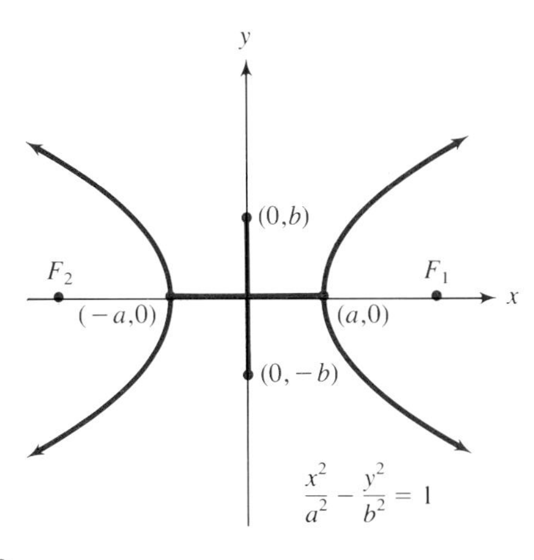

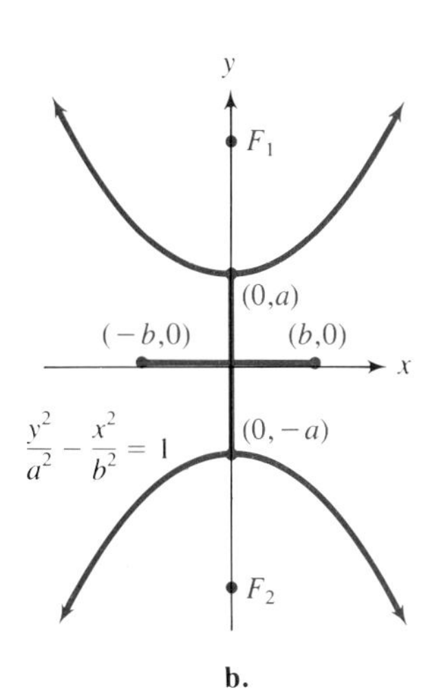

FIGURE 10.12 a. b.

In both orientations in Figure 10.12 the line segment of length $2a$ is called the **transverse axis** and the line segment of length $2b$ is called the **conjugate axis**. The endpoints of the transverse axis are called the **vertices**.

Asymptotes of hyperbolas

We can often simplify the process of graphing a hyperbola by first graphing two straight lines that are approached by the branches of the hyperbola. These lines are called **asymptotes** of the graph. The asymptotes can be obtained by first forming a rectangle centered at the origin with $2a$ as one dimension and $2b$ as the other dimension. Then the asymptotes are the two straight lines that intersect at the origin and pass through the corners of the rectangle.

EXAMPLE 1 Graph $\frac{x^2}{9} - \frac{y^2}{16} = 1$.

Solution The graph is a hyperbola with center at the origin. Since

$$a^2 = 9 \quad\text{and}\quad b^2 = 16,$$

then

$$a = 3 \quad\text{and}\quad b = 4.$$

The x-intercepts are 3 and -3. The central rectangle is constructed with dimensions $2a = 6$ and $2b = 8$. The asymptotes are drawn through the vertices of the rectangle, and the hyperbola is sketched as shown.

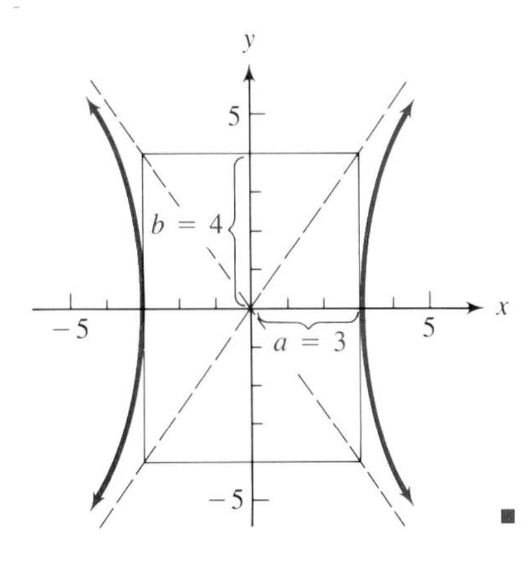

■

EXAMPLE 2 Graph $\frac{y^2}{9} - \frac{x^2}{4} = 1$.

Solution The graph is a hyperbola with center at the origin. Since

$$a^2 = 9 \quad \text{and} \quad b^2 = 4,$$

then

$$a = 3 \quad \text{and} \quad b = 2.$$

The y-intercepts are 3 and -3. The central rectangle is constructed with dimensions $2a = 6$ and $2b = 4$. The asymptotes are drawn through the vertices of the central rectangle, and the hyperbola is sketched as shown.

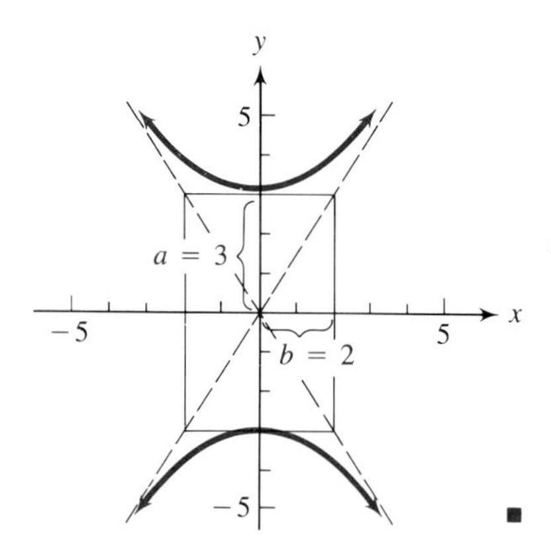

■

Translated graphs

It can be shown that Equation (1) can take the form

$$\frac{(x-h)^2}{a^2} - \frac{(y-k)^2}{b^2} = 1 \tag{1'}$$

for a hyperbola with center at (h, k), foci in a translated x'-axis, and intercepts a and $-a$ on this axis.

It can also be shown that Equation (2) can take the form

$$\frac{(y-k)^2}{a^2} - \frac{(x-h)^2}{b^2} = 1 \tag{2'}$$

for a hyperbola with center (h, k), foci in a translated y'-axis, and intercepts a and $-a$ on this axis. (See Figure 10.13.)

The Equations (1), (1′), (2), and (2′) are the **standard forms** for a hyperbola. Similar to the other conic sections, hyperbolas are **symmetric** about their axes.

The asymptotes of the graph of (1′) and (2′) in relation to the translated axes x' and y' can be obtained by forming a central rectangle about the origin of the translated axes.

EXAMPLE 3 Graph $\frac{(x-3)^2}{8} - \frac{(y+2)^2}{10} = 1$.

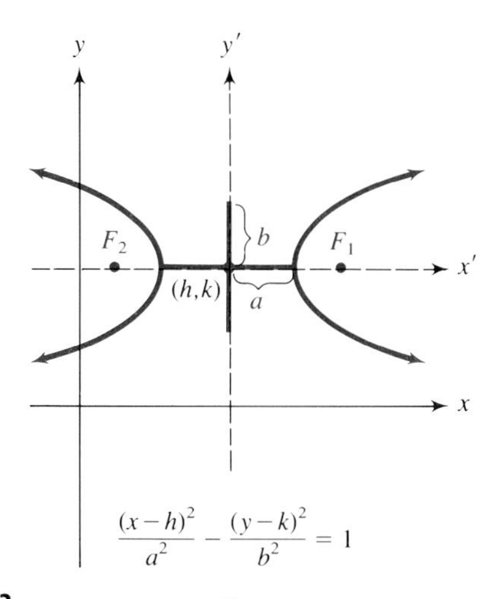

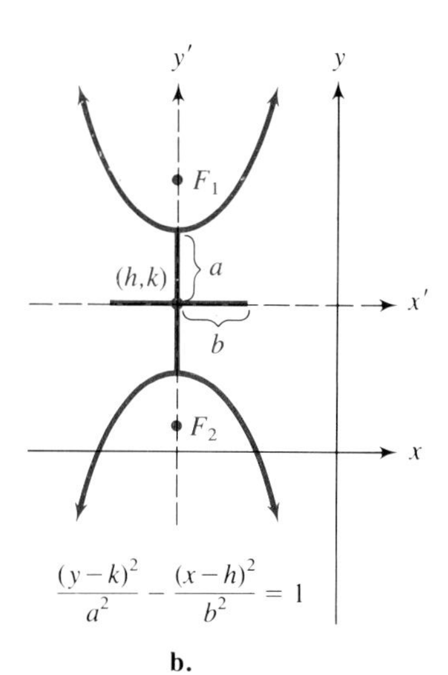

FIGURE 10.13 a. b.

Solution The graph is a hyperbola with center $(3, -2)$. The vertices are in the x'-axis with intercepts $\sqrt{8}$ and $-\sqrt{8}$. Since

$$a = \sqrt{8} \quad \text{and} \quad b = \sqrt{10},$$

the central rectangle can be constructed with dimensions

$$2a = 2\sqrt{8} \quad \text{and} \quad 2b = 2\sqrt{10}.$$

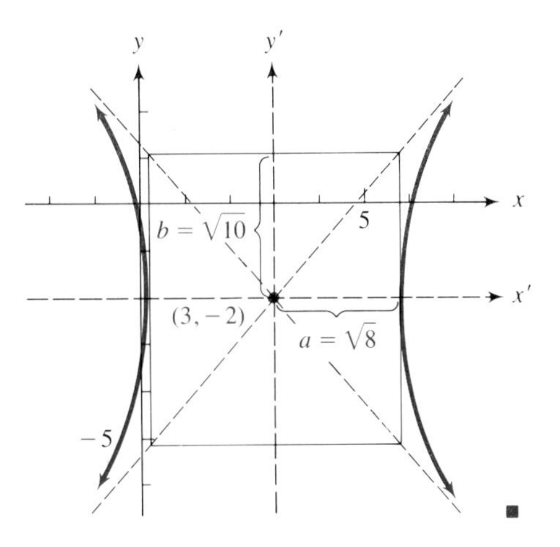

■

We can use the same procedure (completing the square) that we used to rewrite other conic sections to write the equation of a hyperbola in standard form.

EXAMPLE 4 a. Write the equation

$$y^2 - 4x^2 + 4y - 8x - 9 = 0$$

in standard form.

b. Graph the equation.

Solution a. We first prepare to complete the square in both x and y and write

$$(y^2 + 4y \qquad) - 4(x^2 + 2x \qquad) = 9.$$

Completing the square in y by adding 4 to each member and completing the square in x by adding -4 $(-4 \cdot 1)$ to each member, we obtain

$$(y^2 + 4y + 4) - 4(x^2 + 2x + 1) = 9 + 4 - 4$$

and then

$$(y + 2)^2 - 4(x + 1)^2 = 9,$$

from which we have the standard form

$$\frac{(y + 2)^2}{9} - \frac{(x + 1)^2}{9/4} = 1.$$

b. The graph is a hyperbola with center $(-1, -2)$. The vertices are in the y'-axis with intercepts 3 and -3. Since $a = 3$ and $b = 3/2$, the central rectangle can be constructed with dimensions $2a = 6$ and $2b = 3$. The asymptotes are drawn through the origin of the x'- and y'-axes and the vertices of the rectangle. The hyperbola is sketched as shown. ■

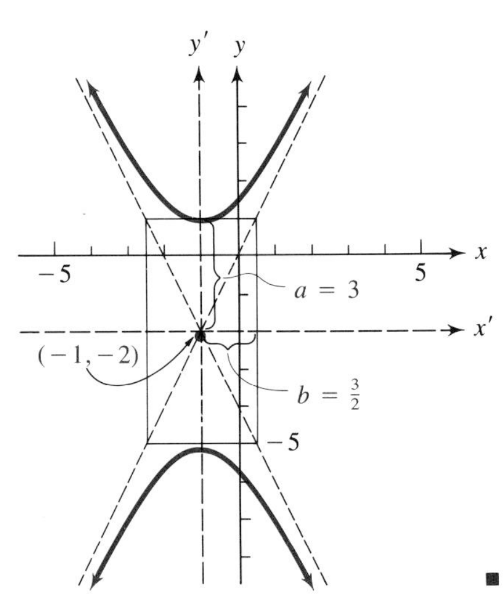

Recognizing basic conic sections

In general, to graph a second-degree equation it is helpful to first write the equation in standard form in order to recognize the form of the graph. The forms of equations that we have considered are summarized in Table 10.1. For a parabola, (h, k) is the vertex, while for the other conics, (h, k) is the center.

EXAMPLE 5 Name the graph of each equation.

a. $x^2 = 9y^2 - 9$ b. $x^2 - y = 2x + 4$ c. $9y^2 = 4 - x^2$

Name of curve	Standard form of equation	Graph
Parabola	$y = a(x - h)^2 + k$	$a > 0$; $a < 0$
Circle	$(x - h)^2 + (y - k)^2 = 1$	(h,k)
Ellipse: Major axis parallel to x-axis Major axis parallel to y-axis	$\dfrac{(x-h)^2}{a^2} + \dfrac{(y-k)^2}{b^2} = 1$ $\dfrac{(x-h)^2}{b^2} + \dfrac{(y-k)^2}{a^2} = 1$	F_1, (h,k), F_2; F_1, (h,k), F_2
Hyperbola: Transverse axis parallel to x-axis Transverse axis parallel to y-axis	$\dfrac{(x-h)^2}{a^2} - \dfrac{(y-k)^2}{b^2} = 1$ $\dfrac{(y-k)^2}{a^2} - \dfrac{(x-h)^2}{b^2} = 1$	F_1, F_2; F_1, F_2

TABLE 10.1

Solution It is helpful to first rewrite the equation in one of the forms shown in the table.

a. $x^2 = 9y^2 - 9$ is equivalent to

$$9y^2 - x^2 = 9,$$

from which $\dfrac{y^2}{1} - \dfrac{x^2}{9} = 1$. The graph is a hyperbola with center at the origin and vertices $(0,1)$ and $(0,-1)$.

b. $x^2 - y = 2x + 4$ is equivalent to

$$y = x^2 - 2x - 4.$$

The graph is a parabola that opens upward.

c. $9y^2 = 4 - x^2$ is equivalent to

$$x^2 + 9y^2 = 4,$$

from which $\dfrac{x^2}{2^2} + \dfrac{y^2}{(2/3)^2} = 1$. The graph is an ellipse with center at the origin and major axis $2a = 4$ in the x-axis. ■

As we have noted, it may be necessary to complete the square in expressions in x and y in order to write an equation in standard form. However, it is not always necessary to complete the process if we only want to determine the type of conic section.

EXAMPLE 6 Name the graph of each equation, assuming the graph exists.

a. $3x^2 + 3y^2 - 2x + 4y - 6 = 0$

b. $4y^2 + 8x^2 - 3y = 0$

c. $4x^2 - 6y^2 + x - 2y = 0$

d. $y + x^2 - 4x + 1 = 0$

Solutions

a. The graph is a circle because the coefficients of x^2 and y^2 are equal.

b. The graph is an ellipse because the coefficients of x^2 and y^2 are both positive and unequal.

c. The graph is a hyperbola because the coefficients of x^2 and y^2 are of opposite signs.

d. The graph is a parabola because y is of first degree and x is of second degree. ■

EXERCISE 10.3

A

■ *Graph each equation. See Examples* 1, 2, *and* 3.

1. $\dfrac{x^2}{25} - \dfrac{y^2}{9} = 1$

2. $\dfrac{y^2}{4} - \dfrac{x^2}{16} = 1$

3. $\dfrac{y^2}{12} - \dfrac{x^2}{8} = 1$

4. $\dfrac{x^2}{15} - \dfrac{y^2}{10} = 1$

5. $\dfrac{(x - 4)^2}{9} - \dfrac{(y + 2)^2}{16} = 1$

6. $\dfrac{(y + 4)^2}{25} - \dfrac{(x - 3)^2}{4} = 1$

7. $\dfrac{x^2}{4} - \dfrac{(y-3)^2}{8} = 1$

8. $\dfrac{y^2}{9} - \dfrac{(x+4)^2}{12} = 1$

9. $\dfrac{(y+2)^2}{6} - \dfrac{(x+2)^2}{10} = 1$

10. $\dfrac{(x-4)^2}{5} - \dfrac{(y-4)^2}{8} = 1$

11. $\dfrac{y^2}{6} - \dfrac{(x-3)^2}{15} = 1$

12. $\dfrac{x^2}{12} - \dfrac{(y+2)^2}{7} = 1$

■ *Graph each equation. See Example 4.*

13. $9x^2 - 4y^2 - 36x - 24y - 36 = 0$

14. $9y^2 - 4x^2 - 72y - 24x + 72 = 0$

15. $16y^2 - 4x^2 + 32x - 128 = 0$

16. $16x^2 - 9y^2 + 54y - 225 = 0$

17. $4x^2 - 6y^2 - 32x - 24y + 16 = 0$

18. $9y^2 - 8x^2 + 72y + 16x + 64 = 0$

19. $12x^2 - 3y^2 + 24y - 84 = 0$

20. $10y^2 - 5x^2 + 30x - 95 = 0$

■ *The graphs of the following equations are circles, ellipses, parabolas, or hyperbolas. Name the graph. See Examples 5 and 6.*

21. $y^2 = 4 - x^2$

22. $y^2 = 6 - 4x^2$

23. $4y^2 = x^2 - 8$

24. $x^2 + 2y - 4 = 0$

25. $4x^2 = 12 - 2y^2$

26. $6x^2 = 8 - 6y^2$

27. $4x^2 = 6 + 4y$

28. $2x^2 = 5 + 4y^2$

29. $6 + \dfrac{x^2}{4} = y^2$

30. $y^2 = 6 - \dfrac{2x^2}{3}$

31. $\dfrac{1}{2}x^2 - y = 4$

32. $\dfrac{x^2}{4} = 4 + 6y^2$

33. $y = (x-3)^2 + 2$

34. $\dfrac{(y-2)^2}{4} - \dfrac{(x+3)^2}{8} = 1$

35. $(x+1)^2 + (x-4)^2 = 16$

36. $\dfrac{(x+3)^2}{4} + \dfrac{y^2}{12} = 1$

37. $4x^2 + 8y^2 + 2x + 6 = 0$

38. $2y^2 - 8x^2 + 4y - 2x = 0$

39. $2x^2 + 2y^2 + 6x - 8 = 0$

40. $y - 2 = \dfrac{(x+4)^2}{4}$

B

41. Graph $x^2 - y^2 = 0$. *Hint:* First write as $(x-y)(x+y) = 0$ and then graph $y = x$ and $y = -x$.

42. Graph $4x^2 - y^2 = 0$. See hint for Exercise 41.

43. Graph $x^2 - y^2 = 4, x^2 - y^2 = 1$, and $x^2 - y^2 = 0$ on the same set of axes.

44. Graph $4x^2 - y^2 = 16, 4x^2 - y^2 = 4$, and $4x^2 - y^2 = 0$ on the same set of axes.

45. Find values for a, b, and c such that the graph of $x^2 + y^2 + ax + by + c = 0$ will contain the points $(3, 0)$, $(3, -8)$, and $(7, -4)$.

46. Find the equation of the circle whose graph contains the points $(2, 3)$, $(3, 2)$, and $(-4, -5)$. *Hint:* See Exercise 45.

10.4

SOLUTION OF SYSTEMS BY SUBSTITUTION

Real solutions obtained from graphs

Approximate solutions of systems of equations in two variables, where one or both of the equations are quadratic, may often be found by graphing both equations and estimating the coordinates of any points they have in common. For example, to find the solution set of the system

$$x^2 + y^2 = 26 \qquad (1)$$

$$x + y = 6, \qquad (2)$$

we graph the equations on the same set of axes, as shown in Figure 10.14, and observe that the graphs appear to intersect at (1, 5) and (5, 1). The solution set of the system (1) and (2) is, in fact, {(1, 5), (5, 1)}.

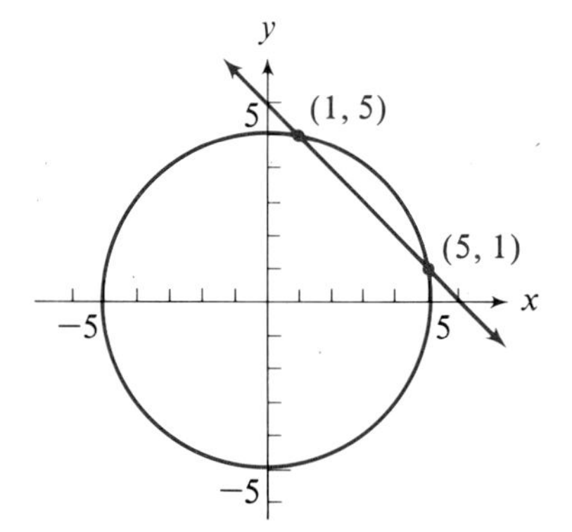

FIGURE 10.14

However, solving second-degree systems graphically on the real plane may produce only approximations to real solutions, and we cannot expect to locate solutions in which one or both of the components are imaginary numbers. It is therefore more practical to concentrate on algebraic methods of solution since the results are exact and we can obtain imaginary solutions. It is suggested that, whenever feasible, you sketch the graphs of the equations as a rough check on an algebraic solution.

Solving a system by substitution

One of the most useful techniques available for finding solution sets for systems of equations is **substitution**. This technique is particularly helpful with systems containing one first-degree and one higher-degree equation.

EXAMPLE 1 Solve

$$x^2 + y^2 = 26 \qquad (1)$$

$$x + y = 6 \qquad (2)$$

using algebraic methods.

Solution Equation (2) can be written in the form

$$y = 6 - x, \qquad (3)$$

and we can argue that for any ordered pair (x, y) in the solution set of both (1) and (2), x and y in (1) represent the same numbers as x and y in (3). Hence, the substitution property may be used to replace y in (1) by its equal $(6 - x)$ from (3). This will produce

$$x^2 + (6 - x)^2 = 26, \tag{4}$$

which will have as a solution set those values of x for which the ordered pair (x, y) is a common solution of (1) and (2). Rewriting (4) equivalently, we have

$$\begin{aligned} x^2 + 36 - 12x + x^2 &= 26 \\ 2x^2 - 12x + 10 &= 0 \\ x^2 - 6x + 5 &= 0 \\ (x - 5)(x - 1) &= 0, \end{aligned}$$

from which x is either 1 or 5. Now, by replacing x in (3) with each of these numbers, we have

$$y = 6 - (1) = 5 \quad \text{and} \quad y = 6 - (5) = 1;$$

so the solution set of the system (1) and (2) is $\{(1, 5), (5, 1)\}$.

Check this solution and notice that these ordered pairs are also solutions of (1). ■

Note that in the above example if we used (1) rather than (2) or (3) to obtain values for the y-component, we would have

$$\begin{aligned} (1)^2 + y^2 &= 26 \\ y &= \pm 5 \end{aligned} \qquad \begin{aligned} (5)^2 + y^2 &= 26 \\ y &= \pm 1 \end{aligned}$$

and the solutions obtained would be $(1, 5)$, $(1, -5)$, $(5, 1)$, and $(5, -1)$. However, $(1, -5)$ and $(5, -1)$ are not solutions of (2). The solution set is again $\{(1, 5), (5, 1)\}$.

The foregoing example suggests: *if the degrees of equations differ, one component of a solution should be substituted in the equation of lower degree in order to find only those ordered pairs that are solutions of both equations.*

EXAMPLE 2 Solve

$$\begin{aligned} y &= x^2 + 2x + 1 && (1) \\ y - x &= 3 && (2) \end{aligned}$$

using algebraic methods.

Solution Solve (2) explicitly for y.

$$y = x + 3 \qquad (2a)$$

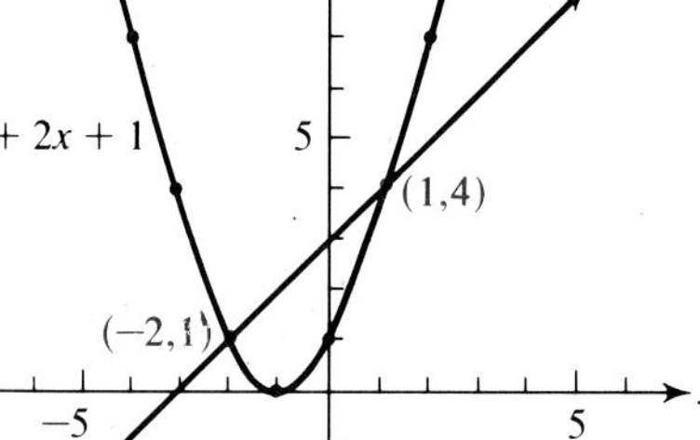

Substitute $(x + 3)$ for y in (1).

$$x + 3 = x^2 + 2x + 1$$

Solve for x.

$$x^2 + x - 2 = 0$$
$$(x + 2)(x - 1) = 0$$
$$x = -2 \quad \text{or} \quad x = 1$$

Substitute each of these values in (2a) to determine values for y. If $x = -2$, then $y = 1$, and if $x = 1$, then $y = 4$. The solution set of the system is $\{(-2, 1), (1, 4)\}$. ■

Imaginary solutions In the foregoing examples the components of each solution are real numbers, and their graphs are the points of intersection of the graphs of each equation. If one or more of the components of the solutions of a system are imaginary numbers, we can find these solutions but, in such cases, the graphs in the real plane do not have points of intersection.

EXAMPLE 3 Solve

$$x^2 + y^2 = 26 \qquad (1)$$
$$x + y = 8. \qquad (2)$$

Solution Solving (6) for y, we have

$$y = 8 - x. \qquad (2a)$$

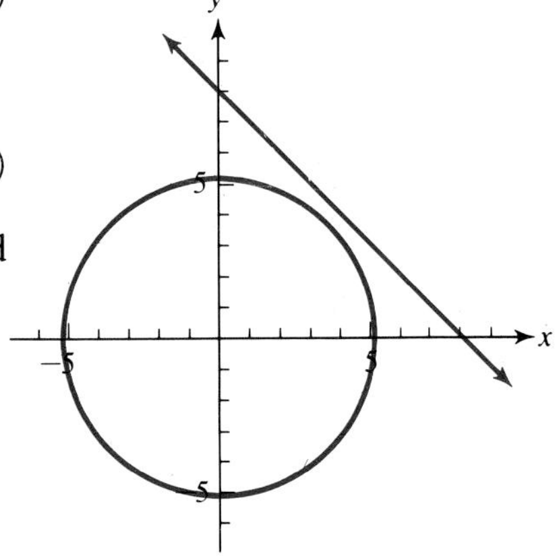

Substituting $(8 - x)$ for y in (1) and simplifying yields

$$x^2 + (8 - x)^2 = 26$$
$$x^2 + 64 - 16x + x^2 = 26$$
$$2x^2 - 16x + 38 = 0$$
$$x^2 - 8x + 19 = 0.$$

Using the quadratic formula to solve for x, we obtain

$$\begin{aligned} x &= \frac{8 \pm \sqrt{64 - 76}}{2(1)} = \frac{8 \pm \sqrt{-12}}{2} \\ &= \frac{8 \pm 2i\sqrt{3}}{2} = \frac{2(4 \pm i\sqrt{3})}{2} \\ &= 4 \pm i\sqrt{3}. \end{aligned}$$

Then, substituting $(4 + i\sqrt{3})$ for x in (2a) gives $y = 4 - i\sqrt{3}$, and substituting $(4 - i\sqrt{3})$ for x in (2a) gives $y = 4 + i\sqrt{3}$. Hence, the solution set is

$$\{(4 + i\sqrt{3}, 4 - i\sqrt{3}), (4 - i\sqrt{3}, 4 + i\sqrt{3})\};$$

the graphs of the equations are shown in the figure. ■

EXERCISE 10.4

A

■ *Solve by the method of substitution. In Problems* 1–10, *sketch the graphs of the equations. See Examples* 1, 2, *and* 3.

1. $y = x^2 - 5$, $y = 4x$

2. $y = x^2 - 2x + 1$, $y + x = 3$

3. $x^2 + y^2 = 13$, $x + y = 5$

4. $x^2 + 2y^2 = 12$, $2x - y = 2$

5. $x + y = 1$, $xy = -12$

6. $2x - y = 9$, $xy = -4$

7. $xy = 4$, $x^2 + y^2 = 8$

8. $x^2 - y^2 = 35$, $xy = 6$

9. $x^2 + y^2 = 9$, $y = 4$

10. $2x^2 - 4y^2 = 12$, $x = 4$

11. $x^2 + y = 4$, $x - y = -1$

12. $x^2 + 9y^2 = 36$, $x - 2y = -8$

13. $x^2 - xy - 2y^2 = 4$, $x - y = 2$

14. $x^2 - 2x + y^2 = 3$, $2x + y = 4$

15. $2x^2 - 5xy + 2y^2 = 5$, $2x - y = 1$

16. $2x^2 + xy + y^2 = 9$, $-x + 3y = 9$

■ *Solve each problem using a system of equations.*

17. The sum of the squares of two positive numbers is 13. If 2 times the first number is added to the second, the sum is 7. Find the numbers.

18. The sum of two numbers is 6 and their product is $35/4$. Find the numbers.
19. The perimeter of a rectangle is 26 inches and the area is 12 square inches. Find the dimensions of the rectangle.
20. The area of a rectangle is 216 square feet. If the perimeter is 60 feet, find the dimensions of the rectangle.

B

21. The annual income from an investment is \$32. If the amount invested were \$200 more and the rate 1/2% less, the annual income would be \$35. What are the amount and rate of the investment?
22. At a constant temperature, the pressure P and volume V of a gas are related by the equation $PV = K$. The product of the pressure (in pounds per square inch) and the volume (in cubic inches) of a certain gas is 30 inch-pounds. If the temperature remains constant as the pressure is increased 4 pounds per square inch, the volume is decreased by 2 cubic inches. Find the original pressure and volume of the gas.

10.5

SOLUTION OF SYSTEMS BY OTHER METHODS

If both the equations in a system are second-degree in both variables, the use of linear combinations of members of the equations often provides a simpler means of solution than does substitution.

EXAMPLE 1 Solve

$$4x^2 + y^2 = 25 \qquad (1)$$

$$x^2 - y^2 = -5. \qquad (2)$$

Solution By forming a linear combination using 1 times Equation (1) and 1 times Equation (2), we have

$$5x^2 = 20,$$

from which

$$x = 2 \quad \text{or} \quad x = -2.$$

We now have the x-components of the members of the solution set of the system (1) and (2). Substituting 2 for x in either (1) or (2)—we shall use (1)—we obtain

$$\begin{aligned} 4(2)^2 + y^2 &= 25 \\ y^2 &= 25 - 16 \\ y^2 &= 9, \end{aligned}$$

from which

$$y = 3 \quad \text{or} \quad y = -3.$$

Thus, the ordered pairs $(2, 3)$ and $(2, -3)$ are in the solution set of the system. Substituting -2 for x in (1) or (2)—this time we shall use (2)—gives us

$$\begin{aligned} (-2)^2 - y^2 &= -5 \\ -y^2 &= -5 - 4 \\ y^2 &= 9, \end{aligned}$$

from which

$$y = 3 \quad \text{or} \quad y = -3.$$

Thus, the ordered pairs $(-2, 3)$ and $(-2, -3)$ are also solutions of the system, and the complete solution set is

$$\{(2, 3), (2, -3), (-2, 3), (-2, -3)\}. \quad \blacksquare$$

The following example shows another means of solving a system of two second-degree equations.

EXAMPLE 2 Solve

$$x^2 + y^2 = 5 \qquad (3)$$

$$x^2 - 2xy + y^2 = 1. \qquad (4)$$

Solution By forming a linear combination using 1 times Equation (3) and -1 times Equation (4), we have

$$2xy = 4$$

or

$$xy = 2, \qquad (5)$$

which has a solution set containing all the ordered pairs that satisfy both (3) and (4). Therefore, forming the new system

$$x^2 + y^2 = 5 \tag{3}$$

$$xy = 2, \tag{5}$$

we can be sure that the solution set of this system is the same as the solution set of the system (3) and (4).

This latter system can be solved by substitution. We have, from (5),

$$y = \frac{2}{x}.$$

Replacing y in (3) by $2/x$, we find

$$x^2 + \left(\frac{2}{x}\right)^2 = 5,$$

from which

$$x^2 + \frac{4}{x^2} = 5. \tag{6}$$

Multiplying each member by x^2, we have

$$x^4 + 4 = 5x^2 \tag{7}$$

$$x^4 - 5x^2 + 4 = 0, \tag{7a}$$

which is quadratic in x^2. Factoring the left-hand member of (7a), we obtain

$$(x^2 - 1)(x^2 - 4) = 0,$$

from which

$$x^2 - 1 = 0 \quad \text{or} \quad x^2 - 4 = 0.$$

Solving these equations (by factorization), we obtain

$$x = 1, \quad x = -1 \quad \text{and} \quad x = 2, \quad x = -2.$$

Since we multiplied Equation (6) by a variable, we are careful to note that these values of x $(x \neq 0)$ all satisfy (6).

Now, substituting 1, -1, 2, and -2 for x in the equation $xy = 2$ or $y = 2/x$, we have:

For $x = 1$, $y = 2$. For $x = -1$, $y = -2$.
For $x = 2$, $y = 1$. For $x = -2$, $y = -1$.

The solution set of either system (3) and (5) or system (3) and (4) is

$$\{(1, 2), (-1, -2), (2, 1), (-2, -1)\}.$$ ■

There are other techniques involving substitution in conjunction with linear combinations that are useful in handling systems of higher-degree equations, but they are all similar to those illustrated. Each system should be scrutinized for some means of finding an equivalent system that will lend itself to solution by linear combination or substitution.

EXERCISE 10.5

A

■ *Solve each system. See Example* 1.

1. $x^2 + y^2 = 10$
$9x^2 + y^2 = 18$

2. $x^2 + 4y^2 = 52$
$x^2 + y^2 = 25$

3. $x^2 + 4y^2 = 17$
$3x^2 - y^2 = -1$

4. $9x^2 + 16y^2 = 100$
$x^2 + y^2 = 8$

5. $x^2 - y^2 = 7$
$2x^2 + 3y^2 = 24$

6. $x^2 + 4y^2 = 25$
$4x^2 + y^2 = 25$

7. $3x^2 + 4y^2 = 16$
$x^2 - y^2 = 3$

8. $4x^2 + 3y^2 = 12$
$x^2 + 3y^2 = 12$

9. $4x^2 - 9y^2 + 132 = 0$
$x^2 + 4y^2 - 67 = 0$

10. $16y^2 + 5x^2 - 26 = 0$
$25y^2 - 4x^2 - 17 = 0$

■ *See Example* 2 *for Problems* 11–20.

11. $2x^2 + xy - 4y^2 = -12$
$x^2 - 2y^2 = -4$

12. $x^2 + 2xy - y^2 = 14$
$x^2 - y^2 = 8$

13. $x^2 + 3xy - y^2 = -3$
$x^2 - xy - y^2 = 1$

14. $2x^2 + xy - 2y^2 = 16$
$x^2 + 2xy - y^2 = 17$

15. $x^2 - xy + y^2 = 7$
$x^2 + y^2 = 5$

16. $3x^2 - 2xy + 3y^2 = 34$
$x^2 + y^2 = 17$

17. $3x^2 + 3xy - y^2 = 35$
$x^2 - xy - 6y^2 = 0$

18. $x^2 - xy + y^2 = 21$
$x^2 + 2xy - 8y^2 = 0$

19. $2x^2 - xy - 6y^2 = 0$
$x^2 + 3xy + 2y^2 = 4$

20. $2x^2 + xy - y^2 = 0$
$6x^2 + xy - y^2 = 1$

B

21. How many *real* solutions are possible for systems of *independent* equations that consist of:

a. Two linear equations in two variables?

b. One linear equation and one quadratic equation in two variables?

c. Two quadratic equations in two variables?

Support your answers with sketches.

22. Consider the system

$$x^2 + y^2 = 8 \qquad (1)$$

$$xy = 4. \qquad (2)$$

We can solve this system by substituting $4/x$ for y in (1) to obtain

$$x^2 + \frac{16}{x^2} = 8,$$

from which we have $x = 2$ or $x = -2$. Now, if we obtain the y-components of the solution from (2), we find that:

For $x = 2,\ y = 2.$ For $x = -2,\ y = -2.$

But if we seek y-components from (1), we have:

For $x = 2,\ y = \pm 2.$ For $x = -2,\ y = \pm 2.$

Graph Equations (1) and (2) and discuss the fact that we seem to obtain more solutions from (1) than from (2). What is the solution set of the system?

10.6

INEQUALITIES

Quadratic inequalities can be graphed in the same way that we graphed linear inequalities in Section 8.5. We first graph the equation that has the same members and then shade the appropriate region as required. We use a solid curve if the coordinates of the points in the graph of the equation *are included* in the solution set of the inequality ($\leqslant$ or $\geqslant$) and a dashed curve if the coordinates of the points in the graph of the equation *are not included* in the solution set of the inequality ($<$ or $>$).

EXAMPLE 1 Graph $4x^2 + 9y^2 \leqslant 36$. (1)

Solution We first graph $4x^2 + 9y^2 = 36$. In standard form the equation is

$$\frac{x^2}{9} + \frac{y^2}{4} = 1, \tag{2}$$

and we determine that the graph is an ellipse centered at the origin with x-intercepts 3 and -3 and y-intercepts 2 and -2.

Substituting the coordinates of the origin (0, 0) in the original inequality, we obtain

$$4(0)^2 + 9(0)^2 = 36,$$

which is true, hence that part of the plane including the origin is shaded. Since the graph of (2) is part of the graph of (1) a solid curve is used.

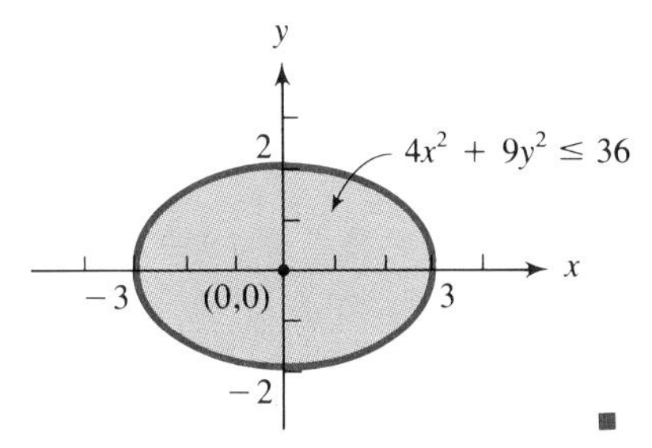

■

EXAMPLE 2 Graph $y < x^2 + 2$. (3)

Solution We first graph

$$y = x^2 + 2. \tag{4}$$

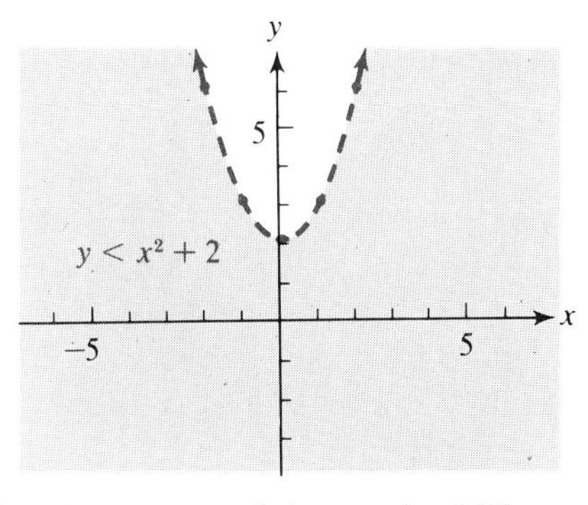

We graphed this equation on page 366 by noting that the parabola opens upward with a vertex at (0, 2) because 2 is the least value for y and this is the value when $x = 0$. Hence, plotting another point and using the fact that the parabola is symmetric with the y-axis we complete the graph of the equation. Since the graph of (4) is not part of the graph of (3) we use a *dashed* curve.

Substituting the coordinates of the origin (0, 0) in Inequality (3) we obtain

$$(0) < (0) + 2,$$

which is true, hence that part of the plane including the origin is shaded. ■

Systems of inequalities

We obtained graphic solutions of systems of linear inequalities in Section 9.1 by using double and triple shading. We can graph systems involving quadratic inequalities in a similar way.

EXAMPLE 3 Graph

$$y > x^2 - 4 \tag{1}$$

$$\frac{x^2}{16} + \frac{y^2}{4} \leqslant 1 \tag{2}$$

on the same coordinate system and use double shading to indicate the region common to both.

Solution We first graph the associated equations

$$y = x^2 - 4 \tag{1'}$$

$$\frac{x^2}{16} + \frac{y^2}{4} = 1, \tag{2'}$$

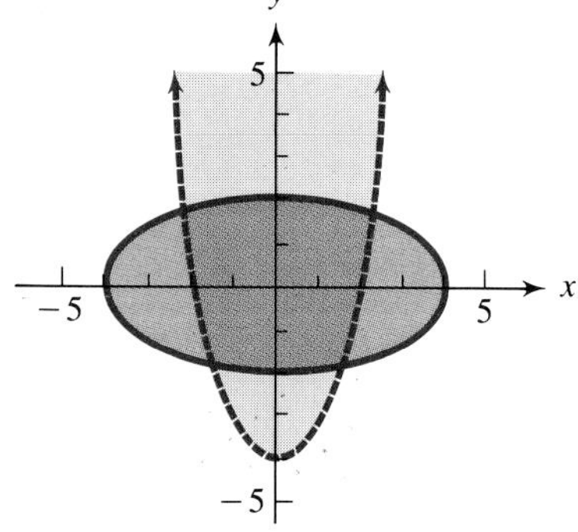

where we use a dashed curve for (1′) because the graph of (1′) *is not* part of the graph of (1), and use a solid curve for (2′) because the graph of (2′) *is* part of the graph of (2).

Substituting (0, 0) in (1) we obtain

$$(0) > (0) - 4,$$

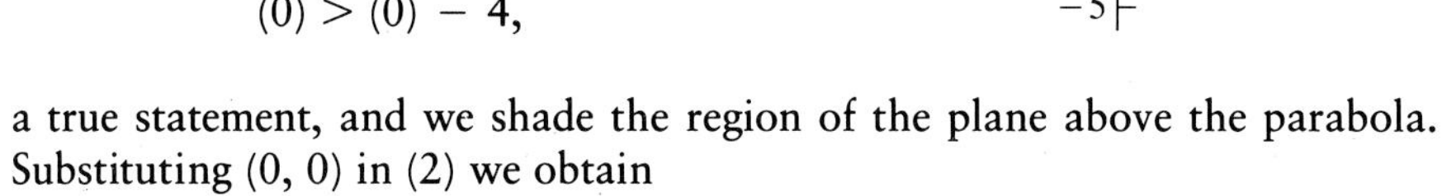

a true statement, and we shade the region of the plane above the parabola. Substituting (0, 0) in (2) we obtain

$$\frac{(0)^2}{16} + \frac{(0)^2}{4} < 1,$$

also a true statement, and we shade the region inside the ellipse.

The double-shaded region is common to Inequalities (1) and (2). ■

EXERCISE 10.6

A

■ *Sketch the graph of each inequality. See Examples* 1 *and* 2.

1. $y > x^2$
2. $y < x^2$
3. $y \leqslant x^2 - 4$
4. $y \geqslant x^2 - 4$
5. $y > x^2 + 4$
6. $y < x^2 + 4$
7. $y \geqslant x^2 - 6x + 8$
8. $y \leqslant x^2 - 6x + 8$
9. $y < x^2 + x - 6$
10. $y > x^2 + x - 6$
11. $y \leqslant x^2 + 3x + 2$
12. $y \geqslant x^2 + 3x + 2$
13. $x^2 + y^2 \leqslant 9$
14. $x^2 + y^2 \geqslant 9$
15. $x^2 + 4y^2 \leqslant 16$
16. $4x^2 + y^2 \geqslant 16$
17. $9x^2 - 4y^2 \leqslant 36$
18. $9y^2 - 4x^2 \geqslant 36$

■ *Graph each pair of inequalities on the same coordinate system and use double shading to indicate the region common to both. See Example* 3.

19. $y > x^2$, $y \leqslant x + 2$
20. $y < x^2 + 4$, $x - y \leqslant 4$
21. $x^2 + y^2 < 25$, $y > x$
22. $x^2 + y^2 \geqslant 9$, $y < x$
23. $4x^2 + y^2 \leqslant 36$, $x \geqslant 2$
24. $4x^2 + 3y^2 \leqslant 24$, $y \leqslant 0$
25. $4x^2 + y^2 \leqslant 16$, $y \geqslant x^2$
26. $x^2 + 2y^2 \geqslant 8$, $y < x^2$
27. $x^2 + y^2 \geqslant 16$, $y \geqslant x^2 - 6$
28. $3y^2 + x^2 < 12$, $y < x^2 - 6$
29. $y > x^2 - 5x + 4$, $y \leqslant 4 - x^2$
30. $y < -x^2 + x + 6$, $y > x^2 - 4$

CHAPTER SUMMARY

[10.1–10.3] The graphs of second-degree equations in two variables are **conic sections.** A quick sketch of a graph can be made by first writing the equation in standard form (see Table 10.1 on page 373) to identify its form.

[10.4–10.5] Systems of equations in two variables in which either or both equations are second-degree in one or both variables may have solutions with real components, imaginary components, or solutions of both kinds. Such systems can be solved by using substitution methods or by using linear combinations of the members of the equations in the system.

[10.6] Quadratic inequalities can be graphed by first graphing the equation that has the same members and then shading the appropriate region.

Systems of inequalities can be graphed by using double shading.

REVIEW EXERCISES

A

[10.1–10.3]

■ *Graph each conic section.*

1. $x^2 + 4y^2 = 24$
2. $y^2 - 4x^2 = 12$
3. $\dfrac{(x-5)^2}{6} + \dfrac{(y+2)^2}{9} = 1$
4. $\dfrac{x^2}{4} - \dfrac{(y+2)^2}{12} = 1$
5. $x^2 + y^2 + 6x - 4y - 2 = 0$
6. $x^2 + 4y^2 - 4x + 8y = 0$
7. $y = x^2 - x - 6$
8. $y = 3x^2 + 5x - 2$
9. $y^2 - 4x^2 - 8x - 16 = 0$
10. $x^2 - 6y^2 - 12y - 18 = 0$

[10.4–10.5]

■ *Solve each system by substitution or linear combination.*

11. $x + 3y^2 = 4$
    $x = 3$
12. $x^2 + 2y^2 = -8$
    $y = -2$
13. $x^2 + y = 3$
    $5x + y = 7$
14. $x^2 + 3xy + x = -12$
    $2x - y = 7$
15. $6x^2 - y^2 = 1$
    $3x^2 + 2y^2 = 13$
16. $2x^2 + 5y^2 - 53 = 0$
    $4x^2 + 3y^2 - 43 = 0$
17. $x^2 - 2xy + 3y^2 = 17$
    $2x^2 + xy + 6y^2 = 24$
18. $x^2 - xy - y^2 = 1$
    $x^2 + 3xy - y^2 = 9$
19. The perimeter of a rectangle is 34 centimeters long and the area is 70 square centimeters. Find the dimensions of the rectangle.
20. A rectangle has a perimeter of 18 feet. If the length is decreased by 5 feet and the width is increased by 12 feet, the area is doubled. Find the dimensions of the original rectangle.

[10.6]

■ *Graph each pair of inequalities on the same coordinate system and use double shading to indicate the region common to both.*

21. $y > x^2 + 1$
    $x^2 + 2y^2 \leqslant 18$

22. $y \geqslant x^2 - 4$
    $y < -x^2$

B

23. Find an equation of the parabola $y = ax^2 + bx + c$ whose graph contains the points $(-1, -4)$, $(0, -6)$, and $(4, 6)$.
24. Find an equation of the circle $x^2 + y^2 + ax + by + c = 0$ whose graph contains the points $(-1, 2)$, $(1, 4)$, and $(3, -2)$.
25. What relationship must exist between the numbers a and b so that the solution set of the system

$$y = x^2 - 4$$
$$y = ax + b$$

will have exactly one solution?

Cumulative Review Chapters 1-10

The numbers in brackets refer to the sections where such problems are first considered.

1. If $Q(x) = -x^4 + 4x^3 - 9x - 2$, find $Q(-2)$. [2.1]
2. Write $\sqrt[4]{3x} + x\sqrt[3]{y^2}$ with exponential notation. [6.5]
3. Write $3x^{1/5}(2x - y)^{-2/5}$ with radical notation. [6.5]
4. Evaluate $\begin{vmatrix} a & b & c \\ 0 & a & b \\ 0 & 0 & a \end{vmatrix}$. [9.4]

■ *Simplify.*

5. $\dfrac{3^3(3^2 - 2^2)}{-3(2 + 3)} - \dfrac{(2^2 + 3)^2}{2(2 + 3) - 3}$ [2.1]

6. $-2x[2x - x(3 - x) - (x + 1)^2]$ [2.4]

7. $\left(\dfrac{-ab}{s^2t}\right)^3 \left(\dfrac{s^3t^5}{a^4bc^2}\right)^2 \left(\dfrac{-b^3c^3}{s}\right)^4$ [6.1]

8. $\dfrac{\sqrt[3]{4y^5x^3z^5}\,\sqrt[3]{10y^4z^{11}}}{\sqrt[3]{9x^2y}}$ [6.6]

■ *Factor completely.*

9. $a^2 - 3a + 2ab - 6b$ [4.3]

10. $a^6b - a^2b^5$ [4.2]

11. $27a^6 + b^3$ [4.3]

12. $5(x - 3)(x - 2)^2 - 3(x - 3)^2(x - 2)$ [4.1]

■ *Write as a single fraction in lowest terms.*

13. $\dfrac{x - 2}{x^2 - 4x + 3} - \dfrac{2x + 1}{x^2 - x - 6} + \dfrac{x + 1}{x^2 + x - 2}$ [5.5]

14. $\dfrac{1}{x - 4} - \dfrac{2x - 6}{3x^2 + 5x - 2} \div \dfrac{4x^2 - 36}{3 - 8x - 3x^2}$ [5.6]

15. $\dfrac{\dfrac{a}{a - 2}}{\dfrac{2}{a^2} - \dfrac{2}{a}}$ [5.7]

16. Divide $\dfrac{3t^3 - 7t + 3}{t - 2}$. [5.2]

17. Write $(yx^{-1} - xy^{-1})^{-2}$ as a single fraction involving only positive exponents. [6.2]

18. Multiply $(x^{-1/4} - x^{1/2})^2$. [6.4]

19. Reduce $\dfrac{\sqrt{2y^2} - y^2\sqrt{8}}{y\sqrt{2}}$. [6.7]

20. Rationalize the denominator:

$$\frac{\sqrt{x} + \sqrt{2}}{\sqrt{x} - \sqrt{3}}. \qquad [6.7]$$

21. Write $(-3 - 2i)^2$ in the form $a + bi$. [6.8]

■ *Solve.*

22. $(3x - 5)^{1/2} = 2 + (x - 3)^{1/2}$ [7.3]

23. $3x^{1/2} - 10x^{1/4} - 8 = 0$ [7.4]

24. Solve by completing the square: $2x^2 + 3 = 3x$. [7.1]

25. Solve for l: $3V = (la + lb + Q)h$. [3.1]

26. Solve for y explicitly in terms of x: $2x^2y + 5y = xy - 2$. [8.1]

■ *Solve, and graph the solution set.*

27. $\dfrac{x - 3}{x + 2} < -4$ [7.5]

28. $x^2 + 12 > 7x$ [7.5]

29. $|6 - 4x| \geq 8$ [3.4]

30. Find an equation with integral coefficients whose solutions are $-1/4$ and $3/4$. [4.4]

31. Solve the system by linear combinations:

$$\frac{1}{3}x + \frac{1}{2}y = 2$$
$$\frac{1}{2}x - \frac{3}{4}y = 9. \qquad [9.1]$$

32. Solve by Cramer's rule:

$$\frac{1}{4}x - \frac{2}{3}y = 6$$
$$2x + 3y = -2. \qquad [9.3]$$

33. Solve

$$3x^2 + 5y^2 = 63$$
$$2x^2 - 3y^2 = 23. \qquad [10.5]$$

34. Determine how many solutions the system has:

$$3x = 2y + 2$$
$$4y - 6x = 4. \qquad [9.1]$$

■ *Solve using linear combinations. If there is no unique solution, so state.*

35.

$$x + 2y = 3z$$
$$x + \frac{1}{2}y + \frac{1}{2}z = \frac{1}{2}$$
$$\frac{2}{3}z + x = \frac{1}{3}y - 1 \qquad [9.2]$$

36.

$$x + 3y - 2z = 2$$
$$2x - 2y + z = 4$$
$$x - 5y + 3z = 2 \qquad [9.2]$$

37. Solve using Cramer's rule:

$$3x + 4y - z = 2$$
$$\frac{1}{2}x + 2y + z = 5$$
$$6x + y - \frac{3}{4}z = 9. \qquad [9.4]$$

■ *Use row operations on the augmented matrix to solve.*

38. $3x + 2y + z = -4$
$x + y - z = 5$
$2x - y - 3z = 5$ [9.5]

39. $x + 3z = 2$
$2y - 3z = 5$
$2x - 7y = 3$ [9.5]

■ *Solve by substitution.*

40. $x^2 - y^2 = 8$
$xy = 3$ [10.4]

41. $2x^2 - xy + 2y^2 = 3$
$2x - y = 1$ [10.4]

■ *Graph.*

42. $2x + 3y = 1$ [8.2]

43. $x^2 + y^2 - 6x + 2y + 6 = 0$ [10.1]

44. $9x^2 + 8y^2 + 72x - 16y + 80 = 0$ [10.1]

45. $y = -2x^2 + x + 3$ [10.2]

46. $x^2 - 2y^2 - 4x - 8y - 8 = 0$ [10.3]

■ *Graph the solution set in the xy-plane.*

47. $3x - y < 2$ [8.5]

48. $-4 \leq x < -1$ [8.5]

49. $x - 3y \geq 3$
$2x + y < 3$ [9.1]

50. $y \leq 4 - x^2$
$x^2 + 4y^2 < 16$ [10.6]

51. Find an equation for the line which includes the points $(-2, 5)$ and $(1, 3)$. [8.4]

52. Find the equation in standard form for the line with x-intercept $a = 1/3$ and y-intercept $b = -3/2$. [8.4]

53. Show that the two line segments whose endpoints are $(-3, 2)$, $(4, -1)$ and $(-2, -5)$, $(4, 9)$ are perpendicular. [8.3]

54. Find the perimeter of the triangle whose vertices are $(-2, 1)$, $(1, 2)$, and $(2, 2)$. [8.3]

55. The length of a rectangle is 5 feet greater than its width. A square whose side is equal to the length of the rectangle has an area 60 square feet greater than the area of the rectangle. Find the dimensions of the rectangle. [3.2]

56. A small sign is 11 inches tall and 8 inches wide. The margin must be twice as wide at top and bottom as it is at the sides. If 52 square inches are needed for the printed message, how wide should the margin be? [4.5]

57. The area of a rectangle is 208 square meters. Its perimeter is 58 meters. Find the dimensions of the rectangle. [10.4]

58. Find the maximum area of a rectangle whose perimeter is 80 incnes. [10.2]

■ *Use a system of equations to solve.*

59. Student tickets to a concert sell for \$5, and general admission tickets for \$16. 350 tickets were sold, bringing in \$4940. How many of each kind of ticket were sold? [9.6]

60. One angle of a triangle measures twice the second angle, and the third angle measures 10° more than the larger of the other two. Find the measure of each angle. [9.6]

B

61. Simplify $-(2x - [3y - (4x - 2) - (2y + 3)]) - [8x - (6y - 1)]$. [2.2]
62. Multiply $(x^{2n} + 2)(x^n - x^{n+1})$. [2.4]
63. Simplify $\dfrac{(a^{2n}b^{4m+2})^{1/2}}{(a^{3n}b^{2n})^{1/n}}$. [6.4]
64. Factor $3x^{2n} - 11x^n - 20$. [4.2]
65. Determine k so that the polynomial $x^3 + kx - 21$ has $x - 3$ as a factor. [5.2].
66. Simplify $\sqrt{\dfrac{4}{4x^2 - 4xy + y^2}}$. [6.5]
67. Evaluate $x^2 - 4x + 5$ for $x = 2 + i$. [6.8]
68. Solve for y in terms of x: $4x^2 + 25y^2 = 100$. [7.1]
69. Solve for c: $\dfrac{1}{a} = \dfrac{1}{b} + \dfrac{1}{c} - \dfrac{1}{d}$. [5.8]
70. Solve $\dfrac{4}{2x + 1} > \dfrac{3}{x - 2}$. [7.5]
71. Find the solution to $y = |x - 1| + |2x - 1|$ for $x = -2$. [8.1]
72. Show that $(-1, -1)$, $(9, 7)$, $(5, 12)$, and $(-5, 4)$ are the vertices of a rectangle. [8.3]
73. Find an equation for the line perpendicular to the graph of $3x + 2y = 6$ and passes through the point $(-1, 6)$. [8.4]
74. Show that $\begin{vmatrix} x & y & 1 \\ x_1 & y_1 & 1 \\ x_2 & y_2 & 1 \end{vmatrix} = 0$ is the equation of the line passing through the points (x_1, y_1) and (x_2, y_2). [9.4]
75. Solve the system
$$\frac{3}{x} + \frac{4}{y} = -1$$
$$\frac{2}{3x} + \frac{5}{2y} = 3. \qquad [9.1]$$
76. Three solutions of the equation $ax + by + cz = 1$ are $(3, 2, 1)$, $(-2, 0, 1)$, and $(1, 3, 4)$. Find the coefficients a, b, and c. [9.6]
77. Find the equation of the circle whose graph contains the points $(4, 2)$, $(-2, -6)$, and $(-3, 1)$. [10.3]
78. Sketch the family of curves $y = (x - k)^2$, $(k = -1, -2, 0, 1, 2)$ on a single set of axes. [10.2]
79. Graph the region described by $x \leqslant 3y$ and $x + 2y < 4$. [8.5]
80. Graph the solution set for the system, and find the coordinates of the vertices:
$$x + y \geqslant 4$$
$$x - 2 \leqslant 0$$
$$x \geqslant 0$$
$$y \geqslant 0. \qquad [9.1]$$

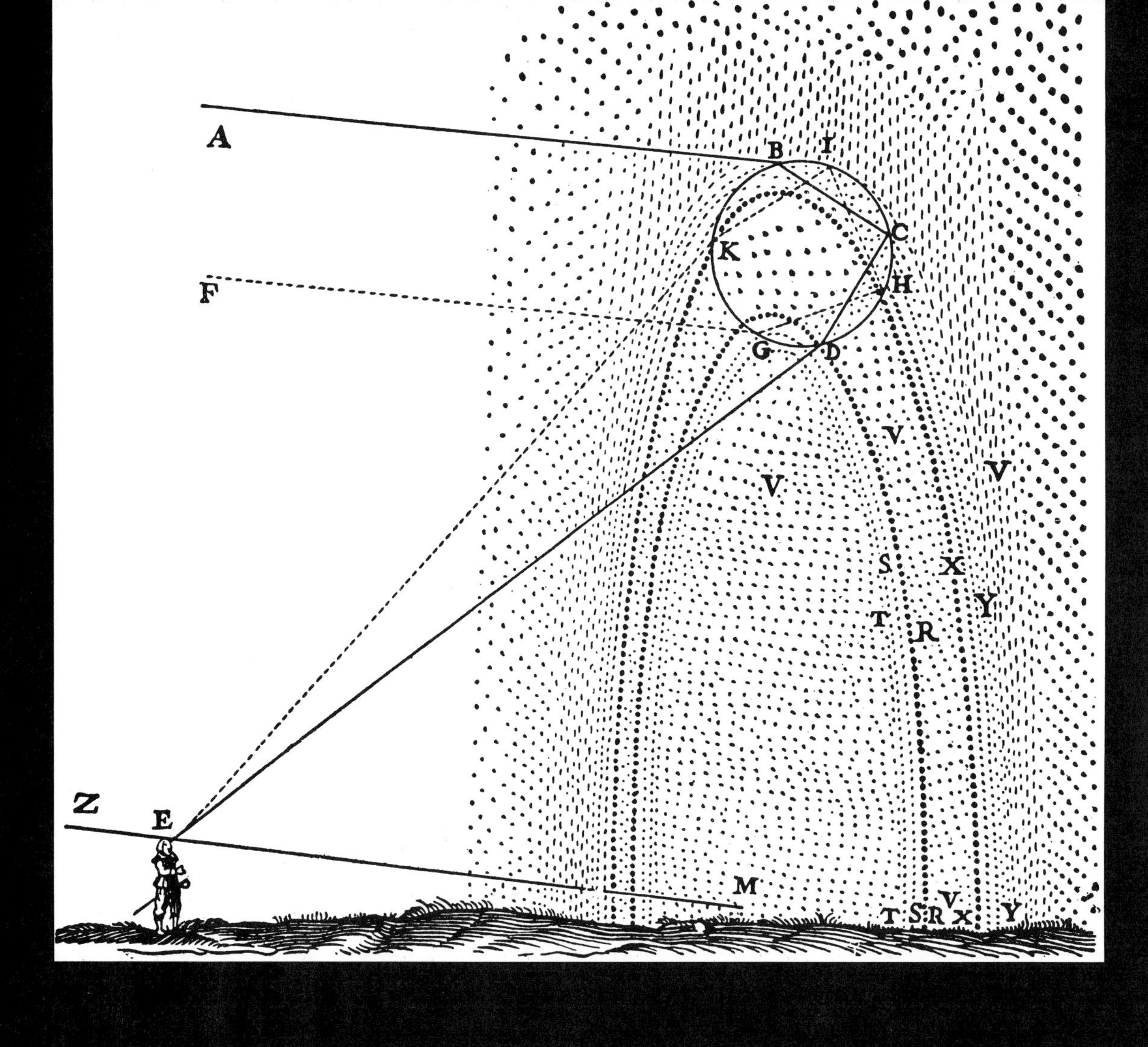
A
B
I
C
K
F
H
G
D
V
V
V
V
S
X
T
Y
R
Z
E
M
T
S
R
V
X
Y

11 Relations and Functions

11.1

DEFINITIONS AND NOTATIONS

In previous chapters we noted that equations and inequalities in two variables are useful models for conditions which describe a relationship between two quantities. In this section we shall consider some general notions concerning such relationships and in particular consider a special kind of relation called a *function.*

Relations

First let us consider the simple equation

$$y = x + 2.$$

If we specify $A = \{1, 2, 3\}$ as the replacement set for x, then $B = \{3, 4, 5\}$ is the replacement set for y, where the individual elements are paired as follows:

$$\begin{array}{rl} A = & \{1,\ 2,\ 3\} \\ & \ \ \updownarrow\ \ \updownarrow\ \ \updownarrow\ . \\ B = & \{3,\ 4,\ 5\} \end{array}$$

Because such associations display a relationship between the elements of the two sets involved, they are called **relations.**

Ordered pairs offer a clear way of specifying a relation. By using each element from set A as a first component for an ordered pair and the associated element in set B as a second component, we can exactly specify any relation between the elements of sets A and B. For example, the relation depicted above can be displayed as the set of ordered pairs

$$\{(1, 3), (2, 4), (3, 5)\}.$$

The set of all first components of the ordered pairs in a relation is the **domain** of the relation, and the set of all second components of the ordered pairs is the **range** of the relation. For example, the relation

$$\{(1, 3), (2, 4), (3, 5)\}$$

elements in the domain (first components); elements in the range (second components)

has domain $\{1, 2, 3\}$ and range $\{3, 4, 5\}$. Each element in the range of a relation is called a **value** of the relation.

Functions

In a relation, each element in the domain is paired with *one or more* elements in the range. Figure 11.1a illustrates a relation in which an element in the domain is paired with two elements in the range; two pairings, (3, 10) and (3, 15), have the same first component. Figure 11.1b illustrates a relation in which each element in the domain is associated with *only one* element in the range. In this relation no two pairings have the same first component. Such a relation is called a **function.**

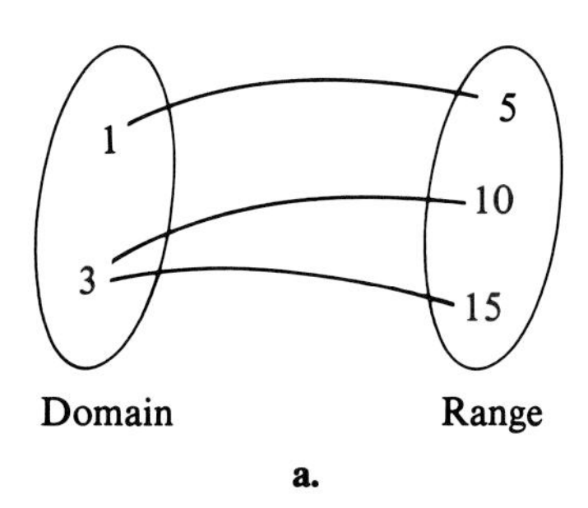

a.
Not a function

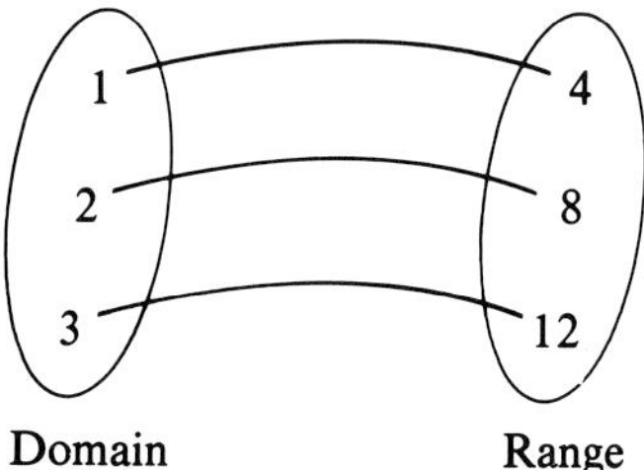

b.
Function

FIGURE 11.1

> A **function** is a relation in which no two distinct ordered pairs have the same first component.

EXAMPLE 1

a. $\{(2, 3), (3, 5), (4, 5)\}$ is a relation; it is also a function because no two ordered pairs have the same first component.

b. $\{(3, 4), (3, 5), (4, 7)\}$ is a relation but not a function, because two ordered pairs, (3, 4) and (3, 5), have the same first component. ■

An equation or inequality in two variables clearly specifies a relation. If each value of x is paired with exactly one value of y by an equation, then the relation specified is a function. In general, inequalities do not define functions.

EXAMPLE 2

a. $y = 2x + 1$ defines a relation that is a function because exactly one value of y is paired with each real number replacement for x.

b. $y^2 - x^2 = 1$, or, equivalently, $y = \pm\sqrt{x^2 + 1}$, defines a relation; it does not define a function because two values of y are paired with each real number replacement for x.

c. $y < x + 3$ defines a relation; it does not define a function, because more than one value of y is paired with each replacement of x. ■

Geometric test for a function

The graph of a relation gives us a means of visually checking to see whether a particular graph does or does not represent a function. If we imagine a line parallel to the y-axis passing across the graph, we can determine whether or not it intersects the graph at more than one point at each position on the x-axis; that is, *whether or not there are two or more different y-coordinates for any given x-coordinate*. If there are, the graph is not the graph of a function. See Figure 11.2

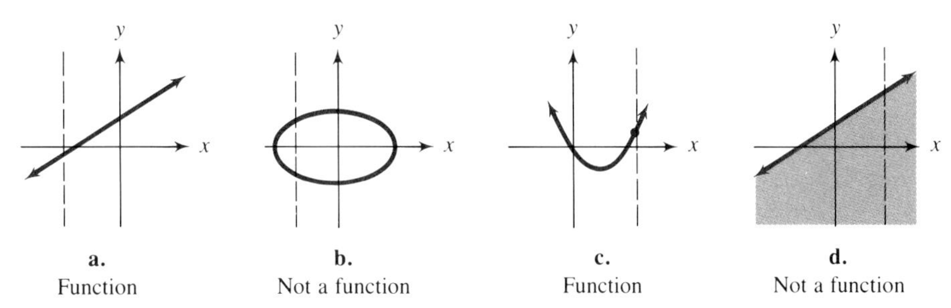

FIGURE 11.2 a. Function; b. Not a function; c. Function; d. Not a function

Note that Figure 11.2a is the graph of a first-degree equation. Because any first-degree equation (except for equations of the form $x = k$) has a graph that is a nonvertical straight line, such equations define functions.

Domain of a function

Let us make the following agreement concerning functions defined by equations:

If the domain of a function (or relation) is not specified, we shall assume that the function (or relation) has as domain the set of real numbers for which real numbers exist in the range.

Thus, values of the variable which make a denominator zero or result in an imaginary value for y are excluded from the domain.

EXAMPLE 3 Specify the domain (the set of replacements for x) for each function.

a. $y = x + 5$ b. $\dfrac{3}{x - 4}$ c. $y = \sqrt{x - 2}$

Solutions a. The entire set of real numbers because y is a real number for all real numbers x.

b. The set of real numbers except $x = 4$, because $3/(x - 4)$ is undefined for this value of x.

c. The set of real numbers $x \geq 2$ because $\sqrt{x - 2}$ is a real number for these values only. (If $x < 2$, then $x - 2$ is negative and $\sqrt{x - 2}$ is imaginary.) ■

Function notation

Functions are usually designated by means of a single symbol, P, R, f, g, or some other letter. The symbol for the function can be used in conjunction with the variable representing an element in the domain to represent an expression in that variable. Thus, we shall use $f(x)$ (read "f of x" or "the value of f at x") in precisely the same way that we used $P(x)$ in Section 2.1 when we discussed expressions and, in particular, polynomial expressions. This notation is commonly called **function notation**.

EXAMPLE 4 a. If f is the function defined by the equation

$$y = x - 3,$$

then we can just as well write

$$f(x) = x - 3,$$

where $f(x)$ plays the same role as y.

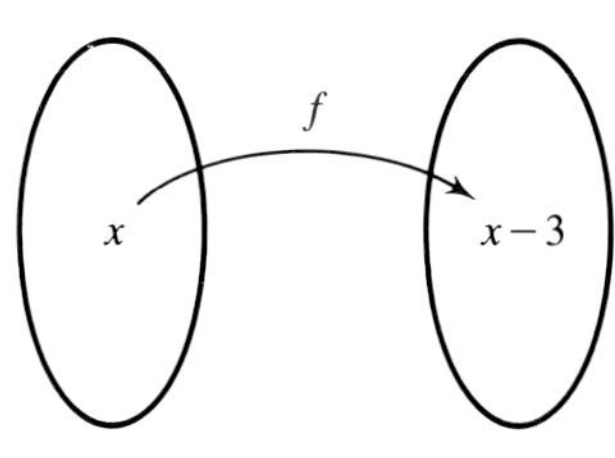

b. If $f(x) = x - 3$, then

$$f(6) = (6) - 3 = 3;$$
$$f(-2) = (-2) - 3 = -5;$$
$$f(t) = (t) - 3 = t - 3.$$

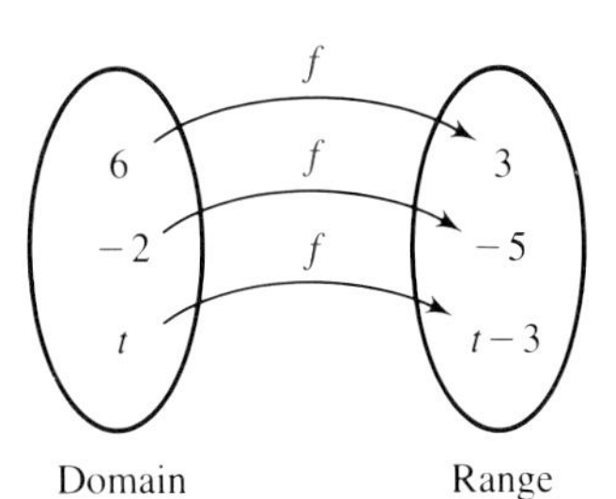

■

When the intent is clear from the context, we shall often use the terminology "the function $y = x - 3$ or $f(x) = x - 3$," rather than the longer phrase "the function defined by . . . ." In Section 8.2 a line was referred to as the "graph of the solution set of the equation" or simply the "graph of the equation." The line is also referred to as the "graph of the function defined by the equation" or simply as the "graph of the function."

The elements in the domain of a function can be other variables or expressions in the same variable.

EXAMPLE 5 Find a. $f(a)$; b. $f(b)$; c. $f(a + b)$ for the function

$$f(x) = x^2 - 2x.$$

Solution a. $f(a) = (a)^2 - 2(a) = a^2 - 2a$

b. $f(b) = (b)^2 - 2(b) = b^2 - 2b$

c. $f(a + b) = (a + b)^2 - 2(a + b) = a^2 + 2ab + b^2 - 2a - 2b$ ■

EXAMPLE 6 Find a. $f(x + h)$; b. $f(x + h) - f(x)$; c. $\dfrac{f(x + h) - f(x)}{h}$ for the function

$$f(x) = x^2 + 1.$$

Solution a. $f(x + h) = (x + h)^2 + 1 = x^2 + 2xh + h^2 + 1$

b. $f(x + h) - f(x) = (x^2 + 2xh + h^2 + 1) - (x^2 + 1) = 2xh + h^2$

c. $\dfrac{f(x + h) - f(x)}{h} = \dfrac{2xh + h^2}{h} = 2x + h$ ■

Graphical interpretation

Function notation can be used to denote the ordinate associated with a given abscissa of a graph. For example, consider the graph of a function $y = f(x)$ shown in Figure 11.3. Specific ordinates are shown as line segments labeled with their "lengths" in function notation. The "length" in this case is called a **directed length** because any function value is simply the ordinate of a point and can be positive, negative, or zero.

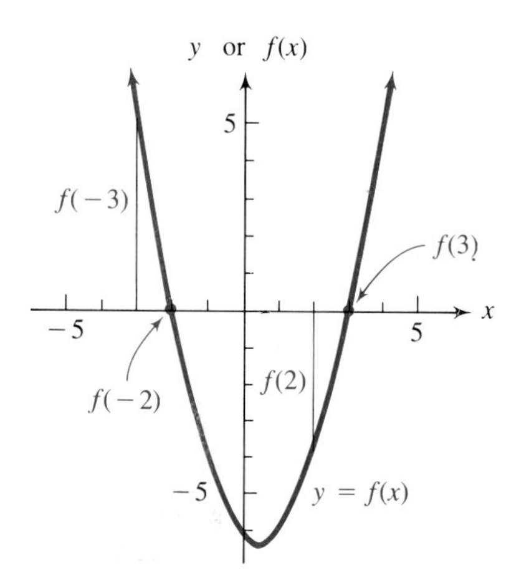

FIGURE 11.3

EXAMPLE 7 We can estimate the ordinates $f(-3)$ and $f(2)$ in Figure 11.3 to be 6 and -4, respectively. We also note that the x-intercepts of the graph are -2 and 3. Hence, $f(-2) = 0$ and $f(3) = 0$. ■

Zeros of functions

For any function f, values of x for which $f(x) = 0$ are called **zeros** of the function. Thus, we have three different names for a single idea. In particular, for a quadratic function:

1. The elements of the solution set of the equation

$$ax^2 + bx + c = 0.$$

2. The zeros of the function defined by

$$y = ax^2 + bx + c.$$

3. The x-intercepts of the graph of

$$y = ax^2 + bx + c.$$

EXAMPLE 8 a. The zeros of the function with graph shown in Figure 11.3 are -2 and 3 because $f(-2) = 0$ and $f(3) = 0$.

b. The zeros of

$$f(x) = x^2 + 3x - 10 = (x + 5)(x - 2)$$

and -5 and 2 because $f(-5) = 0$ and $f(2) = 0$. ■

If f is a function and x is some element of the domain of f, then the corresponding element y of the range depends on the element x. Thus, we call x the **independent variable** and y the **dependent variable**. In such cases we say "y is a function of x."

EXERCISE 11.1

A

■ a. *Specify the domain and the range of each relation.*

b. *State whether or not the relation is a function. See Example* 1.

1. $\{(-2, 3), (-1, 4), (0, 5)\ (1, 6)\}$
2. $\{(-1, -1), (0, 0), (1, 1), (2, 2)\}$
3. $\{(5, 1), (6, 1), (7, 1), (8, 1)\}$
4. $\{(4, 1), (4, 2), (4, 3), (4, 4)\}$
5. $\{(2, 3), (2, 4), (3, 3), (3, 4)\}$
6. $\{(-5, 2), (2, -5), (3, 6), (6, 3)\}$
7. $\{(0, 0), (2, 1), (2, 3), (3, 3)\}$
8. $\{(2, -1), (3, -1), (3, 1), (4, 2)\}$

■ a. *Solve each equation or inequality explicitly for y in terms of x.*

b. *State whether the relation is or is not a function. See Example* 2.

c. *Specify the domain of the relation defined by the equation or inequality. See Example* 3.

9. $2x + y = 6$
10. $2x + 3y = 12$
11. $x - 2y = 8$
12. $3x - 2y = 8$
13. $(x - 2)y = 4$
14. $(x + 3)y = 6$
15. $xy - 4y = 6$
16. $xy + 2y = 8$
17. $2x^2 + y^2 = 8$
18. $x^2 + 2y^2 = 8$
19. $4x^2 - y^2 = 16$
20. $y^2 - 4x^2 = 16$
21. $y - x^2 > 3$
22. $y + x^2 \leq 4$

■ *Find the value of each expression. See Example* 4.

23. $f(4)$, if $f(x) = x^2 - 2x + 1$
24. $g(3)$, if $g(x) = 2x^2 + 3x - 1$
25. $g(5) - g(2)$, if $g(x) = x + 3$
26. $f(2) - f(0)$, if $f(x) = 3x - 1$
27. $f(3) - f(-3)$, if $f(x) = x^2 - x + 1$
28. $f(0) - f(-2)$, if $f(x) = x^2 + 3x - 2$

■ *In Problems 29–32, find:*

a. $f(a)$; b. $f(b)$; c. $f(a + b)$

for each function. See Example 5.

29. $f(x) = 5x - 3$
30. $f(x) = 3x + 2$
31. $f(x) = x^2 + 3x$
32. $f(x) = x^2 - 2x$

■ *In Problems 33–38, find:*

a. $f(x + h)$; b. $f(x + h) - f(x)$; c. $\dfrac{f(x + h) - f(x)}{h}$

for each function. See Examples 5 and 6.

33. $f(x) = 3x - 4$
34. $f(x) = 2x + 5$
35. $f(x) = x^2 - 3x + 5$
36. $f(x) = x^2 + 2x$
37. $f(x) = x^3 + 2x - 1$
38. $f(x) = x^3 + 3x^2 - 1$

■ *Estimate each ordinate specified and the zeros of the function. See Example 7.*

39. $f(-2)$, $f(3)$, and $f(5)$

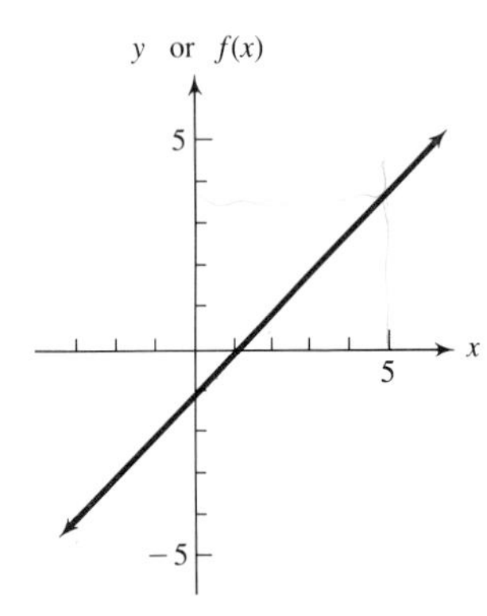

40. $f(-3)$, $f(2)$, and $f(5)$

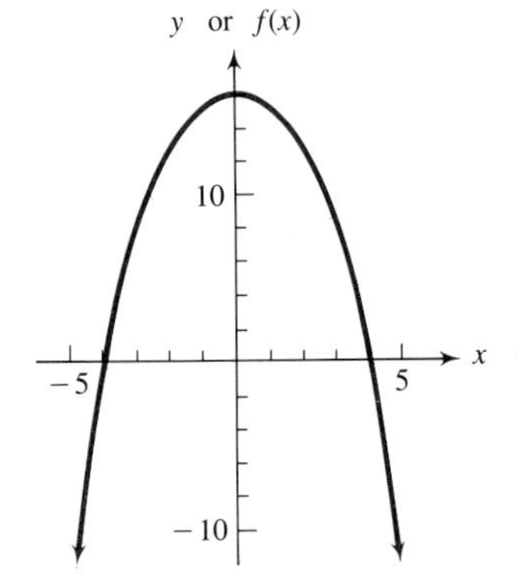

■ *Find the zeros of each function. See Examples 7 and 8.*

41. $f(x) = 3x + 6$
42. $f(x) = 2x - 8$
43. $f(x) = x^2 - 5x + 4$
44. $f(x) = x^2 - 6x + 8$
45. $f(x) = -x^2 + x + 6$
46. $f(x) = -x^2 + 2x + 15$

11.2

FUNCTIONS AS MATHEMATICAL MODELS

In applications of mathematics to other sciences, mathematical models are often used to answer questions posed or to make predictions. The mathematical models generally involve one or more functions. In this section we construct several such models.

Direct variation

There are two types of functional relationships, widely used in the sciences, to which custom has assigned special names. First, the variable y is said to **vary directly** as the variable x if

$$y = kx^n \qquad (k \text{ a positive constant, } n \text{ a natural number}). \tag{1}$$

Note that since Equation (1) associates one and only one y with each x, a direct variation defines a polynomial function.

EXAMPLE 1 a. The circumference of a circle varies directly as the radius since

$$C = 2\pi r.$$

b. The area of a circle varies directly as the square of the radius since

$$A = \pi r^2. \quad ■$$

Inverse variation

The second important type of variation arises from the equation

$$y = \frac{k}{x^n} \qquad (k \text{ a positive constant, } n \text{ a natural number, } x \neq 0) \tag{2}$$

in which y is said to **vary inversely** as x.

Since Equation (2) associates only one y with each x ($x \neq 0$), an inverse variation defines a rational function, with domain $\{x \mid x \neq 0\}$.

EXAMPLE 2 a. For an ideal gas at constant absolute temperature (T), the volume (V) and pressure (P) vary inversely since

$$VP = kT \qquad (k, T \text{ constants}),$$

or

$$V = \frac{kT}{P}.$$

b. For a right circular cylinder with constant volume (V), the height (h) and the square of the radius (r) vary inversely since

$$V = \pi r^2 h \qquad (V \text{ a constant}),$$

or

$$h = \frac{V}{\pi r^2}. \qquad ■$$

In Example 2a above, we can also express the relationship thus:

"V varies directly with T and inversely as P."

In 2b:

"h varies directly with V and inversely as r^2."

The names "direct" and "inverse," as applied to variation, arise from the fact that in direct variation an assignment of increasing absolute values of x results in an association with increasing absolute values of y, whereas in inverse variation an assignment of increasing absolute values of x results in an association with decreasing absolute values of y.

The constant involved in an equation defining a direct or inverse variation is called the **constant of variation.**

For example, $C = 2\pi r$ has 2π as the constant of variation. Also, $h = \dfrac{V}{\pi r^2}$ has $\dfrac{1}{\pi}$ as the constant of variation.

If we know that one variable varies directly or inversely as another, and if we have one set of associated values for the variables, we can find the constant of variation involved. We can then use this constant to express one of the variables as a function of the other.

EXAMPLE 3 The speed (v) at which a particle falls in a certain medium varies directly with the time (t) it falls. The particle is falling at a speed of 20 feet per second 4 seconds after being dropped. Express v as a function of t.

Solution Since v varies directly with t, we know there is a positive constant k so that

$$v = kt.$$

Since $v = 20$ when $t = 4$, we have

$$20 = k(4),$$

from which

$$k = 5.$$

Thus, the functional relationship between v and t is given by

$$v = 5t. \quad ■$$

Solving for unknown values

Once we have constructed a model, values for one of the variables can readily be obtained for known values of the other variable. For example, in Example 3 above, we can obtain the velocity of the particle at any time by substituting for t in the equation $v = 5t$. Thus for t equal to 6, 8, and 10 seconds:

$$v_1 = 5(6) = 30; \qquad v_2 = 5(8) = 40; \qquad v_3 = 5(10) = 50.$$

Other functional relationships

The following examples illustrate other functional relationships that occur in mathematical models. Note that these examples are not direct or inverse variations as defined above.

EXAMPLE 4 A company that produces computer chips finds that after a start-up cost of \$10,000, each new chip costs \$200 to manufacture. Express the cost (C) of producing x new chips as a function of x.

Solution The cost C is the sum of the initial or start-up cost and the cost of manufacturing x chips. Since the cost of manufacturing x chips is

$$\underset{\text{[Cost per chip]}}{\$200} \cdot \underset{\text{[Number of chips]}}{x}$$

and the start-up cost is \$10,000, we have

$$C = 10{,}000 + 200x.$$

Since the company cannot produce a negative number of chips, x is always non-negative, and thus we restrict x to the interval $[0, \infty)$. ■

In the above example, we can now find the cost of any specified number of new chips by simply making a substitution for x and solving for C.

The following example illustrates a case in which the value of the independent variable is restricted to intervals other than $[0, \infty)$.

EXAMPLE 5 A rope 12 feet long is to be cut into two pieces of unequal length, and each piece is to be shaped to enclose a square. Express the total area (A) enclosed as a function of the length (x) of the shorter piece of rope.

Solution We first draw and label a figure as shown.

12 ft

x | $12 - x$

A_1 $\frac{x}{4}$ A_2 $\frac{12 - x}{4}$

Length of short piece: x

Length of long piece: $12 - x$

Smaller area: A_1

Larger area: A_2

Total area enclosed: $A_1 + A_2$

Since the perimeter of the smaller square is x, the side length of this square is $x/4$, as shown in figure. Similarly, the side length of the larger square is $(12 - x)/4$.

Since the area of a square is the square of the length of its side,

$$A_1 = \left(\frac{x}{4}\right)^2 \quad \text{and} \quad A_2 = \left(\frac{12 - x}{4}\right)^2,$$

and the total area is given by

$$A = \left(\frac{x}{4}\right)^2 + \left(\frac{12 - x}{4}\right)^2.$$

Since x represents a length, it is always positive (we assume a cut is made). Further, x represents the shorter of the two lengths of rope and hence it is less than or equal to one-half the total length of the rope. Thus we restrict x to the interval $(0, 6)$. ■

In the above example, we can now find the total area enclosed for any specified length x by simply making a substitution for x and solving for A.

EXERCISE 11.2

A

■ *In Problems* 1–8: **a.** *Construct a mathematical model;* **b.** *Solve for the specified value. See Examples* 1–3.

1. y varies directly with x, and $y = 20$ when $x = 12$. Find y when $x = 20$.

2. y varies inversely with x, and $y = 16$ when $x = 4$. Find y when $x = 12$.
3. y varies directly with the square of x and inversely with z, and $y = 12$ when $x = 3$ and $z = 2$. Find y when $x = 6$ and $z = 5$.

4. y varies directly with x and inversely with the square of z, and $y = 20$ when $x = 4$ and $z = 6$. Find y when $x = 12$ and $z = 10$.
5. The distance (s) a particle falls in a certain medium varies directly with the time (t) it falls. **a.** If the particle falls 16 feet in 2 seconds, express s as a function of t. **b.** Find the distance the particle falls in 7 seconds.
6. The pressure (P) exerted by a liquid at a given point varies directly as the depth (d) of the point beneath the surface of the liquid. **a.** If the liquid exerts a pressure of 40 pounds per square foot at a depth of 10 feet, express P as a function of d. **b.** Find the pressure exerted by a liquid 18 feet beneath the surface.
7. The maximum safe uniformly distributed load (L) for a horizontal beam with breadth $b = 2$ feet and depth $d = 4$ feet varies inversely with the length (l). **a.** If an 8-foot long beam will support up to 750 pounds, express L as a function of l. **b.** Find the safe load for a 3-foot long beam.
8. The resistance (R) of wire 50 feet long varies inversely as the square of its diameter (d). **a.** If a wire with diameter 0.012 inch has a resistance of 10 ohms, express R as a function of d. **b.** Find the resistance of a wire with diameter 0.018 inch.

■ *In Problems 9–22:* **a.** *Construct a mathematical model;* **b.** *Solve. See Examples* 4 *and* 5. *(See Appendix C for appropriate formulas from geometry.)*

9. The set-up cost for printing a book is \$5000, after which it costs \$10 for each book printed. Express the printing costs (C) for a book as a function of the number (n) of the books printed. Find the printing costs for 640 books.
10. The set-up cost to produce a certain kind of computer chip is \$12,000, after which it costs \$24 to produce each chip. Express the cost (C) of production as a function of the number (n) of chips produced. Find the cost of producing 8000 chips.
11. The cost of fencing is \$7.50 per foot. Express the cost (C) of fencing a square field as a function of the length (s) of the side of the field. Find the cost of fencing a square field with sides 300 feet in length.
12. The cost of fencing material is \$5.50 per foot. Express the cost (C) of fencing a rectangular field with its length twice its width as a function of its width (w). Find the cost of fencing a rectangular field with width 140 feet.
13. The cost of building a brick wall is \$15 per foot and the cost of fencing material is \$7.50 per foot. A rancher is building a square corral with three sides brick wall and one side fencing. Express the cost (C) as a function of the side length (s) of the corral. Find the cost of a square corral with side 60 feet in length.
14. Assume the same costs as in Problem 13. A farmer wishes to build a rectangular pen using the wall of his barn as one length of the pen and with the opposite side fencing. The other two sides are to be brick walls. The length is to be twice the width. Express the cost (C) of building the pen as a function of the width (w). Find the cost of a rectangular pen with width 30 feet.
15. Express the volume (V) of a cone as a function of its height (h) if the height of the cone is equal to three times the radius (r). Find the volume of a cone with height 12 centimeters.

16. Express the area (A) of a circle as a function of the circumference (C) of the circle. Find the area of a circle with circumference 24 centimeters.
17. Express the area (A) of a square as a function of the length (d) of its diagonal. Find the area of a square with a 10-inch diameter.
18. A square is inscribed in a circle. Express the area (A) of the square as a function of the radius (r) of the circle. Find the area of a square inscribed in a circle with a 4-inch radius.
19. A circular metal plate is to be cut from a square metal sheet as shown in the figure. Express the area (A) of the waste metal as a function of the radius (r) of the circle. Find the area of the waste metal for an inscribed circle with a 6-inch radius.

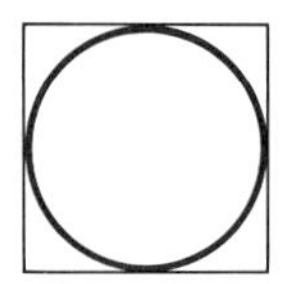

20. For the situation described in Problem 19, express the area (A) of the waste metal as a function of the length (s) of a side of the square. Find the area of the waste metal for a square with 8-inch sides.
21. A printer is to leave a uniform margin of 1 inch on top, bottom, and both sides of a square poster. Express the area (A) of printed matter on the poster as a function of the side length (x) of the poster. Find the area of the printed matter of the poster with side 30 inches.
22. A printer is to leave a margin of 1 inch on the top and bottom of a square poster and ½ inch on the sides. Express the area (A) of the printed matter on the poster as a function of the side length x. Find the area of the printed matter of the poster with side 40 inches.

B

■ *In Problems 23–30, set up a mathematical model for the functional relationship.*

23. A rope l feet long is to be cut in half and each half is to be shaped to enclose a square. Express the total area (A) enclosed by the two pieces as a function of l.
24. A rope 200 feet long is to be cut into two pieces. The shorter piece is to be shaped to enclose a square, and the longer piece is to be shaped to enclose a rectangle with length equal to four times the width. Express the total area (A) enclosed by the two pieces as a function of the length (x) of the shorter piece.
25. A rope 50 feet long is to be cut into two pieces. Each piece is to be shaped to enclose a circle. Express the total area (A) enclosed by the two pieces as a function of the length (x) of the shorter piece.
26. A rope 100 feet long is to be cut into two pieces. The shorter piece is to be shaped to enclose a square and the longer piece is to be shaped to enclose a circle. Express the total area (A) enclosed by the two pieces as a function of the length (x) of the shorter piece.
27. The cost of fencing material is \$7.50 per foot. A rancher is to enclose a rectangular area of 100 square feet with this fencing. Express the cost (C) of the fencing material as a function of the width (w) of the rectangle.
28. The cost of fencing material is \$10 per foot. A rancher is to enclose a rectangular area of 125 square feet adjacent to his barn. The barn wall will act as one length

of the enclosure, and fencing material will be used for the other three sides. Express the cost (C) of the fencing material as a function of the width (w) of the enclosure.

29. The telephone company is laying a cable from an island 12 miles offshore to a relay station 18 miles from the point on the shore directly opposite the island, as shown in the figure. It costs \$50 per foot to lay cable underwater and \$30 per foot to lay cable underground. Express the cost (C) of laying the cable as a function of the distance (x) from the relay station to where the cable leaves the water.

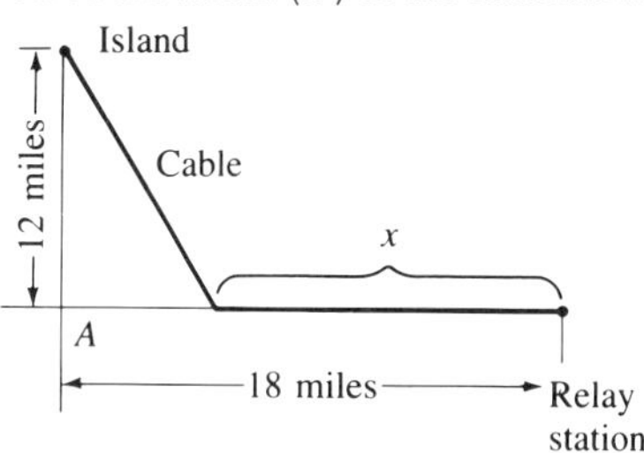

30. A cable TV company is laying a cable from Island A, which is 12 miles off-shore, to a relay station 15 miles from the point on the shore directly opposite the island, and then to Island B, which is 6 miles directly opposite the relay station, as shown in the figure. The cost of laying cable underwater is \$50 per foot and the cost of laying cable underground is \$30 per foot. Express the cost (C) of laying the cable as a function of the distance (x) from the point O in the figure to the point where the cable leaves the water.

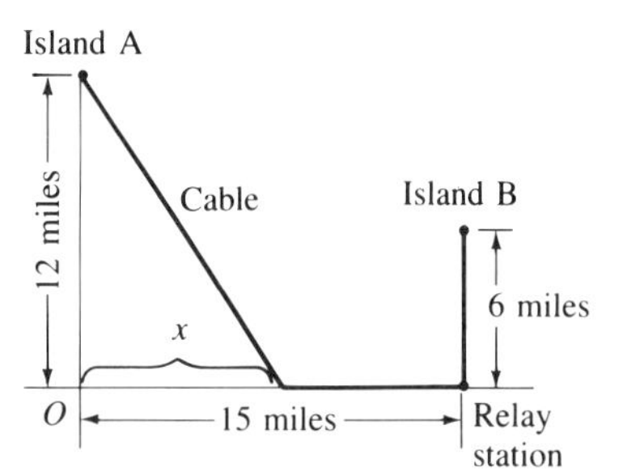

11.3

EXPONENTIAL FUNCTIONS

Powers b^x, x a real number

In Chapter 2, powers a^x were defined for any real number a and natural number x. In Chapter 6, the definition was extended to allow x to be a negative integer, or zero; also a rational number if $a > 0$. We now inquire whether we can interpret powers with irrational exponents, such as

$$b^{\pi}, \quad b^{\sqrt{2}}, \quad \text{or} \quad b^{-\sqrt{3}},$$

to be real numbers.

In Section 6.5, we observed that irrational numbers may be approximated by rational numbers to as great a degree of accuracy as desired. That is, $\sqrt{2} \approx 1.4$ or $\sqrt{2} \approx 1.414$, and so on. Also, although we do not prove it here, if x and y are rational numbers such that $x > y$,

$$\text{if} \quad b > 1, \quad \text{then} \quad b^x > b^y,$$

and

$$\text{if} \quad 0 < b < 1, \quad \text{then} \quad b^x < b^y.$$

Now, because 2^x is defined for any rational number x, the following sequence of inequalities should seem plausible even though 2^x has not been defined for an irrational number x.

$$2^1 < 2^{\sqrt{2}} < 2^2$$
$$2^{1.4} < 2^{\sqrt{2}} < 2^{1.5}$$
$$2^{1.41} < 2^{\sqrt{2}} < 2^{1.42}$$
$$2^{1.414} < 2^{\sqrt{2}} < 2^{1.415},$$

and so on, where $2^{\sqrt{2}}$ is a number lying between the number on the left and that on the right.

It is clear that this process can be continued indefinitely and that the difference between the number on the left and that on the right can be made as small as we please. This being the case, we assume that there is just one number, $2^{\sqrt{2}}$, that will satisfy each inequality if this process is carried on indefinitely. Since we can produce the same type of argument for any irrational exponent x, we shall assume that b^x $(b > 0)$ is defined for all real values of x.

We can find values for some powers b^x by inspection when x is an integer and n is a natural number.

EXAMPLE 1 a. $2^4 = 16$ b. $3^{-2} = \dfrac{1}{3^2} = \dfrac{1}{9}$ c. $4^0 = 1$ ■

If you have a scientific calculator you can obtain approximations to powers b^x for all values of b and x for which b^x is defined.*

EXAMPLE 2* Find an approximation for each power to four places.

a. $2^{1.2}$ b. $2.4^{1.7}$ c. $1.6^{-1.3}$

*Examples and exercises marked with an asterisk require a scientific calculator. They are not prerequisite for work in the following sections and may be omitted.

Solutions The following keystrokes can be used.

a. 2 $\boxed{x^y}$ 1.2 $\boxed{=}$ *ans.* 2.2974

b. 2.4 $\boxed{x^y}$ 1.7 $\boxed{=}$ *ans.* 4.4295

c. 1.6 $\boxed{x^y}$ 1.3 $\boxed{=}$ *ans.* 0.5428 ■

Since for each real x there is one and only one number b^x, the equation

$$\boldsymbol{f(x) = b^x \qquad (b > 0)} \tag{1}$$

defines a function. Because $1^x = 1$ for all real values of x, Equation (1) defines a constant function if $b = 1$. If $b \neq 1$, we say that (1) defines an **exponential function.**

Note that b is restricted to be greater than 0. One reason is that if $b < 0$, b^x is *imaginary* for $x = 1/n$, n even. For example, $(-4)^{1/2}$, $(-9)^{1/2}$, and $(-16)^{1/2}$ are imaginary numbers.

Graphs of exponential functions

Exponential functions can be visualized more clearly by considering their graphs. We illustrate two typical examples in which $0 < b < 1$ and $b > 1$, respectively. Assigning values to x in the equations

$$f(x) = \left(\frac{1}{2}\right)^x \quad \text{and} \quad f(x) = 2^x,$$

we find ordered pairs in each function and sketch the graphs in Figure 11.4 on page 414.

If $f(x) = \left(\frac{1}{2}\right)^x$, then

$$f(-3) = \left(\frac{1}{2}\right)^{-3} = 8;$$

$$\vdots \qquad \vdots \qquad \vdots$$

$$f(0) = \left(\frac{1}{2}\right)^0 = 1;$$

$$\vdots \qquad \vdots \qquad \vdots$$

$$f(3) = \left(\frac{1}{2}\right)^3 = \frac{1}{8}.$$

If $f(x) = 2^x$, then

$$f(-3) = 2^{-3} = \frac{1}{8};$$

$$\vdots \qquad \vdots \qquad \vdots$$

$$f(0) = 2^0 = 1;$$

$$\vdots \qquad \vdots \qquad \vdots$$

$$f(3) = 2^3 = 8.$$

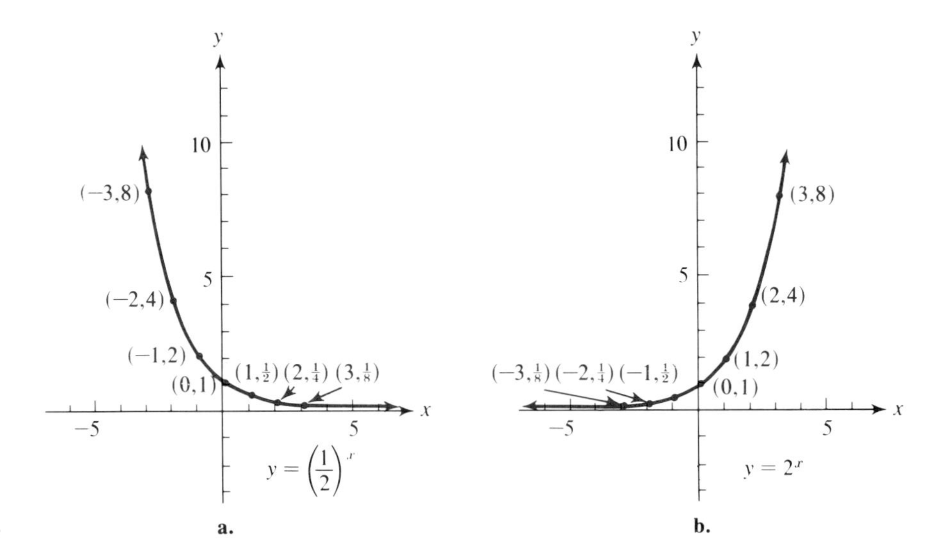

FIGURE 11.4

Additional ordered pairs can be obtained by using a calculator to obtain function values for non-integral values in the domain.

Notice that the graph of the function determined by $f(x) = (1/2)^x$ goes *down* to the right, and the graph of the function determined by $f(x) = 2^x$ goes *up* to the right. For this reason, we say that the former function is a **decreasing function** and the latter is an **increasing function.** In each case, the domain is the set of real numbers and the range is the set of positive real numbers.

The above properties are summarized as follows:

For the Exponential Function $f(x) = b^x$ $(b > 0, b \neq 1)$

1. Domain: x a real number.
2. Range: $b^x > 0$.
3. If $b > 1$, the function is increasing;
   if $0 < b < 1$, the function is decreasing.

EXAMPLE 3 Graph $y = 3^x$. Use selected integral values for x.

Solution For convenience, arbitrarily select integral values of x, say, -2, -1, 0, 1, 2, and 3.

Determine the y-components of each ordered pair.

$$\left(-2, \frac{1}{9}\right), \quad \left(-1, \frac{1}{3}\right), \quad (0, 1), \quad (1, 3), \quad (2, 9), \quad (3, 27).$$

Plot the points and connect them with a smooth curve.

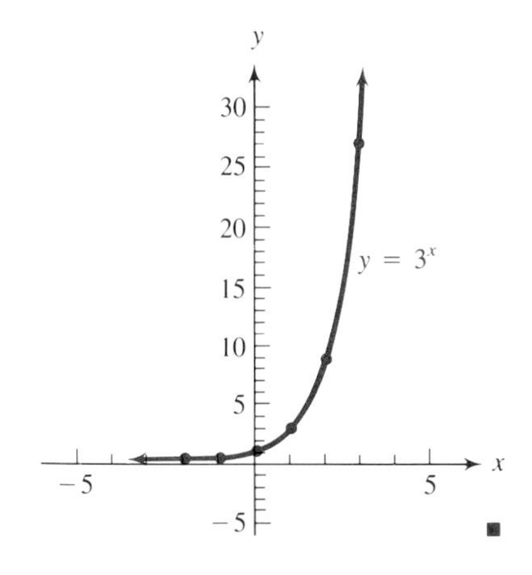

■

EXAMPLE 4* Graph $f(x) = 2.31^x$.

Solution Using a calculator with arbitrarily selected integral values of x—say, -2, -1, 0, 1, 2, and 3—we obtain an approximation for each corresponding function value.

$$(-2, 0.2), \quad (-1, 0.4), \quad (0, 1.0), \quad (1, 2.3), \quad (2, 5.3), \quad (3, 12.3)$$

Plot the points and connect with a smooth curve.

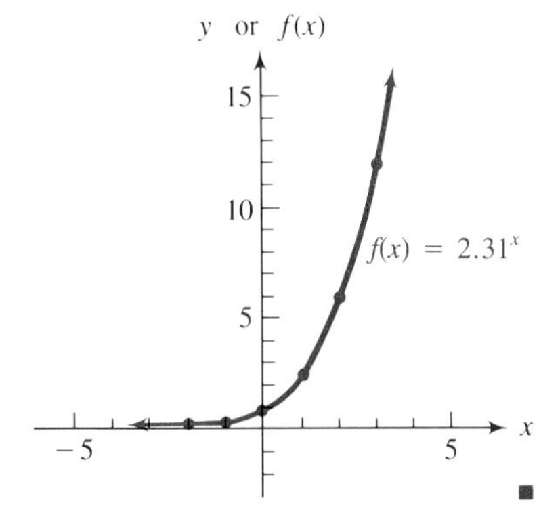

■

Solving $y = b^x$ for x

In this section we obtained values (or approximations) for function values (elements in the range) of exponential functions for given values in their domain. Thus, we obtained values for b^x ($b > 0$), for real values of x. We shall need some additional mathematical "tools" to be able to find values of x for given values of b^x. The required mathematics is developed in detail in the next chapter.

EXERCISE 11.3

A

■ *Evaluate the power for each given value for x. See Example* 1.

1. 3^x; for x equal to 0, 1, and 2.

2. 4^x; for x equal to $-\frac{1}{2}$, 0, and $\frac{1}{2}$.

3. 2^x; for x equal to -4, 0, and 4.

4. 5^x; for x equal to -2, 0, and 2.

5. $(\frac{1}{2})^x$; for x equal to -4, 0, and 4.
6. $(\frac{1}{3})^x$; for x equal to -3, 0, and 3.
7. 10^x; for x equal to -2, -1, and 0.
8. 10^x; for x equal to 0, 1, and 2.

■ **Evaluate the power, to the nearest hundredth, for each given value for x. See Example 2.*

9. 2.7^x; for x equal to -2 and 3.
10. 1.8^x; for x equal to -3 and 2.
11. 0.4^x; for x equal to 0.2 and 2.3.
12. 0.3^x; for x equal to 0.3 and 1.6.

■ *Graph each equation. Use selected integral values for x. See Example 3.*

13. $y = 4^x$
14. $y = 6^x$
15. $y = 10^x$
16. $y = 2^{-x}$
17. $y = 3^{-x}$
18. $y = -3^x$
19. $y = -2^x$
20. $y = (\frac{1}{3})^x$
21. $y = (\frac{1}{4})^x$
22. $y = (\frac{1}{10})^x$
23. $y = (\frac{1}{2})^{-x}$
24. $y = (\frac{1}{3})^{-x}$

■ **Graph each equation. See Example 4.*

25. $f(x) = 1.3^x$
26. $f(x) = 2.4^x$
27. $f(x) = 0.8^x$
28. $f(x) = 0.7^x$

B

■ *In Problems 29 and 30, use graphical methods to estimate the solutions of the exponential equations.*

29. $2^x = 5$ [*Hint:* Graph $y_1 = 2^x$ and $y_2 = 5$ and approximate the value of x at their point of intersection.]
30. $3^x = 4$ [See hint for Problem 29.]
31. For what set of positive real numbers a will $y = a^x$ define an increasing function? A decreasing function?
32. For what set of positive real numbers a will $y = a^{-x}$ define an increasing function? A decreasing function?

■ *Solve each system by graphing. Approximate components of the solution of the system.*

33. $y = 10^x$
 $x + y = 6$
34. $y = 2^x$
 $y - x = 2$

11.4

THE INVERSE OF A FUNCTION

One-to-one functions

A function f has been viewed as a relation in which each element in the domain is associated with just one element in its range. If each element in its range is also associated with just one element in its domain, the function f is called a **one-to-one function.**

Whether or not a function defined by an equation in two variables is a one-to-one function can be readily determined from its graph. Recall (page 399) that any vertical line will intersect the graph of a function in at most one point. If the function f is one-to-one, any horizontal line will also intersect its graph in at most one point, as shown in Figure 11.5a. If it is not one-to-one, a horizontal line will intersect the graph at more than one point, as shown in Figure 11.5b. A function is one-to-one provided each of its y-values maps back to exactly one x-value.

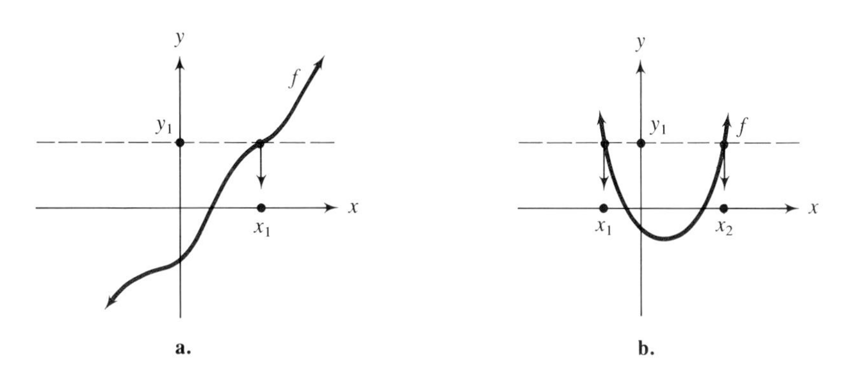

FIGURE 11.5

Inverse functions

If the components of each ordered pair in a one-to-one function f are interchanged, the result will be a function, denoted by f^{-1} (read "f inverse" or "the inverse of f") and called the **inverse function** of f.

EXAMPLE 1 a. If $f = \{(1, 2), (3, 4), (5, 6)\}$, then $f^{-1} = \{(2, 1), (4, 3), (6, 5)\}$, a function.

b. If $f = \{(1, 2), (2, 5), (3, 5)\}$, then its inverse $\{(2, 1), (5, 2), (5, 3)\}$ is not a function because f is not a one-to-one function. ■

It is evident from the definition of an inverse function that the domain and range of f^{-1} are the range and domain, respectively, of f. Hence, if a one-to-one function f is defined by an equation in two variables, the inverse f^{-1} can be obtained by interchanging the variables.

EXAMPLE 2 The inverse of the function defined by

$$y = 4x - 3 \tag{1}$$

is defined by

$$x = 4y - 3, \tag{2}$$

where the variables in (1) have been interchanged. When y is expressed in terms of x,

$$y = \frac{1}{4}(x + 3). \tag{2a}$$

Equations (2) and (2a) are equivalent. ■

Graphs of inverse functions

The graphs of a function and its inverse are related in an interesting way. To see this, we first observe in Figure 11.6 that the graphs of the ordered pairs (a, b) and (b, a) are always located symmetrically with respect to the graph of $y = x$. Therefore, because for every ordered pair (a, b) in f, the ordered pair (b, a) is in f^{-1}, the graphs of $y = f^{-1}(x)$ and $y = f(x)$ are reflections of each other about the graph of the equation $y = x$. Figure 11.7 shows the graphs of

$$y = 4x - 3$$

and its inverse,

$$x = 4y - 3 \quad \text{or} \quad y = \frac{1}{4}(x + 3),$$

together with the graph of $y = x$.

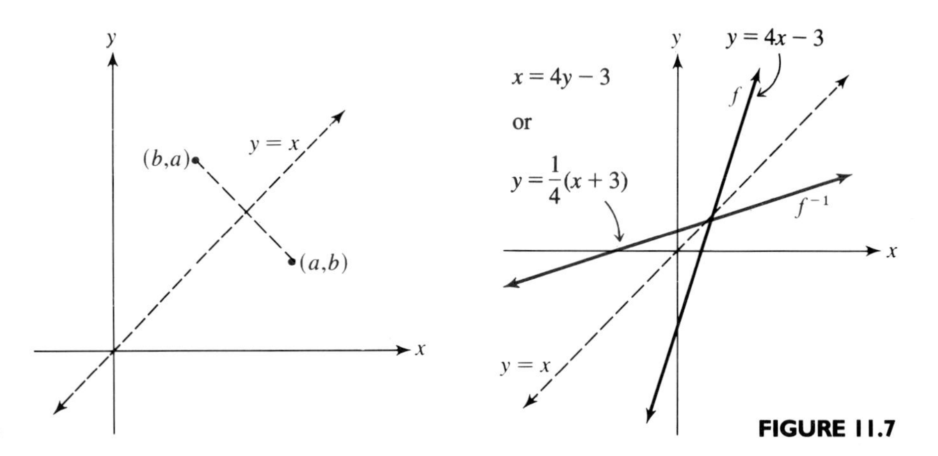

FIGURE 11.6

FIGURE 11.7

Of course, if a function defined by an equation in two variables is not a one-to-one function, its inverse is not a function.

EXAMPLE 3 The figure shows the graph of the function

$$y = x^2,$$

together with the graph of its inverse,

$$x = y^2 \quad \text{or} \quad y = \pm\sqrt{x}.$$

Since, for all but one value in its domain ($x = 0$), the inverse associates two different values of y with each x, the inverse is not a function.

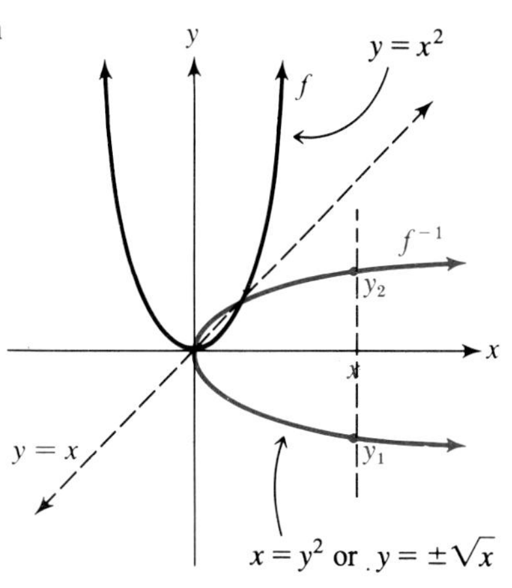

■

$f^{-1}[f(x)] = x$; $f[f^{-1}(x)] = x$

If f and f^{-1} are both functions, f associates the number a with the *unique* number b, and f^{-1} associates the number b with the *unique* number a. Therefore, it must be true that, for every x in the domain of f,

$$f^{-1}[f(x)] = x$$

(read "f inverse of f of x is equal to x"). And, for every x in the domain of f^{-1},

$$f[f^{-1}(x)] = x$$

(read "f of f inverse of x is equal to x"). For example, if $f = \{(2, 5)\}$, then $f(2) = 5$, $f^{-1} = \{(5, 2)\}$, and $f^{-1}(5) = 2$. Note that

$$f^{-1}[f(2)] = f^{-1}(5) = 2$$

and

$$f[f^{-1}(5)] = f(2) = 5.$$

EXAMPLE 4 Consider Equation (1) on page 417 and note that if f is the linear function defined by

$$f(x) = 4x - 3,$$

then f is a one-to-one function. Now f^{-1} is defined by

$$f^{-1}(x) = \frac{1}{4}(x + 3),$$

(Equation (2a) on page 418) and we see that

$$f^{-1}[f(x)] = \frac{1}{4}[(4x - 3) + 3] = x$$

and

$$f[f^{-1}(x)] = 4\left[\frac{1}{4}(x + 3)\right] - 3 = x. \quad ■$$

EXERCISE 11.4

A

■ *Find the inverse of each function and state whether the inverse is also a function. See Example 1.*

1. $f = \{(-2, -2), (2, 2)\}$
2. $f = \{(-5, 1), (5, 2)\}$
3. $q = \{(1, 3), (2, 3), (3, 4)\}$
4. $q = \{(2, 2), (3, 3), (4, 3)\}$
5. $g = \{(1, 1), (2, 3), (3, 3)\}$
6. $g = \{(-2, 0), (0, 0), (4, -2)\}$

■ *In Problems 7–16, each equation defines a function, f.*
a. *Write the equation defining its inverse.*
b. *Sketch the graphs of f and its inverse on the same set of axes.*
c. *State whether the inverse is a function.*
See Examples 2 and 3.

7. $2x + 4y = 7$
8. $3x - 2y = 5$
9. $x - 3y = 6$
10. $4x + y = 4$
11. $y = x^2 - 4x$
12. $y = x^2 - 4$
13. $y = \sqrt{4 + x^2}$
14. $y = -\sqrt{x^2 - 4}$
15. $y = |x|$
16. $y = |x| + 1$

■ *Sketch the graph of the inverse of the function whose graph is given. See Example 3.*

17.

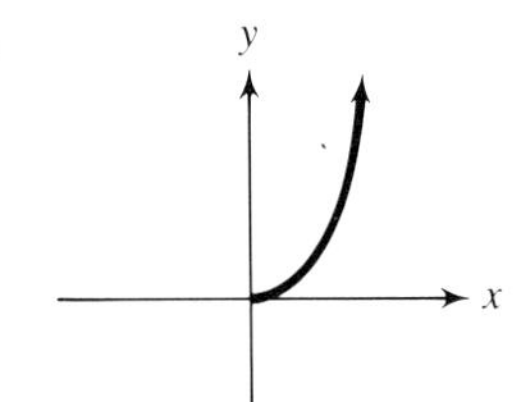

18.

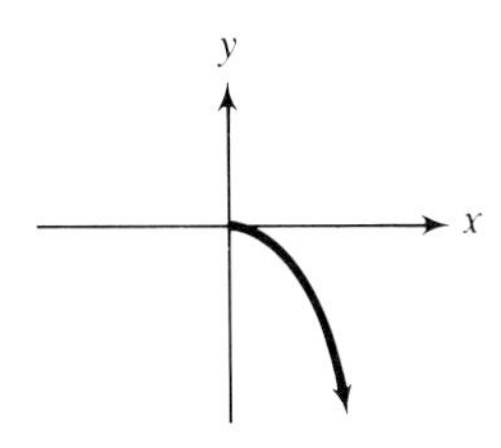

19.

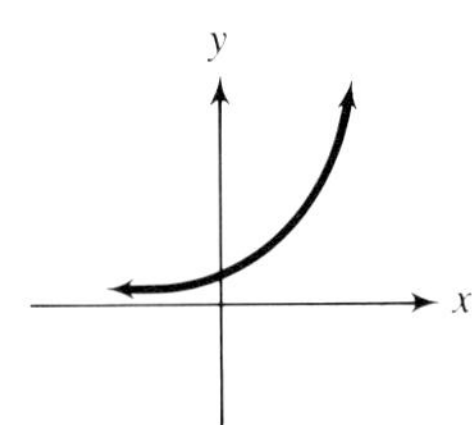

20.

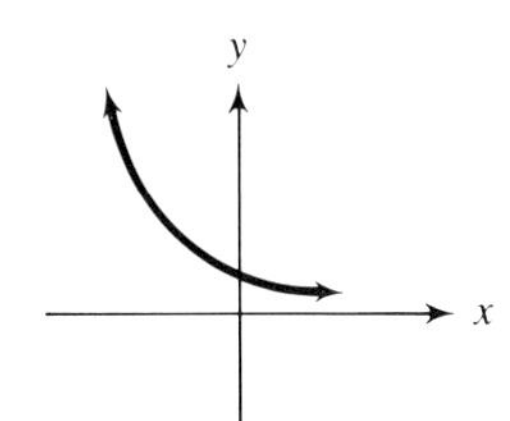

21.

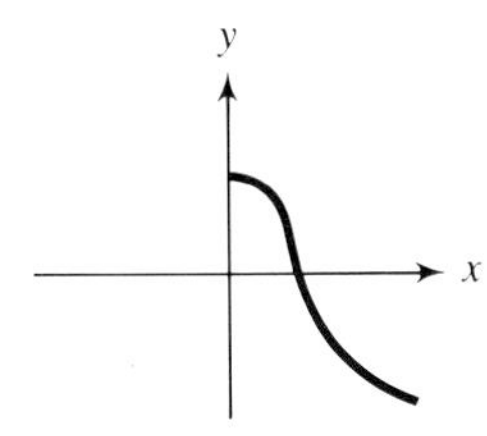

22.

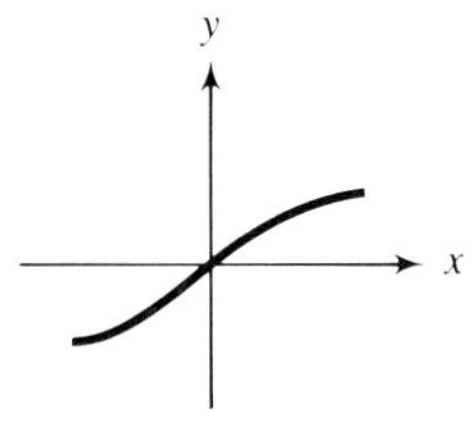

B

■ *In Problems 23–28, each equation defines a one-to-one function f. Find the equation defining f^{-1} and show that $f[f^{-1}(x)] = f^{-1}[f(x)] = x$. [Hint: Solve explicitly for y and let $y = f(x)$.] See Example 4.*

23. $y = x$
24. $y = -x$
25. $2x + y = 4$
26. $x - 2y = 4$
27. $3x - 4y = 12$
28. $3x + 4y = 12$

CHAPTER SUMMARY

[11.1] A **relation** is a set of ordered pairs. A **function** is a relation in which no two ordered pairs have the same first component.

The **domain** of a relation (or function) is the set of all first components in the ordered pairs in the relation (or function). The **range** of a relation (or function) is the set of all second components in the ordered pairs in the relation (or function).

Symbols such as $f(x)$, $P(x)$, and $R(x)$ are called **function notations.**

[11.2] The equation $y = kx$ (k a positive constant) defines a function called a **direct variation.** The equations $xy = k$ (k a positive constant) and $y = \frac{k}{x}$ define a function called an **inverse variation.** In each case the constant k is called the **constant of variation.**

[11.3] A function defined by an equation of the form

$$f(x) = b^x \qquad (b > 0, b \neq 1),$$

is called an **exponential function.** If $b > 1$, it is an **increasing function**; if $0 < b < 1$, it is a **decreasing function.** The domain of the function is the set of real numbers; the range is the set of positive real numbers.

[11.4] The **inverse** of the function f can be obtained by interchanging the components of each ordered pair in f. If f is a one-to-one function, then the inverse f^{-1} is also a function. The graphs of f^{-1} and f are reflections of each other in the graph of $y = x$.

REVIEW EXERCISES

A

[11.1]

1. Specify the domain and range of each relation. Is the relation a function?

   a. $[(3, 5), (4, 5), (4, 6)\}$ b. $\{(2, 3), (3, 4), (4, 5), (5, 5)\}$

2. Specify the domain of each relation and state whether the relation is a function.

   a. $xy - y = 6$ b. $4x^2 + y^2 = 16$

3. Given that $f(x) = x - 4$, find each of the following:

   a. $f(6)$ b. $f(x + h) - f(x)$

4. Given that $f(x) = 2x^2 - 3x + 1$, find each of the following:

   a. $f(3)$ b. $\dfrac{f(x + h) - f(x)}{h}$

[11.2]

5. If y varies inversely with t^2, and $y = 16$ when $t = 3$, **a.** express y as a function of t; **b.** find y when $t = 8$.
6. If y varies directly with s^2 and inversely with t, and $y = 4$ when $s = 2$ and $t = 5$, **a.** express y as a function of s and t; **b.** find y when $s = 3$ and $t = 4$.
7. The distance s a particle falls in a certain medium varies directly with the square of the length of time t it falls. If the particle falls 28 centimeters in 4 seconds, **a.** express the distance a particle will fall as a function of time it falls; **b.** find the distance a particle falls in 6 seconds.
8. The volume V of a gas varies directly with the absolute temperature T and inversely with the pressure P of the gas. If $V = 40$ when $T = 300$ and $P = 30$,

a. express the volume of the gas as a function of the absolute temperature and pressure of the gas; **b.** find the volume when $T = 320$ and $P = 40$.

9. **a.** Express the area (A) of an equilateral triangle as a function of the length of a side (s); **b.** Find the area of an equilateral triangle with sides 4 centimeters in length.

10. The hypotenuse (c) of a right triangle is 12 centimeters long. **a.** Express the area (A) of the triangle as a function of the shortest side; **b.** Find the area of a right triangle with short side 4 centimeters.

[11.3]

11. Find each value.

    a. 10^{-3} **b.** $(\frac{1}{2})^{-3}$

12. Sketch the graph of each equation.

    a. $y = 5^x$ **b.** $y = 5^{-x}$

[11.4]

13. Find the inverse of each function and state whether the inverse is also a function.

    a. $f = \{(3, 7), (4, 8), (8, 4)\}$ **b.** $g = \{(2, 6), (3, 8), (5, 8)\}$

14. Find the inverse of the function defined by $y = x^2 + 9x$ and state whether the inverse is a function.

15. Graph the function defined by $y = 2x + 6$ and its inverse on the same set of axes.

16. Graph the function defined by $y = x^2 + 1$ and its inverse on the same set of axes.

B

17. Approximate the solution of the system

$$y = 2^x$$
$$x + y = 5$$

graphically.

*18. Sketch the graph of $y = 2.2^x$.

19. A function f is defined by $y = x + 3$. Find f^{-1} and show that $f[f^{-1}(x)] = x$ and $f^{-1}[f(x)] = x$.

20. A function f is defined by $y - 2x = 4$. Find f^{-1} and show that $f[f^{-1}(x)] = x$ and $f^{-1}[f(x)] = x$.

歳ノ五月トニアリ

第二巻

上弦圖

西

東

南中

地

日入

改圖

天心

地平

日入

12 Logarithmic Functions

12.1

DEFINITIONS AND NOTATIONS

In Section 11.3, we were concerned with finding values for powers b^x $(b > 0, b \neq 1)$ for given values of the exponent x. In this section, we shall first consider the inverse of the exponential function $y = b^x$. Then we shall see how the notion of this inverse will enable us to find values for the exponent x for given values of the power b^x.

Inverse of the exponential function

Recall from Section 11.4 that the inverse of a function can be obtained by interchanging the components in each ordered pair of the function, and that this may be accomplished by interchanging the variables in the defining equation of the function. Recall also that the graph of the inverse of a function is symmetric to the graph of the function about the line with equation $y = x$.

Because the exponential function

$$f: y = b^x \qquad (b > 0, b \neq 1) \tag{1}$$

with domain $\{x \mid x \in R\}$ and range $\{y \mid y > 0\}$ is a one-to-one function, there is only one y associated with each x as well as only one x associated with each y. Therefore, the inverse of function (1) is also a function, namely,

$$f^{-1}: x = b^y \qquad (b > 0, b \neq 1). \tag{2}$$

Since the domain of (1) is the same as the range of (2) and the range of (1) is the same as the domain of (2), we have $\{x \mid x > 0\}$ for the domain of (2), while the range of (2) is the set of real numbers.

The graphs of functions of the form (2) can be illustrated by the example

$$x = 10^y \qquad (x > 0).$$

We assign arbitrary values to x, say, 0.01, 0.1, 1, 10, and 100, and obtain the ordered pairs which can be plotted and connected with a smooth curve as in Figure 12.1a.

$$\begin{array}{llllllll}
\text{If} & x = 0.01, & \text{then} & 0.01 = 10^{-2} & = 10^y; & \text{so} & y = -2. \\
\text{If} & x = 0.1, & \text{then} & 0.1 = 10^{-1} & = 10^y; & \text{so} & y = -1. \\
\text{If} & x = 1, & \text{then} & 1 = 10^{0} & = 10^y; & \text{so} & y = 0. \\
\text{If} & x = 10, & \text{then} & 10 = 10^{1} & = 10^y; & \text{so} & y = 1. \\
\text{If} & x = 100, & \text{then} & 100 = 10^{2} & = 10^y; & \text{so} & y = 2.
\end{array}$$

Alternatively, we can reflect the graph of $y = 10^x$ about the graph of $y = x$ and obtain the same result as shown in Figure 12.1b.

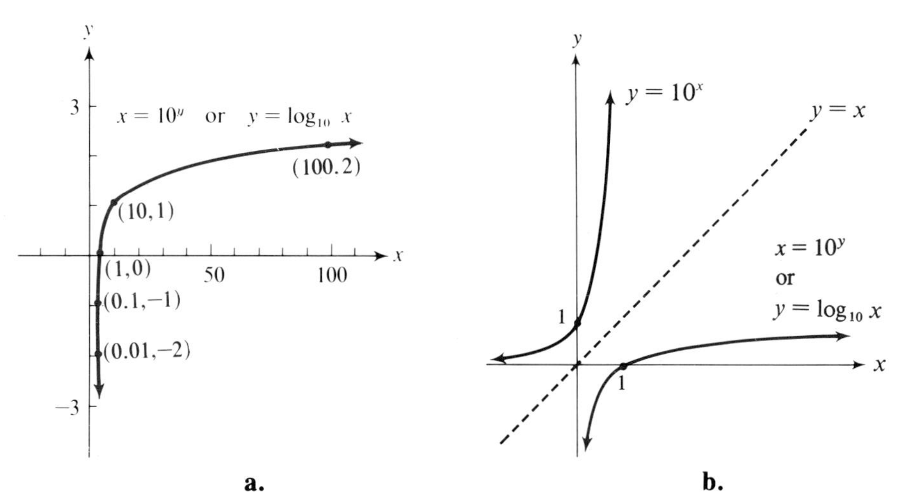

FIGURE 12.1

Logarithmic notation

It is always useful to be able to write an equation in the two variables x and y so that the variable y is expressed explicitly in terms of the variable x. To do this in equations such as (2), we use the notation

$$\mathbf{y = \log_b x \qquad (x > 0,\ b > 0,\ b \neq 1),} \tag{3}$$

where $\log_b x$ is read "logarithm to the base b of x" or "logarithm of x to the base b." The functions defined by such equations are called **logarithmic functions.**

It should be recognized that

$$\mathbf{x = b^y \qquad \text{and} \qquad y = \log_b x} \tag{4}$$

are two different forms of an equation defining the same function—in the same way that $x - y = 4$ and $y = x - 4$ define the same function—and that we may use whichever equation suits our purpose. The two equations in (4), which define $\log_b x$ as an *exponent*, enable us to write exponential equations in logarithmic form.

EXAMPLE 1 a. $5^2 = 25$ is equivalent to $\log_5 25 = 2$.

b. $8^{1/3} = 2$ is equivalent to $\log_8 2 = \frac{1}{3}$.

c. $3^{-2} = \frac{1}{9}$ is equivalent to $\log_3 \frac{1}{9} = -2$. ■

Also, logarithmic statements may be written in exponential form.

EXAMPLE 2 a. $\log_{10} 100 = 2$ is equivalent to $10^2 = 100$.

b. $\log_3 81 = 4$ is equivalent to $3^4 = 81$.

c. $\log_2 \frac{1}{2} = -1$ is equivalent to $2^{-1} = \frac{1}{2}$. ■

Note that because $x = b^y$ and $y = \log_b x$ are equivalent equations, $\log_b x$ (equal to y) is an *exponent*. In particular, $\log_b x$ is the exponent on b such that the power equals x. Thus,

$$\boldsymbol{b^{\log_b x} = x.}$$

EXAMPLE 3 Express each logarithm as an integer.

a. $\log_{10} 100$

b. $\log_2 16$

Solutions a. $\log_{10} 100$ is the *exponent* on 10 such that the power equals 100. Hence,

$$\log_{10} 100 = 2.$$

b. $\log_2 16$ is the *exponent* on 2 such that the power equals 16. Hence,

$$\log_2 16 = 4. \quad ■$$

Note that because $b^0 = 1$,

$$\boldsymbol{\log_b 1 = 0 \qquad (b > 0,\ b \neq 1).}$$

Furthermore, because $b^1 = b$,

$$\mathbf{\log_b b = 1 \qquad (b > 0,\ b \neq 1).}$$

For example,

$$\log_{10} 1 = 0, \qquad \log_5 1 = 0,$$
$$\log_5 5 = 1, \quad \text{and} \quad \log_{10} 10 = 1.$$

If two of the three variables in an equation of the form $y = \log_b x$ are known, we can either determine the third variable directly by using the definition of a logarithm, or by first writing the equation in exponential form.

EXAMPLE 4 Solve for the unknown value in each equation.

a. $\log_2 x = 3$

b. $\log_b 2 = \dfrac{1}{2}$

Solutions

a. Write in exponential form.

$$2^3 = x$$

Solve for the variable.

$$x = 8$$

b. Write in exponential form.

$$b^{1/2} = 2$$

Solve for the variable.

$$(b^{1/2})^2 = 2^2$$
$$b = 4$$

■

An expression involving more than one logarithm can sometimes be simplified by "parts."

EXAMPLE 5 Simplify each expression.

a. $\log_2(\log_3 3)$

b. $\log_5(\log_2 32)$

Solutions

a. Since $\log_3 3 = 1$,

$$\log_2(\log_3 3) = \log_2 1$$
$$= 0.$$

b. Since $\log_2 32 = 5$,

$$\log_5(\log_2 32) = \log_5 5$$
$$= 1.$$

■

EXERCISE 12.1

A

■ *Express each equation in logarithmic notation. See Example 1.*

1. $4^2 = 16$
2. $5^3 = 125$
3. $3^3 = 27$
4. $8^2 = 64$
5. $\left(\frac{1}{2}\right)^2 = \frac{1}{4}$
6. $\left(\frac{1}{3}\right)^2 = \frac{1}{9}$
7. $8^{-1/3} = \frac{1}{2}$
8. $64^{-1/6} = \frac{1}{2}$
9. $10^2 = 100$
10. $10^0 = 1$
11. $10^{-1} = 0.1$
12. $10^{-2} = 0.01$

■ *Express each equation in exponential notation. See Example 2.*

13. $\log_2 64 = 6$
14. $\log_5 25 = 2$
15. $\log_3 9 = 2$
16. $\log_{16} 256 = 2$
17. $\log_{1/3} 9 = -2$
18. $\log_{1/2} 8 = -3$
19. $\log_{10} 1000 = 3$
20. $\log_{10} 1 = 0$
21. $\log_{10} 0.01 = -2$
22. $\log_{10} 0.0001 = -4$

■ *Find the value of each logarithm, using the fact that $\log_b x$ is the exponent on b such that the power is equal to x. See Example 3.*

23. $\log_7 49$
24. $\log_2 32$
25. $\log_4 64$
26. $\log_3 27$
27. $\log_3 \sqrt{3}$
28. $\log_5 \sqrt{5}$
29. $\log_5 \frac{1}{5}$
30. $\log_3 \frac{1}{3}$
31. $\log_2 2$
32. $\log_{10} 10$
33. $\log_{10} 100$
34. $\log_{10} 1$
35. $\log_{10} 0.1$
36. $\log_{10} 0.01$

■ *Solve for the unknown value. See Example 4.*

37. $\log_3 9 = y$
38. $\log_5 125 = y$
39. $\log_b 8 = 3$
40. $\log_b 625 = 4$
41. $\log_4 x = 3$
42. $\log_{1/2} x = -5$
43. $\log_2 \frac{1}{8} = y$
44. $\log_5 \frac{1}{5} = y$
45. $\log_b 10 = \frac{1}{2}$
46. $\log_b 0.1 = -1$
47. $\log_2 x = 2$
48. $\log_{10} x = -3$

B

■ *Simplify each expression. See Example 5.*

49. $\log_2(\log_4 16)$
50. $\log_5(\log_5 5)$
51. $\log_{10}[\log_3(\log_5 125)]$
52. $\log_{10}[\log_2(\log_3 9)]$
53. $\log_2[\log_2(\log_2 16)]$
54. $\log_4[\log_2(\log_3 81)]$
55. $\log_b(\log_b b)$
56. $\log_b(\log_a a^b)$
57. For what values of x is $\log_b(x - 9)$ defined?
58. For what values of x is $\log_b(x^2 - 4)$ defined?

12.2

PROPERTIES OF LOGARITHMS

Because a logarithm is an exponent by definition, the three laws given below are valid for positive real numbers b ($b \neq 1$), x_1, x_2, and all real numbers m.

$$\log_b(x_1x_2) = \log_b x_1 + \log_b x_2 \qquad (1)$$

For example,

$$\begin{array}{ccccccc} \log_2 32 & = & \log_2(4 \cdot 8) & = & \log_2 4 & + & \log_2 8 \\ \downarrow & & & & \downarrow & & \downarrow \\ 5 & & & = & 2 & + & 3 \end{array}$$

The validity of (1) is established as follows: Since

$$x_1 = b^{\log_b x_1} \quad \text{and} \quad x_2 = b^{\log_b x_2},$$

then

$$\begin{aligned} x_1x_2 &= b^{\log_b x_1} \cdot b^{\log_b x_2} \\ &= b^{\log_b x_1 + \log_b x_2}, \end{aligned}$$

and by the definition of a logarithm

$$\log_b (x_1x_2) = \log_b x_1 + \log_b x_2.$$

The validity of Laws (2) and (3) below can be established in a similar way.

$$\log_b \frac{x_2}{x_1} = \log_b x_2 - \log_b x_1 \qquad (2)$$

For example,

$$\begin{array}{ccccccc} \log_2 8 & = & \log_2 \dfrac{16}{2} & = & \log_2 16 & - & \log_2 2 \\ \downarrow & & & & \downarrow & & \downarrow \\ 3 & & & = & 4 & - & 1 \end{array}$$

$$\log_b (x_1)^m = m \log_b x_1 \qquad (3)$$

For example,

$$\begin{array}{ccccc} \log_2 64 & = \log_2 (4)^3 & = & 3 & \log_2 4 \\ \downarrow & & & \downarrow & \downarrow \\ 6 & & = & 3 & \cdot\ 2 \end{array}$$

We shall refer to the above equations as the first, second, and third laws of logarithms, respectively. These laws can be applied to rewrite expressions involving products, quotients, powers, and roots in forms that are sometimes more useful.

EXAMPLE 1 Simplify $\log_b \left(\frac{x^3 y^{1/2}}{z} \right)^{1/5}$.

Solution We can first write

$$\frac{1}{5} \log_b \left(\frac{x^3 y^{1/2}}{z} \right)$$

by using the third law of logarithms. Then, using the first and second laws, we have that

$$\frac{1}{5} \log_b \left(\frac{x^3 y^{1/2}}{z} \right) = \frac{1}{5} \left[\log_b x^3 + \log_b y^{1/2} - \log_b z \right],$$

from which, by using the third law again, we have that

$$\log_b \left(\frac{x^3 y^{1/2}}{z} \right)^{1/5} = \frac{1}{5} \left[3 \log_b x + \frac{1}{2} \log_b y - \log_b z \right]. \qquad ■$$

EXAMPLE 2 Simplify $\log_b \sqrt{\frac{xy}{z}}$.

Solution First express $\sqrt{\frac{xy}{z}}$ using a fractional exponent.

$$\log_b \sqrt{\frac{xy}{z}} = \log_b \left(\frac{xy}{z} \right)^{1/2}$$

By the third law of logarithms,

$$\log_b\left(\frac{xy}{z}\right)^{1/2} = \frac{1}{2}\log_b\left(\frac{xy}{z}\right).$$

Now, by the first and second laws of logarithms,

$$\frac{1}{2}\log_b\left(\frac{xy}{z}\right) = \frac{1}{2}(\log_b x + \log_b y - \log_b z).$$

Therefore,

$$\log_b \sqrt{\frac{xy}{z}} = \frac{1}{2}(\log_b x + \log_b y - \log_b z). \quad \blacksquare$$

EXAMPLE 3 Given that $\log_b 2 = 0.3010$ and $\log_b 3 = 0.4771$, find the value of:

a. $\log_b 9$

b. $\log_b 12$

Solutions First express $\log_b 9$ and $\log_b 12$ in terms of the known logarithmic quantities.

a.
$$\begin{aligned} \log_b 9 &= \log_b 3^2 \\ &= 2\log_b 3 \\ &= 2(0.4771) \\ &= 0.9542 \end{aligned}$$

b.
$$\begin{aligned} \log_b 12 &= \log_b 4 + \log_b 3 \\ &= \log_b 2^2 + \log_b 3 \\ &= 2\log_b 2 + \log_b 3 \\ &= 2(0.3010) + 0.4771 \\ &= 1.0791 \quad \blacksquare \end{aligned}$$

We can also use the three laws of logarithms to write sums and differences of logarithmic quantities as a single term.

EXAMPLE 4 Express $\frac{1}{2}(\log_b x - \log_b y)$ as a single logarithm with a coefficient of 1.

Solution By the second law of logarithms,

$$\frac{1}{2}(\log_b x - \log_b y) = \frac{1}{2}\log_b\left(\frac{x}{y}\right).$$

By the third law of logarithms,

$$\frac{1}{2}\log_b\left(\frac{x}{y}\right) = \log_b\left(\frac{x}{y}\right)^{1/2}.$$

Therefore,

$$\frac{1}{2}(\log_b x - \log_b y) = \log_b\left(\frac{x}{y}\right)^{1/2}. \quad ■$$

Common Errors

Note that

$$\log_b (x + y) \neq \log_b x + \log_b y$$

and

$$\log_b \frac{x}{y} \neq \frac{\log_b x}{\log_b y}.$$

Solving logarithmic equations

We sometimes may want to use the laws of logarithms to write the sum or difference of two logarithms as a single term in the solution of an equation, as illustrated in the following example.

EXAMPLE 5 Solve $\log_{10}(x + 1) + \log_{10}(x - 2) = 1$.

Solution Using the first law of exponents to rewrite the left-hand member, we obtain

$$\log_{10}(x + 1)(x - 2) = 1.$$

Then, writing this equation in exponential form, we have

$$(x + 1)(x - 2) = 10^1,$$

from which

$$\begin{aligned} x^2 - x - 2 &= 10, \\ x^2 - x - 12 &= 0, \\ (x - 4)(x + 3) &= 0. \end{aligned}$$

Thus,

$$x = 4 \quad \text{or} \quad x = -3.$$

The number -3 is not a solution of the original equation: The terms $\log_{10}(x + 1)$ and $\log_{10}(x - 2)$ are not defined for $x = -3$. The solution set is $\{4\}$. ■

So far in this chapter, we have generalized the discussion of the properties of the exponential and logarithmic functions for any base $b > 0$, $b \neq 1$. In Sections 12.3 and 12.4, we shall find values for the powers and logarithms with base 10 and base e that occur most often in applied problems. In Section 12.3, we shall use tables to find such values, and, in Section 12.4, we shall see how a scientific calculator can be used for the same purpose.

EXERCISE 12.2

A

■ *Use Properties (1), (2), and (3) on page 430 to write each expression in terms of simpler logarithmic quantities. Assume that all variables denote positive real numbers. See Examples 1 and 2.*

1. $\log_b(2x)$
2. $\log_b(xy)$
3. $\log_b(3xy)$
4. $\log_b(4yz)$
5. $\log_b\left(\frac{x}{y}\right)$
6. $\log_b\left(\frac{y}{x}\right)$
7. $\log_b\left(\frac{xy}{z}\right)$
8. $\log_b\left(\frac{x}{yz}\right)$
9. $\log_b x^3$
10. $\log_b x^{1/3}$
11. $\log_b \sqrt{x}$
12. $\log_b \sqrt[5]{y}$
13. $\log_b \sqrt[3]{x^2}$
14. $\log_b \sqrt{x^3}$
15. $\log_b(x^2y^3)$
16. $\log_b(x^{1/3}y^2)$
17. $\log_b\left(\frac{x^{1/2}y}{z^2}\right)$
18. $\log_b\left(\frac{xy^3}{z^{1/2}}\right)$
19. $\log_{10}\sqrt[3]{\frac{xy^2}{z}}$
20. $\log_{10}\sqrt[5]{\frac{x^2y}{z^3}}$
21. $\log_{10}\left(x\sqrt{\frac{x}{y}}\right)$
22. $\log_{10}\left(2y\sqrt[3]{\frac{x}{y}}\right)$
23. $\log_{10}\left(2\pi\sqrt{\frac{l}{g}}\right)$
24. $\log_{10}\sqrt{\frac{2L}{R^2}}$
25. $\log_{10}\sqrt{(s-a)(s-b)}$
26. $\log_{10}\sqrt{s^2(s-a)^3}$

■ *Given that* $\log_b 2 = 0.3010$, $\log_b 3 = 0.4771$, *and* $\log_b 5 = 0.6990$, *find the value of each expression. See Example 3.*

27. $\log_b 6$
28. $\log_b 10$
29. $\log_b \frac{2}{5}$
30. $\log_b \frac{3}{2}$
31. $\log_b 9$
32. $\log_b 25$
33. $\log_b \frac{15}{2}$
34. $\log_b \frac{6}{5}$
35. $\log_b(0.002)^3$
36. $\log_b \sqrt{50}$
37. $\log_b 75$
38. $\log_b \frac{0.08}{15}$

■ *Express as a single logarithm with a coefficient of 1. See Example 4.*

39. $\log_b 8 - \log_b 2$
40. $\log_b 5 + \log_b 2$

41. $\frac{1}{2}\log_b 16 + 2\log_b 2 - \log_b 8$

42. $\frac{1}{2}(\log_b 6 + 2\log_b 4 - \log_b 2)$

43. $2\log_b x - 3\log_b y$

44. $\frac{1}{4}\log_b x + \frac{3}{4}\log_b y$

45. $3\log_b x + \log_b y - 2\log_b z$

46. $\frac{1}{3}(\log_b x + \log_b y - 2\log_b z)$

47. $\frac{1}{2}(\log_{10} y + \log_{10} x - 2\log_{10} z)$

48. $\frac{1}{2}(\log_{10} x - 3\log_{10} y - \log_{10} z)$

49. $-2\log_b x$

50. $-\log_b x$

■ *Solve each logarithmic equation. See Example 5.*

51. $\log_{10} x + \log_{10} 2 = 3$

52. $\log_{10}(x - 1) - \log_{10} 4 = 2$

53. $\log_{10} x + \log_{10}(x + 21) = 2$

54. $\log_{10}(x + 3) + \log_{10} x = 1$

55. $\log_{10}(x + 2) + \log_{10}(x - 1) = 1$

56. $\log_{10}(x + 3) - \log_{10}(x - 1) = 1$

57. Show by a numerical example that $\log_{10}(x + y)$ is not equivalent to $\log_{10} x + \log_{10} y$.

58. Show by a numerical example that $\log_{10}\frac{x}{y}$ is not equivalent to $\frac{\log_{10} x}{\log_{10} y}$.

B

■ *Verify that each statement is true.*

59. $\log_b 4 + \log_b 8 = \log_b 64 - \log_b 2$

60. $\log_b 24 - \log_b 2 = \log_b 3 + \log_b 4$

61. $2\log_b 6 - \log_b 9 = 2\log_b 2$

62. $4\log_b 3 - 2\log_b 3 = \log_b 9$

63. $\frac{1}{2}\log_b 12 - \frac{1}{2}\log_b 3 = \frac{1}{3}\log_b 8$

64. $\frac{1}{4}\log_b 8 + \frac{1}{4}\log_b 2 = \log_b 2$

12.3

USING TABLES*

For $b > 0$ and $b \neq 1$, values for b^x $(x \in R)$ and $\log_b x$ $(x > 0)$ can be obtained by using prepared tables in conjunction with the laws of logarithms. Because we are familiar with the number 10 as the base of our numeration system, we shall first give our attention to logarithms and powers to the base 10. This was the base first used in the invention of logarithms to perform computations in astronomy and navigation.

*The topics covered in this section are also covered in Section 12.4 using a calculator instead of tables.

Common logarithms

Recall from Equation (4) on page 426 that $\log_{10} x$ is the exponent that must be placed on 10 so that the resulting power is x. Values for $\log_{10} x$ are sometimes called **common logarithms**.

Values for $\log_{10} 10^k$, $k \in J$

Some values of $\log_{10} x$ can be obtained simply by considering the definition of a logarithm, while other values require tables. Let us first consider values of $\log_{10} x$ for all values of x that are integral powers of 10. These can be obtained by inspection.

$$
\begin{aligned}
&\text{Since} \quad 10^3 = 1000, && \log_{10} 1000 = 3;\\
&\text{since} \quad 10^2 = 100, && \log_{10} 100 = 2;\\
&\text{since} \quad 10^1 = 10, && \log_{10} 10 = 1;\\
&\text{since} \quad 10^0 = 1, && \log_{10} 1 = 0;\\
&\text{since} \quad 10^{-1} = 0.1, && \log_{10} 0.1 = -1;\\
&\text{since} \quad 10^{-2} = 0.01, && \log_{10} 0.01 = -2;\\
&\text{since} \quad 10^{-3} = 0.001, && \log_{10} 0.001 = -3.
\end{aligned}
$$

Notice that the logarithm of a power of 10 is simply the exponent on the base 10. For example,

$$
\begin{aligned}
\log_{10} 100 &= \log_{10} 10^2 = 2,\\
\log_{10} 0.01 &= \log_{10} 10^{-2} = -2,
\end{aligned}
$$

and so on.

EXAMPLE 1 Find each logarithm.

a. $\log_{10} 10^6$ b. $\log_{10} 10^{-5}$

Solutions By definition:

a. $\log_{10} 10^6$ is the exponent on 10 so that the power equals 10^6. Hence,

$$\log_{10} 10^6 = 6.$$

b. $\log_{10} 10^{-5}$ is the exponent on 10 so that the power equals 10^{-5}. Hence,

$$\log_{10} 10^{-5} = -5. \quad ■$$

Values for $\log_{10} x$, $1 < x < 10$

Table II in Appendix B gives values for $\log_{10} x$ for $1 < x < 10$. Consider the following excerpt from the table. Each number in the column headed x represents the first two digits of the numeral for x, while each of the other column-head numbers represents the third significant digit of the numeral for x. The number located at the intersection of a row and a column forms the logarithm of x. For example, to find $\log_{10} 4.25$, we look at the intersection of the row containing 4.2 under x and the column containing 5. Thus,

$$\log_{10} 4.25 = 0.6284.$$

Similarly,

$$\log_{10} 4.02 = 0.6042,$$
$$\log_{10} 4.49 = 0.6522,$$

and so on.

x	0	1	2	3	4	5	6	7	8	9
3.8	.5798	.5809	.5821	.5832	.5843	.5855	.5866	.5877	.5888	.5899
3.9	.5911	.5922	.5933	.5944	.5955	.5966	.5977	.5988	.5999	.6010
4.0	.6021	.6031	.6042	.6053	.6064	.6075	.6085	.6096	.6107	.6117
4.1	.6128	.6138	.6149	.6160	.6170	.6180	.6191	.6201	.6212	.6222
4.2	.6232	.6243	.6253	.6263	.6274	.6284	.6294	.6304	.6314	.6325
4.3	.6335	.6345	.6355	.6365	.6375	.6385	.6395	.6405	.6415	.6425
4.4	.6435	.6444	.6454	.6464	.6474	.6484	.6493	.6503	.6513	.6522
4.5	.6532	.6542	.6551	.6561	.6571	.6580	.6590	.6599	.6609	.6618
4.6	.6628	.6637	.6646	.6656	.6665	.6675	.6684	.6693	.6702	.6712

The values in the tables are rational-number approximations of irrational numbers. We shall follow customary usage and write $=$ instead of $\approx$.

EXAMPLE 2 Find an approximation for each logarithm using Table II in Appendix B.

a. $\log_{10} 1.68 = 0.2253$ b. $\log_{10} 4.3 = 0.6335$ ■

Values for $\log_{10} x$, $x > 10$

Now suppose we wish to find $\log_{10} x$ for values of x outside the range of the table—that is, for $x > 10$ or $0 < x < 1$ (see Figure 12.2). This can be done quite readily by first representing the number in scientific notation and then applying the first law of logarithms.

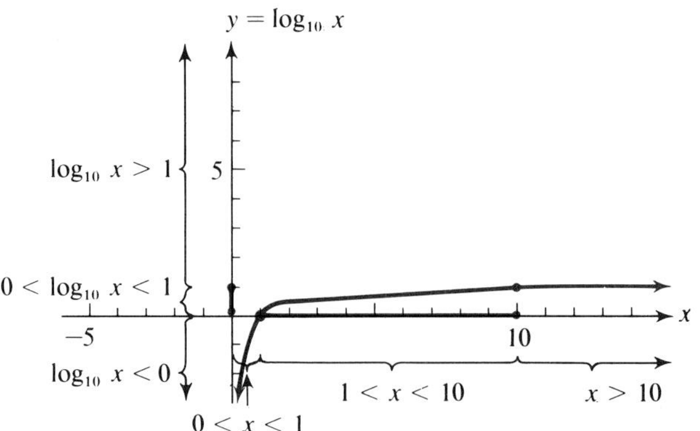

FIGURE 12.2

EXAMPLE 3

a. $\log_{10} 42.5$

$= \log_{10}(4.25 \times 10^1)$

$= \log_{10} 4.25 + \log_{10} 10^1$

$= 0.6284 + 1$

$= 1.6284$

b. $\log_{10} 425$

$= \log_{10}(4.25 \times 10^2)$

$= \log_{10} 4.25 + \log_{10} 10^2$

$= 0.6284 + 2$

$= 2.6284$

c. $\log_{10} 4250$

$= \log_{10}(4.25 \times 10^3)$

$= \log_{10} 4.25 + \log_{10} 10^3$

$= 0.6284 + 3$

$= 3.6284$

d. $\log_{10} 42{,}500$

$= \log_{10}(4.25 \times 10^4)$

$= \log_{10} 4.25 + \log_{10} 10^4$

$= 0.6284 + 4$

$= 4.6284$ ■

Observe that the decimal portion of the logarithms in Examples a–d is always 0.6284 and *the integral portion is the exponent on* 10 *when the number is written in scientific notation.*

This process can be reduced to a mechanical one by considering $\log_{10} x$ to consist of two parts, an *integral part* (called the **characteristic**) and a *nonnegative decimal fraction part* (called the **mantissa**). Thus, the table of values for $\log_{10} x$ for $1 < x < 10$ can be looked upon as a table of mantissas for $\log_{10} x$ for all $x > 0$.

In Examples a, b, c, and d above, where $x > 10$, the mantissa in each case is 0.6284 and the characteristics are 1, 2, 3, and 4, respectively.

Values for $\log_{10} x$, $0 < x < 1$

Now consider an example of the form $\log_{10} x$ for $0 < x < 1$ (see Figure 12.2). To find $\log_{10} 0.00425$, we write

$$\begin{aligned}\log_{10} 0.00425 &= \log_{10}(4.25 \times 10^{-3})\\ &= \log_{10} 4.25 + \log_{10} 10^{-3}.\end{aligned}$$

We find from the table that $\log_{10} 4.25 = 0.6284$. Upon adding 0.6284 to the characteristic -3, we obtain

$$\begin{aligned}\log_{10} 0.00425 &= 0.6284 + (-3)\\ &= -2.3716,\end{aligned}$$

where the decimal part of the logarithm is no longer 0.6284. The decimal part is -0.3716, a negative number.

If we want to use the table, which contains only positive entries, it is customary to write the logarithm in a form in which the decimal part is positive. In the above example, we write

$$\log_{10} 0.00425 = 0.6284 - 3,$$

where the decimal part is positive. Because -3 can be written $1 - 4$, $2 - 5$, $3 - 6$, $7 - 10$, and so on, the forms $1.6284 - 4$, $2.6284 - 5$, $3.6284 - 6$, $7.6284 - 10$, and so on, are equally valid representations of the desired logarithm. It will sometimes be convenient to use these alternative forms.

EXAMPLE 4 a. $\log_{10} 0.294$

$$\begin{aligned}&= \log_{10}(2.94 \times 10^{-1})\\ &= \log_{10} 2.94 + \log_{10} 10^{-1}\\ &= 0.4683 - 1\end{aligned}$$

b. $\log_{10} 0.00294$

$$\begin{aligned}&= \log_{10}(2.94 \times 10^{-3})\\ &= \log_{10} 2.94 + \log_{10} 10^{-3}\\ &= 0.4683 - 3\end{aligned}$$

■

Antilog$_{10}$ *N*

Given a value for an exponent, $\log_{10} x$, we can use Table II to find the power x by reversing the process described to find the logarithm of a number. In this case, the power x is called the **antilogarithm** of $\log_{10} x$. For example, if

$$\log_{10} x = 0.4409,$$

then

$$x = \text{antilog}_{10} 0.4409,$$

which can be obtained by locating 0.4409 in the body of Table II and observing that

$$\log_{10} 2.76 = 0.4409,$$

or

$$\text{antilog}_{10} 0.4409 = 2.76.$$

If the $\log_{10} x$ is greater than one, it can first be written as the sum of a positive decimal (the mantissa) and a positive integer (the characteristic). $\text{Antilog}_{10}\, x$ can then be written as the product of a number between one and ten and a power of ten.

EXAMPLE 5 a. If $\log_{10} x = 2.7364$, then

$$x = \text{antilog}_{10}\, 2.7364 = \text{antilog}_{10}(0.7364 + 2)$$

Locate the mantissa 0.7364 in the body of Table II and determine the associated antilog_{10} (a number between 1 and 10). Write the characteristic 2 as an exponent on the factor with base 10.

$$\begin{aligned} x &= \text{antilog}_{10}(\underbrace{0.7364} + 2) \\ &= 5.45 \times 10^2 \\ &= 545 \end{aligned}$$

b. If $\log_{10} x = 0.4409 - 3$, then

$$\begin{aligned} x &= \text{antilog}_{10}\, 0.4409 - 3 \\ &= \text{antilog}_{10}(\underbrace{0.4409} - 3) \\ &= 2.76 \times 10^{-3} \\ &= 0.00276 \end{aligned}$$ ■

If the decimal part of $\log_{10} x$ is negative and we wish to use the table to obtain x, we cannot use the table directly. However, we can first write $\log_{10} x$ equivalently with a positive decimal part. For example, to find

$$\text{antilog}_{10}(-0.4522) \quad \text{or} \quad \text{antilog}_{10}(-2.4522),$$

we can first add $(+1 - 1)$ to write -0.4522 as

$$\underbrace{-0.4522 + 1} - 1 = 0.5478 - 1$$

and add $(+3 - 3)$ to write -2.4522 as

$$\underbrace{-2.4522 + 3} - 3 = 0.5478 - 3,$$

and then use the tables. Thus,

$$\text{antilog}_{10}(-0.4522) = \text{antilog}_{10}(\underbrace{0.5478} - 1)$$
$$= 3.53 \times 10^{-1} = 0.353$$

and

$$\text{antilog}_{10}(-2.4522) = \text{antilog}_{10}(\underbrace{0.5478} - 3)$$
$$= 3.53 \times 10^{-3} = 0.00353.$$

EXAMPLE 6 Use Table II to find the value of x.

a. $\log_{10} x = -0.7292$ b. $\log_{10} x = -1.4634$

Solutions

a.
$$\begin{aligned} x &= \text{antilog}_{10}(-0.7292) \\ &= \text{antilog}_{10}(\underbrace{-0.7292 + 1} - 1) \\ &= \text{antilog}_{10}(0.2708 - 1) \\ &= 1.87 \times 10^{-1} = 0.187 \end{aligned}$$

b.
$$\begin{aligned} x &= \text{antilog}_{10}(-1.4634) \\ &= \text{antilog}_{10}(\underbrace{-1.4634 + 2} - 2) \\ &= \text{antilog}_{10}(0.5366 - 2) \\ &= 3.44 \times 10^{-2} = 0.0344 \end{aligned}$$ ■

In the above examples, the mantissas, 0.2718 and 0.5366, were listed in Table II. If we seek the common logarithm of a number that is not an entry in the table (for example, $\log_{10} 23.42$) or if we seek x when $\log_{10} x$ is not an entry in the table, we shall simply use the entry in the table which is closest to the value that we seek.

Powers to the base 10

By the definition of a logarithm,

$$P = 10^E$$

can be written in logarithmic form as

$$\log_{10} P = E,$$

from which we see that the power P is the antilogarithm of the exponent E,

$$P = \text{antilog}_{10} E.$$

Since $P = 10^E$,

$$\mathbf{10^E = antilog_{10}\ E}$$

and we can obtain a power 10^E simply by finding the antilogarithm of the exponent E.

EXAMPLE 7 Compute each power.

a. $10^{0.2148}$ b. $10^{-1.6345}$

Solutions

a. $10^{0.2148}$

$= \text{antilog}_{10} 0.2148$

$= 1.64$

b. $10^{-1.6345}$

$= \text{antilog}_{10}(-1.6345)$

$= \text{antilog}_{10}(\underbrace{-1.6345 + 2} - 2)$

$= \text{antilog}_{10}(0.3645 - 2)$

$= 2.32 \times 10^{-2}$ ■

Powers to the base *e*

The number $e \approx 2.7182818$* is an irrational number that has applications in business, biological and physical sciences, and in engineering. Because of its importance, special tables have been prepared for both e^x and $\log_e x$.

Table III in Appendix B gives approximations for powers e^x and e^{-x} for $0 \leqslant x \leqslant 1.00$ in 0.01 intervals and for $1.00 < x \leqslant 10.00$ in 0.1 intervals.

EXAMPLE 8 Using Table III:

a. $e^{2.4} = 11.023$ b. $e^{-4.7} = 0.0091$ ■

Although we can obtain values for e^x and e^{-x} outside the interval 0 to 10 by using the table along with the first law of exponents, at this time the function values in the table over this interval will be adequate for our work.

Natural logarithms, ln *x*

The most-used values for $\log_e x$ $(x > 0)$ are printed in Table IV of Appendix B. These values, like those for e^x, e^{-x}, and $\log_{10} x$, are *approximations* accurate to the number of decimals shown. The symbol $\log_e x$ is often written as **ln *x*** and read as "**natural logarithm of *x*.**" Unlike Table II for common logarithms, which only provides the decimal part of $\log_{10} x$, the table for natural logarithms gives the entire value for ln x, both the integral and decimal portions.

*The number e is discussed in more detail in Section 13.4.

EXAMPLE 9 Using Table IV:

a. $\ln 6.6 = 1.8871$ b. $\ln 0.7 = -0.3567$ ■

If we seek a value for e^x or $\ln x$ for a value of x between two entries in the tables, we shall simply use the entry in the table which is closest to the value that we seek.

Powers to the base *e*, antilogarithms Given an exponent $\ln x$, we can obtain the power x by using the definition of a logarithm and Table III.

EXAMPLE 10 a. $\ln x = 1.3$ b. $\ln x = -0.47$

Solutions a. $\ln x = 1.3$ is equivalent to

$$x = e^{1.3}$$
$$= 3.6693$$

b. $\ln x = -0.47$ is equivalent to

$$x = e^{-0.47}$$
$$= 0.6250$$ ■

As noted above, a power to the base b is called the antilog_b of the exponent. Thus in Example a,

$$x = e^{1.3} = \text{antilog}_e\ 1.3$$
$$= 3.6693.$$

EXERCISE 12.3

A

■ *Find each logarithm by inspection. See Example* 1.

1. $\log_{10} 10^2$
2. $\log_{10} 10^4$
3. $\log_{10} 10^{-4}$
4. $\log_{10} 10^{-6}$
5. $\log_{10} 10^0$
6. $\log_{10} 10^n$

■ *Find an approximation for each logarithm using Table II. See Examples* 2, 3, *and* 4.

7. $\log_{10} 6.73$
8. $\log_{10} 891$
9. $\log_{10} 83.7$

10. $\log_{10} 21.4$
11. $\log_{10} 317$
12. $\log_{10} 219$
13. $\log_{10} 0.813$
14. $\log_{10} 0.00214$
15. $\log_{10} 0.08$
16. $\log_{10} 0.000413$
17. $\log_{10}(2.48 \times 10^2)$
18. $\log_{10}(5.39 \times 10^{-3})$

■ *Solve for x using Table II. See Example 5.*

19. $\log_{10} x = 0.6128$
20. $\log_{10} x = 0.2504$
21. $\log_{10} x = 1.5647$
22. $\log_{10} x = 3.9258$
23. $\log_{10} x = 0.8075 - 2$
24. $\log_{10} x = 0.9722 - 3$
25. $\log_{10} x = 7.8562 - 10$
26. $\log_{10} x = 1.8155 - 4$

■ *For Problems 27–32 see Example 6.*

27. $\log_{10} x = -0.5272$
28. $\log_{10} x = -0.4123$
29. $\log_{10} x = -1.2984$
30. $\log_{10} x = -1.0545$
31. $\log_{10} x = -2.6882$
32. $\log_{10} x = -2.0670$

■ *Compute each power. See Example 7.*

33. $10^{0.8762}$
34. $10^{1.6405}$
35. $10^{2.8943}$
36. $10^{4.3766}$
37. $10^{-1.4473}$
38. $10^{-2.0958}$

■ *Compute each power. See Example 8.*

39. $e^{0.43}$
40. $e^{0.62}$
41. $e^{-0.57}$
42. $e^{-0.08}$
43. $e^{1.5}$
44. $e^{2.6}$
45. $e^{-2.4}$
46. $e^{-1.2}$

■ *Find each logarithm. See Example 9.*

47. $\ln 3.9$
48. $\ln 6.3$
49. $\ln 16$
50. $\ln 55$
51. $\ln 0.4$
52. $\ln 0.7$

■ *Find each value of x. See Example 10.*

53. $\ln x = 0.16$
54. $\ln x = 0.25$
55. $\ln x = 1.8$
56. $\ln x = 2.4$
57. $\ln x = 4.5$
58. $\ln x = 6.0$

12.4

USING CALCULATORS

Scientific calculators are useful in a variety of computations. Furthermore, a calculator enables us to obtain function values for exponential and logarithmic functions more efficiently than we could obtain them by using tables.

A great variety of calculators exist. We shall need one that has at least one of the keys [LOG], [LN], [e^x] and the inverse operation connected with that key. Many scientific calculators contain all three keys and may contain $\boxed{10^x}$ and $\boxed{y^x}$ keys also.*

Base 10

$\log_{10} 10^k$, $k \in J$

Values of $\log_{10} x$, called **logarithms to the base 10** or **common logarithms** can readily be obtained for all $x > 0$ by using a calculator. However, values for $\log_{10} x$, where *x is an integral power of* 10, can be obtained directly from the definition of a logarithm. You should try to obtain such values *without using your calculator.*

$$\begin{array}{llll}
\text{Since} & 10^3 = 1000, & \log_{10} 1000 = 3; \\
\text{since} & 10^2 = 100, & \log_{10} 100 = 2; \\
\text{since} & 10^1 = 10, & \log_{10} 10 = 1; \\
\text{since} & 10^0 = 1, & \log_{10} 1 = 0; \\
\text{since} & 10^{-1} = 0.1, & \log_{10} 0.1 = -1; \\
\text{since} & 10^{-2} = 0.01, & \log_{10} 0.01 = -2; \\
\text{since} & 10^{-3} = 0.001, & \log_{10} 0.001 = -3.
\end{array}$$

Notice that the logarithm of a power of 10 is simply the exponent on the base 10. For example,

$$\log_{10} 100 = \log_{10} 10^2 = 2,$$
$$\log_{10} 0.01 = \log_{10} 10^{-2} = -2,$$

and so on.

EXAMPLE 1 Find each logarithm.

a. $\log_{10} 10^6$

b. $\log_{10} 10^{-5}$

Solutions By definition:

a. $\log_{10} 10^6$ is the exponent on 10 so that the power equals 10^6. Hence,

$$\log_{10} 10^6 = 6.$$

*The calculator instructions shown in this section apply to calculators with algebraic logic. See the instruction manual for your calculator if it is programmed with a different logic.

b. $\log_{10} 10^{-5}$ is the exponent on 10 so that the power equals 10^{-5}. Hence,

$$\log_{10} 10^{-5} = -5. \quad ■$$

Your ability to determine the logarithms of powers of 10 by inspection will help you estimate function values $\log_{10} x$ when x is not a power of 10.

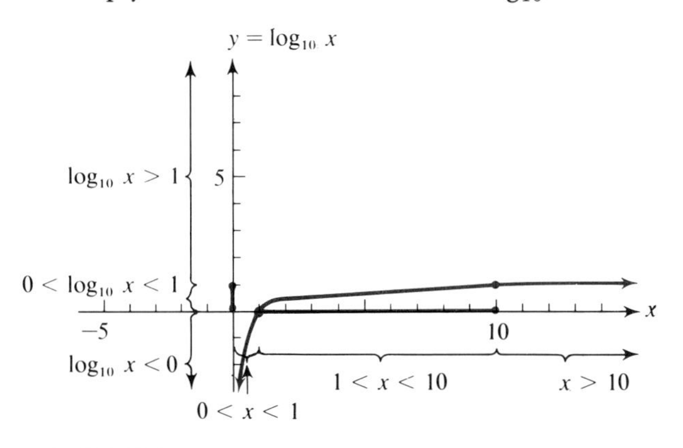

FIGURE 12.2

Note from Figure 12.2 that is reproduced here from Section 12.3:

$$\text{if} \quad 0 < x \leqslant 1, \quad \text{then} \quad \log_{10} x \leqslant 0;$$
$$\text{if} \quad 1 < x \leqslant 10, \quad \text{then} \quad 0 < \log_{10} x \leqslant 1;$$
$$\text{if} \quad 10 < x \leqslant 100, \quad \text{then} \quad 1 < \log_{10} x \leqslant 2;$$
$$\text{if} \quad 100 < x \leqslant 1000, \quad \text{then} \quad 2 < \log_{10} x \leqslant 3;$$

and so on.

$\log_{10} x$, $x > 0$

If a calculator is used to find values for $\log_{10} x$ $(x > 0)$ when x is not an integral power of 10, the function key LOG is ordinarily used to obtain these values. The values are generally produced to five or more significant digits. We shall round off readings of $\log_{10} x$ to five significant digits.

Some calculators do not have a LOG key but do have a LN key. In such cases, function values for $\log_{10} n$ can be obtained by finding the quotient $\ln n/\ln 10$. The reason why this is valid is discussed in Section 12.5.

EXAMPLE 2 Find each logarithm.

a. $\log_{10} 23.4$ **b.** $\log_{10} 0.00402$

Solutions a. Using the [LOG] key,

$$23.4 \boxed{\text{LOG}} = 1.3692.$$

Or, using the fact that $\log_{10} 23.4 = \ln 23.4/\ln 10$,

$$23.4 \boxed{\text{LN}} \boxed{\div} 10 \boxed{\text{LN}} \boxed{=} 1.3692.$$

b. Using the [LOG] key,

$$0.00402 \boxed{\text{LOG}} = -2.3958.$$

Or, using the fact that $\log_{10} 0.00402 = \ln 0.00402/\ln 10$,

$$0.00402 \boxed{\text{LN}} \boxed{\div} 10 \boxed{\text{LN}} \boxed{=} -2.3958.$$ ■

Note that in Example b the integer part of the logarithm (-2) is negative and *the decimal part* (-0.3958) *is also negative*. Contrary to our procedure when using tables, when we use a calculator it is not necessary to maintain a positive decimal part of a logarithm.

Furthermore, it should be understood that the values obtained by using a calculator are, in general, approximations for irrational numbers even though we shall, as is customary, continue to use the "=" sign.

10^x, $x \in R$

Some calculators have a [10^x] key. In this case, the power is readily computed. If a calculator does not have this key, the power can be computed by using the [LOG] key in conjunction with the [INV] key (or [2ND] function key). This is possible because the logarithmic function is the inverse of the exponential function. Alternatively, the [y^x] key can be used. In this case, the [=] key must be pressed as shown in the following example. We shall round off readings of powers of 10^x to five significant digits.

EXAMPLE 3 Compute $10^{2.34}$.

Solution Using the [10^x] key,

$$2.34 \boxed{10^x} = 218.78.$$

Or, using the [INV] (or second function) key,

$$2.34 \boxed{\text{INV}} \boxed{\text{LOG}} = 218.78.$$

Or, using the $\boxed{y^x}$ key,

$$10 \boxed{y^x} 2.34 \boxed{=} 218.78. \quad ■$$

Given a value for $\log_{10} x$, we can find x by first expressing the logarithm in exponential form.

EXAMPLE 4 If $\log_{10} x = 2.34$, then, by the definition of a logarithm,

$$x = 10^{2.34} = 218.78. \quad ■$$

A power to the base b is sometimes called the **antilog$_b$** of the exponent. Hence, in the above example,

$$x = 10^{2.34} = \text{antilog}_{10} 2.34.$$

EXAMPLE 5 Find each value of x.

a. $\log_{10} x = 2.4211$

b. $\log_{10} x = -1.2147$

Solutions

a. $\log_{10} x = 2.4211$
is equivalent to $x = 10^{2.4211}$;
$\text{antilog}_{10} 2.4211 = 263.69$.

b. $\log_{10} x = -1.2147$
is equivalent to $x = 10^{-1.2147}$;
$\text{antilog}_{10} -1.2147 = 0.06100$.

■

Base *e*

The second base in general use with exponential and logarithmic functions is the irrational number $e \approx 2.7182818$.* This number has applications in business, biological and physical sciences, and engineering.

ln *x*, *x* > 0

Approximations for $\log_e x$, commonly written as **ln *x***, are called **natural logarithms** and can be obtained on most calculators by using the $\boxed{\text{LN}}$ key. We shall round off the readings to five significant digits.

EXAMPLE 6 Find each logarithm.

a. $\ln 6.6$

b. $\ln 0.7$

*This number is considered further in Section 13.4.

Solutions a. 6.6 [LN] = 1.8871   b. 0.7 [LN] = −0.35667 ■

e^x, $x \in R$ Some calculators have a special key for e^x. If your calculator does not, this power can be obtained by using the [LN] key in conjunction with the [INV] key (or [2ND] function key).

EXAMPLE 7 Find each power.

a. $e^{2.4}$ b. $e^{-4.7}$

Solutions a. 2.4 [e^x] = 11.023
or
2.4 [INV] [LN] = 11.023

b. −4.7 [e^x] = 0.0090953
or
−4.7 [INV] [LN] = 0.0090953

Given a value for an exponent in the form of a natural logarithm, we can obtain the power by first expressing the logarithm in exponential form. This is the same procedure we used on page 448 for exponents that were written as common logarithms.

EXAMPLE 8 Find each value of x.

a. $\ln x = 2.4$ b. $\ln x = -4.7$

Solutions a. $\ln x = 2.4$ is equivalent to
$x = e^{2.4} = 11.023.$

b. $\ln x = -4.7$ is equivalent to
$x = e^{-4.7} = 0.0090953.$ ■

EXERCISE 12.4

A

■ *Find each logarithm by inspection. See Example* 1.

1. $\log_{10} 10^2$
2. $\log_{10} 10^4$
3. $\log_{10} 10^{-4}$
4. $\log_{10} 10^{-6}$
5. $\log_{10} 10$
6. $\log_{10} 1$
7. $\log_{10} 10{,}000$
8. $\log_{10} 0.001$

■ *Use a calculator in all of the following exercises. Round off each reading to five significant digits.*

■ *Find each logarithm. See Example 2.*

9. $\log_{10} 54.3$ **10.** $\log_{10} 27.9$ **11.** $\log_{10} 2344$

12. $\log_{10} 1476$ **13.** $\log_{10} 0.073$ **14.** $\log_{10} 0.00614$

15. $\log_{10} 0.6942$ **16.** $\log_{10} 0.0104$

■ *Find each power. See Example 3.*

17. $10^{1.62}$ **18.** $10^{0.43}$ **19.** $10^{-0.87}$

20. $10^{-1.31}$ **21.** $10^{2.113}$ **22.** $10^{3.141}$

23. $10^{-0.2354}$ **24.** $10^{-2.0413}$

■ *Solve for x. See Examples 4 and 5.*

25. $\log_{10} x = 1.41$ **26.** $\log_{10} x = 2.3$ **27.** $\log_{10} x = 0.52$

28. $\log_{10} x = 0.8$ **29.** $\log_{10} x = -1.3$ **30.** $\log_{10} x = -1.69$

■ *Find each logarithm. See Example 6.*

31. $\ln 3.9$ **32.** $\ln 6.3$ **33.** $\ln 16$

34. $\ln 55$ **35.** $\ln 6.4$ **36.** $\ln 0.7$

■ *Find each power. See Example 7.*

37. $e^{0.4}$ **38.** $e^{0.73}$ **39.** $e^{2.34}$

40. $e^{3.16}$ **41.** $e^{-1.2}$ **42.** $e^{-2.3}$

43. $e^{-0.4}$ **44.** $e^{-0.62}$

■ *Solve for x. See Example 8.*

45. $\ln x = 1.42$ **46.** $\ln x = 2.03$ **47.** $\ln x = 0.63$

48. $\ln x = 0.59$ **49.** $\ln x = -2.6$ **50.** $\ln x = -3.4$

12.5

SOLVING EXPONENTIAL EQUATIONS

We can now use the definition of a logarithm and Table II or Table IV or a calculator to obtain solutions to some simple equations in which the variable is part of the exponent. Such equations are called **exponential equations.** We shall express the solutions to such equations in the examples in this section to three significant digits.

EXAMPLE 1 Solve each equation.

a. $10^x = 2.73$ b. $e^x = 0.24$

Solutions a. $10^x = 2.73$ is equivalent to $x = \log_{10} 2.73 = 0.436.$

b. $e^x = 0.24$ is equivalent to $x = \ln 0.24 = -1.43.$ ■

In more complicated equations, it is sometimes necessary to first rewrite the equation in an equivalent form so that the power is the only term in one member and has a coefficient of one. The equation can then be written in logarithmic form.

EXAMPLE 2 Solve $4.31 = 1.73 + 2 \cdot 10^{1.2x}$.

Solution Adding -1.73 to each member, we have

$$2.58 = 2 \cdot 10^{1.2x},$$

from which by dividing each member by 2,

$$1.29 = 10^{1.2x}.$$

The equivalent logarithmic equation is

$$1.2\,x = \log_{10} 1.29.$$

Hence, by dividing each member by 1.2 and using Table II or a calculator,

$$x = \frac{\log_{10} 1.29}{1.2} = 0.0922.$$

This result may be obtained on some calculators as follows:

$$1.29 \; \boxed{\text{LOG}} \; \boxed{\div} \; 1.2 \; \boxed{=} \; 0.092158 \approx 0.0922.$$ ■

EXAMPLE 3 Solve $140 = 20e^{0.4x}$.

Solution We first divide each member by 20 to obtain

$$7 = e^{0.4x},$$

from which the equivalent logarithmic equation is

$$0.4x = \ln 7.$$

Hence, by dividing each member by 0.4 and using Table IV or a calculator,

$$x = \frac{\ln 7}{0.4} = 4.86.$$

This result may be obtained on some calculators as follows:

$$7 \boxed{\text{LN}} \quad \boxed{\div} \; 0.4 \; \boxed{=} \; 4.8648 \approx 4.86. \qquad ■$$

Changing bases A procedure similar to the one used in the previous two examples would be applicable to any exponential equation involving the base e or the base 10 because values for $\ln x$ and $\log_{10} x$ can be obtained from tables or a scientific calculator. However, the solution of an exponential equation that involves a base other than e or 10 and for which a table of logarithmic function values is not available requires other methods.

One method involves writing a logarithm to one base in terms of logarithms to a different base. To express $\log_b N$ in terms of another base, say a, we first let

$$y = \log_b N. \tag{1}$$

Then, from the definition of a logarithm, we have that

$$N = b^y.$$

Equating the logarithm of each member using the base a, we have

$$\log_a N = \log_a b^y.$$

From the third law of logarithms,

$$\log_a N = y \log_a b,$$

from which

$$y = \frac{\log_a N}{\log_a b}.$$

Hence, substituting $\log_b N$ for y from (1) above, we obtain the following equation:*

$$\log_b N = \frac{\log_a N}{\log_a b}. \tag{2}$$

EXAMPLE 4 Compute a value for $\log_5 16$.

Solution From Equation (2) above, with $a = 10$,

$$\begin{aligned}\log_5 16 &= \frac{\log_{10} 16}{\log_{10} 5}\\ &= \frac{1.2041}{0.69897} = 1.72.\end{aligned}$$

Calculator sequence:

16 [LOG] [÷] 5 [LOG] [=] $1.7227 \approx 1.72.$

Alternatively, with $a = e$,

$$\begin{aligned}\log_5 16 &= \frac{\ln 16}{\ln 5}\\ &= \frac{2.7726}{1.6094} = 1.72.\end{aligned}$$

Calculator sequence:

16 [LN] [÷] 5 [LN] [=] $1.7227 \approx 1.72.$ ■

Equation (2) above can be used to solve exponential equations in which the base is a positive number other than 10 or e.

*A special case of Equation (2) with $b = 10$ is the relationship that we used on page 446 to find values of $\log_{10} N$ where we used a calculator that did not have a [LOG] key.

EXAMPLE 5 Solve $5^x = 7$.

Solution We can first write this equation in logarithmic form as

$$x = \log_5 7.$$

Now, from Equation (2) and using the base 10, we have

$$\begin{aligned} x &= \log_5 7 \\ &= \frac{\log_{10} 7}{\log_{10} 5} = \frac{0.84510}{0.69897} = 1.21. \end{aligned}$$

The use of natural logarithms will yield the same result. ■

Note that the quotient $\log_{10} 7/\log_{10} 5$ in the above example could have been obtained by first equating the logarithm of each member of $5^x = 7$ to the base 10 to obtain

$$\log_{10} 5^x = \log_{10} 7.$$

Then, from the third law of logarithms,

$$x \log_{10} 5 = \log_{10} 7,$$

from which

$$x = \frac{\log_{10} 7}{\log_{10} 5}.$$

Common Error Note that when we seek a numerical approximation to the solution by the procedure in the above example, the logarithms are *divided*, not subtracted:

$$\frac{\log_{10} 7}{\log_{10} 5} \neq \log_{10} 7 - \log_{10} 5.$$

Equations with several variables We can use the definition of a logarithm in exponential equations and logarithmic equations of more than one variable to solve for one of the variables in terms of the others.

EXAMPLE 6 a. Solve $P = Cb^{kt}$ for t.

Solutions a. First express the power b^{kt} in terms of the other variables.

$$b^{kt} = \frac{P}{C} \qquad (C \neq 0)$$

Write the exponential equation in logarithmic form.

$$kt = \log_b \frac{P}{C}$$

Multiply each member by $^1/_k$.

$$t = \frac{1}{k} \log_b \frac{P}{C} \qquad (k \neq 0)$$

b. Solve $N = N_0 \log_b(ks)$ for s.

b. First express $\log_b(ks)$ in terms of the other variables.

$$\log_b(ks) = \frac{N}{N_0} \qquad (N_0 \neq 0)$$

Write the logarithmic equation in exponential form.

$$ks = b^{N/N_0}$$

Multiply each member by $^1/_k$.

$$s = \frac{1}{k} b^{N/N_0} \qquad (k \neq 0)$$ ■

EXERCISE 12.5

A

■ *Compute solutions of Problems 1–30 to three significant digits. A calculator was used to obtain the answers for this section. Answers obtained by using tables may differ slightly from those given.*

■ *Solve each exponential equation. See Example 1.*

1. $10^x = 4.93$
2. $10^x = 8.07$
3. $10^x = 23.4$
4. $10^x = 182.4$
5. $10^x = 6832.3$
6. $10^x = 9480.2$
7. $e^x = 1.9$
8. $e^x = 2.1$
9. $e^x = 45$
10. $e^x = 60$
11. $e^x = 0.3$
12. $e^x = 0.9$

■ *For Problems 13–20 see Example 2.*

13. $26.1 = 1.4(10^{1.3x})$
14. $140 = 63.1(10^{0.2x})$
15. $14.8 = 1.72 + 10^{-0.3x}$
16. $180 = 64 + 10^{-1.3x}$
17. $12.2 = 2(10^{1.4x}) - 11.6$
18. $163 = 3(10^{0.7x}) - 49.3$
19. $3(10^{-1.5x}) - 14.7 = 17.1$
20. $4(10^{-0.6x}) + 16.1 = 28.2$

■ *For Problems 21–30 see Example 3.*

21. $6.21 = 2.3\, e^{1.2x}$
22. $22.26 = 5.3\, e^{0.4x}$
23. $7.74 = 1.72\, e^{0.2x}$
24. $14.105 = 4.03\, e^{1.4x}$

25. $6.4 = 20\, e^{0.3x} - 1.8$

26. $4.5 = 4\, e^{2.1x} + 3.3$

27. $46.52 = 3.1\, e^{1.2x} + 24.2$

28. $1.23 = 1.3\, e^{2.1x} - 17.1$

29. $16.24 = 0.7\, e^{-1.3x} - 21.7$

30. $55.68 = 0.6\, e^{-0.7x} + 23.1$

■ *Use Equation* (2) *on page* 453 *to find the value of each of the following logarithms to the nearest hundredth. Use base* 10 *or base e. See Example* 4.

31. $\log_3 18$

32. $\log_3 24$

33. $\log_2 7.43$

34. $\log_2 14.3$

35. $\log_4 17.3$

36. $\log_4 28.1$

■ *Solve using logarithms to the base* 10. *See Example* 5.

37. $2^x = 7$

38. $3^x = 4$

39. $3^{x+1} = 8$

40. $2^{x-1} = 9$

41. $4^{x^2} = 15$

42. $3^{x^2} = 21$

43. $3^{-x} = 10$

44. $2.13^{-x} = 8.1$

B

■ *Solve each exponential or logarithmic equation for the specified variable. Leave the results in the form of an equation equivalent to the given equation. See Example* 6.

45. $y = e^{kt}$, for t using the base e

46. $y = k(1 - e^{-t})$, for t using the base e

47. $\dfrac{T}{R} = e^{t/2}$, for t using the base e

48. $B - 2 = (A + 3)e^{-t/3}$, for t using the base e

49. $T = T_0 \ln(k + 10)$, for k

50. $P = P_0 + \ln 10k$, for k

51. Show that $\ln N \approx 2.303 \log_{10} N$.

52. Show that $\ln 10 = \dfrac{1}{\log_{10} e}$.

12.6

APPLICATIONS

Exponential equations involving powers with base 10 or base e can be used as models for a variety of real-world phenomena. In equations such as

$$A = Be^{kt} + C \quad \text{or} \quad A = B \cdot 10^{kt} + C,$$

we may want to find values for a particular variable in an expression that may or may not be "part" of the exponent.

The first case, in which we want to find a value for a variable that is *not* part of the exponent, is illustrated by the following example.

EXAMPLE 1 A scientist starts an experiment with 25 grams of a radioactive element. The number of grams remaining at any time t is given by $y = 25e^{-0.5t}$, where t is in seconds. How much of the element (to the nearest hundredth of a gram) is remaining after three seconds?

Solution Substituting 3 for t, we have

$$\begin{aligned} y &= 25e^{-0.5(3)} \\ &= 25e^{-1.5} = 5.57825. \end{aligned}$$

There are approximately 5.58 grams of material remaining after three seconds. ■

The second case, in which we want to find a value for a variable that is in the exponent, is illustrated by the following example.

EXAMPLE 2 The atmospheric pressure P, in inches of mercury, is given approximately by

$$P = 30(10)^{-0.09a}, \tag{1}$$

where a is the altitude in miles above sea level. How high above the earth (to the nearest hundredth of a mile) is the atmospheric pressure 26.4 inches of mercury?

Solution Substituting 26.4 for P, we obtain

$$26.4 = 30(10)^{-0.09a},$$

which is equivalent to

$$\frac{26.4}{30} = 10^{-0.09a}.$$

Writing the equation in logarithmic form, we have

$$-0.09a = \log_{10} \frac{26.4}{30},$$

from which

$$a = -\frac{1}{0.09}\log_{10}\frac{26.4}{30}$$
$$= 0.61686.$$

Hence, the pressure is 26.4 inches of mercury at approximately 0.62 miles above the earth. ■

EXERCISE 12.6

A *A calculator was used to obtain the answers for this section: Answers obtained by using tables may differ slightly from those given. See Examples 1 and 2.*

■ *Solve Problems 1–6 using the relationship* $P = 30(10)^{-0.09a}$ *between altitude* (a) *in miles and atmospheric pressure* (P) *in inches of mercury. Round off results to the nearest hundredth.*

1. The elevation of Mt. Everest, the highest mountain in the world, is 29,028 feet. What is the atmospheric pressure at the top? [*Hint:* One mile equals 5280 feet.]
2. What is the atmospheric pressure at sea level? 50,000 feet? 100,000 feet?
3. How high above sea level is the atmospheric pressure 20.2 inches of mercury?
4. How high above sea level is the atmospheric pressure 16.1 inches of mercury?
5. Find the height above sea level at which the pressure is equal to one-half of the pressure at sea level. (Pressure at sea level is approximately 30 inches of mercury.)
6. Find the height above sea level at which the pressure is equal to one-fourth of the pressure at sea level.

■ *Solve Problems 7–12 using the fact that population growth is given approximately by* $P = P_0e^{rt}$, *where an initial population* P_0 *increases at an annual rate* r *(expressed as a decimal) to a population* P *after* t *years.*

7. The population of the state of California increased from 1960 to 1970 at a rate of approximately 2.39% per year. The population in 1960 was 15,717,000.
   a. Approximately what was the population in 1970?
   b. Assuming the same rate of growth, estimate the population in the years 1980, 1990, and 2000.
8. The population of the state of New York increased from 1960 to 1970 at a rate of approximately 0.83% per year. The population in 1960 was 16,782,000.
   a. Approximately what was the population in 1970?
   b. Assuming the same rate of growth, estimate the population in the years 1980, 1990, and 2000.

9. The population of the state of Texas in 1960 was 9,579,700. In 1970 the population was 11,196,700. What was the annual rate of growth to the nearest hundredth of a percent?
10. The population of the state of Florida in 1960 was 4,951,600. In 1970 the population was 6,789,400. What was the annual rate of growth to the nearest hundredth of a percent?
11. If the annual rate of growth of a country is 3.7%, how long will it take for the population to double?
12. The population of a country doubled in 20 years. What was the annual rate of growth (to the nearest hundredth of a percent)?
13. The amount of a radioactive element present at any time t is given by $y = y_0 e^{-0.4t}$, where t is measured in seconds and y_0 is the amount present initially. How much of the element (to the nearest hundredth of a gram) would remain after 3 seconds if 40 grams were present initially?
14. The number N of bacteria present in a culture is given by $N = N_0 e^{0.04t}$, where N_0 is the number of bacteria present at time $t = 0$, and t is time in hours. If 6000 bacteria were present at $t = 0$, how many were present 10 hours later?
15. The voltage V across a capacitor in a certain circuit is given by $V = 100(1 - e^{-0.5t})$, where t is the time in seconds. What is the voltage (to the nearest tenth of a volt) after 10 seconds?
16. The intensity I (in lumens) of a light beam after passing through a thickness t (in centimeters) of a medium having an absorption coefficient of 0.1 is given by $I = 1000e^{-0.1t}$. What is the intensity (to the nearest tenth) of a light beam passing through 0.6 centimeters of the medium?
17. The voltage V across a capacitor in a certain circuit is given by $V = 100(1 - e^{-0.5t})$, where t is the time in seconds. How much time must elapse (to the nearest hundredth of a second) for the voltage to reach 75 volts?
18. The amount of a radioactive element present at any time t is given by $y = y_0 e^{-0.4t}$, where t is measured in seconds and y_0 is the amount present initially. How much time must elapse (to the nearest hundredth of a second) for 40 grams to be reduced to 12 grams?
19. The number N of bacteria present in a culture is given by $N = N_0 e^{0.04t}$, where N_0 is the number of bacteria present at time $t = 0$, and t is time in hours. How much time must elapse (to the nearest tenth of an hour) for 2500 bacteria to increase to 10,000?
20. The intensity I (in lumens) of a light beam after passing through a thickness t (in centimeters) of a medium having an absorption coefficient of 0.1 is given by $I = 1000e^{-0.1t}$. How many centimeters (to the nearest tenth) of the material would reduce the illumination to 800 lumens?

B

■ *Solve Problems 21–28 using the following information: P dollars invested at an annual interest rate r (expressed as a decimal) compounded yearly yields an amount A after n years given by* $A = P(1 + r)^n$. *If the interest is compounded t times yearly, the amount is given by*

$$A = P\left(1 + \frac{r}{t}\right)^{tn}.$$

21. One dollar compounded annually for 10 years yields \$5.12. What is the rate of interest to the nearest ½%?
22. How many years (to the nearest year) would it take for \$1000 to "grow" to \$2000 if compounded annually at 12%?
23. What rate of interest (to the nearest ½%) is rquired so that \$100 would yield \$190 after 5 years if the money were compounded semiannually?
24. What rate of interest (to the nearest ½%) is required so that \$40 would yield \$60 after 3 years if the money were compounded quarterly?
25. Find the compounded amount of \$5000 invested at 12% for 10 years when compounded annually. When compounded semiannually.
26. How many years (to the nearest year) would it take for a sum of money to double if invested at 10% and compounded quarterly?
27. How many years (to the nearest year) would it take for a sum of money to increase fivefold if invested at 10% and compounded quarterly?
28. Two investors, A and B, each invested \$10,000 at 8% for 20 years with a bank that computed interest quarterly. Investor A withdrew interest at the end of each 3 month period, but B allowed the investment to be compounded. How much more than A did B earn over the period of 20 years?

CHAPTER SUMMARY

[12.1] The inverse of an exponential function is called a **logarithmic function** and is defined by an equation of the form

$$x = b^y \quad \text{or} \quad y = \log_b x \qquad (b > 0, b \neq 1).$$

$\text{Log}_b\, x$ is the **exponent** on b such that the power equals x. The domain of a logarithmic function is the set of positive real numbers; the range is the set of real numbers.

[12.2] The properties of logarithms given below follow from the definition of a logarithm and the properties of exponents developed in Chapter 6.

$$\log_b(x_1 x_2) = \log_b x_1 + \log_b x_2 \qquad (1)$$

$$\log_b \frac{x_2}{x_1} = \log_b x_2 - \log_b x_1 \qquad (2)$$

$$\log_b(x_1)^m = m \log_b x_1 \qquad (3)$$

[12.3] Values of $\log_{10} x$, $x > 0$ are known as **common logarithms.**

Values of $\log_{10} x$, where x is a power of 10 with an integer exponent, can be obtained by inspection directly from the definition of a logarithm.

Values of $\log_{10} x$ $(1 < x < 10)$ are between 0 and 1 and can be obtained directly from Table II.

Values of $\log_{10} x$ $(x > 10$ and $0 < x < 1)$ can be obtained from Table II in conjunction with the first law of logarithms. The *integral part* of a logarithm to the base 10 is the **characteristic** of the logarithm, and the *positive decimal part* is the **mantissa.**

Values for e^x can be obtained by using Table III; values for $\log_e x$ or $\ln x$, called **natural logarithms,** can be obtained by using Table IV.

A power b^x is called the **antilogarithm** of x. Thus,

$$10^x = \text{antilog}_{10}\, x \quad \text{and} \quad e^x = \text{antilog}_e\, x.$$

[12.4] A scientific calculator can be used to find values for the powers 10^x and e^x $(x \in R)$ and the exponents $\log_{10} x$ and $\ln x$ $(x > 0)$.

[12.5] Exponential equations can be solved by using the properties of logarithms plus a table of logarithms or a calculator.

[12.6] Exponential equations can be used as models for many real-world problems.

The following table summarizes the different ways of finding the exponential and logarithmic function values that have been considered in this chapter.

To find the power for a given value of an exponent:

	10^x ($\text{antilog}_{10}\, x$)	e^x ($\text{antilog}_e\, x$)	b^x $(b > 0, b \neq 1)$
Table	II*	III	
Calculator	[10^x] or [y^x] or [INV]/[2ND] [LOG]	[e^x] or [y^x] or [INV]/[2ND] [LN]	[y^x]

To find the exponent for a given value of a power:

	$10^x = N$ $x = \log_{10} N$	$e^x = N$ $x = \ln N$	$b^x = N\ (b > 0, b \neq 1)$
Table	II*	IV*	II* or IV*
Calculator	LOG or $x = \log_{10} N = \dfrac{\ln N}{\ln 10}$ N LN ÷ 10 LN = x	LN	$x = \log_b N = \dfrac{\ln N}{\ln b}$ N LN ÷ b LN = x

**Note:* When using Table II or IV, the decimal part (mantissa) of a logarithm must be positive; the characteristic must be an integer.

REVIEW EXERCISES

A

[12.1]

1. Write each statement in logarithmic notation.

   a. $9^{3/2} = 27$ b. $\left(\frac{4}{9}\right)^{1/2} = \frac{2}{3}$

2. Write each statement in exponential notation.

   a. $\log_5 625 = 4$ b. $\log_{10} 0.0001 = -4$

■ *Solve for x.*

3. a. $\log_4 16 = x$ b. $\log_2 x = 3$

4. a. $\log_{10} x = -2$ b. $\log_3 \frac{1}{3} = x$

■ *Find the value of each logarithm by inspection.*

5. a. $\log_3 9$ b. $\log_4 \frac{1}{4}$

6. a. $\log_{10} 1000$ b. $\log_{10} 0.001$

[12.2]

7. Express each as the sum or difference of simpler logarithmic quantities.

   a. $\log_b 3x^2y$ b. $\log_b \dfrac{y\sqrt{x}}{z^2}$

8. Given that $\log_b 2 = 0.3010$ and $\log_b 3 = 0.4771$, find the value for each logarithm.

   a. $\log_b 36$ b. $\log_b \sqrt{18}$

9. Write each expression as a single logarithm with a coefficient of 1.

   a. $5 \log_b x - 2 \log_b y$ b. $\dfrac{1}{3}(\log_b x - 4 \log_b y + 2 \log_b z)$

10. a. Solve $\log_{10}(x + 9) - \log_{10} x = 1$.   b. Solve $\log_{10} x + \log_{10}(x - 3) = 1$.

■ *In Problems 11–26, use the tables or a calculator (round off readings to three significant digits).*

[12.3–12.4]

11. Find a value for each logarithm.

    a. $\log_{10} 0.713$ b. $\log_{10} 1810$

12. Solve for x.

    a. $\log_{10} x = 2.6345$ b. $\log_{10} x = -1.4214$

13. Compute each power.

    a. $10^{1.2347}$ b. $10^{-0.5453}$

14. Find the value of each power.

    a. $e^{0.83}$ b. $e^{-1.3}$

15. Find the value of each logarithm.

    a. $\ln 7$ b. $\ln 0.4$

16. Solve for x.

    a. $\ln x = 0.73$ b. $\ln x = 2.7$

[12.5]

■ *Solve each exponential equation.*

17. $10^x = 1.3$
18. $e^x = 62$
19. $7.35 = 2.1(10)^{1.2x}$
20. $12.4 = 2e^{0.3x} - 4.2$
21. Given that $N = N_0 10^{0.4t}$, find t if $N = 280$ and $N_0 = 4$.
22. Given that $y = y_0 e^{-0.2t}$, find t if $y = 4$ and $y_0 = 20$.

■ *Use Equation* (2) *on page* 453 *to find each logarithm.*

23. $\log_2 23.1$

24. $\log_3 7.04$

■ *Solve for x using logarithms to the base* 10 *or the base e.*

25. $3^x = 15$

26. $2^{x-4} = 10$

[12.6]

27. The concentration C of a certain drug in the bloodstream at any time t is given by $C(t) = 10 - 10e^{-0.5t}$, where C is in milligrams and t is in minutes. Determine the concentration to the nearest tenth of a milligram at $t = 0$ and $t = 1$.
28. The intensity I of a light beam after passing through a thickness t of a certain medium is given by $I = 500e^{-0.2t}$, where I is in lumens and t is in centimeters. What is the intensity (to the nearest tenth of a lumen) of a light beam passing through 0.4 centimeters of the medium?
29. Using the formula for population growth on page 458, how long (to the nearest tenth of a year) will it take for a city with an annual rate of growth of 3% to grow from a population of 200,000 to 300,000?
30. Using the formula for radioactive decay in Problem 13 on page 459, how much time (to the nearest tenth of a second) must elapse for 100 grams to be reduced to 50 grams?

B

31. Simplify $\log_{10} \log_2(\log_5 25)$.
32. Verify that $2 \log_b 8 - \log_b 4 = 4 \log_b 2$ is a true statement.
33. Solve $N = N_0 e^{-kt}$ for t, using natural logarithms.
34. Solve $N = N_0 \ln(t/k) + c$ for t.

■ *Chemists define the* pH (*hydrogen potential*) *of a solution by* $\text{pH} = \log_{10} \dfrac{1}{[H^+]}$, *where* $[H^+]$ *represents a numerical value for the concentration of hydrogen ions in aqueous solution in moles per liter.*

35. Calculate to the nearest tenth the pH of a solution with hydrogen ion concentration 6.3×10^{-7}.
36. Calculate the hydrogen ion concentration of a solution with pH 5.6.

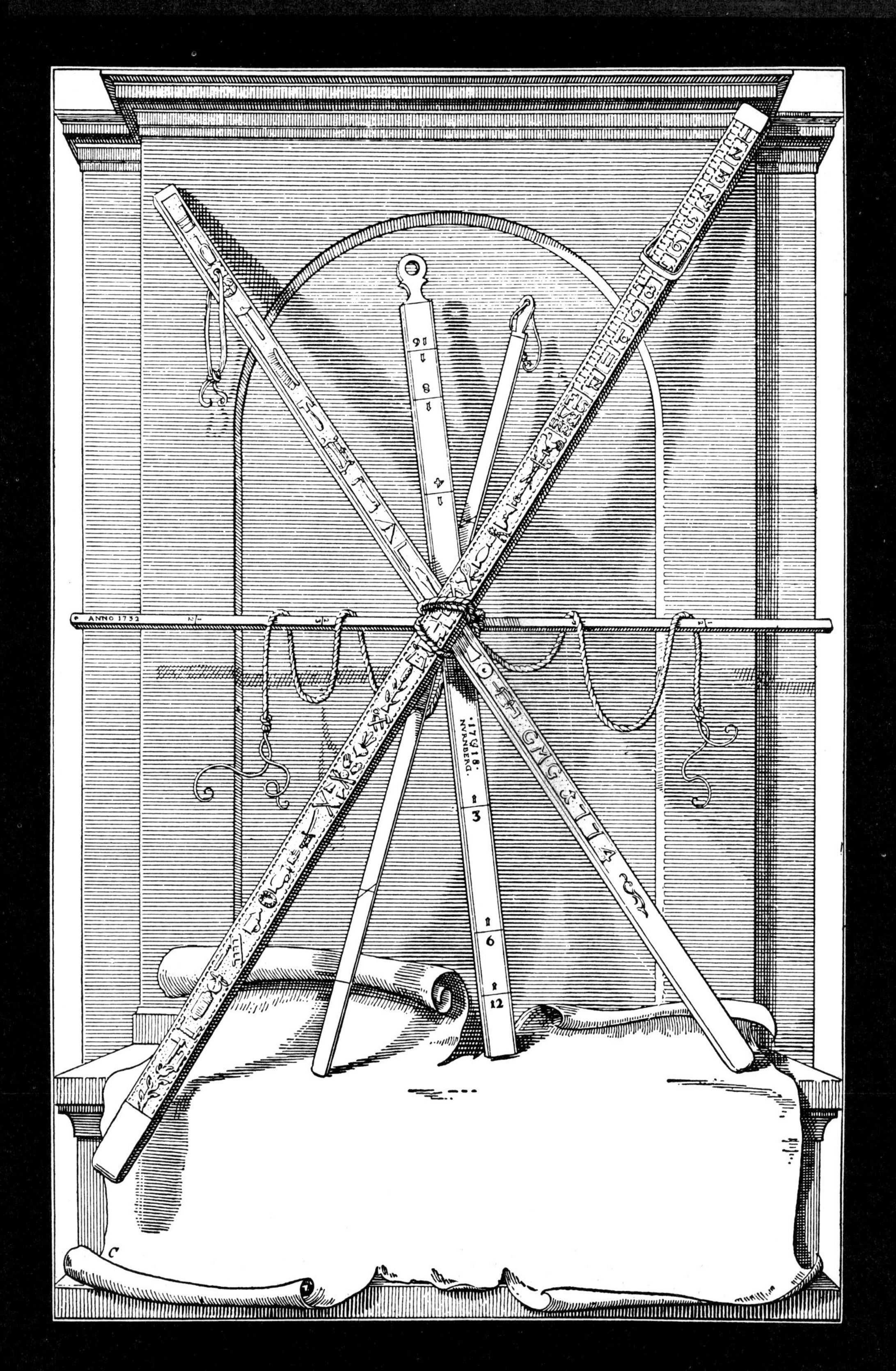
ANNO 1732
NVRNBERG.
3
6
12
8

13 Natural-number Functions

13.1

SEQUENCES AND SERIES

Sequences

A function whose domain is a set of successive positive integers, for example, a function defined by an equation such as

$$s(n) = 2n - 1 \qquad (n \in \{3, 4, 5\}) \tag{1}$$

or

$$s(n) = n + 3 \qquad (n \in \{1, 2, 3, \ldots\}), \tag{2}$$

is called a **sequence function**. The function defined by (1) is called a **finite sequence**, and the function defined by (2) is called an **infinite sequence**. The elements in the range of such functions arranged in the order

$$s(3), s(4), s(5) \quad \text{or} \quad s(1), s(2), s(3), \ldots$$

are said to form a **sequence**, and the elements are referred to as the **terms** of the sequence. Thus, the sequence associated with (2) is found by successively substituting the numbers 1, 2, 3, . . . for n:

$$\begin{aligned} s(1) &= (1) + 3 = 4 \\ s(2) &= (2) + 3 = 5 \\ s(3) &= (3) + 3 = 6; \end{aligned}$$

$s(1)$ or 4 is called the first term, $s(2)$ or 5 is called the second term, $s(3)$ or 6 is called the third term, etc. The expression $n + 3$ is called the **general term** or ***n*th term**.

EXAMPLE 1 The first three terms of the sequence with the general term $\frac{3}{2n-1}$ are:

$$s(1) = \frac{3}{2(1)-1} = \frac{3}{1}$$

$$s(2) = \frac{3}{2(2)-1} = \frac{3}{3}$$

$$s(3) = \frac{3}{2(3)-1} = \frac{3}{5}.$$

The twenty-fifth term is

$$s(25) = \frac{3}{2(25)-1} = \frac{3}{49}.$$ ■

The notation ordinarily used for the terms in a sequence is not function notation as such; rather, it is customary to denote a term in a sequence by means of a subscript. Thus, we will use s_n rather than $s(n)$, and the sequence $s(1), s(2), s(3), \ldots$ will appear as $s_1, s_2, s_3, \ldots$.

EXAMPLE 2 Find the first four terms in a sequence with the given general term.

a. $s_n = \frac{n(n+1)}{2}$

b. $s_n = (-1)^n 2^n$

Solutions

a. $s_1 = \frac{1(1+1)}{2} = 1$

$s_2 = \frac{2(2+1)}{2} = 3$

$s_3 = \frac{3(3+1)}{2} = 6$

$s_4 = \frac{4(4+1)}{2} = 10$

The first four terms are 1, 3, 6, 10.

b. $s_1 = (-1)^1 2^1 = -2$

$s_2 = (-1)^2 2^2 = 4$

$s_3 = (-1)^3 2^3 = -8$

$s_4 = (-1)^4 2^4 = 16$

The first four terms are $-2, 4, -8, 16$. ■

Series

Associated with any sequence is a **series**; the series is defined as the sum of the terms in the sequence and is denoted by S_n.

EXAMPLE 3 a. Associated with the finite sequence

$$4, 7, 10, \ldots, 3n + 1 \tag{3}$$

is the finite series

$$S_n = 4 + 7 + 10 + \cdots + (3n + 1). \tag{4}$$

b. Associated with the sequence

$$x, x^2, x^3, x^4, \ldots, x^n$$

is the series

$$S_n = x + x^2 + x^3 + x^4 + \cdots + x^n. \quad ■$$

Since the terms in the series are the same as those in the corresponding sequence, we can refer to the first term or the second term or the general term of a series in the same manner as we do for a sequence.

Sigma notation

A series with a general term that is known can be represented in a very convenient, compact way by using the symbol Σ (the Greek letter **sigma**) in conjunction with the general term in sigma or summation notation; this denotes the sum of all the terms in the series. For example, series (4) can be written

$$S_n = \sum_{i=1}^{n} (3i + 1),$$

where we understand that S_n is the series with terms obtained by successively replacing i in the expression $3i + 1$ with the numbers, $1, 2, 3, \ldots, n$. Thus,

$$S_6 = \sum_{i=1}^{6} (3i + 1)$$

appears in **expanded form** as

$$[3(1) + 1] + [3(2) + 1] + [3(3) + 1] + [3(4) + 1] + [3(5) + 1] + [3(6) + 1]$$
$$= 4 + 7 + 10 + 13 + 16 + 19.$$

The variable used in conjunction with summation notation (in the above case, i) is called the **index of summation**; the set of integers over which we sum in this case is $\{1, 2, 3, 4, 5, 6\}$. The use of the symbol i as an index of summation should not be confused with its use as an imaginary unit in the set of complex numbers. Alternatively, the summation index can be any letter such as j, k, l, and so on:

$$\sum_{i=1}^{6} (3i + 1) = \sum_{j=1}^{6} (3j + 1) = \sum_{k=1}^{6} (3k + 1) = \cdots.$$

The first member of the replacement set for the index of summation is not necessarily 1. For example,

$$\begin{aligned}\sum_{i=3}^{6} (3i + 1) &= [3(3) + 1] + [3(4) + 1] + [3(5) + 1] + [3(6) + 1] \\ &= 10 + 13 + 16 + 19,\end{aligned}$$

where the first replacement for i is 3; the set of integers over which we sum is $\{3, 4, 5, 6\}$.

Note that the series in the above example contains four terms, which is one more than the difference between the last and the first replacement for i. In general, the series $\sum_{i=a}^{b} s_i$ contains $(b - a + 1)$ terms.

To show that a series has an infinite number of terms—that is, has no last term—we adopt a special notation. For example,

$$S_\infty = \sum_{i=4}^{\infty} (3i + 1)$$

denotes the series that would appear in expanded form as

$$S_\infty = 13 + 16 + 19 + 22 + \cdots, \tag{5}$$

where, in this case, i has been replaced by 4, 5, 6, 7, ...

EXAMPLE 4 Write in expanded form.

a. $\sum_{i=2}^{4} (i^2 + 1)$

b. $\sum_{k=1}^{\infty} (-1)^k 2^{k+1}$

Solutions a. i takes values 2, 3, 4.

$i = 2, \quad (2)^2 + 1 = 5$
$i = 3, \quad (3)^2 + 1 = 10$
$i = 4, \quad (4)^2 + 1 = 17$

Expanded form: $5 + 10 + 17$.

b. k takes values 1, 2, 3, ...

$k = 1, \quad (-1)^1 2^{1+1} = (-1)(4) = -4$
$k = 2, \quad (-1)^2 2^{2+1} = (1)(8) = 8$
$k = 3, \quad (-1)^3 2^{3+1} = (-1)(16) = -16$

Expanded form: $-4 + 8 - 16 + \cdots$. ■

Finding general terms As we have seen in the examples above, for any given general term s_n, we can obtain the elements of an associated sequence by successively substituting the numbers 1, 2, 3, . . . for n. However, it is usually difficult to obtain a general term for a given sequence. In Sections 13.2 and 13.3 we shall consider some formulas that will enable us to find general terms for several particular sequences. For now, we might attempt to find general terms by inspection (trial and error). A series can be represented in summation notation by various general terms and different ranges. For example, both

$$\sum_{i=5}^{\infty} (3i - 2) \quad \text{and} \quad \sum_{i=6}^{\infty} (3i - 5)$$

also represent series (5) above. This can be verified by writing the first few terms in each series.

EXAMPLE 5 Write each series in sigma notation.

a. $5 + 8 + 11 + 14$

b. $x^2 + x^4 + x^6 + \cdots + x^{2n}$

Solutions By inspection (trial and error), we first find a general term for the sequence associated with each series.

a. $3i + 2$

Hence,

$$5 + 8 + 11 + 14 = \sum_{i=1}^{4} (3i + 2).$$

b. x^{2i}

Hence,

$$x^2 + x^4 + x^6 + \cdots + x^{2n} = \sum_{i=1}^{n} x^{2i}.$$ ■

EXERCISE 13.1

A

■ *Find the first four terms in a sequence with the general term as given. See Examples 1 and 2.*

1. $s_n = n - 5$
2. $s_n = 2n - 3$
3. $s_n = \dfrac{n^2 - 2}{2}$
4. $s_n = \dfrac{3}{n^2 + 1}$
5. $s_n = 1 + \dfrac{1}{n}$
6. $s_n = \dfrac{n}{2n - 1}$
7. $s_n = \dfrac{n(n - 1)}{2}$
8. $s_n = \dfrac{5}{n(n + 1)}$
9. $s_n = (-1)^n$
10. $s_n = (-1)^{n+1}$
11. $s_n = \dfrac{(-1)^n(n - 2)}{n}$
12. $s_n = (-1)^{n-1}3^{n+1}$

■ *Write in expanded form. See Example 4.*

13. $\sum_{i=1}^{4} i^2$
14. $\sum_{i=1}^{3} (3i - 2)$
15. $\sum_{j=5}^{7} (j - 2)$
16. $\sum_{j=2}^{6} (j^2 + 1)$
17. $\sum_{k=1}^{4} k(k + 1)$
18. $\sum_{i=2}^{6} \dfrac{i}{2}(i + 1)$
19. $\sum_{i=1}^{4} \dfrac{(-1)^i}{2^i}$
20. $\sum_{i=3}^{5} \dfrac{(-1)^{i+1}}{i - 2}$
21. $\sum_{i=1}^{\infty} (2i - 1)$
22. $\sum_{j=1}^{\infty} \dfrac{1}{j}$
23. $\sum_{k=0}^{\infty} \dfrac{1}{2^k}$
24. $\sum_{k=0}^{\infty} \dfrac{k}{1 + k}$

B

■ *Write each series in sigma notation. See Example 5. (There are no unique solutions.)*

25. $1 + 2 + 3 + 4$
26. $2 + 4 + 6 + 8$
27. $x + x^3 + x^5 + x^7$
28. $x^3 + x^5 + x^7 + x^9 + x^{11}$
29. $1 + 4 + 9 + 16 + 25$
30. $1 + 8 + 27 + 64 + 125$
31. $\dfrac{1}{2} + \dfrac{2}{3} + \dfrac{3}{4} + \dfrac{4}{5} + \cdots$
32. $\dfrac{2}{1} + \dfrac{3}{2} + \dfrac{4}{3} + \dfrac{5}{4} + \cdots$
33. $\dfrac{1}{1} + \dfrac{2}{3} + \dfrac{3}{5} + \dfrac{4}{7} + \cdots$
34. $\dfrac{3}{1} + \dfrac{5}{3} + \dfrac{7}{5} + \dfrac{9}{7} + \cdots$
35. $\dfrac{1}{1} + \dfrac{2}{2} + \dfrac{4}{3} + \dfrac{8}{4} + \cdots$
36. $\dfrac{1}{2} + \dfrac{3}{4} + \dfrac{9}{6} + \dfrac{27}{8} + \cdots$

13.2

ARITHMETIC PROGRESSIONS

Any sequence with a general term that is linear in n has the property that each term except the first can be obtained from the preceding term by adding a common number called the **common difference.** A sequence with this property is called an **arithmetic progression,** or **arithmetic sequence.** We can state the definition for such a sequence symbolically:

$$\begin{aligned} s_1 &= a, \\ s_{n+1} &= s_n + d, \end{aligned}$$

where d is the common difference. Definitions of this sort are called **recursive definitions.** It is customary to denote the first term in such a sequence by the letter a, the common difference between successive terms by d, the number of terms in the sequence (when finite) by n, and the nth term by s_n.

We can verify that a finite sequence is an arithmetic progression simply by subtracting each term from its successor and noting that the difference in each case is the same. For example, the sequence 7, 18, 29, 40 is an arithmetic progression, because

$$18 - 7 = 11, \quad 29 - 18 = 11, \quad \text{and} \quad 40 - 29 = 11.$$

If at least two consecutive terms in an arithmetic progression are known, we can determine the common difference and generate as many terms as we wish.

EXAMPLE 1 The third and fourth terms of an arithmetic progression are 6 and 11, respectively. Write the next three terms of the sequence.

Solution Given $s_3 = 6$ and $s_4 = 11$, we can obtain

$$\begin{aligned} d &= s_4 - s_3 \\ &= 11 - 6 = 5. \end{aligned}$$

Thus, $s_5 = 11 + 5 = 16$, $s_6 = 16 + 5 = 21$, and $s_7 = 21 + 5 = 26$. Hence, the next three terms following 6 and 11 are 16, 21, and 26. ■

***n*th term of an arithmetic progression**

Now consider the general arithmetic progression with first term a and common difference d. The

$$\begin{array}{ll}
\text{first term is} & a, \\
\text{second term is} & a + d, \\
\text{third term is} & a + d + d = a + 2d, \\
\text{fourth term is} & a + d + d + d = a + 3d, \\
\vdots & \vdots \\
n\text{th term is} & a + d + d + \cdots + d = a + (n - 1)d.
\end{array}$$

Thus, we have the following property.

$$\mathbf{s_n = a + (n - 1)d.} \qquad (1)$$

Here we have used an informal inductive process to obtain Equation (1). We shall assume its validity for all natural numbers n.

Equation (1) provides us with a formula to find the nth term of any arithmetic progression when the first term and common difference are known. For example, if the first term of an arithmetic progression is 7 and the common difference is 2, we have from Equation (1) that the nth term is

$$\begin{aligned}
s_n &= 7 + (n - 1)2 \\
&= 7 + 2n - 2 \\
&= 2n + 5.
\end{aligned}$$

EXAMPLE 2 Write the next three terms in each arithmetic progression. Find an expression for the general term.

a. $5, 9, \ldots$

b. $x, x - a, \ldots$

Solutions

a. Find the common difference and then continue the sequence.

$$d = 9 - 5 = 4;$$
$$13, 17, 21$$

b. Find the common difference and then continue the sequence.

$$d = (x - a) - x = -a;$$
$$x - 2a,\ x - 3a,\ x - 4a$$

Use $s_n = a + (n - 1)d$ to find an expression for the general term.

$$s_n = 5 + (n - 1)4$$
$$s_n = 4n + 1$$

Use $s_n = a + (n - 1)d$ to find an expression for the general term.

$$s_n = x + (n - 1)(-a)$$
$$s_n = x - a(n - 1) \quad ■$$

Observe that Equation (1) defines a *linear function* for any values of a and d. As we have noted, the domain (replacement set of n) of this function is a set of natural numbers. Equation (1) can be used directly to find a particular term if the first term and common difference of an arithmetic progression are known.

EXAMPLE 3 Find the fourteenth term of the arithmetic progression $-6, -1, 4, \ldots$.

Solution Find the common difference.

$$d = -1 - (-6) = 5$$

Use $s_n = a + (n - 1)d$.

$$s_{14} = -6 + (14 - 1)5 = 59 \quad ■$$

Equation (1) can also be used to find a particular term in an arithmetic expression in which two terms are known.

EXAMPLE 4 Find the first term in an arithmetic progression in which the third term is 7 and the eleventh term is 55.

Solution A diagram of the situation is helpful here.

n: 1 , 2 , 3 , 4 , 5 , 6 , 7 , 8 , 9 , 10, 11
s_n: <u>?</u>, __, <u>7</u>, __, __, __, __, __, __, __, <u>55</u>

Find a common difference by considering an arithmetic progression with first term 7 and ninth term 55. Use $s_n = a + (n - 1)d$.

$$s_9 = 7 + (9 - 1)d$$
$$55 = 7 + 8d$$
$$d = 6$$

Use this difference to find the first term in an arithmetic progression in which the third term is 7. Use $s_n = a + (n - 1)d$.

$$s_3 = a + (3 - 1)6$$
$$7 = a + 12$$
$$a = -5$$

The first term is -5.

Alternative Solution Use $s_n = a + (n - 1)d$, with $s_3 = 7$.

$$7 = a + (3 - 1)d$$
$$7 = a + 2d \qquad (2)$$

Use $s_n = a + (n - 1)d$ with $s_{11} = 55$.

$$55 = a + (11 - 1)d$$
$$55 = a + 10d \qquad (3)$$

Solve the system (2) and (3) to obtain $a = -5$. ■

Arithmetic means Terms between given terms in an arithmetic progression are called **arithmetic means** of the given terms.

EXAMPLE 5 Insert three arithmetic means between 4 and -8.

Solution If there are three terms between 4 and -8, then the difference between 4 and -8 must be 4 times the common difference (d), as suggested by

$$\underline{4}\ \underline{?}\ \underline{?}\ \underline{?}\ \underline{-8}.$$
$$d \quad d \quad d \quad d$$

Therefore,

$$4d = (-8) - (4) = -12$$
$$d = -3.$$

The three requested arithmetic means can then be obtained by successive additions of -3. We obtain 1, -2, and -5. ■

Sum of n terms

The problem of finding an explicit representation for the sum of n terms of a sequence in terms of n is, in general, very difficult. However, we can obtain such a representation for the sum of n terms in an arithmetic progression. Consider the series of n terms associated with the general arithmetic progression

$$a, (a + d), (a + 2d), \ldots, a + (n - 1)d;$$

that is,

$$S_n = a + (a + d) + (a + 2d) + \cdots + [a + (n - 1)d]. \tag{4}$$

Then consider the same series written as

$$S_n = s_n + (s_n - d) + (s_n - 2d) + \cdots + [s_n - (n - 1)d], \tag{5}$$

where the terms are in reverse order. Adding (4) and (5) term-by-term, we have

$$S_n + S_n = (a + s_n) + (a + s_n) + (a + s_n) + \cdots + (a + s_n),$$

where the term $(a + s_n)$ occurs n times. It follows that

$$2S_n = n(a + s_n),$$

from which we have the following property.

$$S_n = \frac{n}{2}(a + s_n) \tag{6}$$

EXAMPLE 6 The sum of the eight terms of an arithmetic progression with first term -7 and last term 14 is given by

$$S_8 = \frac{8}{2}(-7 + 14) = 28. \quad ■$$

If (6) is rewritten as

$$S_n = n\left(\frac{a + s_n}{2}\right),$$

we observe that the sum is given by the product of the number of terms in the series and the average of the first and last terms.

An alternative form for (6) is obtained by substituting in (6) the value for s_n equal to $a + (n - 1)d$ as given by (1) to obtain

$$S_n = \frac{n}{2}(a + [a + (n - 1)d]),$$

from which we have the following property.

$$\boldsymbol{S_n = \frac{n}{2}[2a + (n - 1)d]} \qquad (7)$$

In (7) the sum is now expressed in terms of a, n, and d. ■

EXAMPLE 7 Compute the sum of the series $\sum_{i=1}^{12}(4i + 1)$.

Solution Write the first two or three terms in expanded form.

$$5 + 9 + 13 + \cdots$$

By inspection, the first term is 5 and the common difference is 4.

Use $S_n = \frac{n}{2}[2a + (n - 1)d]$ with $n = 12$.

$$S_{12} = \frac{12}{2}[2(5) + (12 - 1)4] = 324$$ ■

EXERCISE 13.2

A

■ *Write the next three terms in each arithmetic progression. Find an expression for the general term. See Examples 1 and 2.*

1. $3, 7, \ldots$
2. $-6, -1, \ldots$
3. $-1, -5, \ldots$
4. $-10, -20, \ldots$
5. $x, x + 1, \ldots$
6. $a, a + 5, \ldots$
7. $x + a, x + 3a, \ldots$
8. $y - 2b, y, \ldots$
9. $2x + 1, 2x + 4, \ldots$
10. $a + 2b, a - 2b, \ldots$
11. $x, 2x, \ldots$
12. $3a, 5a, \ldots$

■ *Solve. See Example 3.*

13. Find the seventh term in the arithmetic progression 7, 11, 15, ...
14. Find the tenth term in the arithmetic progression $-3, -12, -21, \ldots$.
15. Find the twelfth term in the arithmetic progression $2, 5/2, 3, \ldots$.
16. Find the seventeenth term in the arithmetic progression $-5, -2, 1, \ldots$.
17. Find the twentieth term in the arithmetic progression $3, -2, -7, \ldots$.
18. Find the tenth term in the arithmetic progression $3/4, 2, 13/4, \ldots$.

■ *Solve. See Example 4.*

19. If the third term in an arithmetic progression is 7 and the eighth term is 17, find the common difference. What is the first term? What is the twentieth term?
20. If the fifth term of an arithmetic progression is -16 and the twentieth term is -46, what is the twelfth term?
21. What term in the arithmetic progression $4, 1, -2, \ldots$ is -77?
22. What term in the arithmetic progression $7, 3, -1, \ldots$ is -81?

■ *Insert the given number of arithmetic means between the given two numbers. See Example 5.*

23. Two between -6 and 15.
24. Four between 10 and 65.
25. One between 12 and 20.
26. One between -11 and 7.
27. Three between 24 and 4.
28. Six between -12 and 23.

■ *Find the sum of each finite series. See Examples 6 and 7.*

29. $\sum_{i=1}^{7} (2i + 1)$
30. $\sum_{i=1}^{21} (3i - 2)$
31. $\sum_{j=3}^{15} (7j - 1)$
32. $\sum_{j=10}^{20} (2j - 3)$
33. $\sum_{k=1}^{8} \left(\frac{1}{2}k - 3\right)$
34. $\sum_{k=1}^{100} k$

B

35. Find the sum of all even integers n, where $13 < n < 89$.
36. Find the sum of all integral multiples of 7 between 8 and 110.
37. How many bricks will there be in a stack one brick deep if there are 27 bricks in the first row, 25 in the second row, . . . , and 1 in the top row?
38. If there is a total of 256 bricks in a stack arranged in the manner of those in problem 37, how many bricks are there in the third row from the bottom?
39. Find three numbers that form an arithmetic sequence such that their sum is 21 and their product is 168.
40. Find three numbers that form an arithmetic sequence such that their sum is 21 and their product is 231.
41. Find k if $\sum_{j=1}^{5} kj = 14$.
42. Find p and q if $\sum_{i=1}^{4} (pi + q) = 28$ and $\sum_{i=2}^{5} (pi + q) = 44$.
43. Show that the sum of the first n odd natural numbers is n^2.
44. Show that the sum of the first n even natural numbers is $n^2 + n$.

13.3

GEOMETRIC PROGRESSIONS

Any sequence in which each term except the first is obtained by multiplying the preceding term by a common multiplier is called a **geometric progression,** or **geometric sequence,** and is defined by the recursive equations

$$s_1 = a,$$
$$s_{n+1} = rs_n,$$

where r is called the **common ratio.**

For example the sequence

$$2, 6, 18, 54, \ldots$$

is a geometric progression in which each term except the first is obtained by multiplying the preceding term by 3.

nth term of a geometric progression

Now, if we designate the first term of a geometric progression by a, then the

second term is	ar,
third term is	$ar \cdot r = ar^2$,
fourth term is	$ar^2 \cdot r = ar^3$,

and it appears that the nth term will take the following form.

$$s_n = ar^{n-1} \qquad (1)$$

EXAMPLE 1 Find the ratio r and general term s_n for the geometric progression

$$2, 6, 18, \ldots.$$

Solution By writing the ratio of any term to its predecessor, say, $18/6$, we obtain $r = 3$. A representation for s_n of this sequence can now be written in terms of n by substituting 2 for a and 3 for r in (1). Thus,

$$s_n = ar^{n-1} = 2(3)^{n-1}. \qquad ■$$

EXAMPLE 2 Write the next three terms in each geometric progression. Find the general term.

a. $3, 6, 12, \ldots$

b. $x, 2, \dfrac{4}{x}, \ldots$

Solutions

a. Find the common ratio.

$$r = \frac{6}{3} = 2$$

Multiply successively by r to determine the following terms:

$$24, \quad 48, \quad 96.$$

b. Find the common ratio.

$$r = \frac{2}{x}$$

Multiply successively by r to determine the following terms:

$$\frac{8}{x^2}, \quad \frac{16}{x^3}, \quad \frac{32}{x^4}.$$

Use $s_n = ar^{n-1}$ to find the general term.

$$s_n = 3(2)^{n-1}$$

Use $s_n = ar^{n-1}$ to find the general term.

$$s_n = x\left(\frac{2}{x}\right)^{n-1} = \frac{2^{n-1}}{x^{n-2}} \quad ■$$

We can use Equation (1) to find a particular term in a geometric progression in which two or more terms are known.

EXAMPLE 3 Find the ninth term of the geometric progression $-24, 12, -6, \ldots$.

Solution Find the common ratio.

$$r = \frac{12}{-24} = -\frac{1}{2}$$

Use $s_n = ar^{n-1}$.

$$s_9 = -24\left(-\frac{1}{2}\right)^8 = -\frac{3}{32} \quad ■$$

Observe that Equation (1) defines an *exponential function* for all values of a and r. As we noted, the domain (replacement set of n) of this function is a set of natural numbers.

Geometric means

Terms between given terms in a geometric progression are called **geometric means** of the given terms.

EXAMPLE 4 Insert two geometric means between 3 and 24.

Solution Since there are three multiplications by the common ratio between 3 and 24, the quotient when 24 is divided by 3 must be the third power of the common ratio, as suggested by the following:

$$\underline{3} \quad \underline{?} \quad \underline{?} \quad \underline{24}.$$
$$\times r \quad \times r \quad \times r$$

Hence, $r^3 = 24/3 = 8$; so $r = \sqrt[3]{8} = 2$. Therefore, the missing terms can be determined by successive multiplications by 2, and we have $3 \times 2 = 6$ and $6 \times 2 = 12$. Thus, the geometric means are 6 and 12. ■

Sum of *n* terms

To find an explicit representation for the sum of a given number of terms in a geometric progression in terms of a, r, and n, we employ a device somewhat similar to the one used in finding the sum of an arithmetic series. Consider the geometric series (2) containing n terms and the series (3) obtained by multiplying both members of (2) by r:

$$S_n = a + ar + ar^2 + ar^3 + \cdots + ar^{n-2} + ar^{n-1}, \quad (2)$$

$$rS_n = ar + ar^2 + ar^3 + ar^4 + \cdots + ar^{n-1} + ar^n. \quad (3)$$

Subtracting (3) from (2), we find that all terms in the right-hand member vanish except the first term in (2) and the last term in (3), and therefore

$$S_n - rS_n = a - ar^n.$$

Factoring S_n from the left-hand member yields

$$(1 - r)S_n = a - ar^n,$$

from which we have the following property.

$$\boldsymbol{S_n = \frac{a - ar^n}{1 - r}} \qquad \boldsymbol{(r \neq 1)} \quad (4)$$

This is a general formula for the sum of n terms of a geometric progression.

EXAMPLE 5 The sum of four terms of a geometric progression with first term 5 and common ratio -3 is given by

$$S_4 = \frac{5 - 5(-3)^4}{1 - (-3)} = \frac{5 - 5(81)}{4} = -100. \quad ■$$

It is helpful to rewrite a geometric progression in sigma notation in expanded form before using Equation (4).

EXAMPLE 6 Compute $\sum_{i=2}^{7} \left(\frac{1}{3}\right)^i$.

Solution Write the first two or three terms in expanded form.

$$\left(\frac{1}{3}\right)^2 + \left(\frac{1}{3}\right)^3 + \cdots$$

By inspection, the first term is 1/9, the ratio is 1/3, and $n = 6$ [recall from page 470 that the series $\sum_{i=a}^{b}$ contains $(b - a + 1)$ terms]. Use $S_n = \dfrac{a - ar^n}{1 - r}$.

$$S_6 = \frac{\frac{1}{9} - \frac{1}{9}\left(\frac{1}{3}\right)^6}{1 - \frac{1}{3}} = \frac{\frac{1}{9}\left(1 - \frac{1}{729}\right)}{\frac{2}{3}} = \frac{1}{9} \cdot \frac{728}{729} \cdot \frac{3}{2} = \frac{364}{2187} \quad ■$$

An alternative expression for (4) can be obtained by noting that it may be written

$$S_n = \frac{a - r(ar^{n-1})}{1 - r},$$

and, since $s_n = ar^{n-1}$, we have the following relationship, where the sum is now given in terms of a, s_n, and r.

$$\mathbf{S_n = \frac{a - rs_n}{1 - r} \qquad (r \neq 1)} \qquad (5)$$

EXAMPLE 7 The sum of the geometric progression

$$6 + 3 + \frac{3}{2} + \frac{3}{4} + \frac{3}{8}$$

with first term 6, common ratio $\frac{1}{2}$, and fifth term $\frac{3}{8}$, is given by

$$S_5 = \frac{6 - \left(\frac{1}{2}\right)\left(\frac{3}{8}\right)}{1 - \frac{1}{2}} = \frac{6 - \frac{3}{16}}{\frac{1}{2}}$$

$$= 2\left(6 - \frac{3}{16}\right) = 12 - \frac{3}{8} = \frac{93}{8}. \qquad \blacksquare$$

EXERCISE 13.3

A

■ *Write the next three terms in each geometric progression. Find the general term. See Examples* 1 *and* 2.

1. $2, 8, 32, \ldots$
2. $4, 8, 16, \ldots$
3. $\frac{2}{3}, \frac{4}{3}, \frac{8}{3}, \ldots$
4. $6, 3, \frac{3}{2}, \ldots$
5. $4, -2, 1, \ldots$
6. $\frac{1}{2}, -\frac{3}{2}, \frac{9}{2}, \ldots$
7. $\frac{a}{x}, -1, \frac{x}{a}, \ldots$
8. $\frac{a}{b}, \frac{a}{bc}, \frac{a}{bc^2}, \ldots$

■ *Solve. See Example* 3.

9. Find the sixth term in the geometric progression $48, 96, 192, \ldots$.
10. Find the eighth term in the geometric progression $-3, \frac{3}{2}, -\frac{3}{4}, \ldots$.
11. Find the seventh term in the geometric progression $-\frac{1}{3}a^2, a^5, -3a^8, \ldots$.
12. Find the ninth term in the geometric progression $-81a, -27a^2, -9a^3, \ldots$.
13. Find the first term of a geometric progression with fifth term 48 and ratio 2.
14. Find the first term of a geometric progression with fifth term 1 and ratio $-\frac{1}{2}$.

■ *Insert the given number of geometric means between the two given numbers. See Example* 4.

15. Two between 1 and 27.
16. Two between -4 and -32.
17. One between 36 and 9. (Two answers are possible.)
18. One between -12 and $-\frac{1}{12}$. (Two answers are possible.)
19. Three between 32 and 2. (Two answers are possible.)
20. Three between -25 and $-\frac{1}{25}$. (Two answers are possible.)

■ *Find each sum. See Examples 5 and 6.*

21. $\sum_{i=1}^{6} 3^i$

22. $\sum_{j=1}^{4} (-2)^j$

23. $\sum_{k=3}^{7} \left(\frac{1}{2}\right)^{k-2}$

24. $\sum_{i=3}^{12} (2)^{i-5}$

25. $\sum_{j=1}^{6} \left(\frac{1}{3}\right)^j$

26. $\sum_{k=1}^{5} \left(\frac{1}{4}\right)^k$

B

27. Graph the geometric progression defined by $s_n = 2^n$ for $1 \leq n \leq 4$. Use the horizontal axis for n and the vertical axis for s_n.

28. Graph the geometric progression defined by $s_n = 2^{n-4}$ for $1 \leq n \leq 7$. Use the horizontal axis for n and the vertical axis for s_n.

29. A culture of bacteria doubles every hour. If there were 10 bacteria in the culture originally, how many are there after 1 hour? After 2 hours? After 3 hours? After 4 hours? After n hours?

30. A certain radioactive substance has a half-life of 2400 years (50% of the original material is present at the end of 2400 years). If 100 grams were produced today, how many grams would be present in 2400 years? In 4800 years? In 7200 years? In 9600 years? In 2400n years?

13.4

INFINITE SERIES

Infinite geometric series

Consider the infinite geometric series

$$\frac{1}{2} + \frac{1}{4} + \frac{1}{8} + \frac{1}{16} + \cdots$$

and the partial sums of terms of the series,

$$S_1 = \frac{1}{2}$$

$$S_2 = \frac{1}{2} + \frac{1}{4} = \frac{3}{4}$$

$$S_3 = \frac{1}{2} + \frac{1}{4} + \frac{1}{8} = \frac{7}{8}$$

$$S_4 = \frac{1}{2} + \frac{1}{4} + \frac{1}{8} + \frac{1}{16} = \frac{15}{16}$$

$$\vdots$$

Note that the nth term of the sequence of partial sums

$$\frac{1}{2}, \frac{3}{4}, \frac{7}{8}, \frac{15}{16}, \ldots, S_n, \ldots$$

appears to be "approaching" 1. That is, as the number n becomes very large, S_n is very close to 1. In fact, we can make the difference between S_n and 1 as small as we like by using a sufficiently large value for n.

Recall from Section 13.3 that the sum of n terms of a geometric progression is given by

$$S_n = \frac{a - ar^n}{1 - r} \qquad (r \neq 1). \tag{1}$$

If $|r| < 1$, that is, if $-1 < r < 1$, then r^n becomes smaller and smaller for increasingly large n. For example, if $r = 1/2$, then

$$r^2 = \left(\frac{1}{2}\right)^2 = \frac{1}{4}, \quad r^3 = \left(\frac{1}{2}\right)^3 = \frac{1}{8}, \quad r^4 = \left(\frac{1}{2}\right)^4 = \frac{1}{16},$$

and so on, and we can make $(1/2)^n$ as small as we please by taking n sufficiently large. Writing (1) in the form

$$S_n = \frac{a}{1 - r}(1 - r^n), \tag{2}$$

we see that the value of the factor $(1 - r^n)$ can be made as close as we please to 1, providing $|r| < 1$ and n is taken large enough. Since this asserts that the sum (2) can be made to approximate

$$\frac{a}{1 - r}$$

as close as we please, we define the sum of an infinite geometric series with $|r| < 1$ as follows.

$$\boldsymbol{S_\infty = \frac{a}{1 - r}} \tag{3}$$

Limit of a sequence

A sequence that has an nth term (and all terms after the nth) that can be made to approximate a fixed number L as closely as desired by simply taking n large enough is said to *approach the limit L* as n increases without bound. We can indicate this in terms of symbols,

$$\lim_{n\to\infty} S_n = L,$$

where S_n is the nth term of the sequence of sums $S_1, S_2, S_3, \ldots, S_n, \ldots$. Thus, (3) might be written

$$S_\infty = \lim_{n\to\infty} S_n = \frac{a}{1-r}.$$

If $|r| \geqslant 1$, then r^n in (2) does not approach 0 and $\lim_{n\to\infty} S_n$ does not exist.

EXAMPLE 1 Find the sum of each series if the sum exists.

a. $3 + 2 + \frac{4}{3} + \cdots$

b. $\frac{1}{81} - \frac{1}{54} + \frac{1}{36} + \cdots$

Solutions a. $r = \frac{2}{3}$; the series has a sum, since $|r| < 1$.

$$S_\infty = \frac{a}{1-r} = \frac{3}{1-\frac{2}{3}} = 9$$

b. $r = -\frac{1}{54} \div \frac{1}{81} = -\frac{3}{2}$; the series does not have a sum, since $|r| > 1$. ■

Repeating decimals

An interesting application of this sum arises in connection with repeating decimals—that is, decimal numerals that, after a finite number of decimal places, have endlessly repeating groups of digits. For example,

$$0.2121\overline{21}, \qquad 0.333\overline{3} \quad \text{and} \quad 0.138512512\overline{512}$$

are repeating decimals, where in each case the bar indicates the repeating digits. Consider the problem of expressing such a decimal numeral as a fraction. We illustrate the process involved with the first repeating decimal above:

$$0.2121\overline{21}. \tag{4}$$

This decimal can be written either as

$$0.21 + 0.0021 + 0.000021 + \cdots \tag{5}$$

or

$$\frac{21}{100} + \frac{21}{10,000} + \frac{21}{1,000,000} + \cdots, \tag{6}$$

which are sums with terms that form a geometric progression with a ratio 0.01 (or $1/100$). Since the ratio is less than 1 in absolute value, we can use (3) to find the sum of an infinite number of terms of (5) or (6). Thus, using (6) we have

$$S_\infty = \frac{a}{1 - r} = \frac{\frac{21}{100}}{1 - \frac{1}{100}} = \frac{\frac{21}{100}}{\frac{99}{100}} = \frac{21}{99} = \frac{7}{33},$$

and the given decimal numeral, $0.2121\overline{21}$, is equivalent to $7/33$.

If we use the decimal form (5) above, we would use $S_\infty = \dfrac{a}{1 - r}$ with $a = 0.21$ and $r = 0.01$.

In the following example three digits are repeated.

EXAMPLE 2 Find a fraction equivalent to $2.045045\overline{045}$.

Solution Rewrite as a series.

$$2 + \frac{45}{1000} + \frac{45}{1,000,000} + \cdots$$

For series beginning with $\dfrac{45}{1000}$, $r = \dfrac{1}{1000}$. Use $S_\infty = \dfrac{a}{1 - r}$.

$$S_\infty = \frac{\frac{45}{1000}}{1 - \frac{1}{1000}} = \frac{\frac{45}{1000}}{\frac{999}{1000}} = \frac{45}{999} = \frac{5}{111}$$

Hence,

$$2.045045\overline{045} = 2\frac{5}{111} = \frac{227}{111}. \quad ■$$

The decimal part of the series in the example above can be written in decimal notation as $0.045 + 0.000045 + \cdots$, with $a = 0.045$ and $r = 0.001$.

The number *e*

The number e that we introduced in Chapter 11 is the limit of the sequence

$$\left(1 + \frac{1}{1}\right)^1, \left(1 + \frac{1}{2}\right)^2, \left(1 + \frac{1}{3}\right)^3, \ldots,$$

and the general term is given by

$$\left(1 + \frac{1}{t}\right)^t,$$

where t is a natural number. We can use the $\boxed{y^x}$ key on a calculator to obtain a decimal approximation for the expression for different values of t. The first several terms of this sequence, rounded to 4 decimal places, are

$$2.0000,\ 2.2500,\ 2.3704,\ 2.4414,\ 2.4883, \cdots,$$

and the 1000th term is

$$\left(1 + \frac{1}{1000}\right)^{1000} = 2.7169.$$

The question arises whether we would approach some number as $t \to \infty$. It is shown in courses in advanced mathematics that

$$\lim_{t \to \infty} \left(1 + \frac{1}{t}\right)^t = e,$$

where, as we have noted, e is an irrational number approximately equal to 2.7182818.

The expression $[1 + (1/t)]^t$ is a special case of the right-hand member of the formula

$$A = P\left(1 + \frac{r}{t}\right)^{tn} \tag{7}$$

for compound interest that we used in Exercise 12.6. This formula is sometimes written in the form

$$A = P(1 + r)^n. \tag{8}$$

Here, r is the rate and n is the number of compounding periods. It can be shown that if $t \to \infty$ in (7), which implies continuous compounding, we obtain

$$A = Pe^{rn}, \tag{9}$$

the formula introduced on page 458.

EXAMPLE 3 Five thousand dollars is invested for 1 year at 10% interest. What is the total value of the investment at the end of the year if the interest is compounded quarterly? If the interest is compounded continuously?

Solution If the interest is compounded quarterly, we have, from Equation (7),

$$\begin{aligned} A &= 5000\left(1 + \frac{0.10}{4}\right)^{4(1)} \\ &= 5000(1.025)^4 \\ &= 5519.06. \end{aligned}$$

The total value is \$5519.06. If the interest is compounded continuously, we have, from Equation (9),

$$A = 5000e^{0.10(1)} = 5525.85.$$

The total value is \$5525.85. ■

The notion of compounding interest continuously does not provide a practical application of Equation (9). However, different forms of this equation do have wide application in a variety of areas. Some of these applications were included in Exercise 12.6.

EXERCISE 13.4

A

■ *Find the sum of each infinite geometric series. If the series has no sum, so state. See Example* 1.

1. $12 + 6 + 3 + \cdots$
2. $2 + 1 + \dfrac{1}{2} + \cdots$
3. $\dfrac{1}{36} + \dfrac{1}{30} + \dfrac{1}{25} + \cdots$

4. $1 + \frac{2}{3} + \frac{4}{9} + \cdots$

5. $\frac{3}{4} - \frac{1}{2} + \frac{1}{3} - \cdots$

6. $\frac{1}{16} - \frac{1}{8} + \frac{1}{4} - \cdots$

7. $\frac{1}{49} + \frac{1}{56} + \frac{1}{64} + \cdots$

8. $2 - \frac{3}{2} + \frac{9}{8} - \cdots$

9. $\sum_{i=1}^{\infty} \left(\frac{2}{3}\right)^i$

10. $\sum_{i=1}^{\infty} \left(-\frac{1}{4}\right)^i$

■ *Find a fraction equivalent to each of the given decimal numerals. See Example 2.*

11. $0.3333\overline{3}$
12. $0.6666\overline{6}$
13. $0.3131\overline{31}$
14. $0.4545\overline{45}$
15. $2.410\overline{410}$
16. $3.027\overline{027}$
17. $0.12888\overline{8}$
18. $0.8333\overline{3}$

■ *Use the formulas for compound interest given on page* 491. *See Example* 3.

19. One hundred dollars is invested for 10 years at 8% interest. What is the total value of the investment if the interest is compounded quarterly? If the interest is compounded continuously?

20. Which investment would produce the greatest annual income: five thousand dollars at 12% compounded semiannually or five thousand dollars at 11% compounded continuously?

B

21. A force is applied to a particle moving in a straight line in such a fashion that each second it moves only one-half of the distance it moved the preceding second. If the particle moves 10 centimeters the first second, approximately how far will it move before coming to rest?

22. The arc length through which the bob on a pendulum moves is nine-tenths of its preceding arc length. Approximately how far will the bob move before coming to rest if the first arc length is 12 inches?

23. A ball returns two-thirds of its preceding height on each bounce. If the ball is dropped from a height of 6 feet, approximately what is the total distance the ball travels before coming to rest?

24. If a ball is dropped from a height of 10 feet and returns three-fifths of its preceding height on each bounce, approximately what is the total distance the ball travels before coming to rest?

25. If P dollars is invested at an interest rate r *compounded annually*, show that the amount A present after two years is given by $A = P(1 + r)^2$ and that the amount present after three years is $A = P(1 + r)^3$. Make a conjecture about the amount that is present after n years.

26. If P dollars is invested at an interest rate r, show that the amount A present in one year when the interest is compounded semiannually for n years is given by $A = P[1 + (r/2)]^{2n}$ and when compounded quarterly for n years is given by $A = P[1 + (r/4)]^{4n}$. Make a conjecture about the amount that is present if the interest is compounded t times yearly.

13.5

THE BINOMIAL EXPANSION

Factorial notation

Sometimes it is necessary to write the product of consecutive positive integers. To do this in some cases, we use a special symbol, $n!$ (read "n factorial" or "factorial n"), which is defined as follows.

$$n! = n(n-1)(n-2) \cdot \cdots \cdot 1$$

For example,

$$5! = 5 \cdot 4 \cdot 3 \cdot 2 \cdot 1 \quad \text{and} \quad 8! = 8 \cdot 7 \cdot 6 \cdot 5 \cdot 4 \cdot 3 \cdot 2 \cdot 1.$$

The factorial symbol applies only to the variable or numeral it follows.

EXAMPLE 1 Write each expression in expanded form.

a. $5n!$ for $n = 4$

b. $(2n-1)!$ for $n = 4$

Solutions

a. $5n! = 5 \cdot (4 \cdot 3 \cdot 2 \cdot 1)$

b. $(2n-1)! = [2(4) - 1]!$
$= 7! = 7 \cdot 6 \cdot 5 \cdot 4 \cdot 3 \cdot 2 \cdot 1$ ■

Since

$$n! = n(n-1)(n-2)(n-3) \cdot \cdots \cdot 5 \cdot 4 \cdot 3 \cdot 2 \cdot 1$$

and

$$(n-1)! = (n-1)(n-2)(n-3) \cdot \cdots \cdot 5 \cdot 4 \cdot 3 \cdot 2 \cdot 1,$$

we can, for $n > 1$, write the following recursive relationship.

$$n! = n(n-1)! \qquad (1)$$

EXAMPLE 2 Write each expression in expanded form and simplify.

a. $\frac{7!}{4!}$ b. $\frac{4!6!}{8!}$

Solutions a. $\frac{7!}{4!} = \frac{7 \cdot 6 \cdot 5 \cdot 4!}{4!} = 210$ b. $\frac{4!6!}{8!} = \frac{4 \cdot 3 \cdot 2 \cdot 1 \cdot 6!}{8 \cdot 7 \cdot 6!} = \frac{3}{7}$ ■

EXAMPLE 3 Write each product in factorial notation.

a. $1 \cdot 2 \cdot 3 \cdot 4 \cdot 5 \cdot 6$ b. $11 \cdot 12 \cdot 13 \cdot 14$

c. 150 d. $149 \cdot 150$

Solutions a. $1 \cdot 2 \cdot 3 \cdot 4 \cdot 5 \cdot 6 = 6!$ b. $11 \cdot 12 \cdot 13 \cdot 14 = \frac{14!}{10!}$

c. $150 = \frac{150!}{149!}$ d. $149 \cdot 150 = \frac{150!}{148!}$ ■

EXAMPLE 4 Write $(2n + 1)!$ in factored form and show the first three factors and the last three factors.

Solution $(2n + 1)! = (2n + 1)(2n)(2n - 1) \cdot \cdots \cdot 3 \cdot 2 \cdot 1$ ■

If $n = 1$ in Equation (1) above, we have

$$1! = 1 \cdot (1 - 1)!$$
$$1! = 1 \cdot 0!.$$

Therefore, for consistency, we shall make the following definition.

$$\mathbf{0! = 1}$$

Note that both 1! and 0! are equal to 1.

Binomial expansion

The series obtained by expanding a binomial of the form

$$(a + b)^n$$

is particularly useful in certain branches of mathematics. Starting with familiar examples where n takes the value 1, 2, 3, 4, and 5, we can show by direct multiplication that

$$
\begin{aligned}
(a + b)^1 &= a + b, \\
(a + b)^2 &= a^2 + 2ab + b^2, \\
(a + b)^3 &= a^3 + 3a^2b + 3ab^2 + b^3, \\
(a + b)^4 &= a^4 + 4a^3b + 6a^2b^2 + 4ab^3 + b^4, \\
(a + b)^5 &= a^5 + 5a^4b + 10a^3b^2 + 10a^2b^3 + 5ab^4 + b^5.
\end{aligned}
$$

We observe that in each case:

1. The first term may be considered to be a^nb^0. The exponent on *a decreases* by 1 and the exponent on b *increases* by 1 in each of the following terms. The last term may be considered to be a^0b^n. For example,

$$
\frac{5}{1!} \quad \frac{5 \cdot 4}{2!} \quad \frac{5 \cdot 4 \cdot 3}{3!} \quad \frac{5 \cdot 4 \cdot 3 \cdot 2}{4!}
$$

$$
(a + b)^5 = a^5 + 5a^4b + 10a^3b^2 + 10a^2b^3 + 5ab^4 + b^5.
$$

Exponents of a decrease by 1
Exponents of b increase by 1

2. The variable factors of the second term are $a^{n-1}b^1$ and the coefficient is n, which can be written in the form

$$\frac{n}{1!}.$$

3. The variable factors of the third term are $a^{n-2}b^2$, and the coefficient can be written in the form

$$\frac{n(n-1)}{2!}.$$

4. The variable factors of the fourth term are $a^{n-3}b^3$, and the coefficient can be written in the form

$$\frac{n(n-1)(n-2)}{3!}.$$

The above results can be generalized to obtain the following **binomial expansion.**

$$(a + b)^n = a^n + \frac{n}{1!}a^{n-1}b + \frac{n(n-1)}{2!}a^{n-2}b^2 + \frac{n(n-1)(n-2)}{3!}a^{n-3}b^3$$
$$+ \cdots + \frac{n(n-1)(n-2)\cdot \cdots \cdot (n-r+2)}{(r-1)!}a^{n-r+1}b^{r-1}$$
$$+ \cdots + b^n, \qquad (2)$$

where r is the number of the term.

EXAMPLE 5 Expand.

a. $(x - 2)^4$ b. $(a - 3b)^4$

Solutions a. In this case, $a = x$ and $b = -2$ in the binomial expansion:

$$(x - 2)^4 = x^4 + \frac{4}{1!}x^3(-2)^1 + \frac{4 \cdot 3}{2!}x^2(-2)^2 + \frac{4 \cdot 3 \cdot 2}{3!}x(-2)^3 + \frac{4 \cdot 3 \cdot 2 \cdot 1}{4!}(-2)^4$$
$$= x^4 - 8x^3 + 24x^2 - 32x + 16$$

b. In this case $a = a$ and $b = -3b$ in the binomial expansion:

$$(a - 3b)^4 = a^4 + \frac{4}{1!}a^3(-3b) + \frac{4 \cdot 3}{2!}a^2(-3b)^2 + \frac{4 \cdot 3 \cdot 2}{3!}a(-3b)^3 + \frac{4 \cdot 3 \cdot 2 \cdot 1}{4!}(-3b)^4$$
$$= a^4 - 12a^3b + 54a^2b^2 - 108ab^3 + 81b^4. \quad ■$$

rth term of an expansion

Note in (2) that the rth term in a binomial expansion is given by

$$\frac{n(n-1)(n-2)\cdot \cdots \cdot (n-r+2)}{(r-1)!}a^{n-r+1}b^{r-1}. \qquad (3)$$

EXAMPLE 6 a. Find the fifth term in the expansion $(x - 2)^{10}$.

b. Find the seventh term in the expansion of $(x - 2)^{10}$.

Solutions a. We can use (3) above, where $n = 10$, $r = 5$, $r - 1 = 4$, and $n - r + 2 = 7$.

$$\frac{10 \cdot 9 \cdot 8 \cdot 7}{4 \cdot 3 \cdot 2 \cdot 1}x^6(-2)^4 = 3360x^6$$

b. We can use (3) above, where $n = 10$, $r = 7$, $r - 1 = 6$, and $n - r + 2 = 5$.

$$\frac{10 \cdot 9 \cdot 8 \cdot 7 \cdot 6 \cdot 5}{6 \cdot 5 \cdot 4 \cdot 3 \cdot 2 \cdot 1} x^4(-2)^6 = 13{,}440x^4$$ ■

When using Equation (3), it is helpful to note that:

1. The exponent on b, $(r - 1)$, is one less than r, the number of the term.
2. The sum of the exponents $(n - r + 1) + (r - 1)$ equals n, the exponent of the power.
3. The factor $(r - 1)$ in the denominator of the coefficient equals the exponent on b.
4. The number of factors in the numerator of the coefficient is the same as the number of factors in the denominator.

EXERCISE 13.5

A

■ *Write each expression in expanded form. See Example* 1.

1. $(2n)!$ for $n = 4$
2. $(3n)!$ for $n = 4$
3. $2n!$ for $n = 4$
4. $3n!$ for $n = 4$
5. $n(n - 1)!$ for $n = 6$
6. $2n(2n - 1)!$ for $n = 2$

■ *Write in expanded form and simplify. See Example* 2.

7. $5!$
8. $7!$
9. $\dfrac{9!}{7!}$
10. $\dfrac{12!}{11!}$
11. $\dfrac{5!7!}{8!}$
12. $\dfrac{12!8!}{16!}$
13. $\dfrac{8!}{2!(8 - 2)!}$
14. $\dfrac{10!}{4!(10 - 4)!}$

■ *Write each product in factorial notation. See Example* 3.

15. $1 \cdot 2 \cdot 3$
16. $1 \cdot 2 \cdot 3 \cdot 4 \cdot 5$
17. $3 \cdot 4 \cdot 5 \cdot 6$
18. 7
19. $8 \cdot 7 \cdot 6$
20. $28 \cdot 27 \cdot 26 \cdot 25 \cdot 24$

■ *Write each expression in factored form and show the first three factors and the last three factors. See Example* 4.

21. $n!$
22. $(n + 4)!$
23. $(3n)!$
24. $3n!$
25. $(n - 2)!$
26. $(3n - 2)!$

■ *Expand. See Example 5.*

27. $(x + 3)^5$
28. $(2x + y)^4$
29. $(x - 3)^4$
30. $(2x - 1)^5$
31. $\left(2x - \frac{y}{2}\right)^3$
32. $\left(\frac{x}{3} + 3\right)^5$
33. $\left(\frac{x}{2} + 2\right)^6$
34. $\left(\frac{2}{3} - a^2\right)^4$

■ *Write the first four terms in each expansion. Do not simplify the terms. See Example 5.*

35. $(x + y)^{20}$
36. $(x - y)^{15}$
37. $(a - 2b)^{12}$
38. $(2a - b)^{12}$
39. $(x - \sqrt{2})^{10}$
40. $\left(\frac{x}{2} + 2\right)^8$

■ *Find each specified term. See Example 6.*

41. $(a - b)^{15}$, sixth term
42. $(x + 2)^{12}$, fifth term
43. $(x - 2y)^{10}$, fifth term
44. $(a^3 - b)^9$, seventh term
45. $(x^2 - y^2)^7$, third term
46. $\left(x - \frac{1}{2}\right)^8$, fourth term
47. $\left(\frac{a}{2} - 2b\right)^9$, fifth term
48. $\left(\frac{x}{2} + 4\right)^{10}$, eighth term

B

49. Given that the binomial formula holds for $(1 + x)^n$, where n is a negative integer:
   a. Write the first four terms of $(1 + x)^{-1}$.
   b. Find the first four terms of the quotient $\frac{1}{1 + x}$ by dividing $(1 + x)$ into 1.
   c. Compare the results of parts a and b.
50. Given that the binomial formula holds as an infinite "sum" for $(1 + x)^n$, where n is a noninteger rational number and $|x| < 1$, find to two decimal places:
   a. $\sqrt{1.02}$
   b. $\sqrt{0.99}$

13.6

PERMUTATIONS

An arrangement of the elements of a set in specified order (where no element is repeated) is called a **permutation** of the elements. For example, the possible permutations of the elements of $\{a, b, c\}$ are

$$abc, \quad bac, \quad cab, \quad acb, \quad bca, \quad \text{and} \quad cba.$$

To count the permutations of a given set containing n elements, think of placing them in order as follows:

1. Select one member to be first. There are n possible selections.
2. Having selected a first element, select a second. Since one element has been placed in the first position, there remain $(n - 1)$ elements to be placed in the second position. The total number of possible first and second selections is then given by the product $n(n - 1)$.
3. Select a third element. Since one has been used for the first position and one for the second, there are $(n - 2)$ possibilities for the third. The total number of possible first, second, and third selections is $n(n - 1)(n - 2)$.
4. Continue this process until the last element is put in the last position. There are then $n(n - 1)(n - 2) \cdot \cdots \cdot 3 \cdot 2 \cdot 1$ possible permutations of n elements.

Formally, we represent the number of permutations of n things by means of the symbol $P(n, n)$(read "the number of permutations of n things taken n at a time"). From the foregoing discussion, we have the following property.

$$\boldsymbol{P(n, n) = n!} \tag{1}$$

For example, suppose we wish to determine how many different signals can be formed with 5 different signal flags on a pole by altering the location of the flags. A helpful way to picture the situation is to first draw 5 dashes to represent the positions of the flags:

$$\underline{\quad}\ \underline{\quad}\ \underline{\quad}\ \underline{\quad}\ \underline{\quad}.$$

Next, since we can place any of the 5 flags in the first position, we write 5 on the first dash:

$$\underline{\ 5\ }\ \underline{\quad}\ \underline{\quad}\ \underline{\quad}\ \underline{\quad}.$$

Having selected 1 flag for the first position, we have a choice from among 4 flags for the second position.

$$\underline{\ 5\ }\ \underline{\ 4\ }\ \underline{\quad}\ \underline{\quad}\ \underline{\quad}.$$

Similarly, 3 flags remain for the third position, 2 for the fourth, and 1 for the fifth. Thus, we fill the dashes:

$$\underline{\ 5\ }\ \underline{\ 4\ }\ \underline{\ 3\ }\ \underline{\ 2\ }\ \underline{\ 1\ }.$$

Then, the total number of different signals is

$$5 \cdot 4 \cdot 3 \cdot 2 \cdot 1 = 120.$$

EXAMPLE 1 Compute each of the following.

a. $P(3, 3)$ b. $P(5, 5)$

Solutions

a. $P(3, 3) = 3! = 3 \cdot 2 \cdot 1 = 6$

b. $P(5, 5) = 5! = 5 \cdot 4 \cdot 3 \cdot 2 \cdot 1 = 120$ ■

EXAMPLE 2 In how many ways can a jury of 12 persons be seated in a jury box containing 12 chairs?

Solution We want to find $P(12, 12)$.

$$P(12, 12) = 12! = 12 \cdot 11 \cdot 10 \cdot 9 \cdot 8 \cdot 7 \cdot 6 \cdot 5 \cdot 4 \cdot 3 \cdot 2 \cdot 1 = 479{,}001{,}600$$

■

Counting principle

Counting permutations of distinct things in this way makes use of a more general counting principle.

> *If one thing can be done in m ways, another thing in n ways, another in p ways, and so on, then the number of ways all the things can be done is*
>
> $$m \cdot n \cdot p \cdot \cdots .$$

EXAMPLE 3 The number of different license plates containing four-digit numerals using the digits 1, 2, 3, 4, 5, 6, 7, 8, and 9, is given by

$$9 \cdot 9 \cdot 9 \cdot 9 = 6561,$$

since nothing prevents the use of the same digit in two, three, or four positions. ■

Frequently, it is important to be able to compute the number of possible permutations of a set contained in a given set. Thus, we might wish to know the number of permutations of n different things taken r ($r \leq n$) at a time. The symbol representing this number is $P(n, r)$, and, by reasoning in exactly the same way as we did to compute $P(n, n)$, we find that:

$$P(n, r) = n(n-1)(n-2) \cdot \cdots \cdot (n-r+1). \tag{2}$$

Now, because

$$n! = n(n-1)(n-2) \cdot \cdots \cdot (n-r+1)(n-r)!,$$

Equation (2) can be written as follows.

$$P(n, r) = \frac{n!}{(n-r)!}. \tag{3}$$

EXAMPLE 4 Find the number of permutations of 5 things taken 3 at a time.

Solution Using Equation (3),

$$P(5, 3) = \frac{5!}{2!} = \frac{5 \cdot 4 \cdot 3 \cdot 2 \cdot 1}{2 \cdot 1}$$
$$= 5 \cdot 4 \cdot 3 = 60.$$

If we wish to use dashes to help visualize the situation in the above example, we could first draw 3 dashes,

$$\_\_ \quad \_\_ \quad \_\_$$

and then fill each dash in order,

$$\underline{5} \quad \underline{4} \quad \underline{3}.$$

The product of these numbers then gives us the total number of possibilities for permuting 5 things taken 3 at a time. ■

EXAMPLE 5 How many different three-digit numerals for whole numbers can be formed from the digits 0, 1, 2, 3, 4, 5, 6, if:

a. No restrictions are placed on the repetition of digits?
b. No digit can be used more than once?
c. The last digit is 6 and no digit can be used more than once?

Solutions a. Since the numeral must contain 3 digits, the first digit cannot be 0. Drawing 3 dashes, we see that the first place can be filled with any 1 of the digits from 1 to 6, a total of 6 digits:

$$\underline{6}\ \underline{\ \ }\ \underline{\ \ }.$$

The remaining places can be filled with any of the 7 digits; so we have

$$\underline{6}\ \underline{7}\ \underline{7}.$$

Then the answer to the original question is

$$6 \cdot 7 \cdot 7 = 294.$$

b. If no digit can be used more than once, then we can select any one of the digits from 1 to 6 for the first digit, but, having fixed the first digit, we have 6 possibilities for the second (since 0 can now be used), and 5 for the third. Our diagram appears as follows:

$$\underline{6}\ \underline{6}\ \underline{5}.$$

The answer to the original question then is

$$6 \cdot 6 \cdot 5 = 180.$$

c. If the last digit is specified, in this case 6, our diagram begins like this:

$$\underline{\ \ }\ \underline{\ \ }\ \underline{1}.$$

Next, having fixed one digit, and being unable to use 0 as the first digit, we have

$$\underline{5}\ \underline{5}\ \underline{1}.$$

The answer to the original question is then

$$5 \cdot 5 \cdot 1 = 25. \quad ■$$

Distinguishable permutations

Sometimes the elements with which we wish to form permutations are not all different. Thus, to find the number of distinguishable permutations of the letters in the word *toast*, we must take into consideration the fact that we cannot distinguish between the 2 *t*'s in any permutation. The number of permutations of the 5 letters in the word is clearly 5!, but since in any one of these the 2 *t*'s can be permuted in 2! ways without producing a different result, the number of *distinguishable* permutations P is given by

$$2!P = 5!.$$

Hence,

$$P = \frac{5!}{2!},$$

where the number of permutations of five letters, 5!, is divided by the number of permutations, 2!, of the repeat letter t. Thus,

$$P = \frac{5!}{2!} = \frac{5 \cdot 4 \cdot 3 \cdot 2 \cdot 1}{2 \cdot 1} = 60.$$

EXAMPLE 6 Find the number of distinguishable permutations of the letters in each word.

a. pineapple

b. seventeen

Solution

a. There is a total of 9! permutations. However, there are 3 *p*'s and 2 *e*'s which can be permuted 3! and 2! ways, respectively, in each permutation of the 9 letters without altering the result. Accordingly, the number of distinguishable permutations P is given by

$$P = \frac{9!}{3!2!} = \frac{9 \cdot 8 \cdot 7 \cdot 6 \cdot 5 \cdot 4 \cdot 3 \cdot 2 \cdot 1}{3 \cdot 2 \cdot 1 \cdot 2 \cdot 1} = 30{,}240.$$

b. There are 9 letters in the word, but there are 2 *n*'s and 4 *e*'s. Hence, the number of distinguishable permutations is given by

$$P = \frac{9!}{2!4!} = \frac{9 \cdot 8 \cdot 7 \cdot 6 \cdot 5 \cdot 4 \cdot 3 \cdot 2 \cdot 1}{2 \cdot 4 \cdot 3 \cdot 2} = 7560.$$ ■

EXERCISE 13.6

A

■ *Solve. See Examples* 1 *and* 2.

1. In how many different ways can 4 books be arranged between bookends?
2. In how many ways can 7 students be seated at 7 desks?
3. In how many ways can 9 players be assigned to the 9 positions on a baseball team?
4. In how many ways can 10 floats be arranged for a parade?

■ *Solve. See Example* 3.

5. In how many ways can six flags be aligned when taken three at a time?
6. In how many ways can seven flags be aligned when taken four at a time?
7. In how many ways can ten books be arranged on a shelf when taken five at a time?
8. In how many ways can twelve books be arranged on a shelf when taken six at a time?

■ *How many different four-digit numerals for whole numbers can be formed from the digits* 0, 1, 2, 3, 4, 5, 6 *in the following exercises? See Example* 4.

9. No restrictions are placed on the repetition of digits?
10. No restrictions are placed on the repetition of digits and the last digit is 3?
11. No digit can be used more than once in each number?
12. No digit may be used more than once in each number, and the last digit is 5?

■ *Find the number of three-digit numerals, using the digits* 0, 1, 2, 3, 4, 5, 6, 7, 8 *for the problem cited. See Example* 4.

13. Problem 9
14. Problem 10
15. Problem 11
16. Problem 12

17. How many three-letter permutations can be formed from the twenty-six letter English alphabet if no restrictions are placed on the repetition of letters?
18. How many five-letter permutations can be formed from the twenty-six letter English alphabet if no restrictions are placed on the repetition of letters?
19. How many license numbers can be formed if each license contains 2 letters of the alphabet followed by 3 digits?
20. How many license numbers can be formed if each license contains 1 digit followed by 1 letter of the alphabet followed by 2 digits?

■ *Find the number of distinguishable permutations of the letters of the given word. See Example* 5.

21. Hurry
22. Sonnet
23. Stress
24. Between
25. Committee
26. Consists
27. Selected
28. Permutation

29. How many distinguishable arrangements can be made by aligning five identically shaped flags, two blue, two red, and one white?

30. How many distinguishable arrangements can be made by aligning seven identically shaped flags, two blue, two red, and three white?

13.7

COMBINATIONS

If the order in which the elements of a set are listed is unimportant, then the set is called a **combination**. Thus, while *abc*, *acb*, *bac*, *bca*, *cab*, and *cba* are six permutations of the elements of $\{a, b, c\}$, the set constitutes a *single* combination, *abc*. Selecting the elements two at a time would yield the following:

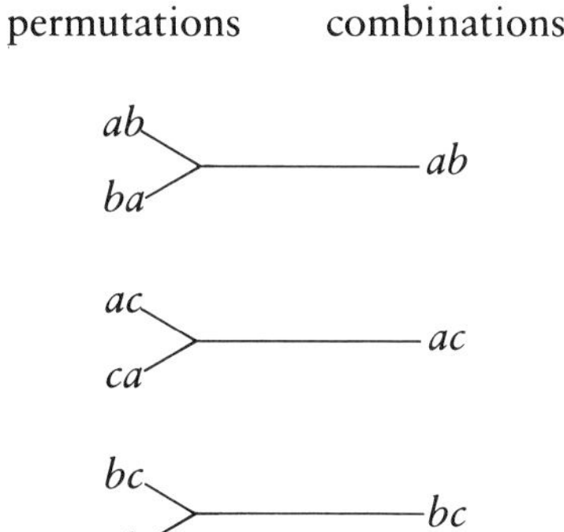

The symbol $\binom{n}{r}$, or sometimes $C(n, r)$, is used to designate the number of combinations of r elements that can be formed from a set of n elements. Thus, in the example above

$$P(3, 2) = 6 \quad \text{and} \quad \binom{3}{2} = 3.$$

We can use permutations to count combinations. Since each r-element combination can be permuted in $P(r, r)$, or $r!$, ways, the total number of permutations of n things taken r at a time can be computed by multiplying $P(r, r)$ by $\binom{n}{r}$. That is,

$$P(r, r)\binom{n}{r} = P(n, r).$$

Multiplying both members by $1/P(r, r)$, we have the following:

$$\binom{n}{r} = \frac{P(n, r)}{P(r, r)}. \tag{1}$$

Since $P(n, r) = n(n - 1)(n - 2) \cdot \cdots \cdot (n - r + 1)$ and $P(r, r) = r!$, it follows that:

$$\binom{n}{r} = \frac{n(n - 1)(n - 2) \cdot \cdots \cdot (n - r + 1)}{r!}. \tag{2}$$

For example, the number of different committees of 3 people that could be appointed in a club having 8 members is the number of combinations of 8 things taken 3 at a time, rather than the number of permutations of 8 things taken 3 at a time, because the order in which the members of a committee are considered is of no consequence. Thus, from Equation (2) we would have

$$\binom{8}{3} = \frac{8 \cdot 7 \cdot 6}{3 \cdot 2 \cdot 1} = 56,$$

and 56 such committees could be appointed in the club. Notice that in computing $\binom{n}{r}$, it is easier to first write $r!$ in the denominator, because we can then simply write the same number of factors in the numerator, rather than actually determining $(n - r + 1)$ as the final factor of the numerator.

In Section 13.6 we observed that

$$P(n, r) = \frac{n!}{(n - r)!}$$

Substituting the right-hand member of this equation for $P(n, r)$ in Equation (1) above, we have

$$\binom{n}{r} = \frac{P(n, r)}{P(r, r)} = \frac{\dfrac{n!}{(n - r)!}}{r!},$$

from which we have the following formula.

$$\binom{n}{r} = \frac{n!}{r!(n - r)!} \tag{3}$$

EXAMPLE 1 Using Equation (3),

$$\binom{8}{3} = \frac{8!}{3!(8-3)!} = \frac{8!}{3!5!} = 56. \qquad \blacksquare$$

As expected, the result in the above example is the same result we obtained using Equation (2).

EXAMPLE 2 How many different amounts of money can be formed from a penny, a nickel, a dime, and a quarter?

Solution Using 1 coin, we can form $\binom{4}{1} = 4$ different amounts; using 2 coins, we obtain $\binom{4}{2} = \frac{4 \cdot 3}{1 \cdot 2} = 6$; using 3 coins, we have $\binom{4}{3} = \frac{4 \cdot 3 \cdot 2}{1 \cdot 2 \cdot 3} = 4$; using all the coins, we obtain $\binom{4}{4} = \frac{4 \cdot 3 \cdot 2 \cdot 1}{1 \cdot 2 \cdot 3 \cdot 4} = 1$; so the total number of different amounts of money is

$$\binom{4}{1} + \binom{4}{2} + \binom{4}{3} + \binom{4}{4} = 4 + 6 + 4 + 1 = 15. \qquad \blacksquare$$

EXERCISE 13.7

A

■ *Compute each combination. See Example* 1.

1. $\binom{7}{2}$
2. $\binom{6}{4}$
3. $\binom{12}{3}$
4. $\binom{10}{4}$
5. $\binom{12}{9}$
6. $\binom{10}{6}$

■ *Solve. See Examples* 1 *and* 2.

7. How many different committees of 4 people can be appointed in a club containing 15 members?

8. How many different committees of 5 people can be appointed from a group containing 9 people?
9. How many different amounts of money can be formed from a nickel, a dime, and a quarter?
10. How many different amounts of money can be formed from a nickel, a dime, a quarter, a penny, and a half-dollar?
11. In how many different ways can a set of 5 cards be selected from a deck of 52 cards?
12. In how many different ways can a set of 13 cards be selected from a deck of 52 cards?
13. How many straight lines are determined by 6 points, no 3 of which are collinear?
14. How many diagonals does a regular octagon (8 sides) have?
15. In how many ways can 3 marbles be drawn from an urn containing 9 marbles?
16. In how many ways can 2 face cards be drawn from a standard deck of 52 cards?
17. In how many ways can a bridge hand consisting of 6 hearts, 3 spades, and 4 diamonds be selected from a deck of 52 cards?
18. In how many ways can a hand consisting of 4 aces and 1 card that is not an ace be selected from a deck of 52 cards?
19. In how many ways can a committee of 2 men and 3 women be selected from a club with 10 men and 11 women members?
20. In how many ways can a committee containing 8 men and 8 women be selected from the club in Problem 13?

B

21. Use Equation (2) or (3) on page 506 to show that $\binom{n}{r} = \binom{r}{n-r}$.
22. Use the results of Problem 21 to compute $\binom{100}{98}$.
23. Given $\binom{n}{3} = \binom{n}{4}$, find n.
24. Given $\binom{n}{7} = \binom{n}{5}$, find n.
25. Show that the coefficients of the terms in the binomial expansion of $(a + b)^4$ are $\binom{4}{0}, \binom{4}{1}, \binom{4}{2}, \binom{4}{3}$, and $\binom{4}{4}$.
26. Show that the coefficients of the first five terms in the binomial expansion of $(a + b)^n$ are $\binom{n}{0}, \binom{n}{1}, \binom{n}{2}, \binom{n}{3}$, and $\binom{n}{4}$.

CHAPTER SUMMARY

[13.1] The elements in the range of a function whose domain is a set of successive positive integers form a **sequence**. The sequence is **finite** if it has a last member; otherwise it is **infinite**.

A **series** is the indicated sum of the terms in a sequence. We can use **sigma,** or **summation,** notation to represent a series.

[13.2] A sequence in which each term after the first is obtained by adding a constant to the preceding term is an **arithmetic progression.** The constant is called the **common difference** of the terms.

The nth term of an arithmetic progression is given by

$$s_n = a + (n - 1)d,$$

and the sum of n terms is given by

$$S_n = \frac{n}{2}(a + s_n) = \frac{n}{2}[2a + (n - 1)d],$$

where a is the first term, n is the number of terms, and d is the common difference.

[13.3] A sequence in which each term after the first is obtained by multiplying its predecessor by a constant is called a **geometric progression.** The constant is called the **common ratio.**

The nth term of a geometric progression is given by

$$s_n = ar^{n-1},$$

and the sum of n terms is given by

$$S_n = \frac{a - ar^n}{1 - r} = \frac{a - rs_n}{1 - r} \qquad (r \neq 1),$$

where a is the first term, n is the number of terms, and r is the common ratio.

[13.4] An infinite geometric series has a sum if the common ratio has an absolute value less than 1. This sum is given by

$$S_\infty = \lim_{n \to \infty} S_n = \frac{a}{1 - r}.$$

The number e is defined by

$$\lim_{t \to \infty} \left(1 + \frac{1}{t}\right)^t.$$

[13.5] Factorial notation is convenient to represent special kinds of products. For n a natural number,

$$\begin{aligned} n! &= n(n-1)(n-2)\cdot \cdots \cdot(3)(2)(1) \\ &= n(n-1)!. \end{aligned}$$

The binomial power $(a+b)^n$ can be expanded into a series containing $(n+1)$ terms for n a natural number:

$$(a+b)^n = a^n + \frac{n}{1!}a^{n-1}b + \frac{n(n-1)}{2!}a^{n-2}b^2 + \frac{n(n-1)(n-2)}{3!}a^{n-3}b^3 + \cdots + \frac{n(n-1)(n-2)\cdot \cdots \cdot(n-r+2)}{(r-1)!}a^{n-r+1}b^{r-1} + \cdots + b^n,$$

where r is the number of the term.

[13.6] An arrangement of the elements in a set (in which no element is repeated) is called a **permutation** of the elements. To count permutations, we can use the general counting principle:

> *If one thing can be done in m ways, another thing in n ways, another in p ways, and so on, then the number of ways all the things can be done is* $m \cdot n \cdot p \cdot \cdots$.

For n, r natural numbers ($r \leq n$),

$$P(n, n) = n!$$

and

$$\begin{aligned} P(n, r) &= n(n-1)(n-2)\cdot \cdots \cdot(n-r+1) \\ &= \frac{n!}{(n-r)!}. \end{aligned}$$

[13.7] Any r-element set contained in an n-element set is called a **combination.** For n, r natural numbers,

$$\binom{n}{r} = \frac{n(n-1)(n-2)\cdot \cdots \cdot(n-r+1)}{r!} = \frac{n!}{r!(n-r)!}.$$

REVIEW EXERCISES

A

[13.1]

1. Find the first four terms in a sequence with the general term $s_n = \dfrac{(-1)^{n-1}}{n}$.
2. Write $\sum_{k=2}^{5} k(k - 1)$ in expanded form.

[13.2]

3. Given that 5 and 9 are the first two terms of an arithmetic progression, find an expression for the general term.
4. a. Find the twenty-third term of the arithmetic progression $-82, -74, -66, \ldots$.

   b. Find the sum of the first twenty-three terms.
5. The first term of an arithmetic progression is 8 and the twenty-eighth term is 89. Find the twenty-first term.

[13.3]

6. Given that 5 and 9 are the first two terms of a geometric progression, find an expression for the general term.
7. a. Find the eighth term of the geometric progression $16/27, -8/9, 4/3, \ldots$.

   b. Find the sum of the first four terms.
8. Find $\sum_{j=1}^{5} \left(\dfrac{1}{3}\right)^j$.

[13.4]

9. Find $\sum_{i=1}^{\infty} \left(\dfrac{1}{3}\right)^i$.
10. Find a fraction equivalent to $0.44\overline{4}$.

[13.5]

11. Simplify $\dfrac{8!}{3!5!}$.
12. Write the first four terms of the binomial expansion of $(x - 2y)^{10}$.
13. Find the eighth term in the expansion of $(x - 2y)^{10}$.

[13.6]

14. How many different three-digit numerals for whole numbers can be formed from the digits 1, 2, 3, 4, 5 when:
    a. No restrictions are placed on the repetition of digits?
    b. No digit can be used more than once?
15. In how many ways can 6 men be assigned to a 4-man relay team?
16. How many distinguishable permutations are there in the letters of the word *fifteen*?

[13.7]

17. How many different doubles combinations could be formed from a tennis team containing 6 members?
18. How many different committees of 4 persons can be formed from a group of 9 persons?
19. In how many ways can 1 ace and 3 kings be selected from a deck of 52 cards?
20. In how many ways can 3 aces and 2 kings be selected from a deck of 52 cards?

B

21. Find the sum of all integer multiples of 3 between 17 and 97.
22. Find three numbers that form an arithmetic sequence such that their sum is 15 and their product is 80.
23. The bacteria in a culture doubles every 4 hours. If there were 2 bacteria in the culture originally, how many are there at the end of n hours?
24. A ball returns three-fourths of its preceding height on each bounce. If the ball is dropped from a height of 8 feet, approximately what is the total distance the ball travels before coming to rest?

■ *Simplify each expression.*

25. $\dfrac{2n!(n+3)}{(n+3)!}$

26. $\dfrac{(2n-1)!(n+1)!}{n!(2n-3)!}$

Cumulative Review Chapters 1-13

The numbers in brackets refer to the sections where such problems are first considered.

1. If $f(x) = x^2 - x + 2$, find $f(3) - f(-3)$. [11.1]
2. Solve the equation $x^2y - 4y = 6$ explicitly for y. Specify the domain of the relation defined by the equation, and state whether or not the relation is a function. [11.1]
3. Find the zeros of the function $f(x) = 14 + 5x - x^2$. [11.1]
4. If $f(x) = -\sqrt{x^2 + 1}$, find an equation defining its inverse. Is the inverse a function? [11.4]
5. Evaluate $\frac{100}{4Lv}(R^2 - r^2)$ for $L = 1$, $v = 5$, $R = 13$, and $r = 2$. [2.1]

■ *Simplify.*

6. $2x^2(xy^2)^3 - x^2y^5(-x^3y) + 3y(-x^2y^3)^3$ [2.3]
7. $\dfrac{(3 \times 10^{-3})^{-2}\,(2 \times 10^{-7})}{(6 \times 10^2)^{-2}}$ [6.3]
8. $\sqrt[4]{64x^3y^7z^{12}}$ [6.6]
9. $\sqrt[3]{250x^3y^2} + \sqrt[3]{64x} - x\sqrt[3]{54y^2}$ [6.7]
10. Find the least common multiple of $4x - 2$, $x - 2x^2$, and $4x^3 - x$. [5.4]

■ *Write as a single fraction in lowest terms.*

11. $\dfrac{x^2 - 8x + 16}{x^2 - 4} \div (x^2 - 16)$ [5.4]
12. $x - \dfrac{x}{x + 3} - \dfrac{x + 2}{x^2 - 9}$ [5.5]
13. Reduce $\dfrac{1 - 64x^3}{64x^3 - 16x^2}$. [5.1]

14. Use synthetic division to divide $\dfrac{x^5 - 2x^3 + x^2 - 3}{x + 2}$. [5.3]

■ *Factor completely.*

15. $6x^2 - x^3 - x^4$ [4.2]
16. $2x^3 - x^2 - 2x + 1$ [4.3]
17. $21x^4y - 56x^3y^2 + 35x^2y^3$ [4.1]

■ *Multiply.*

18. $(3a^2 - 2a - 1)(2a^2 + 3a - 2)$ [2.4]
19. $(\sqrt{2} + 2\sqrt{3})(3\sqrt{2} - 2\sqrt{3})$ [6.7]
20. Write $\sqrt[5]{\dfrac{2}{xy^3}}$ with positive fractional exponents. [6.5]
21. Write as a product or quotient involving only positive exponents:
$\left(\dfrac{a^{1/6}b^{-3/2}c}{a^{3/4}b^3c^{-5/2}}\right)^{-6}$ [6.4]

■ *Write in the form $a + bi$.*

22. $\dfrac{3 + i}{3 - 2i}$ [6.8]

23. $\dfrac{2 - \sqrt{-5}}{2\sqrt{-3}}$ [6.8]

24. Evaluate $\log_4 12.6$. [12.5]
25. Compute $10^{-3.625}$. [12.3, 12.4]
26. Compute $e^{-1.3}$. [12.3, 12.4]
27. Express as a single logarithm with coefficient 1:
$\frac{1}{2}\log_b x - (2\log_b y + 3 \log_b z)$. [12.2]
28. Find the first four terms of the sequence $S_n = \dfrac{(-1)^{n-1}2^n}{n^2}$. [13.1]
29. Write in expanded form: $\displaystyle\sum_{k=2}^{8} (-1)^k \frac{2k - 1}{k - 1}$. [13.1]
30. Find $\displaystyle\sum_{k=10}^{100} (3k - 1)$. [13.2]
31. Find $\displaystyle\sum_{i=4}^{8} (-3)^{i-4}$. [13.3]
32. Find the eighth term of $(2x - y^3)^{11}$. [13.5]
33. Find a fraction equivalent to $1.6\overline{25}$. [13.4]
34. Solve $2 = \dfrac{3x}{3y^2 + 1}$ explicitly for y. [8.1]

■ *Solve.*

35. $-2(x^2 - 7x + 8) = (x + 1)^2 - 3x^2$. [3.1]

36. $2x^2 + 3x + 3 = 0$ [6.8]

37. $(x - 3/4)^2 = 3/16$ [7.1]

38. $2/3x^2 - x + 1/4 = 0$ [7.2]

39. $\dfrac{8}{x-3} - \dfrac{5}{x-2} = \dfrac{19}{x^2 - 5x + 6}$ [5.8]

40. $x^2 - 2 - 3\sqrt{x^2 - 2} - 18 = 0$ [7.4]

41. $\sqrt{3x + 1} - \sqrt{x - 4} = 3$ [7.3]

42. $\log_x 8 = -3$ [12.1]

43. $\log_{64} x = 1/3$ [12.1]

44. $\ln x = -0.26$ [12.3, 12.4]

45. $16.2 = 3(10^{2.3x}) - 8.7$ [12.5]

46. $38.6 = 12.2e^{-1.6x} + 13.2$ [12.5]

47. $5^{x+2} = 7$ [12.5]

48. $\log_{10} x + \log_{10} (x - 3) = 1$ [12.2]

■ *Solve each inequality and graph the solution set.*

49. $\dfrac{2x - 3}{-3} < x - 2$ [3.3]

50. $|6x - 9| > 2$ [3.4]

51. $\dfrac{x - 3}{x + 1} \leqslant 3$ [7.5]

52. Solve by linear combinations: $\begin{aligned} 4x &= 6 - 2y \\ 3y &= 17 + 2x \end{aligned}$ [9.1]

■ *Solve the systems.*

53. $\begin{aligned} x^2 - 3xy + 2y^2 &= 0 \\ 3x + 4y &= -10 \end{aligned}$ [10.4]

54. $\begin{aligned} 2x^2 - 5xy + 2y^2 &= 0 \\ x^2 + xy + y^2 &= 28 \end{aligned}$ [10.5]

■ *Solve by using matrices.*

55. $\begin{aligned} x - 2y &= -7 \\ 3x + 2z &= -9 \\ 2y - 3z &= 4 \end{aligned}$ [9.5]

56. $\begin{aligned} \tfrac{1}{2}x - \tfrac{1}{2}y + z &= -5 \\ \tfrac{2}{3}x + y - \tfrac{1}{5}z &= 3 \\ \tfrac{5}{4}x - \tfrac{1}{4}y + z &= 2 \end{aligned}$ [9.5]

57. Solve by using Cramer's rule: $\begin{aligned} x - 3y &= -4 \\ 2x - y + z &= 5 \\ 2y + 3z &= 0 \end{aligned}$ [9.4]

58. Graph on a number line $-5/3$, $\sqrt{12}$, $-\sqrt{57}$, and $\sqrt{1}$. [6.5]

■ *Graph.*

59. $3x = 4y - 6$ [8.2]

60. $3x - 4y < 0$ [8.5]

61. $y = -(1/2)^x$ [11.3]

62. $9x^2 + 18x + 4y^2 - 24y + 9 = 0$ [10.1]

63. $y = -2x^2 - 9x + 5$ [10.2]

64. $4y^2 + 16y - x^2 + 6x + 3 = 0$ [10.3]

65. $\begin{aligned} x - 3y &< 6 \\ -4x + 2y &\geqslant 8 \end{aligned}$ [9.1]

66. Find an equation for the line which contains the points $(-1, -6)$ and $(-3, 5)$. [8.4]

67. z varies directly with x and inversely with the square of y. If $z = 6$ when $x = 72$ and $y = 6$, find z when $x = 80$ and $y = 4$. [11.2]

68. A rectangular tool chest has a square base, and its height is half its length. Express its surface area (S) as a function of its length (x). Find the surface area of a tool chest whose base measures 20 inches on a side. [11.2]

69. If the third term of an arithmetic progression is $-9/2$, and the thirteenth term is 3, what is the eighth term? [13.2]

70. What is the total value of $350 invested for 6 years at 9% interest compounded monthly? [13.4]

71. The population of a city doubled in 25 years. What was the annual rate of growth? [12.6]

72. A radioactive isotope decays exponentially so that the amount present after t years is $N = N_0 e^{-0.002t}$, where N_0 is the initial amount. How long will it take for 90% of the isotope to decay? [12.6]

73. How many telephone exchanges can be formed using three letters followed by four digits if Q and Z are omitted, and the exchange cannot begin with 0? [13.6]

74. In how many ways can a committee of two Republicans and three Democrats be chosen from a legislature made up of 20 Republicans and 30 Democrats? [13.7]

75. A company ships its product to three cities: Boston, Chicago, and Los Angeles. The cost of shipping is $5 per crate to Chicago, $10 per crate to Boston, and $12 per crate to Los Angeles. The company's shipping bill for April was $445. It shipped 55 crates in all, with twice as many crates going to Boston as to Los Angeles. How many crates were shipped to each destination? [8.6]

B

76. If $P(x) = x^3 - 1$, $Q(x) = x^2 + 1$, and $R(x) = x^3 - x^2$, evaluate $Q(4)[R(-2) - P(2)]$. [2.1]

77. Write as a single interval: $[(-2, 3] \cap (2, 4)] \cup (-1, 2]$ [3.4]

■ *Simplify.*

78. $x^{2n-2} \cdot x^{n^2-3n}$ [2.3]

79. $\left(\dfrac{x^{-n+1}\,y^{2n}}{x^{-2}\,y^{-n}}\right)^{-2}$ [6.2]

80. $\dfrac{\dfrac{1}{x-2} - \dfrac{2}{x+1}}{\dfrac{2}{x+2} - \dfrac{1}{x-2}}$ [5.7]

81. $\sqrt[4]{\dfrac{x^2y^8 - y^6}{x^2y^2}}$ [6.6]

■ *Factor completely.*

82. $y^{n-1} - 3y^n + 6y^{2n}$ [4.1]

83. $36a^{6n} + 6a^{3n} - 12$ [4.2]

84. $2x + 4y = 8$ defines y as a function of x, $y = f(x)$. Find $f^{-1}(x)$ and show that $f^{-1}[f(x)] = f[f^{-1}(x)] = x$. [11.4]

85. For what values of x is $\log_3 (9 - x^2)$ defined? [12.1]

86. Write in sigma notation: $1 + 4x + 9x^2 + 16x^3 + \cdots$ [13.1]

87. Use the binomial formula to estimate $\sqrt{4.1}$ to two decimal places. [13.5]

88. Show that $\begin{vmatrix} a_1 & b_1 \\ a_2 & b_2 \end{vmatrix} = \begin{vmatrix} a_1 & b_1 \\ a_1 + a_2 & b_1 + b_2 \end{vmatrix}$ [9.3]

89. Solve $x^2 + kx + k - 4 = 0$ for x in terms of k. [7.2]

90. Solve $x^{-2} - 2x^{-1} - 15 = 0$. [7.4]

91. Solve $P = P_0 e^{-kt} + C$ for K. [12.5]

92. Solve for n: $\binom{n}{4} = \binom{n}{2}$ [13.7]

93. Solve the system $\frac{1}{x} - \frac{1}{y} + \frac{2}{z} = -1$ [9.2]

$$\frac{2}{x} + \frac{1}{y} - \frac{2}{z} = -2$$

$$\frac{5}{x} + \frac{1}{y} + \frac{4}{z} = 4$$

94. Graph $x = 2y + 6$ and $y = 2x - 6$ on the same coordinate system and estimate the point of intersection. [8.2]

95. Graph $4x^2 - 9y^2 = 0$. [10.3]

96. Graph the geometric progression defined by $s_n = (1/2)^n$ for $1 \leq n \leq 4$. Use the horizontal axis for n and the vertical axis for s_n. [13.3]

97. Show that the triangle whose vertices are $(1, 1)$, $(-6, 2)$, and $(-2, 5)$ is an isosceles triangle. [8.3]

98. Write an equation for the line perpendicular to the graph of $3x + y = 6$ that passes through the point $(-2, -4)$. [8.4]

99. What rate of interest is required for an amount of money to double in 8 years if the interest is compounded monthly? [12.6]

100. Find the sum of all integral multiples of 14 between 25 and 500. [13.2]

A More about Functions

A.1

POLYNOMIAL FUNCTIONS

In Section 8.2 we graphed linear functions defined by

$$y = a_1x + a_0,$$

and in Section 10.2 we graphed quadratic functions defined by

$$y = a_1x^2 + a_1x + a_0.$$

Remainder theorem

We can graph any **polynomial function** defined by

$$y = a_nx^n + a_{n-1}x^{n-1} + \cdots + a_0,$$

where a_n, a_{n-1}, $\cdots + a_0$, and x are real numbers, by obtaining a number of solutions (ordered pairs) sufficient to determine the behavior of its graph. We can obtain the ordered pairs (x, y) by direct substitution of values of x, as we did earlier, or by another method that can sometimes be more efficient. To do this, we first consider an important property of quotients called the **remainder theorem**, which we state without proof as follows.

> *If $P(x)$ is divided by $(x - a)$, the remainder r is $P(a)$.*

Since synthetic division (see section 5.3) offers a means of finding values of $r = P(a)$, we can sometimes find such values more quickly by synthetic division than by direct substitution. For example, if

$$P(x) = 2x^3 - 3x^2 + 2x + 1,$$

we can find $P(2)$ by synthetically dividing $2x^3 - 3x^2 + 2x + 1$ by $x - 2$:

$$\begin{array}{r|rrrr} 2 & 2 & -3 & 2 & 1 \\ & & 4 & 2 & 8 \\ \hline & 2 & 1 & 4 & 9 \end{array}$$

By inspection, we note that $r = P(2) = 9$.

EXAMPLE 1 If $P(x) = 4x^4 - 2x^3 + 3x - 2$, find $P(-1)$ and $P(2)$.

Solution

$$\begin{array}{r|rrrrr} -1 & 4 & -2 & 0 & 3 & -2 \\ & & -4 & 6 & -6 & 3 \\ \hline & 4 & -6 & 6 & -3 & 1 \end{array} \qquad \begin{array}{r|rrrrr} 2 & 4 & -2 & 0 & 3 & -2 \\ & & 8 & 12 & 24 & 54 \\ \hline & 4 & 6 & 12 & 27 & 52 \end{array}$$

$P(-1) = 1$ $\qquad$ $P(2) = 52$ ■

Graphs of polynomial functions

In the following example we graph a third-degree polynomial by finding a sufficient number of ordered pairs to suggest the appearance of its graph.

EXAMPLE 2 Graph $P(x) = x^3 - 2x^2 - 5x + 6$.

Solution We obtain solutions of the equation for selected values of x, say, -3, -2, -1, 0, 1, 2, 3, and 4, by dividing $x^3 - 2x^2 - 5x + 6$ by $x + 3$, $x + 2$, and so on. Dividing synthetically by $x + 3$, we have

$$\begin{array}{r|rrrr} -3 & 1 & -2 & -5 & 6 \\ & & -3 & 15 & -30 \\ \hline & 1 & -5 & 10 & -24 \end{array}$$

and $(-3, -24)$ is a solution of (3). Dividing by $x + 2$, we have

$$\begin{array}{r|rrrr} -2 & 1 & -2 & -5 & 6 \\ & & -2 & 8 & -6 \\ \hline & 1 & -4 & 3 & 0 \end{array}$$

and $(-2, 0)$ is a solution of (3). Dividing by $x + 1$, we have

$$\begin{array}{r|rrrr} -1 & 1 & -2 & -5 & 6 \\ & & -1 & 3 & 2 \\ \hline & 1 & -3 & -2 & 8 \end{array}$$

and $(-1, 8)$ is a solution of (3). Similarly, we find that $P(0) = 6$, $P(1) = 0$, $P(2) = -4$, $P(3) = 0$, and $P(4) = 18$. Hence, $(0, 6)$, $(1, 0)$, $(2, -4)$, $(3, 0)$, and $(4, 18)$ are solutions, and their graphs are on the graph of the function. Note that -2, 1, and 3 are zeros of the function because

$$f(-2) = 0, \quad f(-1) = 0, \quad \text{and} \quad f(3) = 0.$$

The points, shown in Figure a below, enable us to complete the graph as shown in Figure b. Any additional values of x less than -3 and greater than 4 would not change the general appearance of the graph.

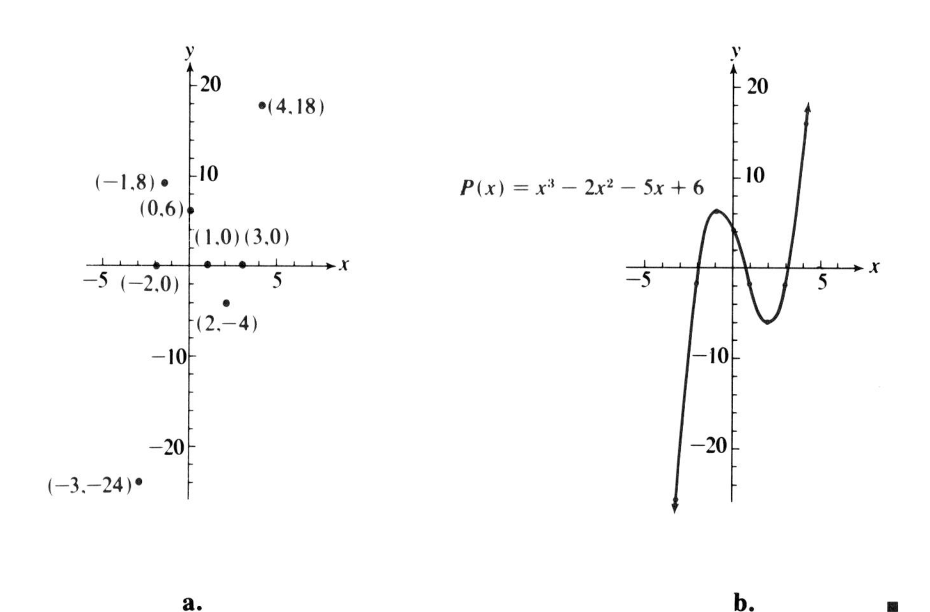

■

Number of direction changes

Note in the above example that the graph of this third-degree polynomial function changes direction twice. In fact, it can be shown, although we will not do so, that the *maximum number of direction changes of the graph of a polynomial function is one less than the degree of the polynomial.*

Observe that the high and low points in the graph of the above example *do not* exactly correspond with the ordered pairs $(-1, 8)$ and $(2, -4)$, respectively. To find the exact values of the components of the ordered pairs that are the high and low points of the graphs of polynomial functions of degree greater than 2 requires methods considered only in more advanced courses. However, we can obtain approximations for these points by plotting ordered pairs as demonstrated in the example above.

Factor theorem

A direct consequence of the remainder theorem is the following result, called the **factor theorem.**

> *If* $P(a) = 0$, *then* $(x - a)$ *is a factor of* $P(x)$, *and if* $(x - a)$ *is a factor, then* $P(x) = 0$.

EXAMPLE 3 For the polynomial

$$P(x) = x^3 - 2x^2 - 5x + 6$$

in Example 2, we obtained $P(-2) = 0$, $P(1) = 0$, and $P(3) = 0$. Hence, by the factor theorem, $x + 2$, $x - 1$, and $x - 3$ are factors of the polynomial. ■

Number of solutions of $P(x) = 0$

We note from the factor theorem that a is a solution of $P(x) = 0$ if and only if $(x - a)$ is a factor of $P(x)$. This suggests that *an equation* $P(x) = 0$, *where* $P(x)$ *is of nth degree, has n solutions.* This is indeed the case.

EXAMPLE 4 We note that the third-degree polynomial $x^3 - 2x^2 - 5x + 6$ is equivalent to $(x + 2)(x - 1)(x - 3)$, and that there are exactly three solutions of

$$x^3 - 2x^2 - 5x + 6 = (x + 2)(x - 1)(x - 3) = 0,$$

namely, -2, 1, and 3. ■

Of course, it may be that one or more factors of such an expression are the same. When this happens, we count the solution as many times as the factor involved occurs. Thus, if

$$\begin{aligned} P(x) &= x^4 + 2x^3 - 2x - 1 \\ &= (x + 1)(x + 1)(x + 1)(x - 1), \end{aligned}$$

we can see that -1 and 1 are the only solutions of $P(x) = 0$, but we say that -1 is a solution of *multiplicity* three.

EXERCISE A.I

A

■ *Find the designated values. See Example* 1.

1. If $P(x) = 3x^3 - 2x^2 + 5x - 4$, find $P(3)$ and $P(-2)$.
2. If $P(x) = 4x^4 - 2x^3 + 3x^2 - 5$, find $P(1)$ and $P(-1)$.
3. If $P(x) = 2x^5 - 3x^3 + x^2 - x + 2$, find $P(-1)$ and $P(2)$.
4. If $P(x) = x^4 - 10x^3 + 5x^2 - 3x + 6$, find $P(-2)$ and $P(3)$.

■ *Use synthetic division and the remainder theorem to find sufficient solutions to graph each function. Specify the zeros of the function. See Example* 2.

5. $P(x) = x^3 + x^2 - 6x$
6. $P(x) = x^3 + 5x^2 + 4x$
7. $P(x) = x^3 - 2x^2 + 1$
8. $P(x) = x^3 - 4x^2 + 3x$
9. $P(x) = 2x^3 + 9x^2 + 7x - 6$
10. $P(x) = x^3 - 3x^2 - 6x + 8$
11. $P(x) = x^4 - 4x^2$
12. $P(x) = x^4 - x^3 - 4x^2 + 4x$

■ *Use the factor theorem to determine whether or not the given binomial is a factor of the given polynomial. See Example* 3.

13. $x - 2$; $x^3 - 3x^2 + 2x + 2$
14. $x - 1$; $2x^3 - 5x^2 + 4x - 1$
15. $x + 3$; $3x^3 + 11x^2 + x - 15$
16. $x + 1$; $2x^3 - 5x^2 + 3x + 3$

B

17. Verify that 1 is a solution of $x^3 + 2x^2 - x - 2 = 0$, and find the other solutions.
18. Verify that 3 is a solution of $x^3 - 6x^2 - x + 30 = 0$, and find the other solutions.
19. Verify that -3 is a solution of $x^4 - 3x^3 - 10x^2 + 24x = 0$, and find the other solutions.
20. Verify that -5 is a solution of $x^4 + 5x^3 - x^2 - 5x = 0$, and find the other solutions.
21. Graph on the same set of axes the equations $y = kx$, $y = kx^2$, and $y = kx^3$, where $k = 2$ and $x \geqslant 0$. What effect does increasing the degree of the equation $y = kx^n$ have on the graph of the equation?

A.2

RATIONAL FUNCTIONS

A function defined by an equation of the form

$$y = \frac{P(x)}{Q(x)},$$

where $P(x)$ and $Q(x)$ are polynomials and $Q(x) \neq 0$, is called a **rational function.** We can graph such a function by obtaining a number of solutions (ordered pairs) sufficient to determine its behavior by direct substitution of arbitrary values of x. It is also helpful to first obtain any asymptotes to the curve that may exist.

Vertical asymptotes

Vertical asymptotes can be found by using the following property.

> *The graph of the function defined by* $y = P(x)/Q(x)$ *has a vertical asymptote* $x = a$ *for each value* a *at which* $Q(x) = 0$ *and* $P(x) \neq 0$.

EXAMPLE 1 Find the vertical asymptotes of the graphs of each function.

a. $y = \dfrac{2}{x - 2}$ b. $y = \dfrac{4}{x^2 - x - 6}$

Solutions

a. $x - 2 = 0$ if $x = 2$.

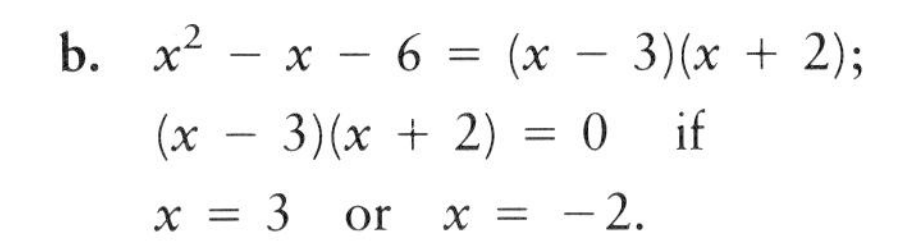

b. $x^2 - x - 6 = (x - 3)(x + 2)$; $(x - 3)(x + 2) = 0$ if $x = 3$ or $x = -2$.

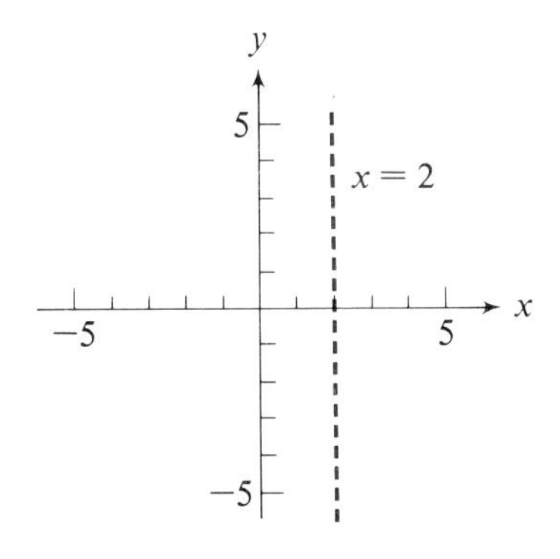

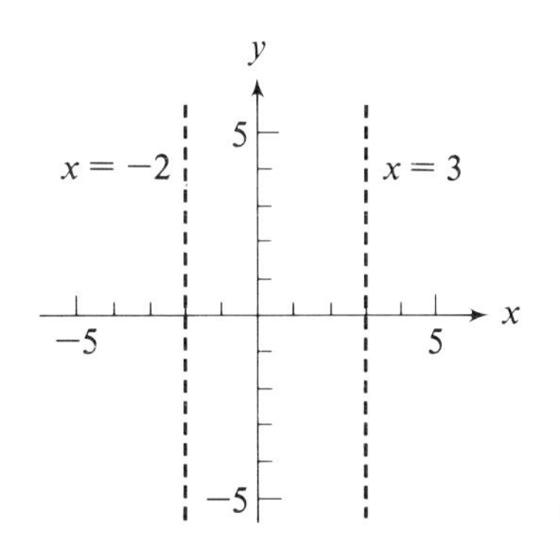

■

Horizontal asymptotes

The graphs of some rational functions have horizontal asymptotes which can be identified by using the following property.

> *If ax^n is the term of highest degree of a polynomial $P(x)$ and bx^m is the term of highest degree of a polynomial $Q(x)$, then the graph of the function $y = P(x)/Q(x)$ has a horizontal asymptote*
>
> *at* $y = 0$ *if* $n < m$;
> *at* $y = a/b$ *if* $n = m$;
> nowhere *if* $n > m$.

EXAMPLE 2 Determine any horizontal asymptotes of the graphs of each function.

a. $y = \dfrac{3x}{x^2 - 5x + 4}$ b. $y = \dfrac{4x^2}{2x^2 - x}$ c. $y = \dfrac{x^4 + 1}{x^2 + 2}$

Solutions a. The degree of $3x$ is less than the degree of x^2. Hence, the graph has a horizontal asymptote at $y = 0$.

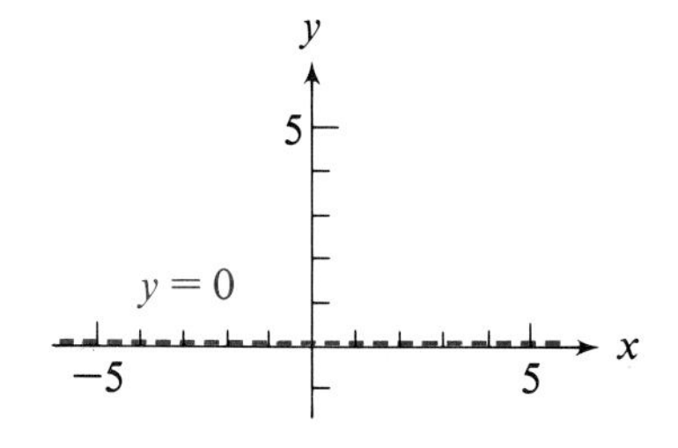

b. $4x^2$ and $2x^2$ are of the same degree. Hence, the graph has a horizontal asymptote at $y = 4/2 = 2$.

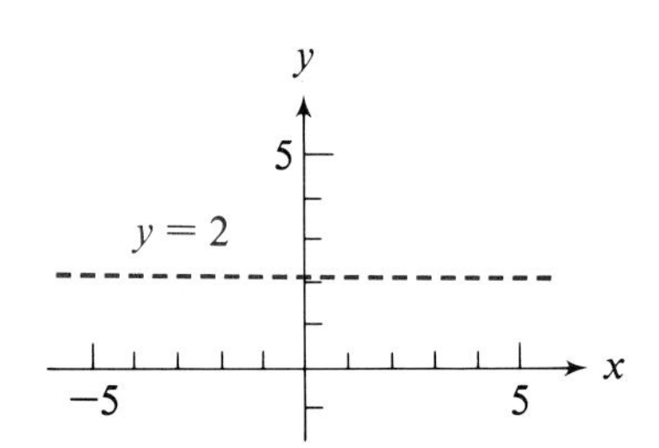

c. The degree of x^4 is greater than the degree of x^2. Hence, the graph does not have a horizontal asymptote. ■

After vertical and horizontal asymptotes have been found, a few additional ordered pairs associated with points on opposite sides of each asymptote will usually be sufficient to complete the graph.

EXAMPLE 3 Graph $y = \dfrac{2}{x - 2}$.

Solution Because $x - 2 = 0$ at $x = 2$, the vertical asymptote is $x = 2$. Because the degree of the numerator is less than the degree of the denominator, the horizontal asymptote is $y = 0$. We note that if $x = 0$, the y-intercept is -1. Several additional ordered pairs,

$$\left(-2, -\frac{1}{2}\right), (1, -2), (3, 2), \text{and } (4, 1),$$

enable us to complete the graph as we use the asymptotes as a guide for the branches of the curve.

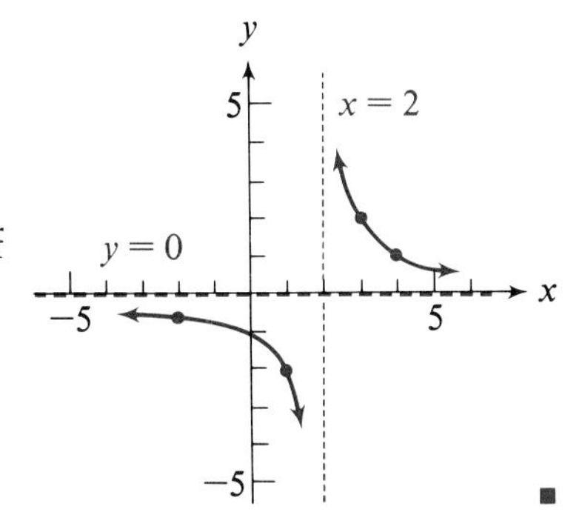

■

EXAMPLE 4 Graph $y = \dfrac{2x - 4}{x^2 - 9}$.

Solution Because $x^2 - 9 = (x - 3)(x + 3)$, which is zero for $x = 3$ and $x = -3$, the vertical asymptotes are $x = 3$ and $x = -3$. Because the degree of $2x$ is less than the degree of x^2, the horizontal asymptote is $y = 0$. If $x = 0$, the y-intercept is $4/9$. If $y = 0$, $2x - 4 = 0$; hence, the x-intercept is 2. Several additional ordered pairs,

$$\left(-4, -\frac{12}{7}\right), \left(-2, \frac{8}{5}\right), \text{ and } \left(4, \frac{4}{7}\right),$$

enable us to complete the graph as we use the asymptotes to direct the branches of the curve.

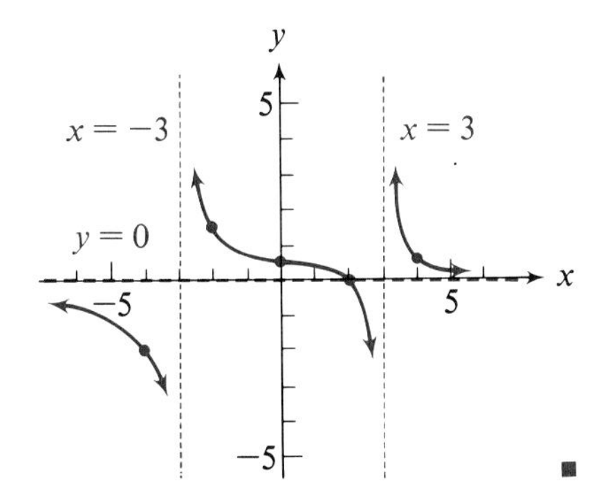

■

Hole The graph of the function defined by $y = P(x)/Q(x)$ has a hole at $x = a$ for each value of a at which both $Q(x)$ and $P(x)$ have $x - a$ as a common factor.

EXAMPLE 5 Graph $y = \dfrac{x^2 - 4}{x + 2}$.

Solution Since $x + 2$ is a common factor of both the numerator and denominator, it can be divided out. But, since the denominator can not equal zero, $x \neq -2$. Thus,

$$y = \frac{x^2 - 4}{x + 2} = \frac{(x + 2)(x - 2)}{x + 2} = x - 2,$$

where $x \neq -2$.

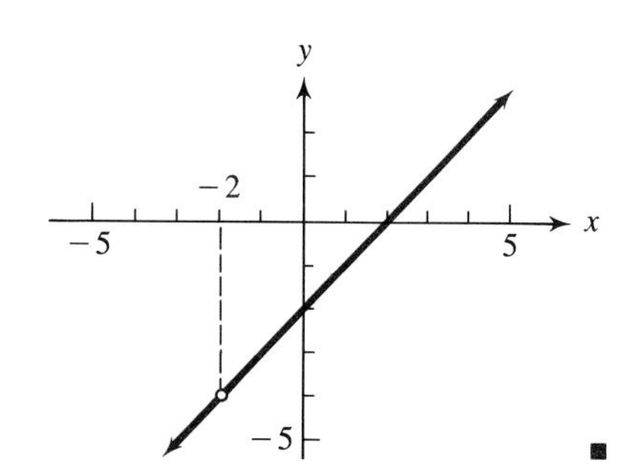

■

EXERCISE A.2

A

■ *Determine the vertical asymptotes of the graph of each function. See Example 1.*

1. $y = \dfrac{2}{x + 3}$
2. $y = \dfrac{1}{x - 4}$
3. $y = \dfrac{3}{(x - 2)(x + 3)}$
4. $y = \dfrac{4}{(x + 1)(x - 4)}$
5. $y = \dfrac{2x}{x^2 - x - 6}$
6. $y = \dfrac{2x + 1}{x^2 - 3x + 2}$

■ *Determine any vertical or horizontal asymptotes of the graphs of each function. See Examples 1 and 2.*

7. $y = \dfrac{x}{x^2 - 9}$
8. $y = \dfrac{2x - 4}{x^2 + 5x + 4}$
9. $y = \dfrac{x - 4}{2x - 1}$
10. $y = \dfrac{2x + 1}{x - 3}$
11. $y = \dfrac{2x^2}{x^2 - 3x - 4}$
12. $y = \dfrac{x^2}{x^2 - x - 12}$

■ *Graph each function after first identifying all asymptotes and intercepts. See Examples 3 and 4.*

13. $y = \dfrac{1}{x + 3}$
14. $y = \dfrac{1}{x - 3}$
15. $y = \dfrac{2}{(x - 4)(x + 1)}$
16. $y = \dfrac{4}{(x + 2)(x - 1)}$
17. $y = \dfrac{2}{x^2 - 5x + 4}$
18. $y = \dfrac{4}{x^2 - x - 6}$

19. $y = \dfrac{x}{x + 3}$

20. $y = \dfrac{x}{x - 2}$

21. $y = \dfrac{x + 1}{x + 2}$

22. $y = \dfrac{x - 1}{x - 3}$

23. $y = \dfrac{2x}{x^2 - 4}$

24. $y = \dfrac{x}{x^2 - 9}$

25. $y = \dfrac{x - 2}{x^2 + 5x + 4}$

26. $y = \dfrac{x + 1}{x^2 - x - 6}$

■ *Graph each function. See Example 5.*

27. $y = \dfrac{x^2 - 9}{x - 3}$

28. $y = \dfrac{x^2 - 1}{x + 1}$

29. $y = \dfrac{x^2 - 5x + 4}{x - 1}$

30. $y = \dfrac{x^2 - x - 6}{x - 3}$

31. $y = \dfrac{(x^2 - 4)(x + 1)}{x + 1}$

32. $y = \dfrac{(x^2 - 1)(x - 3)}{x - 3}$

B

33. Graph $xy = k$ for $k = 4$ and $k = 12$ on the same set of axes.

34. Graph $xy = k$ for $k = -4$ and $k = -12$ on the same set of axes.

A.3

SPECIAL FUNCTIONS

Functions defined piecewise

It is possible to define a function using different rules for different parts of the domain. Such a function is said to be **piecewise defined.** To find the range value paired with a given domain value, we use that part of the function in which the given domain value lies.

EXAMPLE 1 Given that $f(x) = \begin{cases} x^2 & \text{if } x < 0 \\ -x + 1 & \text{if } x \geq 0 \end{cases}$, find the range value.

a. $f(-1)$

b. $f(1)$

Solutions

a. Since $-1 < 0$,
use $f(x) = x^2$.
$f(-1) = (-1)^2 = 1$

b. Since $1 \geq 0$,
use $f(x) = -x + 1$.
$f(1) = -1 + 1 = 0$ ■

The graph of a function that is defined piecewise can be obtained by graphing it separately over each part of the domain determined by the definition of the function.

EXAMPLE 2 Graph

$$f(x) = \begin{cases} x^2 & \text{if} \quad x \leqslant 1 \\ -x^2 + 2 & \text{if} \quad x > 1. \end{cases}$$

Solution Graph the function separately over each part of the domain. First,

$$f(x) = x^2 \quad \text{for} \quad x \leqslant 1$$

is graphed, as shown in Figure a. Then,

$$f(x) = -x^2 + 2 \quad \text{for} \quad x > 1$$

is graphed to obtain the complete graph shown in Figure b.

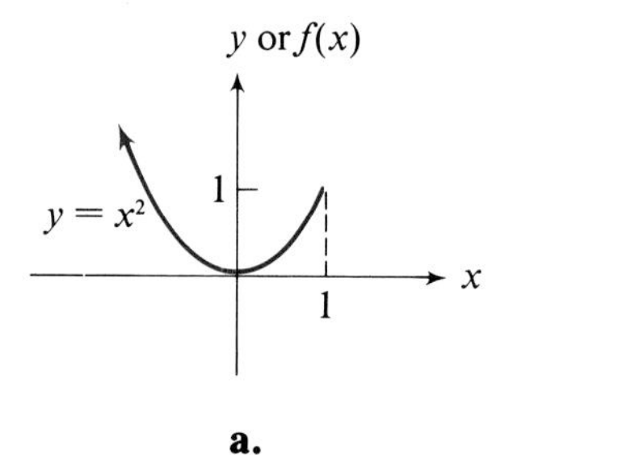

a.

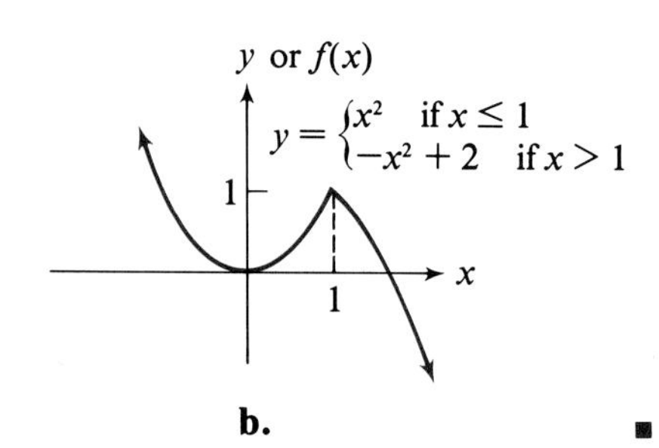

b. ■

Absolute-value functions Functions involving the absolute value of the independent variable are piecewise-defined functions. Consider the function defined by

$$y = |x|. \tag{1}$$

From the definition of $|x|$, Equation (1) is equivalent to

$$y = \begin{cases} x & \text{for} \quad x \geqslant 0 \quad (2) \\ -x & \text{for} \quad x < 0. \quad (3) \end{cases}$$

If we graph (2) and (3) on the same set of axes, we have the graph of $y = |x|$ shown in Figure A.1 on page 529.

y

$y = -x$ and $x < 0$ $y = x$ and $x \geq 0$

x

FIGURE A.1

We can graph any equation involving $|x|$ or $|f(x)|$ by first writing the equation in piecewise-defined form and then using the method described above.

EXAMPLE 3 Graph $y = |x - 2|$.

Solution The definition of absolute value implies that $y = |x - 2|$ is equivalent to

$$y = \begin{cases} x - 2 & \text{if } x - 2 \geqslant 0 \\ -(x - 2) & \text{if } x - 2 < 0. \end{cases}$$

Thus, we graph

$$y = x - 2, \quad x \geqslant 2$$

and

$$y = -x + 2, \quad x < 2$$

on the same set of coordinate axes to obtain the figure shown.

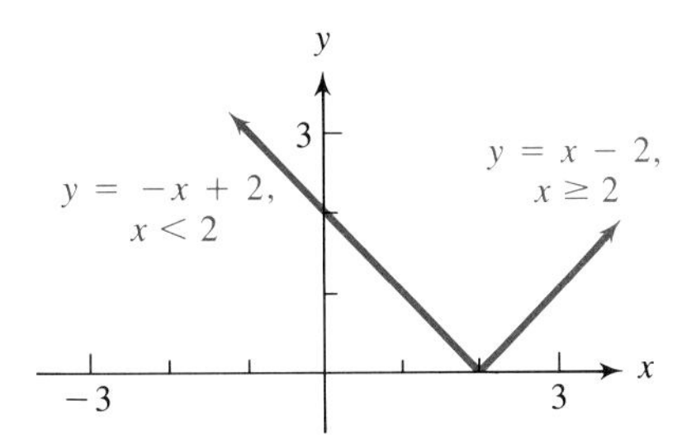

■

EXAMPLE 4 Graph $y = |x| + 1$.

Solution The definition of absolute value implies that $y = |x| + 1$ is equivalent to:

$$y = \begin{cases} x + 1 & \text{if } x \geqslant 0 \\ -x + 1 & \text{if } x < 0 \end{cases}$$

Thus, we graph $y = x + 1,\ x \geqslant 0$ and $y = -x + 1,\ x \leqslant 0$ on the same set of coordinate axes to obtain the figure shown.

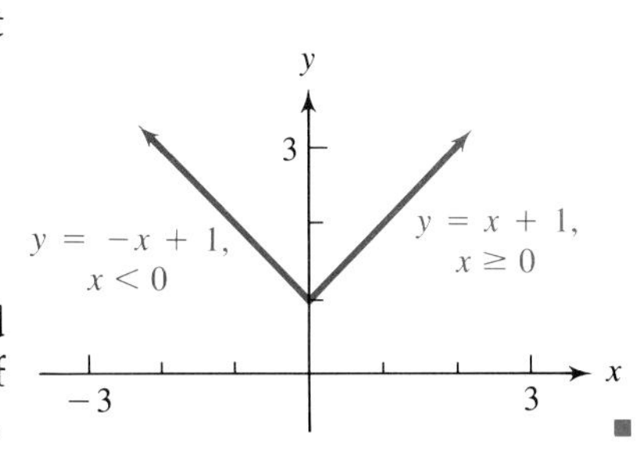

■

Equations of the form

$$y = |f(x)| + c \tag{4}$$

(see Example 4) can be graphed by observing that for each x, $|f(x)| + c$ differs from $|f(x)|$ by c units. Thus, the graph of (4) is simply the graph of $y = |f(x)|$ shifted by c units. The graph is shifted upward if $c > 0$ and downward if $c < 0$.

EXAMPLE 5 Graph $y = |x| - 2$.

Solution We first sketch $y = |x|$ as shown in Figure a. We then shift the graph 2 units down, as shown in Figure b to obtain the graph of $y = |x| - 2$.

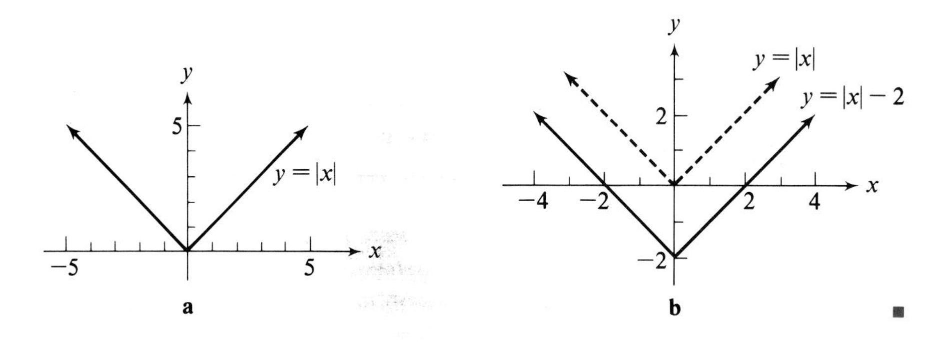

■

Note that $y = |x| - 2$ in Example 5 and $y = |x - 2|$ in Example 3 define different functions.

Bracket function

Another interesting piecewise function (sometimes called the **greatest-integer function**) is defined as follows:

For each x, $f(x)$ is the greatest integer less than or equal to x.

The greatest-integer function is sometimes denoted by

$$f(x) = [x],$$

and hence this function is also called the **bracket function.**

EXAMPLE 6 $[2] = 2,\quad \left[\frac{7}{4}\right] = 1,\quad \left[\frac{-3}{2}\right] = -2,\quad$ and $\quad \left[\frac{-5}{2}\right] = -3.$

$(2 = 2)\quad \left(1 < \frac{7}{4}\right)\quad \left(-2 < \frac{-3}{2}\right)\quad \left(-3 < \frac{-5}{2}\right)$ ■

To graph the bracket function, we consider unit intervals along the x-axis, as shown in Table A.1. The graph is shown in Figure A.2. The heavy dot on the left-hand endpoint of each line segment indicates that the endpoint is a part of the graph. The open dot on the right-hand endpoint indicates that the endpoint is not part of the graph. The bracket function is also called a "step function," for an obvious reason.

x-interval	$[x]$
$[-2, -1)$	-2
$[-1, 0)$	-1
$[0, 1)$	0
$[1, 2)$	1
$[2, 3)$	2

TABLE A.1

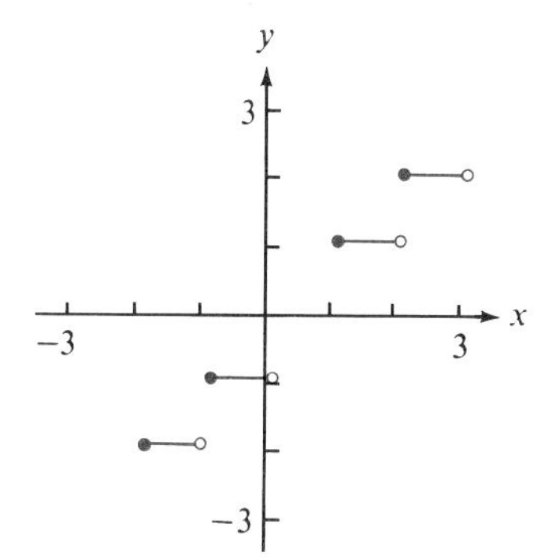

FIGURE A.2

EXERCISE A.3

A

■ *Find the indicated range value for the given function. See Example 1.*

1. $f(3)$ for $f(x) = \begin{cases} x - 2 & \text{if } x \geq 2 \\ 2 - x & \text{if } x < 2 \end{cases}$

2. $f(1)$ for $f(x) = \begin{cases} -3x - 1 & \text{if } x < -1 \\ 2x + 1 & \text{if } x \geq -1 \end{cases}$

3. $f(2)$ for $f(x) = \begin{cases} x & \text{if } x < 0 \\ 2x + 1 & \text{if } 0 \leq x < 2 \\ 3x + 2 & \text{if } x \geq 2 \end{cases}$

4. $f(-1)$ for $f(x) = \begin{cases} -2x & \text{if } x \leq -1 \\ x & \text{if } -1 < x < 1 \\ 2x & \text{if } x \geq 1 \end{cases}$

■ *Graph each function. See Example 2.*

5. $f(x) = \begin{cases} x^2 & \text{if } x \leq 2 \\ x + 2 & \text{if } x > 2 \end{cases}$

6. $f(x) = \begin{cases} x^2 - 4 & \text{if } x \leq 2 \\ 2 - x & \text{if } x > 2 \end{cases}$

7. $f(x) = \begin{cases} x^2 & \text{if } x \leq 0 \\ -x^2 & \text{if } x > 0 \end{cases}$

8. $f(x) = \begin{cases} -x^2 + 1 & \text{if } x \leq 0 \\ x^2 - 1 & \text{if } x > 0 \end{cases}$

9. $f(x) = \begin{cases} x^2 & \text{if } x \leq 2 \\ 4 & \text{if } x > 2 \end{cases}$

10. $f(x) = \begin{cases} x^2 - 4 & \text{if } x \leq -2 \\ 0 & \text{if } x > -2 \end{cases}$

■ *Graph each function. See Examples 3, 4, and 5.*

11. $y = |x| + 2$
12. $y = -|x| + 3$
13. $f(x) = |x + 1|$
14. $f(x) = |x - 2|$
15. $y = -|2x - 1|$
16. $y = -|3x + 2|$
17. $g(x) = |2x| - 3$
18. $y = |3x| + 2$

■ *Graph each function over the interval* $-5 \leq x \leq 5$. *See Example 6.*

19. $y = 2[x]$
20. $y = [2x]$
21. $f(x) = [x + 1]$
22. $f(x) = [x] + 1$

B

■ *Graph each function. For Problems 33–34, graph the function over the interval* $-5 \leq x \leq 5$.

23. $f(x) = \begin{cases} -1 & \text{if } x < -1 \\ 1 & \text{if } -1 \leq x \leq 1 \\ \frac{1}{2} & \text{if } 1 < x \end{cases}$

24. $f(x) = \begin{cases} x + 1 & \text{if } x \leq 0 \\ 1 - x & \text{if } 0 < x \leq 1 \\ x - 1 & \text{if } 1 < x \end{cases}$

25. $f(x) = \begin{cases} -x^2 & \text{if } x < -1 \\ -1 & \text{if } -1 \leq x < 1 \\ x^2 - 2 & \text{if } 1 \leq x \end{cases}$

26. $f(x) = \begin{cases} x & \text{if } x < 0 \\ x^2 & \text{if } 0 \leq x < 2 \\ -x + 6 & \text{if } x \geq 2 \end{cases}$

27. $f(x) = |2x| + |x|$
28. $f(x) = |3x| - |x|$
29. $y = -2|x| + x$
30. $y = 3|x| - x$
31. $y = |x|^2$
32. $y = |x^2|$
33. $y = [x] + x$
34. $y = [x] - x$

APPENDIX SUMMARY

[A.1] The equation

$$y = a_n x^n + a_{n-1} x^{n-1} + \cdots + a_0$$

defines a **polynomial function** of nth degree.

A quotient of the form $\dfrac{\boldsymbol{P(x)}}{\boldsymbol{(x - a)}}$, where $P(x)$ is a polynomial of degree $n \geq 1$ with real coefficients, can be expressed in the form $Q(x) + \dfrac{\boldsymbol{r}}{\boldsymbol{(x - a)}}$, where $Q(x)$ is a polynomial of degree $(n - 1)$ with real coefficients and r is a real number. Furthermore, $r = P(a)$ (**remainder theorem**).

Synthetic division can be used to find values $P(a)$ of a polynomial $P(x)$ in order to determine the behavior of its graph.

If $\dfrac{P(x)}{(x - a)}$ yields a remainder $r = P(a) = 0$, then $(x - a)$ is a factor of $P(x)$ (**factor theorem**).

An equation of the form $P(x) = 0$, where $P(x)$ is of nth degree, has n solutions.

[A.2] A function defined by $y = \dfrac{P(x)}{Q(x)}$, where $P(x)$ and $Q(x)$ are polynomials and $Q(x) \neq 0$, is called a **rational function**.

The process of graphing a rational function can be facilitated by first finding any vertical and/or horizontal asymptotes that exist.

[A.3] Functions involving the absolute value of the independent variable, called **absolute value functions**, are piecewise functions. The **greatest-integer function,** also called the **bracket function,** is also a piecewise function.

REVIEW EXERCISES

A

[A.1]

■ *Use synthetic division and the remainder theorem in Problems 1 and 2.*

1. If $P(x) = 2x^3 - x^2 + 3x + 1$, find $P(2)$ and $P(-2)$.
2. If $P(x) = x^4 + 3x^2 - 2x + 2$, find $P(1)$ and $P(-1)$.

■ *Use synthetic division and the remainder theorem to find solutions to each equation, and then graph the equation.*

3. $y = x^3 + x^2 - 2x$
4. $y = x^4 + 2x^3 - 5x^2 - 6x$
5. Is $(x + 2)$ a factor of $3x^3 - 2x^2 - x + 4$?
6. Verify that $(x - 2)$ is a factor of $x^3 - 4x^2 + x + 6$, and find the other factors.

[A.2]

■ *Graph each function.*

7. $y = \dfrac{1}{x - 4}$
8. $y = \dfrac{2}{x^2 - 3x - 10}$
9. $y = \dfrac{x - 2}{x + 3}$
10. $y = \dfrac{x - 1}{x^2 - 2x - 3}$

[A.3]

■ *Graph each function.*

11. $f(x) = \begin{cases} x^2 & \text{if } x \leq 0 \\ 2x & \text{if } x > 0 \end{cases}$

12. $f(x) = \begin{cases} x^2 & \text{if } x \leq 2 \\ 8 - x^2 & \text{if } x > 2 \end{cases}$

13. $f(x) = |2x + 3|$

14. $f(x) = |2x| + 3$

15. $f(x) = |2x - 3|$

16. $f(x) = |2x| - 3$

17. $f(x) = [x] + 2$

18. $f(x) = [x + 2]$

B

19. Verify that -2 is a solution of $x^3 - x^2 - 4x + 4 = 0$, and find the other solutions.

20. Find the zeros of the function $f(x) = x^5 - 13x^3 + 36x$.

B Tables

TABLE I

SQUARES, SQUARE ROOTS, AND PRIME FACTORS

Number	Square	Square root	Prime factors
1	1	1.000	
2	4	1.414	2
3	9	1.732	3
4	16	2.000	2^2
5	25	2.236	5
6	36	2.449	$2 \cdot 3$
7	49	2.646	7
8	64	2.828	2^3
9	81	3.000	3^2
10	100	3.162	$2 \cdot 5$
11	121	3.317	11
12	144	3.464	$2^2 \cdot 3$
13	169	3.606	13
14	196	3.742	$2 \cdot 7$
15	225	3.873	$3 \cdot 5$
16	256	4.000	2^4
17	289	4.123	17
18	324	4.243	$2 \cdot 3^2$
19	361	4.359	19
20	400	4.472	$2^2 \cdot 5$
21	441	4.583	$3 \cdot 7$
22	484	4.690	$2 \cdot 11$
23	529	4.796	23
24	576	4.899	$2^3 \cdot 3$
25	625	5.000	5^2
26	676	5.099	$2 \cdot 13$
27	729	5.196	3^3
28	784	5.292	$2^2 \cdot 7$
29	841	5.385	29
30	900	5.477	$2 \cdot 3 \cdot 5$
31	961	5.568	31
32	1,024	5.657	2^5
33	1,089	5.745	$3 \cdot 11$
34	1,156	5.831	$2 \cdot 17$
35	1,225	5.916	$5 \cdot 7$
36	1,296	6.000	$2^2 \cdot 3^2$
37	1,369	6.083	37
38	1,444	6.164	$2 \cdot 19$
39	1,521	6.245	$3 \cdot 13$
40	1,600	6.325	$2^3 \cdot 5$
41	1,681	6.403	41
42	1,764	6.481	$2 \cdot 3 \cdot 7$
43	1,849	6.557	43
44	1,936	6.633	$2^2 \cdot 11$
45	2,025	6.708	$3^2 \cdot 5$
46	2,116	6.782	$2 \cdot 23$
47	2,209	6.856	47
48	2,304	6.928	$2^4 \cdot 3$
49	2,401	7.000	7^2
50	2,500	7.071	$2 \cdot 5^2$
51	2,601	7.141	$3 \cdot 17$
52	2,704	7.211	$2^2 \cdot 13$
53	2,809	7.280	53
54	2,916	7.348	$2 \cdot 3^3$
55	3,025	7.416	$5 \cdot 11$
56	3,136	7.483	$2^3 \cdot 7$
57	3,249	7.550	$3 \cdot 19$
58	3,364	7.616	$2 \cdot 29$
59	3,481	7.681	59
60	3,600	7.746	$2^2 \cdot 3 \cdot 5$
61	3,721	7.810	61
62	3,844	7.874	$2 \cdot 31$
63	3,969	7.937	$3^2 \cdot 7$
64	4,096	8.000	2^6
65	4,225	8.062	$5 \cdot 13$
66	4,356	8.124	$2 \cdot 3 \cdot 11$
67	4,489	8.185	67
68	4,624	8.246	$2^2 \cdot 17$
69	4,761	8.307	$3 \cdot 23$
70	4,900	8.367	$2 \cdot 5 \cdot 7$
71	5,041	8.426	71
72	5,184	8.485	$2^3 \cdot 3^2$
73	5,329	8.544	73
74	5,476	8.602	$2 \cdot 37$
75	5,625	8.660	$3 \cdot 5^2$
76	5,776	8.718	$2^2 \cdot 19$
77	5,929	8.775	$7 \cdot 11$
78	6,084	8.832	$2 \cdot 3 \cdot 13$
79	6,241	8.888	79
80	6,400	8.944	$2^4 \cdot 5$
81	6,561	9.000	3^4
82	6,724	9.055	$2 \cdot 41$
83	6,889	9.110	83
84	7,056	9.165	$2^2 \cdot 3 \cdot 7$
85	7,225	9.220	$5 \cdot 17$
86	7,396	9.274	$2 \cdot 43$
87	7,569	9.327	$3 \cdot 29$
88	7,744	9.381	$2^3 \cdot 11$
89	7,921	9.434	89
90	8,100	9.487	$2 \cdot 3^2 \cdot 5$
91	8,281	9.539	$7 \cdot 13$
92	8,464	9.592	$2^2 \cdot 23$
93	8,649	9.644	$3 \cdot 31$
94	8,836	9.695	$2 \cdot 47$
95	9,025	9.747	$5 \cdot 19$
96	9,216	9.798	$2^5 \cdot 3$
97	9,409	9.849	97
98	9,604	9.899	$2 \cdot 7^2$
99	9,801	9.950	$3^2 \cdot 11$
100	10,000	10.000	$2^2 \cdot 5^2$

TABLE II

VALUES OF $\text{LOG}_{10}\, x$ AND $\text{ANTILOG}_{10}\, x$ OR (10^x)

x	0	1	2	3	4	5	6	7	8	9
1.0	.0000	.0043	.0086	.0128	.0170	.0212	.0253	.0294	.0334	.0374
1.1	.0414	.0453	.0492	.0531	.0569	.0607	.0645	.0682	.0719	.0755
1.2	.0792	.0828	.0864	.0899	.0934	.0969	.1004	.1038	.1072	.1106
1.3	.1139	.1173	.1206	.1239	.1271	.1303	.1335	.1367	.1399	.1430
1.4	.1461	.1492	.1523	.1553	.1584	.1614	.1644	.1673	.1703	.1732
1.5	.1761	.1790	.1818	.1847	.1875	.1903	.1931	.1959	.1987	.2014
1.6	.2041	.2068	.2095	.2122	.2148	.2175	.2201	.2227	.2253	.2279
1.7	.2304	.2330	.2355	.2380	.2405	.2430	.2455	.2480	.2504	.2529
1.8	.2553	.2577	.2601	.2625	.2648	.2672	.2695	.2718	.2742	.2765
1.9	.2788	.2810	.2833	.2856	.2878	.2900	.2923	.2945	.2967	.2989
2.0	.3010	.3032	.3054	.3075	.3096	.3118	.3139	.3160	.3181	.3201
2.1	.3222	.3243	.3263	.3284	.3304	.3324	.3345	.3365	.3385	.3404
2.2	.3424	.3444	.3464	.3483	.3502	.3522	.3541	.3560	.3579	.3598
2.3	.3617	.3636	.3655	.3674	.3692	.3711	.3729	.3747	.3766	.3784
2.4	.3802	.3820	.3838	.3856	.3874	.3892	.3909	.3927	.3945	.3962
2.5	.3979	.3997	.4014	.4031	.4048	.4065	.4082	.4099	.4116	.4133
2.6	.4150	.4166	.4183	.4200	.4216	.4232	.4249	.4265	.4281	.4298
2.7	.4314	.4330	.4346	.4362	.4378	.4393	.4409	.4425	.4440	.4456
2.8	.4472	.4487	.4502	.4518	.4533	.4548	.4564	.4579	.4594	.4609
2.9	.4624	.4639	.4654	.4669	.4683	.4698	.4713	.4728	.4742	.4757
3.0	.4771	.4786	.4800	.4814	.4829	.4843	.4857	.4871	.4886	.4900
3.1	.4914	.4928	.4942	.4955	.4969	.4983	.4997	.5011	.5024	.5038
3.2	.5051	.5065	.5079	.5092	.5105	.5119	.5132	.5145	.5159	.5172
3.3	.5185	.5198	.5211	.5224	.5237	.5250	.5263	.5276	.5289	.5302
3.4	.5315	.5328	.5340	.5353	.5366	.5378	.5391	.5403	.5416	.5428
3.5	.5441	.5453	.5465	.5478	.5490	.5502	.5514	.5527	.5539	.5551
3.6	.5563	.5575	.5587	.5599	.5611	.5623	.5635	.5647	.5658	.5670
3.7	.5682	.5694	.5705	.5717	.5729	.5740	.5752	.5763	.5775	.5786
3.8	.5798	.5809	.5821	.5832	.5843	.5855	.5866	.5877	.5888	.5899
3.9	.5911	.5922	.5933	.5944	.5955	.5966	.5977	.5988	.5999	.6010
4.0	.6021	.6031	.6042	.6053	.6064	.6075	.6085	.6096	.6107	.6117
4.1	.6128	.6138	.6149	.6160	.6170	.6180	.6191	.6201	.6212	.6222
4.2	.6232	.6243	.6253	.6263	.6274	.6284	.6294	.6304	.6314	.6325
4.3	.6335	.6345	.6355	.6365	.6375	.6385	.6395	.6405	.6415	.6425
4.4	.6435	.6444	.6454	.6464	.6474	.6484	.6493	.6503	.6513	.6522
4.5	.6532	.6542	.6551	.6561	.6571	.6580	.6590	.6599	.6609	.6618
4.6	.6628	.6637	.6646	.6656	.6665	.6675	.6684	.6693	.6702	.6712
4.7	.6721	.6730	.6739	.6749	.6758	.6767	.6776	.6785	.6794	.6803
4.8	.6812	.6821	.6830	.6839	.6848	.6857	.6866	.6875	.6884	.6893
4.9	.6902	.6911	.6920	.6928	.6937	.6946	.6955	.6964	.6972	.6981
5.0	.6990	.6998	.7007	.7016	.7024	.7033	.7042	.7050	.7059	.7067
5.1	.7076	.7084	.7093	.7101	.7110	.7118	.7126	.7135	.7143	.7152
5.2	.7160	.7168	.7177	.7185	.7193	.7202	.7210	.7218	.7226	.7235
5.3	.7243	.7251	.7259	.7267	.7275	.7284	.7292	.7300	.7308	.7316
5.4	.7324	.7332	.7340	.7348	.7356	.7364	.7372	.7380	.7388	.7396
x	0	1	2	3	4	5	6	7	8	9

Table II *(continued)*

x	0	1	2	3	4	5	6	7	8	9
5.5	.7404	.7412	.7419	.7427	.7435	.7443	.7451	.7459	.7466	.7474
5.6	.7482	.7490	.7497	.7505	.7513	.7520	.7528	.7536	.7543	.7551
5.7	.7559	.7566	.7574	.7582	.7589	.7597	.7604	.7612	.7619	.7627
5.8	.7634	.7642	.7649	.7657	.7664	.7672	.7679	.7686	.7694	.7701
5.9	.7709	.7716	.7723	.7731	.7738	.7745	.7752	.7760	.7767	.7774
6.0	.7782	.7789	.7796	.7803	.7810	.7818	.7825	.7832	.7839	.7846
6.1	.7853	.7860	.7868	.7875	.7882	.7889	.7896	.7903	.7910	.7917
6.2	.7924	.7931	.7938	.7945	.7952	.7959	.7966	.7973	.7980	.7987
6.3	.7993	.8000	.8007	.8014	.8021	.8028	.8035	.8041	.8048	.8055
6.4	.8062	.8069	.8075	.8082	.8089	.8096	.8102	.8109	.8116	.8122
6.5	.8129	.8136	.8142	.8149	.8156	.8162	.8169	.8176	.8182	.8189
6.6	.8195	.8202	.8209	.8215	.8222	.8228	.8235	.8241	.8248	.8254
6.7	.8261	.8267	.8274	.8280	.8287	.8293	.8299	.8306	.8312	.8319
6.8	.8325	.8331	.8338	.8344	.8351	.8357	.8363	.8370	.8376	.8382
6.9	.8388	.8395	.8401	.8407	.8414	.8420	.8426	.8432	.8439	.8445
7.0	.8451	.8457	.8463	.8470	.8476	.8482	.8488	.8494	.8500	.8506
7.1	.8513	.8519	.8525	.8531	.8537	.8543	.8549	.8555	.8561	.8567
7.2	.8573	.8579	.8585	.8591	.8597	.8603	.8609	.8615	.8621	.8627
7.3	.8633	.8639	.8645	.8651	.8657	.8663	.8669	.8675	.8681	.8686
7.4	.8692	.8698	.8704	.8710	.8716	.8722	.8727	.8733	.8739	.8745
7.5	.8751	.8756	.8762	.8768	.8774	.8779	.8785	.8791	.8797	.8802
7.6	.8808	.8814	.8820	.8825	.8831	.8837	.8842	.8848	.8854	.8859
7.7	.8865	.8871	.8876	.8882	.8887	.8893	.8899	.8904	.8910	.8915
7.8	.8921	.8927	.8932	.8938	.8943	.8949	.8954	.8960	.8965	.8971
7.9	.8976	.8982	.8987	.8993	.8998	.9004	.9009	.9015	.9020	.9025
8.0	.9031	.9036	.9042	.9047	.9053	.9058	.9063	.9069	.9074	.9079
8.1	.9085	.9090	.9096	.9101	.9106	.9112	.9117	.9122	.9128	.9133
8.2	.9138	.9143	.9149	.9154	.9159	.9165	.9170	.9175	.9180	.9186
8.3	.9191	.9196	.9201	.9206	.9212	.9217	.9222	.9227	.9232	.9238
8.4	.9243	.9248	.9253	.9258	.9263	.9269	.9274	.9279	.9284	.9289
8.5	.9294	.9299	.9304	.9309	.9315	.9320	.9325	.9330	.9335	.9340
8.6	.9345	.9350	.9355	.9360	.9365	.9370	.9375	.9380	.9385	.9390
8.7	.9395	.9400	.9405	.9410	.9415	.9420	.9425	.9430	.9435	.9440
8.8	.9445	.9450	.9455	.9460	.9465	.9469	.9474	.9479	.9484	.9489
8.9	.9494	.9499	.9504	.9509	.9513	.9518	.9523	.9528	.9533	.9538
9.0	.9542	.9547	.9552	.9557	.9562	.9566	.9571	.9576	.9581	.9586
9.1	.9590	.9595	.9600	.9605	.9609	.9614	.9619	.9624	.9628	.9633
9.2	.9638	.9643	.9647	.9652	.9657	.9661	.9666	.9671	.9675	.9680
9.3	.9685	.9689	.9694	.9699	.9703	.9708	.9713	.9717	.9722	.9727
9.4	.9731	.9736	.9741	.9745	.9750	.9754	.9759	.9763	.9768	.9773
9.5	.9777	.9782	.9786	.9791	.9795	.9800	.9805	.9809	.9814	.9818
9.6	.9823	.9827	.9832	.9836	.9841	.9845	.9850	.9854	.9859	.9863
9.7	.9868	.9872	.9877	.9881	.9886	.9890	.9894	.9899	.9903	.9908
9.8	.9912	.9917	.9921	.9926	.9930	.9934	.9939	.9943	.9948	.9952
9.9	.9956	.9961	.9965	.9969	.9974	.9978	.9983	.9987	.9991	.9996
x	0	1	2	3	4	5	6	7	8	9

TABLE III

VALUES OF e^x

x	e^x	e^{-x}	x	e^x	e^{-x}
0.00	1.0000	1.0000	0.50	1.6487	0.6065
0.01	1.0101	0.9901	0.51	1.6653	0.6005
0.02	1.0202	0.9802	0.52	1.6820	0.5945
0.03	1.0305	0.9705	0.53	1.6990	0.5886
0.04	1.0408	0.9608	0.54	1.7160	0.5827
0.05	1.0513	0.9512	0.55	1.7333	0.5769
0.06	1.0618	0.9418	0.56	1.7507	0.5712
0.07	1.0725	0.9324	0.57	1.7683	0.5655
0.08	1.0833	0.9231	0.58	1.7860	0.5599
0.09	1.0942	0.9139	0.59	1.8040	0.5543
0.10	1.1052	0.9048	0.60	1.8221	0.5488
0.11	1.1163	0.8958	0.61	1.8404	0.5434
0.12	1.1275	0.8869	0.62	1.8590	0.5380
0.13	1.1388	0.8781	0.63	1.8776	0.5326
0.14	1.1503	0.8694	0.64	1.8965	0.5273
0.15	1.1618	0.8607	0.65	1.9155	0.5220
0.16	1.1735	0.8521	0.66	1.9348	0.5169
0.17	1.1853	0.8437	0.67	1.9542	0.5117
0.18	1.1972	0.8353	0.68	1.9739	0.5066
0.19	1.2092	0.8270	0.69	1.9937	0.5016
0.20	1.2214	0.8187	0.70	2.0138	0.4966
0.21	1.2337	0.8106	0.71	2.0340	0.4916
0.22	1.2461	0.8025	0.72	2.0544	0.4868
0.23	1.2586	0.7945	0.73	2.0751	0.4819
0.24	1.2712	0.7866	0.74	2.0959	0.4771
0.25	1.2840	0.7788	0.75	2.1170	0.4724
0.26	1.2969	0.7711	0.76	2.1383	0.4677
0.27	1.3100	0.7634	0.77	2.1598	0.4630
0.28	1.3231	0.7558	0.78	2.1815	0.4584
0.29	1.3364	0.7483	0.79	2.2034	0.4538
0.30	1.3499	0.7408	0.80	2.2255	0.4493
0.31	1.3634	0.7334	0.81	2.2479	0.4449
0.32	1.3771	0.7261	0.82	2.2705	0.4404
0.33	1.3910	0.7190	0.83	2.2933	0.4360
0.34	1.4050	0.7118	0.84	2.3164	0.4317
0.35	1.4191	0.7047	0.85	2.3396	0.4274
0.36	1.4333	0.6977	0.86	2.3632	0.4232
0.37	1.4477	0.6907	0.87	2.3869	0.4190
0.38	1.4623	0.6839	0.88	2.4109	0.4148
0.39	1.4770	0.6771	0.89	2.4351	0.4107
0.40	1.4918	0.6703	0.90	2.4596	0.4066
0.41	1.5068	0.6636	0.91	2.4843	0.4025
0.42	1.5220	0.6570	0.92	2.5093	0.3985
0.43	1.5373	0.6505	0.93	2.5345	0.3946
0.44	1.5527	0.6440	0.94	2.5600	0.3906
0.45	1.5683	0.6376	0.95	2.5857	0.3867
0.46	1.5841	0.6313	0.96	2.6117	0.3829
0.47	1.6000	0.6250	0.97	2.6379	0.3791
0.48	1.6160	0.6188	0.98	2.6645	0.3753
0.49	1.6323	0.6126	0.99	2.6912	0.3716

Table III (*continued*)

x	e^x	e^{-x}
1.0	2.7183	0.3679
1.1	3.0042	0.3329
1.2	3.3201	0.3012
1.3	3.6693	0.2725
1.4	4.0552	0.2466
1.5	4.4817	0.2231
1.6	4.9530	0.2019
1.7	5.4739	0.1827
1.8	6.0496	0.1653
1.9	6.6859	0.1496
2.0	7.3891	0.1353
2.1	8.1662	0.1225
2.2	9.0250	0.1108
2.3	9.9742	0.1003
2.4	11.023	0.0907
2.5	12.182	0.0821
2.6	13.464	0.0743
2.7	14.880	0.0672
2.8	16.445	0.0608
2.9	18.174	0.0550
3.0	20.086	0.0498
3.1	22.198	0.0450
3.2	24.533	0.0408
3.3	27.113	0.0369
3.4	29.964	0.0334
3.5	33.115	0.0302
3.6	36.598	0.0273
3.7	40.447	0.0247
3.8	44.701	0.0224
3.9	49.402	0.0202
4.0	54.598	0.0183
4.1	60.340	0.0166
4.2	66.686	0.0150
4.3	73.700	0.0136
4.4	81.451	0.0123
4.5	90.017	0.0111
4.6	99.484	0.0101
4.7	109.95	0.0091
4.8	121.51	0.0082
4.9	134.29	0.0074
5.0	148.41	0.0067
5.1	164.02	0.0061
5.2	181.27	0.0055
5.3	200.34	0.0050
5.4	221.41	0.0045

x	e^x	e^{-x}
5.5	244.69	0.0041
5.6	270.43	0.0037
5.7	298.87	0.0034
5.8	330.30	0.0030
5.9	365.04	0.0027
6.0	403.43	0.0025
6.1	445.86	0.0022
6.2	492.75	0.0020
6.3	544.57	0.0018
6.4	601.85	0.0017
6.5	665.14	0.0015
6.6	735.10	0.0014
6.7	812.41	0.0012
6.8	897.85	0.0011
6.9	992.27	0.0010
7.0	1096.6	0.0009
7.1	1212.0	0.0008
7.2	1339.5	0.0007
7.3	1480.3	0.0007
7.4	1636.0	0.0006
7.5	1808.0	0.0006
7.6	1998.2	0.0005
7.7	2208.4	0.0005
7.8	2440.6	0.0004
7.9	2697.3	0.0004
8.0	2981.0	0.0003
8.1	3294.5	0.0003
8.2	3641.0	0.0003
8.3	4023.9	0.0002
8.4	4447.1	0.0002
8.5	4914.8	0.0002
8.6	5431.7	0.0002
8.7	6002.9	0.0002
8.8	6634.2	0.0002
8.9	7332.0	0.0001
9.0	8103.1	0.0001
9.1	8955.3	0.0001
9.2	9897.1	0.0001
9.3	10938	0.0001
9.4	12088	0.0001
9.5	13360	0.0001
9.6	14765	0.0001
9.7	16318	0.0001
9.8	18034	0.0001
9.9	19930	0.0001

TABLE IV

VALUES OF ln x

x	$\ln x$
0.1	−2.3026
0.2	−1.6094
0.3	−1.2040
0.4	−0.9163
0.5	−0.6931
0.6	−0.5108
0.7	−0.3567
0.8	−0.2231
0.9	−0.1054
1.0	0.0000
1.1	0.0953
1.2	0.1823
1.3	0.2624
1.4	0.3365
1.5	0.4055
1.6	0.4700
1.7	0.5306
1.8	0.5878
1.9	0.6419
2.0	0.6931
2.1	0.7419
2.2	0.7885
2.3	0.8329
2.4	0.8755
2.5	0.9163
2.6	0.9555
2.7	0.9933
2.8	1.0296
2.9	1.0647
3.0	1.0986
3.1	1.1314
3.2	1.1632
3.3	1.1939
3.4	1.2238
3.5	1.2528
3.6	1.2809
3.7	1.3083
3.8	1.3350
3.9	1.3610
4.0	1.3863
4.1	1.4110
4.2	1.4351
4.3	1.4586
4.4	1.4816

x	$\ln x$
4.5	1.5041
4.6	1.5261
4.7	1.5476
4.8	1.5686
4.9	1.5892
5.0	1.6094
5.1	1.6292
5.2	1.6487
5.3	1.6677
5.4	1.6864
5.5	1.7047
5.6	1.7228
5.7	1.7405
5.8	1.7579
5.9	1.7750
6.0	1.7918
6.1	1.8083
6.2	1.8245
6.3	1.8405
6.4	1.8563
6.5	1.8718
6.6	1.8871
6.7	1.9021
6.8	1.9169
6.9	1.9315
7.0	1.9459
7.1	1.9601
7.2	1.9741
7.3	1.9879
7.4	2.0015
7.5	2.0149
7.6	2.0281
7.7	2.0412
7.8	2.0541
7.9	2.0669
8.0	2.0794
8.1	2.0919
8.2	2.1041
8.3	2.1163
8.4	2.1282
8.5	2.1401
8.6	2.1518
8.7	2.1633
8.8	2.1748
8.9	2.1861

x	$\ln x$
9.0	2.1972
9.1	2.2083
9.2	2.2192
9.3	2.2300
9.4	2.2407
9.5	2.2513
9.6	2.2618
9.7	2.2721
9.8	2.2824
9.9	2.2925
10	2.3026
11	2.3979
12	2.4849
13	2.5649
14	2.6391
15	2.7081
16	2.7726
17	2.8332
18	2.8904
19	2.9444
20	2.9957
25	3.2189
30	3.4012
35	3.5553
40	3.6889
45	3.8067
50	3.9120
55	4.0073
60	4.0943
65	4.1744
70	4.2485
75	4.3175
80	4.3820
85	4.4427
90	4.4998
100	4.6052
110	4.7005
120	4.7875
130	4.8676
140	4.9416
150	5.0106
160	5.0752
170	5.1358
180	5.1930
190	5.2470

C Formulas from Geometry

PLANE FIGURES

1. Triangle ABC with sides of lengths a and c, base b, and altitude h.

   Perimeter: $P = a + b + c$

   Area: $A = \frac{1}{2}bh$

   $\angle A + \angle B + \angle C = 180°$

   a. Isoceles triangle.
      Two sides of equal length.
      Two angles with equal measure.
   b. Equilateral triangle.
      Three sides of equal length.
      Three angles with equal measure.
   c. Right triangle with hypotenuse c:

   $$c^2 = a^2 + b^2$$

2. Square with side of length s.

   Perimeter: $P = 4s$
   Area: $A = s^2$

3. Rectangle with length l and width w.

   Perimeter: $P = l + l + w + w$
   $P = 2l + 2w$
   Area: $A = lw$

4. Circle with radius r.

   Diameter: $d = 2r$
   Circumference: $C = 2\pi r$ or πd
   Area: $A = \pi r^2$

SOLID FIGURES

1. Right circular cylinder with height h and radius r of the base.

   Volume: $V = \pi r^2 h$
   Lateral area: $S = 2\pi rh$

2. Rectangular prism with length l, width w, and height h.

   Volume: $V = lwh$

Answers to Odd-numbered Exercises in Each Section; Answers to All Exercises in Reviews

Exercise 1.1 [page 6] **1.** $\{3, 4, 5\}$ **3.** $\{0, 1, 2\}$ **5.** $\{5, 6, 7, \ldots\}$ **7.** $\{5, 7, 9\}$ **9.** finite **11.** finite **13.** infinite **15.** $\in$ **17.** $\in$ **19.** $\notin$ **21.** $\{0, 8\}$ **23.** $\{-\sqrt{15}\}$ **25.** $\{-5, -\sqrt{15}, -3.44, -2/3\}$ **27.** no **29.** no

Exercise 1.2 [page 12] **1.** $3r$ **3.** 6 **5.** y **7.** r **9.** $6 + x$ **11.** $8 > 5$ **13.** $-6 < -4$ **15.** $x + 1 < 0$ **17.** $x - 4 \leq 0$ **19.** $-2 < y < 3$ **21.** $1 \leq x < 7$ **23.** $-2 < 8$ **25.** $-7 > -13$ **27.** $-6 < -3$ **29.** $1\frac{1}{2} = \frac{3}{2}$ **31.** $3 < 5 < 7$ **33.** $-7 < 0 < 2$

35. 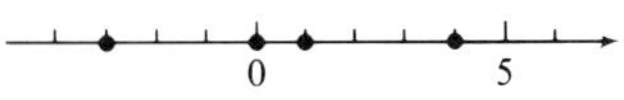

37. 

39. 

41. 

43. 

45. 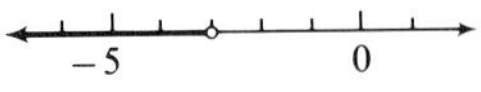

47. 

49. 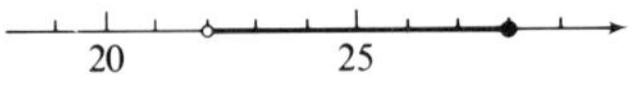

51. 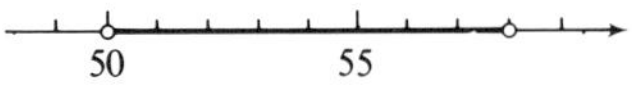

53. 

Exercise 1.3 [page 17] **1.** 12 **3.** $3y$ **5.** $t \cdot 4$ **7.** 1 **9.** 1 **11.** $3x$; $3y$ **13.** 5 **15.** -3 **17.** x **19.** 3 **21.** $-x$ **23.** positive **25.** 3 **27.** 4 **29.** -2 **31.** -5 **33.** x, if $x \geq 0$; $-x$, if $x < 0$ **35.** $x - 2$, if $x \geq 2$; $-(x - 2)$, if $x < 2$ **37.** no; yes **39.** no; no **41.** yes **43.** no

Exercise 1.4 [page 21] **1.** 11 **3.** 6 **5.** 3 **7.** -7 **9.** 5 **11.** -3 **13.** -9 **15.** 6 **17.** 2 **19.** 11 **21.** -6 **23.** 3 **25.** 4 **27.** -10 **29.** 10 **31.** 3 **33.** 7 **35.** 8 **37.** 11

39. Substituting 5 and 3 for a and b, we get $5 - 3 \stackrel{?}{=} 3 - 5$; $2 \neq -2$. Other values can be used for a and b. **41.** integers

Exercise 1.5 [page 26] 1. -12 3. 12 5. 24 7. -20 9. 24 11. 0 13. -24 15. 24 17. -24 19. 24 21. $2 \cdot 2 \cdot 2$ 23. $7 \cdot 7$ 25. prime 27. $-1 \cdot 2 \cdot 2 \cdot 3$ 29. $2 \cdot 2 \cdot 2 \cdot 7$ 31. $-1 \cdot 2 \cdot 2 \cdot 2 \cdot 2 \cdot 3$ 33. -4 35. -13 37. 3 39. 0 41. undefined 43. 4 45. $7\left(\frac{1}{8}\right)$ 47. $3\left(\frac{1}{8}\right)$ 49. $82\left(\frac{1}{11}\right)$ 51. $7\left(\frac{1}{100}\right)$ 53. $\frac{3}{2}$ 55. $\frac{3x}{y}$ 57. $\frac{4x}{4}$ 59. $\frac{2(x+y)}{2}$

61. Substituting 1 for y gives $2(3 \cdot 1) \overset{?}{=} 2 \cdot 3(2 \cdot 1)$; $6 \neq 12$. 63. $x = 0, y \neq 0$

65. $x, y > 0$ or $x, y < 0$ 67. Substituting 8 and 4 for a and b gives $8 \div 4 \overset{?}{=} 4 \div 8$; $2 \neq \frac{1}{2}$.

69. 1. $b \neq 0$ (hypothesis)

2. $\frac{a}{b} = q$ implies $bq = a$ (definition of quotient)

3. $\frac{1}{b}(bq) = \frac{1}{b} \cdot a$ (multiplication property)

4. $\left(\frac{1}{b} \cdot b\right)q = a \cdot \frac{1}{b}$ (associative and commutative properties of multiplication)

5. $1 \cdot q = a \cdot \frac{1}{b}$ (reciprocal property)

6. $q = a \cdot \frac{1}{b}$ (identity element for multiplication)

7. $\frac{a}{b} = a \cdot \frac{1}{b}$ (substitution of $\frac{a}{b}$ for q)

Exercise 1.6 [page 29] 1. 20 3. 0 5. -17 7. -13 9. -2 11. 6 13. -44 15. 5 17. -2 19. -2 21. 60 23. 10 25. 2 27. undefined 29. 1 31. 0 33. 100 35. 4 37. 1080 39. 56 centimeters 41. 1016

Review Exercises [page 32] 1. $\{-3, 0, 1\}$ 2. $\{0, 1\}$ 3. $a < c$ 4. $a \leq 7$ 5. $a < b < c$ 6. $y \geq 6$ 7. 

8. 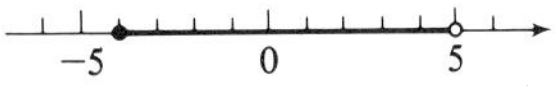

9. $2 + x$ 10. $18 + t$ 11. $\frac{1}{3}$ 12. 1 13. a. 14 b. 8 14. a. -6 b. 5

15. a. -36 b. 0 c. 21 d. 24

16. a. $2 \cdot 2 \cdot 2 \cdot 3$ b. $2 \cdot 2 \cdot 7$ c. $-1 \cdot 2 \cdot 3 \cdot 7$ d. $-1 \cdot 2 \cdot 2 \cdot 2 \cdot 3 \cdot 3$

17. a. 12 b. 0 c. -8 d. -8

18. a. $-7 \cdot \frac{1}{5}$ b. $24 \cdot \frac{1}{7}$ c. $-4 \cdot \frac{1}{5}$ d. $8 \cdot \frac{1}{3}$ 19. a. $\frac{2x}{3}$ b. $\frac{3y}{4}$ c. $\frac{x}{3}$ d. $\frac{y}{4}$

20. a. 3 b. 6 21. a. 5 b. 1 22. a. 16 b. -11 23. 212 24. 8

25. 6 26. -26 27. 2 28. -3 29. $x \geq 0$ 30. a. $x > y$ b. $x < y$

Exercise 2.1 [page 41] 1. binomial; degree 3; 2 and -1 3. monomial; degree 4; 5 5. trinomial; degree 2; 3 and -1 7. trinomial; degree 3; 1, -2, and -1 9. -25 11. 9 13. -2 15. 50 17. -2 19. -5 21. 7 23. 13 25. 5 27. 0 29. 1 31. 64 33. 8 35. 54 37. -1; -21 39. -4; 16 41. 19; 7 43. 37; 11 45. 14 47. -4 49. 4 51. 15 53. -1 55. 2 57. Closure for multiplication; closure for both addition and multiplication.

Exercise 2.2 [page 46] 1. $7x^2$ 3. $-3y^2$ 5. z^2 7. $7x^2y - 2x$ 9. $6r^2 + 4r$ 11. $s^2 - 4s$ 13. $t^2 + 2t - 1$ 15. $-2u^2 - u - 3$ 17. $-x^2 - 4x + 7$ 19. $-t^3 - 4t^2 + t + 1$ 21. $a^2 - 6a - 3$ 23. $8x^2y + 3xy^2 - xy$ 25. $-2x^2 + x - 3$ 27. $b^2 + 3b + 1$ 29. $-2y - 1$ 31. $2 - x$ 33. $x - 1$ 35. $-x^2 - 3x - 1$ 37. Substituting 1 for x gives $-(1 + 1) \stackrel{?}{=} -1 + 1$; $-2 \neq 0$ 39. $2x - y$ 41. $-4x - 5$ 43. $2x - 1$ 45. -1 47. $2x - 1$ 49. -1

Exercise 2.3 [page 50] 1. $-14t^3$ 3. $-40a^3b^3c$ 5. $44x^3y^4z^2$ 7. $6x^5y^5$ 9. $-2r^6s^4t^2$ 11. $3x^2y^6z^4$ 13. $6r^3t^4$ 15. $6x^4$ 17. $6z^5$ 19. x^6 21. x^4y^4 23. y^6z^3 25. $8x^3z^6$ 27. $4x^2y^4z^2$ 29. $-8x^3y^9z^3$ 31. $x^4y^2 + x^3y^3$ 33. $4x^2y^4z^2 - x^2y^2z^4$ 35. $x^2y^2 - x^3y^4$ 37. $4x^5y^3 + xy^2$ 39. $8x^2y^2 - 3x^5y^2$ 41. $2x^4y^2$ 43. a^{3n-3} 45. x^n 47. a^{3n+1} 49. $x^{6n}y^3$ 51. $x^{3n-6}y^3$ 53. $x^{6n+3}y^{3n-3}$

Exercise 2.4 [page 54] 1. $4xy - 8y^2$ 3. $-12x^3 + 6x^2 - 6x$ 5. $-x^2 + 4x + 1$ 7. $x^2 + 6x + 9$ 9. $4y^2 - 20y + 25$ 11. $x^2 - 9$ 13. $n^2 + 10n + 16$ 15. $r^2 + 3r - 10$ 17. $y^2 - 7y + 6$ 19. $2z^2 - 5z - 3$ 21. $8r^2 + 2r - 3$ 23. $4x^2 - a^2$ 25. $y^3 - y + 6$ 27. $x^3 + 2x^2 - 21x + 18$ 29. $x^3 - 7x + 6$ 31. $z^3 - 7z - 6$ 33. $6x^3 + x^2 - 8x + 6$ 35. $6a^4 - 5a^3 - 5a^2 + 5a - 1$ 37. 6 39. $-2a^2 - 2a$ 41. $4a + 4$ 43. $4a + 4$ 45. $-4x^2 - 11x$ 47. $2x^2 + 12x + 2$ 49. $-2x^3 + 12x^2 - 11x$ 51. $-8x^3 + 2x^2 + 6x$ 53. $8x - 16$

55. $$\begin{aligned}(x + a)(x + b) &= x(x + a) + b(x + a)\\ &= x^2 + ax + bx + ab\\ &= x^2 + (a + b)x + ab\end{aligned}$$

57. $$\begin{aligned}(x + a)(x - a) &= x(x + a) - a(x + a)\\ &= x^2 + ax - ax - a^2\\ &= x^2 - a^2\end{aligned}$$

59. $$\begin{aligned}(x + a)(x^2 - ax + a^2) &= x(x^2 - ax + a^2) + a(x^2 - ax + a^2)\\ &= (x^3 - ax^2 + a^2x) + (ax^2 - a^2x + a^3)\\ &= x^3 - ax^2 + a^2x + ax^2 - a^2x + a^3\\ &= x^3 + a^3\end{aligned}$$

61. Substituting 1 for x and 2 for y gives $(1 + 2)^2 \stackrel{?}{=} 1^2 + 2^2$; $9 \neq 5$ 63. $2x^{2n} - x^n$ 65. $a^{2n+1} - a^{n+1}$ 67. $a^{3n+1} + a^{2n+2}$ 69. $1 - a^{2n}$ 71. $a^{6n} + a^{3n} - 2$ 73. $2a^{2n} + 3a^nb^n - 2b^{2n}$

Review Exercises [page 56] 1. a. monomial; degree 3 b. trinomial; degree 2
2. a. binomial; degree 5 b. trinomial; degree 4 3. $1/5$ 4. 7 5. a. 8 b. 13
6. a. -2 b. -14 7. a. $x + y - z$ b. $2x^2 - 2z^2 - x + 3y$
8. a. $3x - 3y + z$ b. $-2x - y - 6z$
9. a. $-2x^2 - 4y - z^2 - z$ b. $-2x^2 + 2x - y^2 - y + 2z$
10. a. $2x - 5$ b. $-x^2 - 2x$ 11. a. $2y^2 - 1$ b. $2y + 1$
12. a. $-z$ b. 2 13. a. $-6x^3y^4$ b. $-6x^2y^3z^3$
14. a. $-8x^6y^9z^3$ b. $x^3y^6 - 4x^6y^2$
15. a. $9x^4y^5$ b. $2x^4y^5$ 16. a. $6x^3y^4z^2$ b. $6x^2y^4z^3$
17. a. $5x^3y^2 - x^4y^3$ b. $5x^3y^3 - 3x^3y^2$ 18. a. $5x^2y^2 + 2x^3y$ b. $6x^2y^3$
19. a. $2x^3 - 4x^2 + 2x$ b. $4x^2 + 10x - 6$
20. a. $y^3 - 3y^2 + 3y - 2$ b. $z^3 + 2z^2 - z - 2$
21. a. $-3x - 15$ b. $7x^2 - 7x$ 22. a. $-y^2 + 6y$ b. $-2y^2 + 4y$
23. a. $-10x + 9$ b. $-x^2 + x - 2$
24. a. $-8x^2 + 26x - 22$ b. $6x^2 + 21x + 39$ 25. 17 26. 0 27. 4 28. 14
29. $x^{3n-3}y^{3n}$ 30. $2x^{2n} - 5x^n - 3$

Exercise 3.1 [page 67] 1. $\{7\}$ 3. $\{7\}$ 5. $\{240\}$ 7. $\{2890\}$ 9. $\{8/3\}$ 11. $\{4\}$
13. $\{-13/2\}$ 15. $\{16/3\}$ 17. $\{-6\}$ 19. $\{8/3\}$ 21. $\{1/2\}$ 23. $\{6\}$ 25. $\{4/3\}$
27. $\{-22/27\}$ 29. $\{0\}$ 31. $\{11\}$ 33. $\{24\}$ 35. $\{15\}$ 37. $\{18\}$ 39. $\{74\}$ 41. $\{4\}$
43. $m = \dfrac{f}{a}$ 45. $p = \dfrac{I}{rt}$ 47. $w = \dfrac{P - 2l}{2}$ 49. $g = \dfrac{v - k}{t}$
51. $h = \dfrac{S}{2\pi} - r$ or $h = \dfrac{S - 2\pi r}{2\pi}$ 53. $n = \dfrac{l - a}{d} + 1$ or $n = \dfrac{l - a + 1}{d}$ 55. $t = \dfrac{A - P}{pr}$

Exercise 3.2 [page 77] 1. $3x + 5 = 26$ 3. $4x - 6 = 22$ 5. $2x + x = 21$
7. $\dfrac{x + 2}{4} = 5$

Note: Equations for Problems 9–59 are not unique. The answers given can be obtained using different equations.

9. a. Number of students that applied: x
 b. $0.75x = 600$
 c. 800 students

11. a. Total sales: x
 b. $0.02x = 300$
 c. $15,000

13. a. Number of students enrolled: x
 b. $0.30x = 12$
 c. 40 students

15. a. Total number of bats: x
 b. $0.95x = 1710$
 (If 5% of bats are discarded, 95% are not discarded.)
 c. 1800 bats

17. a. Total number of employees: x
    b. $0.06x = 9$
    (If 94% were present, 6% were absent.)
    c. 150 employees

19. a. Length of shorter piece: x
    Length of longer piece: $x + 6$
    b. $x + (x + 6) = 24$
    c. 9 feet and 15 feet

21. a. Number of votes of loser: x
    Number of votes of winner: $x + 140$
    b. $x + (x + 140) = 620$
    c. Loser: 240 votes
    Winner: 380 votes

23. a. Number of votes of winner: x
    Number of votes of second candidate: $x - 210$
    Number of votes of third candidate: $x - 490$
    b. $x + (x - 210) + (x - 490) = 6560$
    c. Winner: 2420 votes
    Second candidate: 2210 votes
    Third candidate: 1930 votes

25. a. Measure of smallest angle: x
    Measure of second angle: $2x$
    Measure of third angle: $3x + 12$
    b. $x + 2x + (3x + 12) = 180$
    c. Smallest angle: 28°
    Second angle: 56°
    Third angle: 96°

27. a. Measure of smallest angle: x
    Measure of second angle: $x + 10$
    Measure of third angle: $2x + 10$
    b. $x + (x + 10) + (2x + 10) = 180$
    c. Smallest angle: 40°
    Second angle: 50°
    Third angle: 90°

29. a. Length of smallest side: x
    Length of each equal side: $x + 15$
    b. $x + (x + 15) + (x + 15) = 66$
    c. Smallest side: 12 cm
    Each equal side: 27 cm

31. a. Length of side of original square: x
    b. $(x + 5)^2 - x^2 = 85$
    c. 6 cm

33. a. Width of table: x
    Length of table: $x + 4$
    b. $2x + 2(x + 4) = 28$
    c. Width: 5 feet
    Length: 9 feet

35. a. Number of quarters: x
    Number of dimes: $2x$
    b. $0.25x + 0.10(2x) = 3.60$
    c. 8 quarters; 16 dimes

37. a. Number of dimes: d
    Number of quarters: $d + 12$
    b. $10d + 25(d + 12) = 1245$
    c. 27 dimes; 39 quarters

39. a. Number of first-class passengers: f
    Number of tourist passengers: $42 - f$

41. a. Pounds of less expensive brand: x
    Pounds of more expensive brand: $x + 2$

b. $80f + 64(42 - f) = 2880$
c. 12 first class; 30 tourist

43. a. Number of children's tickets: x
Number of adult tickets: $82 - x$
b. $2.50x + 4.00(82 - x) = 310$
c. 12 children's tickets; 70 adult tickets

47. a. Amount invested at 9%: x
Amount invested at 11%: $42{,}000 - x$
b. $0.09x = 0.11(42{,}000 - x)$
c. \$23,100 at 9% and \$18,900 at 11%

51. a. Money invested at 13%: x
b. $0.08(3000) + 0.13x = 0.10(3000 + x)$
c. \$2000

55. a. Number of ounces: x
b. $0.40x + 0.60(60 - x) = 0.50(60)$
c. 30 ounces

59. a. Liters of pure alcohol: x
b. $x + 0.45(12) = 0.60(x + 12)$
c. 4.5 liters

b. $1.20x + 1.40(x + 2) = 28.80$
c. Less expensive brand: 10 pounds;
More expensive brand: 12 pounds

45. a. Amount invested at 10%: x
Amount invested at 12%: $x + 4000$
b. $0.10x + 0.12(x + 4000) = 920$
c. \$2000 at 10%; \$6000 at 12%

49. a. Amount invested at 10%: x
Amount invested at 12%: $8000 - x$
b. $0.10x + 0.12(8000 - x) = 844$
c. \$5800 at 10% and \$2200 at 12%

53. a. Number of quarts of 10% solution: x
b. $0.10x + 0.40(20) = 0.30(x + 20)$
c. 10 quarts

57. a. Liters of 30% solution: x
b. $0.30x + 0.12(40) = 0.20(x + 40)$
c. 32 liters

Exercise 3.3 [page 88]

1. $\{x|x < 2\}$ or $(-\infty, 2)$

3. $\{x|x \leq 12\}$ or $(-\infty, 12]$

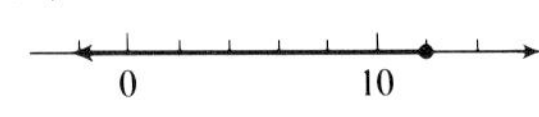

5. $\{x|x > 3\}$ or $(3, +\infty)$

7. $\{x|x > 3\}$ or $(3, +\infty)$

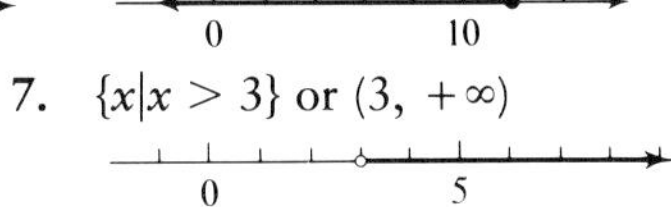

9. $\{x|x < -6\}$ or $(-\infty, -6)$

11. $\{x|x \geq -14/3\}$ or $[-14/3, +\infty)$

13. $\{x|x > -11\}$ or $(-11, +\infty)$

15. $\{x|x \geq 0\}$ or $[0, +\infty)$

17. $\{x|6 \leq x < 10\}$ or $(6, 10)$

19. $\{x|-2 < x \leq 3\}$ or $(-2, 3]$

21. $\{x|-6 < x < -2\}$ or $(-6, -2)$

23. $\{x|-2 < x < 2\}$ or $(-2, 2)$

25. $\{x|-4 \leq x \leq 2\}$ or $[-4, 2]$

27. $\{x|-7 < x \leq -4\}$ or $(-7, -4]$

29. a. Grade on fifth test: x

b. $80 \leqslant \dfrac{78 + 64 + 88 + 76 + x}{5} < 90$

c. 94% or more

31. a. Range in degrees Fahrenheit: F

b. $30 < \dfrac{5F - 160}{9} < 40$

c. Between 86°F and 104°F

33. $\{x | -8 \leqslant x \leqslant 2\}$; $\{x | 3 < x \leqslant 7\}$

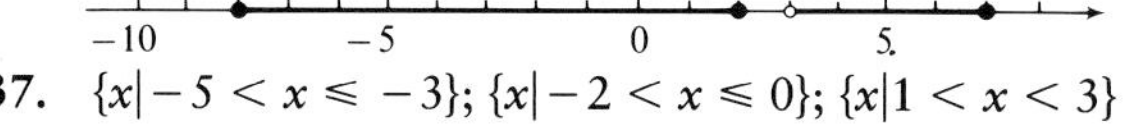

35. $\{x | -7 \leqslant x \leqslant -3\}$; $\{x | 0 < x \leqslant 4\}$

37. $\{x | -5 < x \leqslant -3\}$; $\{x | -2 < x \leqslant 0\}$; $\{x | 1 < x < 3\}$

39. $[-6, -4]$ 41. disjoint 43. disjoint 45. $(-3, 2)$

Exercise 3.4 [page 93] 1. $\{5, -5\}$ 3. $\{13, -5\}$ 5. $\{6, -7\}$ 7. $\{1, 5/3\}$ 9. $\{-3/4\}$

11. $\{0, 3\}$

13. $\{x | -2 < x < 2\}$ or $(-2, 2)$

15. $\{x | -7 \leqslant x \leqslant 1\}$ or $[-7, 1]$

17. $\{x | 1 < x < 4\}$ or $(1, 4)$

19. $\{x | -4 \leqslant x \leqslant 12\}$ or $[-4, 12]$

21. $\{x | x < -3\} \cup \{x | x > 3\}$ or $(-\infty, -3) \cup (3, +\infty)$

23. $\{x | x < -3\} \cup \{x | x > 7\}$ or $(-\infty, -3) \cup (7, \infty)$

25. $\{x | x \leqslant -2\} \cup \{x | x \geqslant 5\}$ or $(-\infty, -2] \cup [5, +\infty)$

27. $[-8, 6]$ 29. $(-\infty, 3/2) \cup (9/2, \infty)$ 31. $\{-2/5\}$ 33. $[-5/4, 7/4]$ 35. $[0, 4]$

37. $[-3, 3)$ 39. $[-2, 0)$

Review Exercises [page 95] 1. $\{1\}$ 2. $\{600\}$ 3. $\{2/3\}$ 4. $\{-5\}$ 5. $\{5\}$ 6. $\{8\}$

7. $\{3/2\}$ 8. $\{-1/2\}$ 9. $\{5\}$ 10. $\{200\}$ 11. $\{3\}$ 12. $\{17\}$ 13. $h = \dfrac{l + d - a}{d}$

14. $a = \dfrac{2s - k}{t}$ 15. $R = \dfrac{S}{2\pi} + r$ or $R = \dfrac{S + 2\pi r}{2\pi}$ 16. $F = \dfrac{9C}{5} + 32$ or $F = \dfrac{9C + 160}{5}$

17. a. Total number of serves: x

b. $0.26x = 52$

c. 200 serves

18. a. Total salary: x

b. $x - 0.20x = 180$

$0.80x = 180$

c. $225

19. a. Cost of computer B: x

Cost of computer A: $x + 120$

b. $x + (x + 120) = 460$

c. Computer B: $170

Computer A: $290

20. a. Number of votes of winner: x

Number of votes of second candidate: $x - 120$

Number of votes of third candidate: $x - 160$

b. $x + (x - 120) + (x - 160) = 440$
c. Winner: 240 votes
Second candidate: 120 votes
Third candidate: 80 votes

21. a. Number of 13¢ bolts: x
Number of 10¢ bolts: $3x$
b. $0.13x + 0.10(3x) = 2.58$
c. Six 13¢ bolts;
eight 10¢ bolts

22. a. Number of 22¢ stamps: x
Number of 2¢ stamps: $x + 6$
b. $0.22x + 0.02(x + 6) = 7.32$
c. Thirty 22¢ stamps;
thirty-six 2¢ stamps

23. a. Amount invested at 10%: x
Amount invested at 11%: $x + 1000$
b. $0.10x + 0.11(x + 1000) = 320$
c. \$1000 at 10%; \$2000 at 11%

24. a. Amount invested at 8%: x
Amount invested at 10%: $3600 - x$
b. $0.08x = 0.10(3600 - x)$
c. \$2000 at 8%; \$1600 at 10%

25. a. Number of quarts of 20% solution: x
b. $0.20x + 0.30(10) = 0.25(x + 10)$
c. 10 quarts

26. a. Number of pounds of alloy containing 60% copper: x
b. $0.60x + 0.20(4 - x) = 0.35(4)$
c. $1\frac{1}{2}$ pounds

27. $\{x|x \leq 27\}$ or $(-\infty, 27]$

28. $\{x|x \leq 3/2\}$ or $(-\infty, 3/2]$

29. $\{x|2 < x \leq 5\}$ or $(2, 5]$

30. $\{x|1 < x < 3\}$ or $(1, 3)$

31. $\{x|-5 < x < 2\}$ or $(-5, 2)$

32. $\{x|0 < x < 4\}$ or $(0, 4)$

33. a. $\{3, -4\}$ b. $\{-5/4, 11/4\}$

34. a. $\{x|x \leq -8\} \cup \{x|x \geq 2\}$ or $(-\infty, -8] \cup [2, +\infty)$ b. $\{x|-3 < x < 7\}$ or $(-3, 7)$

35. a. $\{x \mid -4 \leq x < 5\}$ b. $\{x \mid -2 < x \leq 6\}$ 36. a. $\{x \mid -5 \leq x \leq -3\}$ b. $\{x \mid 4 < x < 8\}$ 37. $[0, 5)$ 38. $(-2, 0)$ 39. $[-1, 3]$ 40. $(-4, 0]$

Exercise 4.1 [page 102] 1. $2(x + 3)$ 3. $4x(x + 2)$ 5. $3x(x - y + 1)$ 7. $6(4a^2 + 2a - 1)$ 9. $2x(x^3 - 2x + 4)$ 11. $3z^2(4z^2 + 5z - 3)$ 13. $a(y^2 + by + b)$ 15. $3mn(m - 2n + 4)$ 17. $3ac(5ac - 4 + 2c^2)$ 19. $(a + b)(a + 3)$ 21. $(2x - y)(x + 3)$ 23. $(2y - x)(a + b)$ 25. $(x + 3)(x + 4)$ 27. $4(x - 2)(x - 4)$ 29. $(2x - 1)^2(x + 1)$ 31. $-5x(x - 5)$ 33. $2(x + 3)(3x - 1)$ 35. $-4(x - 1)$ 37. $-(r - 7)$

39. $-(b - 2a)$ 41. $-2(x - 1)$ 43. $-a(b + c)$ 45. $-(-2x + 1)$
47. $-(-x + y - z)$ 49. $x^n(x^n - 1)$ 51. $x^n(x^{2n} - x^n - 1)$ 53. $x^n(x^2 + 1)$
55. $-x^n(x^n + 1)$ 57. $-x^a(x + 1)$

Exercise 4.2 [page 108] 1. $(x + 3)(x + 2)$ 3. $(y - 4)(y - 3)$ 5. $(x - 3)(x + 2)$
7. $(y - 5)(y + 2)$ 9. $(2x - 1)(x + 2)$ 11. $(4x - 1)(x + 2)$ 13. $(3x - 1)(x - 1)$
15. $(3x - 8)(3x + 1)$ 17. $(5x - 3)(2x + 1)$ 19. $(2x + 3)(2x + 3)$ or $(2x + 3)^2$
21. $(3x - a)(x - 2a)$ 23. $(4xy - 1)(4xy + 1)$ 25. $(x - 5)(x + 5)$ 27. $(xy - 1)(xy + 1)$
29. $(y^2 - 3)(y^2 + 3)$ 31. $(x - 2y)(x + 2y)$ 33. $(2x - 5y)(2x + 5y)$
35. $(4xy - 1)(4xy + 1)$ 37. $3(x + 2)(x + 2)$ or $3(x + 2)^2$ 39. $2a(a - 5)(a + 1)$
41. $4(a - b)(a - b)$ or $4(a - b)^2$ 43. $4y(x - 3)(x + 3)$ 45. $x(4 + x)(3 - x)$
47. $x^2y^2(x - 1)(x + 1)$ 49. $(y^2 + 1)(y^2 + 2)$ 51. $(3x^2 + 1)(x^2 + 2)$
53. $(x^2 + 4)(x - 1)(x + 1)$ 55. $(x - 2)(x + 2)(x - 1)(x + 1)$ 57. $(2a^2 + 1)(a - 1)(a + 1)$
59. $(x^2 + 2a^2)(x - a)(x + a)$ 61. $(x^{2n} + 1)(x^n + 1)(x^n - 1)$
63. $(x^{2n} + y^{2n})(x^n + y^n)(x^n - y^n)$ 65. $(3x^{2n} - 1)(x^{2n} - 3)$

Exercise 4.3 [page 112] 1. $(a + b)(x + 1)$ 3. $(ax + 1)(x + a)$ 5. $(x + a)(x + y)$
7. $(3a - c)(b - d)$ 9. $(3x + y)(1 - 2x)$ 11. $(a^2 + 2b^2)(a - 2b)$ 13. $(x + 2y)(x - 1)$
15. $(2a^2 - 1)(b + 3)$ 17. $(x^3 - 3)(y^2 + 1)$ 19. $(x + 2)(x^2 + 4)$ 21. $(2x - 3)(x^2 + 1)$
23. $(x - 1)(x^2 + 1)$ 25. $(x + 2 - y)(x + 2 + y)$ 27. $(x + 1 - y)(x + 1 + y)$
29. $(y - x + 1)(y + x - 1)$ 31. $(2x + 1 - 2y)(2x + 1 + 2y)$ 33. $(x - 1)(x^2 + x + 1)$
35. $(2x + y)(4x^2 - 2xy + y^2)$ 37. $(a - 2b)(a^2 + 2ab + 4b^2)$ 39. $(xy - 1)(x^2y^2 + xy + 1)$
41. $(3a + 4b)(9a^2 - 12ab + 16b^2)$ 43. $(4ab - 1)(16a^2b^2 + 4ab + 1)$
45. $(2x - y)(x^2 - xy + y^2)$ 47. $[(x + 1) - 1][(x + 1)^2 + (x + 1) + 1] = x(x^2 + 3x + 3)$
49. $[(x + 1) - (x - 1)][(x + 1)^2 + (x + 1)(x - 1) + (x - 1)^2] = 2(3x^2 + 1)$
51. $ac - ad + bd - bc$
$= (ac - ad) - (bc - bd)$
$= a(c - d) - b(c - d)$
$= (a - b)(c - d)$;

$ac - ad + bd - bc$
$= (bd - ad) - (bc - ac)$
$= (b - a)d - (b - a)c$
$= (b - a)(d - c)$

53. $(x^2 + 2y^2 + xy)(x^2 + 2y^2 - xy)$ 55. $(a^2 + 5b^2 - 2ab)(a^2 + 5b^2 + 2ab)$

Exercise 4.4 [page 116] 1. $\{-2, 5\}$ 3. $\{-5/2, 2\}$ 5. $\{0, -1/2\}$ 7. $\{6, -3/2\}$ 9. $\{2, -1/2\}$
11. $\{5/2, -2/3\}$ 13. $\{0, 3\}$ 15. $\{0, 3]$ 17. $\{3, -3\}$ 19. $\{3, -3\}$ 21. $\{2/3, -2/3\}$
23. $\{3/2, -3/2\}$ 25. $\{4, 1\}$ 27. $\{7, -2\}$ 29. $\{1\}$ 31. $\{1/2, 1\}$ 33. $\{3, -2\}$
35. $\{1, -6\}$ 37. $\{1, -5\}$ 39. $\{1/2, -3\}$ 41. $\{6, -3\}$ 43. $\{3\}$ 45. $\{1, 4\}$
47. $\{3, -3\}$ 49. $x^2 + x - 2 = 0$ 51. $x^2 + 5x = 0$ 53. $x^2 + 6x + 9 = 0$
55. $x^2 - ax - bx + ab = 0$ 57. $x = \pm 2b$ 59. $x = 4a,\ x = -a$ 61. $x = a,\ x = b$
63. $x = -a,\ x = -b/2$

Exercise 4.5 [page 120]

1. a. Smaller positive number: x
   Larger positive number: $x + 3$
   b. $x(x + 3) = 40$
   c. 5 and 8

3. a. A negative integer: x
   Next consecutive integer: $x + 1$
   b. $x^2 + (x + 1)^2 = 61$
   c. -6 and -5

5. a. A positive integer: x
   Next consecutive integer: $x + 1$
   Next consecutive integer: $x + 2$
   b. $x^2 + (x + 1)^2 + (x + 2)^2 = 149$
   c. 6, 7, and 8

7. a. Width: x
   Length: $x + 2$
   b. $x(x + 2) = 63$
   c. Width: 7 meters; length: 9 meters

9. a. Length of altitude: x
   Length of base: $x - 3$
   b. $\frac{1}{2}x(x - 3) = 27$
   c. Altitude: 9 inches;
   base: 6 inches

11. a. Shorter side: x
    Longer side: $x + 2$
    b. $x^2 + (x + 2)^2 = 10^2$
    c. Shorter side: 6 cm;
    longer side: 8 cm

13. a. Time to reach 28 feet on the way down: t
    b. $28 = 64t - 16t^2$
    c. 3½ seconds (reaches 28 feet on the way up in ½ second)

15. a. Time to move 24 cm in a positive direction: t
    b. $24 = t^2 - 5t$
    c. 8 seconds

17. a. Width: x
    Length: $2x$
    b. $(x - 4)(2x - 4) = 48$
    c. Width: 8 in.; length: 16 in.

19. a. Width of border: x
    b. $2x(25 + x) + 50x = 310$
    c. 3 meters

21. a. Width of tray: x
    Length of tray: $x + 2$
    Height of tray: 2
    b. $2(x - 4)(x + 2 - 4) = 160$
    c. Width: 12 cm; length: 14 cm

Review Exercises [page 122] 1. a. $4(3x^2 - 2x + 1)$ b. $x(x^2 - 3x - 1)$
2. a. $4y^2(y - 2)$ b. $2xy^2(2x^2 - x + 3)$ 3. a. $-(y - 3x)$ b. $-(-2x + y - z)$
4. a. $-x(x - 2)$ b. $-3xy(2x - 1 + y)$ 5. a. $(a + 2)(x - y)$ b. $(x - y)(2a + b)$
6. a. $(x + 2y)(1 - x)$ b. $(x + y)(y + 1)$ 7. a. $(x - 7)(x + 5)$ b. $(y + 8)(y - 4)$

8. a. $(xy - 6)(xy + 6)$ b. $(a - 7b)(a + 7b)$
9. a. $(3y - 1)(y + 4)$ b. $x(x + 5)(x - 2)$
10. a. $9(x - 2)(x + 2)$ b. $3(2x - y)(2x + y)$
11. a. $(2x - y)(x + 2y)$ b. $(3x + y)(2x - y)$
12. a. $(3a + 2b)(5a + 6b)$ b. $6(2a - b)(a - b)$
13. a. $(x + y)(2x + 1)$ b. $(y - 3)(x - 1)$
14. a. $(x + y)(a - 2b)$ b. $(x - 2y)(2a + b)$
15. a. $(2a - 1)(a - 2)(a + 2)$ b. $(a + 2)(a + 2)(a - 2)$
16. a. $(2y - 3 - 2x)(2y - 3 + 2x)$ b. $(2y - 1 - x)(2y - 1 + x)$
17. a. $(y - 2x - 1)(y + 2x + 1)$ b. $(x - y + 4)(x + y - 4)$
18. a. $(2x - y)(4x^2 + 2xy + y^2)$ b. $(x + 4y)(x^2 - 4xy + 16y^2)$
19. a. $(3y + z)(9y^2 - 3yz + z^2)$ b. $(x - 2a)(x^2 + 2ax + 4a^2)$
20. a. $(2xy - 5)(4x^2y^2 + 10xy + 25)$ b. $(1 + 4xy)(1 - 4xy + 16x^2y^2)$
21. a. $\{0, 2\}$ b. $\{2, 3\}$ 22. a. $\{-3, 5\}$ b. $\{-3, 2\}$ 23. a. $\{2\}$ b. $\{1, 3\}$
24. a. $x^2 - 3x - 10 = 0$ b. $x^2 + 3x = 0$
25. a. Length of altitude: x
Length of base: $2x + 1$
b. $\frac{1}{2}x(2x + 1) = 18$
c. Altitude: 4 inches; base: 9 inches

26. a. Length: x
Width: $x - 4$
b. $x(x - 4) = 60$
c. Length: 10 cm; width: 6 cm

27. a. An integer: x
Next consecutive integer: $x + 1$
Next consecutive integer: $x + 2$
b. $x^2 + (x + 1)^2 + (x + 2)^2 = 194$
c. 7, 8, and 9

28. a. Time to travel 63 meters: t
b. $63 = 15t + 2t^2$
c. 3 seconds

29. $x^{2n}(x^{2n} + 1)$ 30. $x^n(x - 1)(x + 1)$ 31. $2x(x^2 + 3y^2)$ 32. $2y(3x^2 + y^2)$

Exercise 5.1 [page 133] 1. $\frac{-1}{4}$ 3. $\frac{3}{5}$ 5. $\frac{2}{5}$ 7. $\frac{-3}{7}$ 9. $\frac{-2x}{y}$ 11. $\frac{3x}{4y}$

13. $\frac{-(x + 1)}{x}$ or $\frac{-x - 1}{x}$ 15. $\frac{-(x - y)}{y + 2}$, $\frac{-x + y}{y + 2}$, or $\frac{y - x}{y + 2}$ 17. $\frac{4}{y - 3}$

19. $\frac{-1}{y - x}$ 21. $\frac{2 - x}{x - 3}$ 23. $\frac{-x - 1}{y - x}$ 25. $\frac{x - 2}{y - x}$ 27. $\frac{a - 1}{3a + b}$

29. Substituting 1 for x gives $-\frac{1 + 3}{2} \stackrel{?}{=} \frac{-1 + 3}{2}$; $-2 \neq 1$. 31. $\frac{2x^2}{y}$ 33. $\frac{2r^2}{t}$ 35. $\frac{-2}{cd^2}$

37. $\frac{1}{a^3c^2}$ 39. $\frac{-1}{6p^2}$ 41. $\frac{2r}{t}$ 43. $\frac{2x + 3}{3}$ 45. $\frac{3x - 1}{3}$ 47. $-a^2 + 3a - 2$

49. $y + 7$ 51. $\frac{5}{3}$ 53. $\frac{a - b}{2}$ 55. $\frac{6t + 6}{t - 1}$ 57. $y - 2$ 59. $\frac{-2}{y + 3}$

61. $\dfrac{-x-y}{x-y}$ 63. $\dfrac{x-4}{x+1}$ 65. $\dfrac{2y-3}{y-1}$ 67. $\dfrac{x+2y}{x+y}$ 69. $\dfrac{x+y}{2}$ 71. $\dfrac{x+y}{a+2b}$

73. $\dfrac{4y^2+6y+9}{2y+3}$ 75. Let $x = 1$ and $y = 1$: $\dfrac{2(1)+(1)}{1} \neq 2(1)$

Exercise 5.2 [page 138] 1. $4a^2 + 2a + \dfrac{1}{2}$ 3. $y^2 - 2 + \dfrac{3}{7y^2}$ 5. $6rs - 5 + \dfrac{2}{rs}$

7. $4ax - 2x + \dfrac{1}{2}$ 9. $-5m^3 + 3 - \dfrac{7}{5m^3}$ 11. $8m^2 - 5 + \dfrac{7}{5m}$ 13. $2y + 5 + \dfrac{2}{2y+1}$

15. $2t - 1 - \dfrac{6}{2t-1}$ 17. $x^2 + 4x + 9 + \dfrac{19}{x-2}$ 19. $a^3 - 3a^2 + 6a - 16 + \dfrac{47}{a+3}$

21. $4z^3 - 2z^2 + 3z + 1 + \dfrac{2}{2z+1}$ 23. $x^3 + 2x^2 + 4x + 8 + \dfrac{15}{x-2}$

25. $x - 1 + \dfrac{-7x+12}{x^2-2x+7}$ 27. $4a^2 - 9a + 31 + \dfrac{-104a+32}{a^2+3a-1}$

29. $t - 1 + \dfrac{-t^2-3t+3}{t^3-2t^2+t+2}$ 31. $k = -2$

Exercise 5.3 [page 143] 1. $x - 2 \quad (x \neq 6)$ 3. $x + 2 \quad (x \neq -2)$

5. $x^3 - x^2 + \dfrac{-1}{x-2} \quad (x \neq 2)$ 7. $2x^2 - 2x + 3 + \dfrac{-8}{x+1} \quad (x \neq -1)$

9. $2x^3 + 10x^2 + 50x + 249 + \dfrac{1251}{x-5} \quad (x \neq 5)$ 11. $x^2 + 2x - 3 + \dfrac{4}{x+2} \quad (x \neq -2)$

13. $x^3 + x + 2 + \dfrac{3}{x-2}$ 15. $x^4 + 2x^3 + 2x^2 + 4x + 4 + \dfrac{3}{x-1}$

17. $x^4 - x^3 + x^2 - 4x + 4 - \dfrac{5}{x+1}$

19. $x^5 + x^4 + 2x^3 + 2x^2 + 2x + 1 + \dfrac{1}{x-1} \quad (x \neq 1)$

21. $x^4 + x^3 + x^2 + x + 1 \quad (x \neq 1)$ 23. $x^5 + x^4 + x^3 + x^2 + x + 1 \quad (x \neq 1)$

Exercise 5.4 [page 147] 1. $\dfrac{6}{9}$ 3. $\dfrac{-30}{14}$ 5. $\dfrac{20}{5}$ 7. $\dfrac{6}{18x}$ 9. $\dfrac{-a^2b}{b^3}$ 11. $\dfrac{xy^2}{xy}$

13. $\dfrac{3a+3b}{a^2-b^2}$ 15. $\dfrac{3xy-9x}{y^2-y-6}$ 17. $\dfrac{-2x-4}{x^2+3x+2}$ 19. 60 21. 120 23. 252

25. $6ab^2$ 27. $24x^2y^2$ 29. $a(a-b)^2$ 31. $(a-b)(a+b)$ 33. $(a+4)(a+1)^2$

35. $(x+4)(x-1)^2$ 37. $x(x-1)^3$ 39. $4(a+1)(a-1)^2$ 41. $x^3(x-1)^2$

43. $\dfrac{4y}{6xy}$ and $\dfrac{1}{6xy}$ 45. $\dfrac{9}{12y^2}$ and $\dfrac{8y}{12y^2}$ 47. $\dfrac{1}{2(x+1)}$ and $\dfrac{x+1}{2(x+1)}$

49. $\dfrac{3y+6}{(y-3)(y+2)}$ and $\dfrac{2y-6}{(y-3)(y+2)}$ 51. $\dfrac{10y+5}{3(y-2)(2y+1)}$ and $\dfrac{y-2}{3(y-2)(2y+1)}$

53. $\dfrac{2}{a^2-1}$ and $\dfrac{3a+3}{a^2-1}$ 55. $\dfrac{a-1}{(a-4)(a-1)^2}$ and $\dfrac{2a-8}{(a-4)(a-1)^2}$

57. $\dfrac{3y^2+6y}{(y+1)(y+2)^2}$ and $\dfrac{y^2+y}{(y+1)(y+2)^2}$ 59. $\dfrac{4x^2-2xy}{(2x-y)^2(2x+y)}$ and $\dfrac{6x^2+3xy}{(2x-y)^2(2x+y)}$

61. $\dfrac{10y}{2(x+y)^2}$ and $\dfrac{3x+3y}{2(x+y)^2}$ 63. $\dfrac{1}{(x-1)(x^2+x+1)}$ and $\dfrac{3x^2+3x+3}{(x-1)(x^2+x+1)}$

65. $\dfrac{4x}{(x+y)(x^2-xy+y^2)}$ and $\dfrac{x^2y-xy^2+y^3}{(x+y)(x^2-xy+y^2}$

67. $\dfrac{3x}{(x-y)(x-y)}$ and $\dfrac{-2x^2+2xy}{(x-y)(x-y)}$

69. $\dfrac{2}{2(x-2y)}$ and $\dfrac{-3}{2(x-2y)}$ or $\dfrac{-2}{2(2y-x)}$ and $\dfrac{3}{2(2y-x)}$

Exercise 5.5 [page 152] 1. $\dfrac{x-3}{2}$ 3. $\dfrac{a+b-c}{6}$ 5. $\dfrac{2x-1}{2y}$ 7. $\dfrac{1-2x}{x+2y}$

9. $\dfrac{6-2a}{a^2-2a+1}$ 11. $\dfrac{7x}{6}$ 13. $\dfrac{-y}{15}$ 15. $\dfrac{5x}{12}$ 17. $\dfrac{7}{6}x$ 19. $\dfrac{1}{12}y$ 21. $\dfrac{3}{4}y$

23. $\dfrac{3xy+y-x}{xy}$ 25. $\dfrac{3x-4y}{6xy}$ 27. $\dfrac{5y-3x-8}{(x+1)(y-1)}$ 29. $\dfrac{4x^2+x}{(3x+2)(x-1)}$

31. $\dfrac{-3y^2+3y}{(2y-1)(y+1)}$ 33. $\dfrac{2x^2+8x+7}{(x+2)(x+3)}$ 35. $\dfrac{y^2-4y+5}{(y+1)(2y-3)}$ 37. $\dfrac{17}{6(x+2)}$

39. $\dfrac{-4}{15(x-2)}$ 41. $\dfrac{4x-2}{(x-2)(x+1)(x+1)}$ 43. $\dfrac{-6y-4}{(y+4)(y-4)(y-1)}$

45. $\dfrac{-x^2-4x-1}{(x-1)(x-1)(x+1)}$ 47. $\dfrac{3y+1}{y(y-3)(y+2)}$ 49. $\dfrac{-x^2-7x+2}{(x-2)(x-2)(x+2)}$

51. $\dfrac{-1}{(z-4)(z-3)(z-2)}$ 53. $\dfrac{6}{y-4}$ 55. $\dfrac{2}{y+5}$ 57. $\dfrac{x^3-2x^2+2x-2}{(x-1)(x-1)}$

59. $\dfrac{y^3-2y^2-y}{(y-1)(y+1)}$ 61. $\dfrac{x^3+2x^2+3x-4}{(x+1)(x+1)(x-1)}$ 63. $\dfrac{x^2+x+1}{x+2}$ 65. $\dfrac{x^2+5x+4}{x+3}$

67. $\dfrac{x^3+2x+2}{x+1}$

69. $$\frac{c}{s}+\frac{s}{c}=\frac{c\cdot c}{c\cdot s}+\frac{s\cdot s}{c\cdot s}$$
$$=\frac{c^2+s^2}{cs}=\frac{1}{sc}$$

71. $$c+\frac{s^2}{c}=\frac{c\cdot c}{c\cdot 1}+\frac{s^2}{c}$$
$$=\frac{c^2+s^2}{c}=\frac{1}{c}$$

73. $$\frac{(c-1)(c+1)}{s^2}+1=\frac{c^2-1}{s^2}+\frac{s^2}{s^2}$$
$$=\frac{c^2+s^2-1}{s^2}$$
$$=\frac{1-1}{s^2}=0$$

75. $$\frac{s-c}{s+c}+\frac{2sc}{s^2-c^2}$$
$$=\frac{(s-c)(s-c)}{(s-c)(s+c)}+\frac{(2sc)}{(s-c)(s+c)}$$
$$=\frac{s^2-2sc+c^2+2sc}{(s-c)(s+c)}$$
$$=\frac{s^2+c^2}{s^2-c^2}=\frac{1}{s^2-c^2}$$

Exercise 5.6 [page 158] **1.** $\frac{2}{3}$ **3.** $\frac{7}{10}$ **5.** $\frac{10}{3}$ **7.** $\frac{1}{8x}$ **9.** $\frac{25p}{n}$ **11.** $\frac{-n}{2}$ **13.** $\frac{2x^5}{7}$

15. $\frac{x^3y^3}{2}$ **17.** $\frac{-x^3y^2z^2}{3}$ **19.** $\frac{-b^2}{a}$ **21.** $\frac{3c}{35ab}$ **23.** $\frac{5}{ab}$ **25.** 5 **27.** $\frac{a(2a-1)}{a+4}$

29. $\frac{x+3}{x-5}$ **31.** $\frac{x-7}{x-5}$ **33.** $\frac{(x-2)(3x+1)}{(3x-1)(x-1)}$ **35.** $\frac{3-a}{a+1}$ **37.** $\frac{1}{5}x^2 - 3x$

39. $x^2 + \frac{2}{3}x + \frac{1}{9}$ **41.** $y^2 - \frac{1}{2}y + \frac{1}{16}$ **43.** $\frac{4}{3}$ **45.** $\frac{1}{ax^2y}$ **47.** $\frac{20ay}{3}$ **49.** $\frac{2}{9}$

51. $\frac{a+1}{a-2}$ **53.** $\frac{x-5}{x+5}$ **55.** $\frac{3x-1}{x-2}$ **57.** $\frac{x+2}{x^2-1}$ **59.** $\frac{x^3-4x^2}{x+1}$ **61.** $\frac{x+3}{6y}$

63. $\frac{6y+11}{(2y-1)(3y+2)}$ **65.** $\frac{5y}{(2y-3)(3y-2)}$ **67.** $\frac{2}{y(y-2)}$

69. $3(x^2 - xy + y^2)$ **71.** $(y-3)(x+2)$ **73.** $\frac{x+1}{x-1}$ **75.** $\frac{x+1}{x}$ **77.** $\frac{1}{x+1}$

Exercise 5.7 [page 163] **1.** $\frac{3}{2}$ **3.** $\frac{2}{21}$ **5.** $\frac{a}{bc}$ **7.** $\frac{4y}{3}$ **9.** $\frac{1}{5}$ **11.** $\frac{1}{10}$

13. $\frac{8}{13}$ **15.** $\frac{11}{35}$ **17.** $\frac{7}{2(5a+1)}$ **19.** x **21.** $\frac{x}{x-1}$ **23.** $\frac{y}{y+2}$

25. $\frac{y^2}{y^3+y^2-y-1}$ **27.** $\frac{x^2y-x^2}{xy^2+y^2}$ **29.** $\frac{-y+2}{2y}$ **31.** $\frac{-y+3}{y}$ **33.** $\frac{y+1}{y-2}$

35. $\frac{a-2b}{a+2b}$ **37.** $\frac{a+6}{a-1}$ **39.** 1

41.
$$\frac{(c+s)^2}{c-s} - \frac{1}{\frac{c-s}{2cs}} = \frac{c^2+2cs+s^2}{c-s} - \frac{2cs}{c-s}$$
$$= \frac{c^2+s^2+2cs-2cs}{c-s}$$
$$= \frac{1}{c-s}$$

43.
$$\frac{\frac{1}{s-c}}{\frac{s}{s^2-c^2}} + \frac{s-c}{s} = \frac{\frac{1}{s-c}\cdot(s-c)(s+c)}{\frac{s}{(s-c)(s+c)}\cdot(s-c)(s+c)} + \frac{s-c}{s}$$
$$= \frac{s+c}{s} + \frac{s-c}{s}$$
$$= \frac{s+c+s-c}{s} = \frac{2s}{s} = 2$$

Exercise 5.8 [page 169] **1.** $\{-27/2\}$ **3.** $\{3\}$ **5.** $\{-30\}$ **7.** $\{-7\}$ **9.** $\emptyset$
11. $\{13\}$ **13.** $\emptyset$ **15.** $\{4\}$ **17.** $\{-2\}$ **19.** $\{4\}$ **21.** $\{40\}$ **23.** $\{-2, 3/2\}$

25. $\{-3, 1\}$ 27. $\{-10/3, 1\}$ 29. 66 inches 31. 132 feet 33. 120 ohms

35. $8\frac{3}{4}$ years; 15 years 37. 62.6° Fahrenheit 39. 15 sides 41. $r = \dfrac{S - a}{S}$

43. $t = \dfrac{2rs}{s - r}$ 45. $t = \dfrac{15c - 15V}{c}$ 47. $n = \dfrac{125b - 50w}{w}$

Exercise 5.9 [page 177]

1. a. A number: x
   b. $\frac{1}{2}x + 3x = \frac{35}{2}$
   c. 5

3. a. An integer: x
   Next consecutive integer: $x + 1$
   b. $\frac{1}{2}x + \frac{2}{3}(x + 1) = 17$
   c. 14 and 15

5. a. A number: x
   b. $x + \frac{1}{x} = \frac{17}{4}$
   c. 4 or $\frac{1}{4}$

7. a. Denominator: x
   Numerator: $x - 1$
   b. $\dfrac{x - 1}{x} + 2\left(\dfrac{x}{x - 1}\right) = \dfrac{41}{12}$
   c. Denominator: 4
   Numerator: 3

9. a. Profit: P
   b. $\frac{2}{3}P = 160$
   c. $240

11. a. Amount greater than $100: A
   b. $\frac{1}{6}(A + 100) = 25$
   c. $50

13. a. Numerator: N
   b. $\dfrac{N}{N + 6} = \dfrac{3}{4}$
   c. 18

15. a. Number of bricks laid in 40 hours: x
   b. $\dfrac{x}{450} = \dfrac{40}{3}$
   c. 6000 bricks

17. a. Number of pounds of tin to make 90 pounds of alloy: x
   b. $\dfrac{x}{2.5} = \dfrac{90}{12}$
   c. $18\frac{3}{4}$ pounds

19. a. Number addressed in 5 hours: N
   b. $\dfrac{N}{144} = \dfrac{5}{3}$
   c. 240 envelopes

21. a. Earnings in 52 weeks: E
   b. $\dfrac{E}{6200} = \dfrac{52}{20}$
   c. $16,120

23. a. Rate of auto: r
   Rate of plane: $r + 120$
   b. $\dfrac{1260}{r + 120} = \dfrac{420}{r}$
   c. Auto: 60 miles per hour
   Plane: 180 miles per hour

25. a. Time for second ship to reach first shp: t
b. $20t + 5 = 30t$
c. ½ hour

27. a. Speed of man: s
Speed of woman: $s + 20$
b. $\frac{120}{s + 20} = \frac{80}{s}$
c. Speed of man: 40 miles per hour
Speed of woman: 60 miles per hour

29. a. Speed of current: s
b. $\frac{63}{18 + r} - \frac{30}{18 - r} = 1$
c. 3 miles per hour

31. a. Rate to the city: r
Rate from the city: $r + 10$
b. $\frac{10}{r} + \frac{10}{r + 10} = \frac{5}{6}$
c. 20 miles per hour; 30 miles per hour

33. a. Time to fill tank with both pipes running together: t
b. $\frac{1}{30}t + \frac{1}{45}t = 1$
c. 18 hours

35. a. Time of the older machine: t
b. $\left(\frac{1}{10}\right)6 + \left(\frac{1}{t}\right)6 = 1$
c. 15 hours

37. a. Time for slow operator alone: t
b. $20\left(\frac{1}{t}\right) + 20\left(\frac{5}{4t}\right) = 1$
c. 45 hours

Review Exercises [page 181]

1. a. $\frac{-1}{-(a - b)}, \quad \frac{-1}{b - a}, \quad -\frac{-1}{a - b}, \quad -\frac{-1}{-(b - a)}, \quad -\frac{1}{-(a - b)}, \quad -\frac{1}{b - a}$

b. $\frac{1}{a - b}$ is a positive number if $a > b$ and a negative number if $a < b$.

2. $\frac{1}{a - 1}, \quad \frac{-1}{1 - a}$ 3. a. $\frac{2x}{5y}$ b. $x - 2$ 4. a. $-2x - 1$ b. $-\frac{1}{2}$

5. a. $6x - 3 + \frac{3}{2x}$ b. $2y + 1 - \frac{1}{3y}$

6. a. $y - 3 + \frac{10}{2y + 3}$ b. $bx^2 + 3x^2 + 3x + 5 + \frac{7}{x - 1}$

7. $y^2 + 2y - 4$ 8. $y^6 + y^5 + y^4 + y^3 + y^2 + y + 1$ 9. a. $\frac{-18}{24}$ b. $\frac{x^2y}{2xy^2}$

10. a. $\frac{2x - 6y}{x^2 - 9y^2}$ b. $\frac{y - y^2}{y^2 - 4y + 3}$ 11. a. $\frac{4x}{3}$ b. $\frac{-x - y}{4x}$

12. a. $\frac{8y - 15x + 7}{20xy}$ b. $\frac{3y + 4}{6y - 18}$ 13. a. $\frac{2x + 21}{(x - 2)(x + 3)}$ b. $\frac{x - 2y + 3}{x^2 - 4y^2}$

14. a. $\frac{5x + 1}{(x^2 - 1)(x - 1)}$ b. $\frac{-6y - 4}{(y - 4)(y + 4)(y - 1)}$ 15. a. $\frac{x^2}{3}$ b. $\frac{2}{y}$

16. a. 1 b. $\frac{2y^2 - 2y}{y + 1}$ 17. a. $\frac{5xy}{3}$ b. 1 18. a. $\frac{(y + 3)(y + 2)}{(y - 2)(y + 1)}$ b. $\frac{1}{1 - x}$

19. a. $\frac{5}{3}$ b. $\frac{2}{11}$ 20. a. $\frac{6x + 9}{3x - 2}$ b. $\frac{y^2 - 1}{y^2 + 1}$ 21. a. $\frac{2x}{x^2 - 1}$ b. 10

22. a. $\frac{y^3 + y^2 - y - 1}{y^2}$ b. $\frac{(x^2 + 1)(x - 1)}{x(x - 2)}$ 23. a. $\left\{\frac{1}{2}\right\}$ b. {12}

24. a. {4} b. $\left\{\frac{-10}{7}\right\}$ 25. a. $\left\{\frac{-4}{3}\right\}$ b. {−17} 26. a. {1/8, 4} b. {1/2, 6}

27. a. Length of larger side: $\frac{x}{11}$

b. $\frac{x}{5} = \frac{11}{4}$

c. $13\frac{3}{4}$ in.

28. a. Rate of car: x

Rate of plane: $3x$

b. $3x - x = 100$

c. Rate of car: 50 miles per hour; rate of plane: 150 miles per hour

29. a. Time after leaving station: t

b. $20t + 60t = 200$

c. $2\frac{1}{2}$ hours

30. a. Number of hits: $-x$

b. $\frac{17 + x}{60 + 20} = 0.3$

c. 7 hits

31. $\frac{x^2 + x + 1}{x + 1}$ 32. $\frac{4y^2 - 2y + 1}{2y - 1}$ 33. $x - 1 + \frac{-2x + 3}{x^2 - x + 2}$ 34. $x + \frac{-x^2 - x + 4}{x^3 - 2x + 1}$

35. $x^2 + 3x + 1$ 36. $x^2 - 5 + \frac{18}{x^2 + 4}$ 37. $\frac{x^2 + xy + y^2}{x}$ 38. $\frac{y^2 + 2y + 4}{2}$

39. $\frac{13}{10}$ 40. $\frac{3x - 7y}{2x + 2y}$ 41. 1 42. $\frac{a^2 - 4a + 1}{a^2 + a - 5}$ 43. 2 44. $2b^2$

45. a. Original cost of each notebook: x

b. $96 = \left(x + \frac{2}{5}\right)\left(\frac{96}{x} - 20\right)$

c. $1.20

46. a. Number of seats now: x

b. $1200 = (x + 60)\left(\frac{1200}{x} - 1\right)$

c. 300 seats

Cumulative Review Exercises, Chapters 1–5 [page 185]

1. 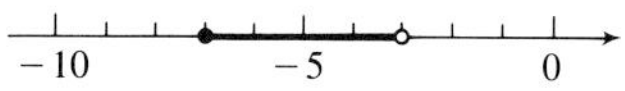

2. $|3 - x| = \begin{cases} 3 - x & \text{if } x \leq 3 \\ x - 3 & \text{if } x > 3 \end{cases}$ 3. 288 4. 14. 5. 56

6. $2x^3 + x^2 - 13x + 6$ 7. $3x(x - 3)^2(x + 2)(x^3 - 6x^2 + 5x - 8)$

8. $(2x^2 - y^3)(4x^4 + 2x^2y^3 + y^6)$ 9. $(x^2 - y)(x + 2y^2)$ 10. −12 11. −18

12. 4 13. $x^2 + 9x - 5$ 14. $-11a^4b^5 - a^4b^4$ 15. $x^4 - 3x^3 + 5x^2 - 2x + 6$

16. {8} 17. no solution 18. {8000} 19. {−3, 6} 20. {0} 21. [1, 13]

22. $(-\infty, 1] \cup [2, \infty)$ 23. $P = \frac{A}{1 + rt}$ 24. $h = \frac{S - 2\pi r^2}{2\pi r}$ 25. $\frac{3 - 2x}{2x(x + 3)}$

26. $x^4 + 2x^3 + x^2 + 2x + 4 + \frac{7}{x - 2}$ 27. $18x^2(x + 2)(x - 2)^2$

28. $\dfrac{2x^2 - 3x - 2}{(x - 3)(x - 1)(x + 2)}$

29. $\dfrac{-x(x^2 + x + 2)}{(x - 1)^2}$

30. \$1200 31. $6x^2 - x - 2 = 0$

32. a. Number of freshmen: x
b. $0.68x = 493$
c. 725 freshmen

33. a. Amount put into Christmas club: x
Amount put into savings: $x + 200$
Amount put into checking: $2(x + 200)$
b. $x + (x + 200) + 2(x + 200) = 1484$
c. \$221 into Christmas club,
\$421 into savings, \$842 into checking

34. a. Amount invested at 14%: x
b. $0.09(16{,}000) + 0.14x = 0.10(x + 16{,}000)$
c. \$4000 must be invested at 14%

35. a. Number of liters of 80% acid: x
b. $0.80x + 0.30(40) = 0.60(x + 40)$
c. 60 liters

36. a. Width of garden: x
Length of garden: $2x$
b. $(x + 6)(2x + 6) = 2x^2 + 972$
c. 52 feet by 104 feet

37. a. Shorter leg: x
Hypotenuse: $2x - 3$
b. $x^2 + 12^2 = (2x - 3)^2$
c. The leg is 9 inches; the hypotenuse is 15 inches

38. a. Time to reach 380 feet: t
b. $380 = 500 - 28t - 16t^2$
c. 2 seconds

39. a. Number of miles: x
b. $\dfrac{5/4}{20} = \dfrac{13/8}{x}$
c. 26 miles

40. a. Speed of train: x
b. $\dfrac{700}{x} = \dfrac{700}{2x - 20} + 6$
c. 35 miles per hour or 33⅓ miles per hour

41. a. Any x and y with the same sign
b. Any x and y with opposite signs

42. $(-2, -1]$ 43. -6

44. $x^{3n-3}y^{6n}$ 45. $x^n(x^{2n} - x^3)$ 46. $(2y^3 - 5)(y^3 + 2)$ 47. $y^2 - 3y + 8 + \dfrac{8 - 27y}{y^2 + 3y - 1}$

48. $\dfrac{x^3 - x^2 + 2x - 1}{x(x - 1)}$ 49. $\dfrac{ab - a^2}{ab + b^2}$ 50. $b = \dfrac{(a + q)(a + c)}{2q - a - c}$

Exercise 6.1 [page 192] 1. x^5 3. a^8 5. a^6 7. x^6 9. x^3y^6 11. $a^4b^4c^8$

13. $-16x^4$ 15. $4a^7b^6$ 17. x^2 19. xy^2 21. $\dfrac{x^3}{y^6}$ 23. $\dfrac{8x^3}{y^6}$ 25. $\dfrac{-8x^3}{27y^6}$

27. $\dfrac{16}{x}$ 29. $\dfrac{y^4}{x}$ 31. x^4y 33. $\dfrac{8x}{9y^2}$ 35. $36y^2$ 37. $\dfrac{x^2s^{14}}{y^2t^2}$ 39. $\dfrac{-1}{x^3a^3b}$ 41. $\dfrac{y + z}{x^4}$

43. Use 1 for x and 2 for y: $(1^2 + 2^2)^3 \stackrel{?}{=} 1^6 + 2^6$; $125 \neq 1 + 64$. 45. x^{2n} 47. x^{3n}

49. x^{2n}

Exercise 6.2 [page 197] 1. $\frac{1}{2}$ 3. 3 5. $-\frac{1}{8}$ 7. $\frac{1}{5}$ 9. $\frac{5}{3}$ 11. $\frac{9}{5}$ 13. $\frac{82}{9}$

15. $\frac{3}{16}$ 17. $\frac{x^2}{y^3}$ 19. $\frac{1}{x^6y^3}$ 21. $\frac{x^2}{y^6}$ 23. $\frac{y^4}{x}$ 25. x^4 27. $\frac{1}{x^6}$

29. $\frac{y}{x}$ 31. $4x^5y^2$ 33. $\frac{z^2}{y^2}$ 35. $\frac{xy}{z}$ 37. $\frac{z}{y^2}$ 39. $\frac{x^2 + y^2}{x^2y^2}$ 41. $\frac{x^2y^2 + 1}{xy}$

43. $\frac{1}{(x - y)^2}$ 45. $\frac{y^2 - x^2}{xy}$ 47. $1 - xy$ 49. $x + y$

51. Substituting 1 for x and 2 for y gives $(1 + 2)^{-2} \stackrel{?}{=} \frac{1}{1^2 + 2^2}$; $\frac{1}{9} \neq \frac{1}{5}$.

53. a^{3-n} 55. a^{2-2n} 57. $b^{-1}c^{-1}$ 59. x^{-1}

Exercise 6.3 [page 202] 1. 2.85×10^2 3. 2.1×10 5. 8.372×10^6

7. 2.4×10^{-2} 9. 4.21×10^{-1} 11. 4×10^{-6} 13. 240 15. 687,000

17. 0.005 19. 0.0202 21. 12,270 23. 0.00235 25. 0.0005 27. 12.5

29. 0.00006 31. 10^{-5} or 0.00001 33. 10^0 or 1 35. 6×10^{-3} or 0.006

37. 1.8×10^6 or 1,800,000 39. four 41. two 43. three 45. three

47. 72×10 or 720 49. 4×10^{-1} or 0.4 51. 8

53. a. 3×10^8 b. 1.18×10^{10} inches per second 55. 3×10^8 57. 1.5×10^{-5}

Exercise 6.4 [page 206] 1. 3 3. 2 5. -2 7. 8 9. 27 11. 16 13. $\frac{1}{4}$

15. $\frac{1}{8}$ 17. 9 19. 81 21. $x^{2/3}$ 23. $x^{1/3}$ 25. $a^{3/2}$ 27. $\frac{1}{x^{1/2}}$ 29. $a^{1/3}b^{1/2}$

31. $\frac{a^4}{b^2}$ 33. $\frac{r^2}{t^3}$ 35. $\frac{t^2}{z}$ 37. $\frac{yz}{x}$ 39. $x^{3/2} + x$ 41. $x - x^{2/3}$ 43. $x^{-1} + 1$

45. $t + 1$ 47. $b + b^{1/4}$ 49. $x^2 + 2x - 1$ 51. $2x + x^{1/2} - 1$ 53. $x - 4x^{1/2} + 4$

55. $x - x^{3/2} - 2x^2$ 57. Use 1 for a and 1 for b: $(1 + 1)^{1/2} \stackrel{?}{=} 1^{1/2} + 1^{1/2}$; $2^{1/2} \neq 1 + 1$.

59. $x^{3n/2}$ 61. $x^{3n/2}$ 63. $x^{5n/2}y^{(3m+2)/2}$ 65. $x^{n/3}y^n$ 67. $16^{1/2} > 16^{1/4}$; $(\frac{1}{16})^{1/4} > (\frac{1}{16})^{1/2}$

Exercise 6.5 [page 212] 1. $7^{1/2}$ 3. $(2x)^{1/3}$ 5. $x^{1/3} - 3y^{1/2}$ 7. $x^{1/3}y^{1/2}$ 9. $\sqrt{3}$

11. $4\sqrt[3]{x}$ 13. $\sqrt[4]{x - 2}$ 15. $3\sqrt[3]{xy}$ 17. 3 19. 2 21. 3 23. 5 25. $\sqrt[3]{x^2}$

27. $3\sqrt[5]{x^3}$ 29. $\sqrt{(x + 2y)^3}$ 31. $\frac{1}{\sqrt{y}}$ 33. $\frac{1}{\sqrt[3]{x^2}}$ 35. $\frac{3}{\sqrt[3]{y^2}}$ 37. $x^{2/3}$ 39. $(xy)^{2/3}$

41. $(xy^3)^{1/2}$ 43. $\frac{1}{x^{1/2}}$ 45. 3 47. -4 49. x 51. x^2 53. $2y^2$ 55. $-x^2y^3$

57. $\frac{2}{3}xy^4$ 59. $\frac{-2}{5}x$ 61. $2xy^2$ 63. $2a^2b^3$ 65. $\sqrt{3}$ 67. $\sqrt{3}$ 69. $\sqrt[3]{9}$ 71. $\sqrt{x}$

73.
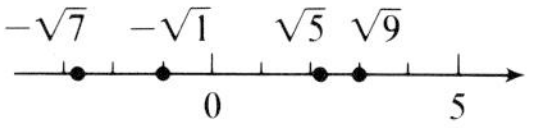

75.
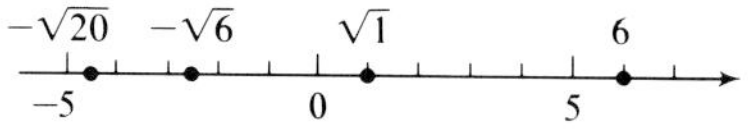

77. $2|x|$ 79. $|x + 1|$ 81. $\frac{2}{|x + y|}$ $(x + y \neq 0)$

83. Substitute a negative number such as -2 for a in the given equation and solve:

$$\sqrt{a^2} \stackrel{?}{=} a$$
$$\sqrt{(-2)^2} \stackrel{?}{=} -2$$
$$\sqrt{4} \stackrel{?}{=} -2$$
$$2 \neq -2.$$

Exercise 6.6 [page 218] 1. $3\sqrt{2}$ 3. $2\sqrt{5}$ 5. $5\sqrt{3}$ 7. x^2 9. $x\sqrt{x}$ 11. $3x\sqrt{x}$ 13. $2x^3\sqrt{2}$ 15. $x\sqrt[4]{x}$ 17. $xyz^2\sqrt[5]{x^2y^4z}$ 19. $ab^2c^2\sqrt[6]{ac^3}$ 21. $3abc\sqrt[7]{ab^2c^3}$ 23. 6 25. x^3y 27. 2 29. $10\sqrt{3}$ 31. $100\sqrt{6}$ 33. $\frac{\sqrt{5}}{5}$ 35. $\frac{-\sqrt{2}}{2}$ 37. $\frac{\sqrt{2x}}{2}$ 39. $\frac{-\sqrt{xy}}{x}$ 41. $\sqrt{x}$ 43. $\frac{\sqrt{2xy}}{2x}$ 45. $\frac{\sqrt[3]{x}}{x}$ 47. $\frac{\sqrt[3]{18y^2}}{3y}$ 49. $\frac{\sqrt[3]{2xy}}{2y}$ 51. $\frac{\sqrt[5]{48x^2}}{2x}$ 53. a^2b 55. $7y\sqrt{2x}$ 57. $\frac{2b}{a^2}\sqrt[3]{b}$ 59. $\sqrt[5]{b}$ 61. $\frac{1}{\sqrt{3}}$ 63. $\frac{x}{\sqrt{xy}}$

65. Substitute two numbers such as 4 and 9 for a and b in the given equation and solve:

$$(\sqrt{a} + \sqrt{b})^2 \stackrel{?}{=} a + b$$
$$(\sqrt{4} + \sqrt{9})^2 \stackrel{?}{=} 4 + 9$$
$$(2 + 3)^2 \stackrel{?}{=} 13$$
$$25 \neq 13.$$

67. $2(x-1)\sqrt{2(x-1)}$ 69. $x(y-2)^2\sqrt{x(y-2)}$ 71. $\frac{y-3}{xy^2}\sqrt{xy(y-3)}$ 73. $x\sqrt[3]{4x^2-1}$ 75. $\frac{(x-1)}{xy}\sqrt[3]{x^2y(x-1)}$ 77. $\frac{1}{x^2}\sqrt{3x(x-1)}$

Exercise 6.7 [page 222] 1. $5\sqrt{7}$ 3. $\sqrt{3}$ 5. $9\sqrt{2x}$ 7. $-6y\sqrt{x}$ 9. $15\sqrt{2a}$ 11. $5\sqrt[3]{2}$ 13. $6 - 2\sqrt{5}$ 15. $2\sqrt{3} + 2\sqrt{5}$ 17. $1 - \sqrt{5}$ 19. $x - 9$ 21. $-4 + \sqrt{6}$ 23. $7 - 2\sqrt{10}$ 25. $2(1 + \sqrt{3})$ 27. $6(\sqrt{3} + 1)$ 29. $4(1 + \sqrt{y})$ 31. $\sqrt{2}(1 - \sqrt{3})$ 33. $1 + \sqrt{3}$ 35. $1 + \sqrt{2}$ 37. $1 - \sqrt{x}$ 39. $x - y$ 41. $2\sqrt{3} - 2$ 43. $\frac{2(\sqrt{7} + 2)}{3}$ 45. $\frac{x(\sqrt{x} + 3)}{x - 9}$ 47. $\frac{\sqrt{6}}{2}$ 49. $\frac{-1}{2(1 + \sqrt{2})}$ 51. $\frac{x - 1}{3(\sqrt{x} + 1)}$ 53. $\frac{x - y}{x(\sqrt{x} + \sqrt{y})}$ 55. $\frac{\sqrt{x + 1}}{x + 1}$ 57. $\frac{-\sqrt{x^2 + 1}}{x(x^2 + 1)}$ 59. $\frac{x - 2}{\sqrt{x(x-1)} - \sqrt{x - 1} + \sqrt{x} - 1}$ 61. $\frac{1}{2x + 1 + 2\sqrt{x(x + 1)}}$

Exercise 6.8 [page 228] 1. $2i$ 3. $4i\sqrt{2}$ 5. $6i\sqrt{2}$ 7. $6i\sqrt{6}$ 9. $40i$ 11. $-4i\sqrt{3}$ 13. $4 + 2i$ 15. $2 + 15i\sqrt{2}$ 17. $2 + 2i$ 19. $5 + 5i$ 21. $-2 + i$ 23. $-1 - 2i$

25. $8 + i$ 27. $13 + 3i$ 29. $21 - 18i$ 31. $3 - 4i$ 33. $5 + 0i$ or 5 35. $\frac{-i}{3}$

37. $\frac{-1}{5} - \frac{3}{5}i$ 39. $1 + i$ 41. $\frac{1}{2} - \frac{1}{2}i$ 43. $\frac{12}{13} - \frac{5}{13}i$ 45. $\frac{9}{34} + \frac{9}{34}i$ 47. $4 + 2i$

49. $15 + 3i$ 51. $-\frac{3}{2}i$ 53. $\frac{3}{5} - \frac{4}{5}i$ 55. $x \geq 5; x < 5$

57. a. -1 b. 1 c. $-i$ d. -1 59. $5 + 4i$

Review Exercises [page 232] 1. a. x^3y^2 b. $27x^6y^3$ 2. a. $\frac{1}{x^3y}$ b. $\frac{y^6}{x^8}$

3. a. $\frac{11}{18}$ b. $\frac{x^2 + y^2}{xy}$ 4. a. $\frac{2x}{4x^2 + 4x + 1}$ b. $\frac{(y^2 - x^2)(x - y)^2}{x^2y^2}$

5. a. 2.3×10^{-11} b. 3.07×10^{11} 6. a. 0.0000349 b. 0.0075

7. a. 10^2 or 100 b. 8×10^1 or 80 8. a. three b. four

9. a. 9 b. ½ 10. a. x^2 b. x^2y^3 11. a. $x - x^{5/3}$ b. $y - y^0$ or $y - 1$

12. a. $x^{1/5}(x^{3/5})$ b. $y^{-1/2}(y^{-1/4})$ 13. a. $\sqrt[3]{(1 - x^2)^2}$ b. $\sqrt[3]{(1 - x^2)^{-2}}$ or $\frac{1}{\sqrt[3]{(1 - x^2)^2}}$

14. a. $(x^2y)^{1/3}$ or $x^{2/3}y^{1/3}$ b. $(a + b)^{-2/3}$ 15. a. $2y$ b. $-2xy^2$

16. a. $\frac{1}{2}x^2y^3$ b. $\frac{1}{2}y$ 17. a. $6\sqrt{5}$ b. $2xy\sqrt[4]{2y}$ 18. a. $\frac{\sqrt{2}}{2}$ b. $\frac{\sqrt{3}}{3}$

19. a. $\frac{\sqrt{xy}}{y}$ b. $x\sqrt{3xy}$ 20. a. $\frac{\sqrt[3]{4}}{2}$ b. $\frac{\sqrt[4]{54x^3}}{3x}$ 21. a. $xy\sqrt[3]{x^2y}$ b. xy

22. a. $2xy\sqrt[3]{y}$ b. $x\sqrt[5]{y^2}$ 23. a. $\frac{3x}{2\sqrt{3x}}$ b. $\frac{2x}{y\sqrt{2x}}$ 24. a. $\frac{6x}{\sqrt{2x}}$ b. $\frac{2x}{\sqrt{x}}$

25. a. $18\sqrt{3}$ b. $19\sqrt{2x}$ 26. a. $3y\sqrt{2x}$ b. $4x\sqrt{3x}$ 27. a. $3\sqrt{2} - 2\sqrt{3}$
b. $5\sqrt{2} - 5$ 28. a. $12 - 7\sqrt{3}$ b. $x - 4$ 29. a. $1 + 2\sqrt{2}$ b. $1 - \sqrt{2}$

30. a. $1 + \sqrt{xy}$ b. $1 - \sqrt{x}$ 31. a. $\frac{\sqrt{x}}{2x}$ b. $\frac{10\sqrt{y}}{3y}$ 32. a. $8 + 4\sqrt{3}$ b. $2\sqrt{5} - 4$

33. a. $\frac{y(\sqrt{y} + 3)}{y - 9}$ b. $\sqrt{x} - \sqrt{y}$ 34. a. $\frac{13}{8 - 2\sqrt{3}}$ b. $\frac{x - 16}{x(\sqrt{x} - 4)}$

35. a. $4 + 6i$ b. $5 - 6i\sqrt{3}$ 36. a. $6 - i$ b. $5 + 3i$ 37. a. $7 + 17i$
b. $4 + 3i$ 38. a. $\frac{-4i}{3}$ b. $1 - i$ 39. a. $x^{4n}y^{2n-2}$ b. $\frac{y^{n-1}}{x^{1-n}}$ or $\frac{x^{n-1}}{y^{1-n}}$

40. a. x^ny^2 b. xy 41. 0.2 42. $|x - 5|$ 43. $|2x - 2|$ 44. $\sqrt[3]{4}$ 45. $-i$

Exercise 7.1 [page 243] 1. $\{10, -10\}$ 3. $\{5/3, -5/3\}$ 5. $\{\sqrt{7}, -\sqrt{7}\}$ 7. $\{i\sqrt{6}, -i\sqrt{6}\}$
9. $\{\sqrt{6}, -\sqrt{6}\}$ 11. $\{9i/2, -9i/2\}$ 13. $\{5, -1\}$ 15. $\{5/2, -3/2\}$

17. $\{-2 + i\sqrt{3}, -2 - i\sqrt{3}\}$ 19. $\left\{\frac{1 + \sqrt{3}}{2}, \frac{1 - \sqrt{3}}{2}\right\}$ 21. $\{-2/9, -4/9\}$ 23. $\{1, 4\}$

25. $\{-2/3, -8/3\}$ 27. $\left\{\frac{1 + i\sqrt{15}}{7}, \frac{1 - i\sqrt{15}}{7}\right\}$ 29. $\left\{\frac{7 + 2i\sqrt{2}}{8}, \frac{7 - 2i\sqrt{2}}{8}\right\}$ 31. $\{2, -6\}$

33. $\{1\}$ 35. $\{-5, -4\}$ 37. $\{1 + \sqrt{2}, 1 - \sqrt{2}\}$ 39. $\left\{\frac{-3 + \sqrt{21}}{2}, \frac{-3 - \sqrt{21}}{2}\right\}$

41. $\left\{\frac{-2 + \sqrt{10}}{2}, \frac{-2 - \sqrt{10}}{2}\right\}$ 43. $\{5/2, -1\}$ 45. $\left\{\frac{1 + i\sqrt{11}}{2}, \frac{1 - i\sqrt{11}}{2}\right\}$

47. $\left\{\frac{-2 + i\sqrt{2}}{2}, \frac{-2 - i\sqrt{2}}{2}\right\}$ 49. $\{\sqrt{a}, -\sqrt{a}\}$ 51. $\left\{\sqrt{\frac{bc}{a}}, -\sqrt{\frac{bc}{a}}\right\}$ 53. $\{a + 4, a - 4\}$

55. $\left\{\frac{3 - b}{a}, \frac{-3 - b}{a}\right\}$ 57. $\{\sqrt{c^2 - a^2}, -\sqrt{c^2 - a^2}\}$ 59. $\left\{-1 + \sqrt{\frac{A}{P}}, -1 - \sqrt{\frac{A}{P}}\right\}$

61. $y = \pm\frac{1}{3}\sqrt{9 - x^2}$ 63. $y = \pm\frac{2}{3}\sqrt{x^2 - 9}$ 65. (See page 245.)

Exercise 7.2 [page 248] 1. $\{4, 1\}$ 3. $\{1, -4\}$ 5. $\left\{\frac{3 + \sqrt{5}}{2}, \frac{3 - \sqrt{5}}{2}\right\}$

7. $\left\{\frac{5 + \sqrt{13}}{6}, \frac{5 - \sqrt{13}}{6}\right\}$ 9. $\{3/2, -2/3\}$ 11. $\{0, 5\}$ 13. $\{i\sqrt{2}, -i\sqrt{2}\}$

15. $\left\{\frac{1 + i\sqrt{7}}{4}, \frac{1 - i\sqrt{7}}{4}\right\}$ 17. $\left\{\frac{3 + \sqrt{13}}{2}, \frac{3 - \sqrt{13}}{2}\right\}$ 19. $\left\{\frac{2 + \sqrt{7}}{3}, \frac{2 - \sqrt{7}}{3}\right\}$

21. $\left\{\frac{1 + i\sqrt{23}}{6}, \frac{1 - i\sqrt{23}}{6}\right\}$ 23. $\left\{\frac{1 + i\sqrt{3}}{4}, \frac{1 - i\sqrt{3}}{4}\right\}$ 25. 1; real and unequal

27. -16; imaginary and unequal 29. 0; one real 31. $x = 2k$; $x = -k$

33. $x = \frac{1 \pm \sqrt{1 - 4ac}}{2a}$ 35. $x = -1 \pm \sqrt{1 + y}$ 37. $x = \frac{-y \pm \sqrt{24 - 11y^2}}{6}$

39. $y = \frac{-x \pm \sqrt{8 - 11x^2}}{2}$

41. If r_1 and r_2 are solutions of a quadratic equation, let

$$r_1 = \frac{-b + \sqrt{b^2 - 4ac}}{2a} \quad \text{and} \quad r_2 = \frac{-b - \sqrt{b^2 - 4ac}}{2a}.$$

Then,

$$r_1 + r_2 = \left(\frac{-b}{2a} + \frac{\sqrt{b^2 - 4ac}}{2a}\right) + \left(\frac{-b}{2a} - \frac{\sqrt{b^2 - 4ac}}{2a}\right) = \frac{-2b}{2a} = -\frac{b}{a}$$

and

$$r_1 \cdot r_2 = \left(\frac{-b}{2a} + \frac{\sqrt{b^2 - 4ac}}{2a}\right)\left(\frac{-b}{2a} - \frac{\sqrt{b^2 - 4ac}}{2a}\right) = \frac{b^2 - b^2 + 4ac}{4a^2} = \frac{c}{a}.$$

Exercise 7.3 [page 252] 1. $\{64\}$ 3. $\{-2\}$ 5. $\{-1/3\}$ 7. $\{2, -1/2\}$ 9. $\{12\}$ 11. $\{5\}$
13. $\{4\}$ 15. $\{-27\}$ 17. $\{17\}$ 19. $\{5\}$ 21. $\{0\}$ 23. $\{4\}$ 25. $\{1, 3\}$

27. $A = \pi r^2$ 29. $S = \frac{1}{R^3}$ 31. $t = \pm\sqrt{r^2 + s^2}$ 33. $E = \pm\sqrt{\left(\frac{A - B}{C}\right)^2 - D}$

35. $A \geq 0$ 37. $2\sqrt{10}$ centimeters 39. 4 feet

Exercise 7.4 [page 256] 1. $\{1, -1, 2, -2\}$ 3. $\{1, -1, \sqrt{3}, -\sqrt{3}\}$

5. $\{\sqrt{2}/2, -\sqrt{2}/2, 3i, -3i\}$ 7. $\{i, -i, \sqrt{3}i, -\sqrt{3}i\}$ 9. $\{25\}$ 11. 0 13. $\{3, -3\}$
15. $\{\sqrt{10}, -\sqrt{10}\}$ 17. $\{64, -8\}$ 19. $\{64, -1\}$ 21. $\{9, 36\}$ 23. $\{1/4, 16\}$
25. $\{1/4, -1/3\}$ 27. $\{626\}$ 29. $\{9, 16\}$ 31. $\{4, 81\}$

Exercise 7.5 [page 262] 1. $\{x|x < -1\} \cup \{x|x > 2\}$ or $(-\infty, -1); (2, +\infty)$

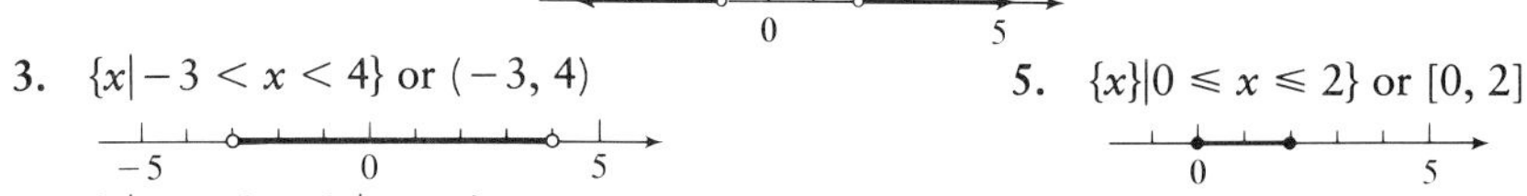

3. $\{x|-3 < x < 4\}$ or $(-3, 4)$

5. $\{x\}|0 \leq x \leq 2\}$ or $[0, 2]$

7. $\{x|x \leq 0\} \cup \{x|x > 5\}$ or $(-\infty, 0); (5, \infty)$

9. $\{x|x < -1\} \cup \{x|x > 4\}$ or $(-\infty, -1); (4, +\infty)$

11. $\{x|-2 \leq x \leq 3\}$ or $[-2, 3]$

13. $\{x|-5 \leq x \leq 3\}$ or $[-5, 3]$

15. $\{x|x < -3\} \cup \{x|x > 4\}$ or $(-\infty, -3); (4, \infty)$

17. $\{x|-\sqrt{5} < x < \sqrt{5}\}$ or $(-\sqrt{5}, \sqrt{5})$ 19. $\varnothing$

21. $\{x|-3 < x < 2\}$ or $(-3, 2)$

23. $\{x|-8/3 < x < -2\}$ or $(-8/3, -2)$

25. $\{x|x \leq -1/2\} \cup \{x|x > 0\}$ or $(-\infty, -1/2]; (0, +\infty)$

27. $\{x|x < 1\} \cup \{x \geq 2\}$ or $(-\infty, 1); [2, +\infty)$

29. $\{x|x < -1\} \cup \{x|0 < x < 3\}$ or $(-\infty, -1); (0, 3)$

31. $\{x|x < 0\} \cup \{x|2 < x \leq 4\}$ or $(-\infty, 0); (2, 4]$

33. $\{x|-3 < x < -2\} \cup \{x|2 < x < 3\}$ or $(-3, -2); (2, 3)$

35. a. Time higher than 1024 feet: t
 b. $16t^2 - 320t + 1024 < 0$
 c. Greater than 4 seconds and less than 16 seconds

Review Exercises [page 264] 1. a. $\{5, -5\}$ b. $\left\{\frac{1}{3}i\sqrt{21}, -\frac{1}{3}i\sqrt{21}\right\}$

2. a. $\{2, -8\}$ b. $\{4 + \sqrt{15}, 4 - \sqrt{15}\}$

3. a. $\left\{\frac{1 + \sqrt{2}}{3}, \frac{1 - \sqrt{2}}{3}\right\}$ b. $\left\{\frac{-2 + \sqrt{5}}{3}, \frac{-2 - \sqrt{5}}{3}\right\}$

4. a. $\left\{\frac{3 + i\sqrt{5}}{2}, \frac{3 - i\sqrt{5}}{2}\right\}$ b. $\left\{\frac{-2 + i\sqrt{7}}{3}, \frac{-2 - i\sqrt{7}}{3}\right\}$

5. a. $\{2+\sqrt{10}, 2-\sqrt{10}\}$ b. $\left\{\frac{-3+\sqrt{33}}{4}, \frac{-3-\sqrt{33}}{4}\right\}$

6. a. $\left\{\frac{-3+\sqrt{21}}{2}, \frac{-3-\sqrt{21}}{2}\right\}$ b. $\left\{\frac{1+\sqrt{13}}{3}, \frac{1-\sqrt{13}}{3}\right\}$

7. a. $\left\{\frac{1+i\sqrt{7}}{2}, \frac{1-i\sqrt{7}}{2}\right\}$ b. $\{-1+i\sqrt{2}, -1-i\sqrt{2}\}$

8. a. $\left\{\frac{1+i\sqrt{5}}{2}, \frac{1-i\sqrt{5}}{2}\right\}$ b. $\left\{\frac{3+i\sqrt{3}}{6}, \frac{3-i\sqrt{3}}{6}\right\}$

9. a. $\{1, 2\}$ b. $\left\{\frac{3+i\sqrt{19}}{2}, \frac{3-i\sqrt{19}}{2}\right\}$ 10. a. $\left\{\frac{3+\sqrt{5}}{2}, \frac{3-\sqrt{5}}{2}\right\}$ b. $\{1, -3/2\}$

11. a. $\left\{\frac{1+i\sqrt{7}}{2}, \frac{1-i\sqrt{7}}{2}\right\}$ b. $\{1+i\sqrt{3}, 1-i\sqrt{3}\}$

12. a. $\left\{\frac{-3+i\sqrt{7}}{4}, \frac{-3-i\sqrt{7}}{4}\right\}$ b. $\left\{\frac{1+i\sqrt{23}}{4}, \frac{1-i\sqrt{23}}{4}\right\}$ 13. $\{1, 4\}$ 14. $\{8\}$

15. $t = \pm\sqrt{\frac{1-p^2s}{2}}$ 16. $p = \pm 2\sqrt{R^2 - R}$ 17. $\{2, -2, i, -i\}$ 18. $\{16\}$

19. $\{16\}$ 20. $\{2\}$

21. a. $\{x|x < 0\} \cup \{x|x > 9\}$ or $(-\infty, 0); (9, +\infty)$

b. $\{x|-3 < x < -2\}$ or $(-3, -2)$

22. a. $\{x|-4 \leq x \leq 2\}$

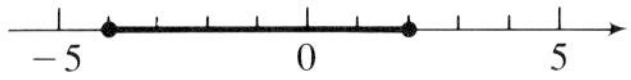

b. $\{x|x \leq -1\} \cup \{x > 4\}$ or $(-x, -1); (4, \infty)$

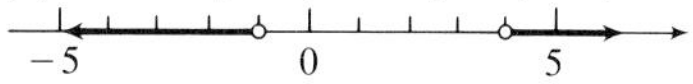

23. a. $\{x|x \leq -1\} \cup \{x|x > 3\}$ or $(-\infty, -1]; (3, +\infty)$

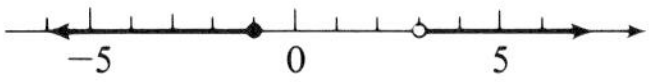

b. $\{x|x \leq -4\} \cup \{x|x > -2\}$ or $(-\infty, -4]; (-2, +\infty)$

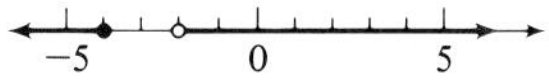

24. $x = \frac{b \pm 2}{a}$ 25. $x = \frac{k \pm \sqrt{k^2 - 8}}{4}$ 26. $y = \frac{-x \pm i\sqrt{7x^2}}{2}$ 27. $k \geq 0$ 28. $\{82\}$

29. $\{x|0 < x < 2\} \cup \{x|x \geq 4\}$ or $(0, 2); [4, +\infty)$

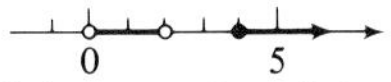

30. $\{x|x < -3\} \cup \{x|-1 < x < 0\}$ or $(-\infty, -3); (-1, 0)$

Cumulative Review Exercises, Chapters 1–7 [page 267]

1. $2t^3 + 3t^2 + 9t - 2$ 2. $6x^4 - 5x^3 - 9x^2 + 14x - 6$ 3.

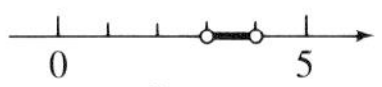

4. $\frac{2-y}{2+y}$ 5. $x^5 - 2x^4 + 4x^3 - 8x^2 + 16x - 32 + \frac{66}{x+2}$ 6. $\frac{-3x^2 - 6x}{2x^3 - 8x}$

7. $\frac{x^2 - x - 2}{x^2 - x}$ 8. $\frac{3}{2x(x-1)}$ 9. $-xy$ 10. $\dfrac{2x - 1 - \frac{14}{x-2}}{x - \frac{8}{x-2}} = \frac{2x+3}{x+2}$

11. $2(3x-1)(9x+2)$ **12.** $x^2y^2(2-y)(2+y)$ **13.** $(3x-2-2y)(3x-2+2y)$

14. $(2x-3y)(4x^2+6xy+9y^2)$ **15.** $17x-14x^2-22x^3$ **16.** $45x^2y^2+13x^3y^4$

17. $\dfrac{a^5b^5c^{12}x^2}{y^4}$ **18.** $\dfrac{8xz^{10}}{9}$ **19.** $\dfrac{x^2}{y-x}$ **20.** 4.8×10^9 **21.** $\dfrac{a^{3/2}}{bc^{5/2}}$ **22.** $\dfrac{1}{256}$

23. $-2x^{-1/3}+2x^{1/3}-4$ **24.** $2\sqrt[3]{\dfrac{(x+y)^2}{x}}$ **25.** $-3xy^{10}$ **26.** $\sqrt[3]{4}$

27. $\sqrt{3}(\sqrt{6}-2\sqrt{2})$ **28.** $3\sqrt{y}-x\sqrt{3}$ **29.** $\dfrac{\sqrt[4]{6x^2}}{2x}$ **30.** $\dfrac{2+3x+7\sqrt{x}}{1-9x}$

31. $-\dfrac{1}{5}-\dfrac{2}{5}i$ **32.** $1+3i\sqrt{3}$ **33.** $\{-5\}$ **34.** $(-\infty, 1/5] \cup [13/5, \infty)$ **35.** $\{-2, 7/2\}$

36. $\{3/2, 5\}$ **37.** $\{1\pm i\sqrt{2}\}$ **38.** $\{5/3\pm i\sqrt{2}\}$ **39.** $\left\{\dfrac{1}{3}\pm i\dfrac{\sqrt{2}}{3}\right\}$ **40.** $\left\{\dfrac{2}{3}\pm i\dfrac{\sqrt{2}}{3}\right\}$

41. $d=\dfrac{t+(s-c)(1-e)}{t+c(1-e)}$ **42.** $\{-6\}$ **43.** $\{9\}$ **44.** $\{-1, 512\}$

45. $(-\infty,-4)\cup(1,\infty)$ **46.** $Q=b+2A^3$ **47.** $D=41>0$; two real solutions

48. a. Number of tulip bulbs: x
Number of daffodil bulbs: $50-x$
b. $0.69x+0.89(50-x)=40.50$
c. 30 tulip bulbs; 20 daffodils

49. a. Amount he must make: x
b. $80\leqslant\dfrac{92+65+103+79+x}{5}\leqslant120$
c. Between \$61 and \$261

50. a. Length of hypotenuse: x
b. $(x-3)^2+9^2=x^2$
c. 15 feet

51. a. Gallons of gas: x
b. $\dfrac{18}{504}=\dfrac{x}{602}$
c. 21.5 gallons

52. a. Speed of the balloon: x
b. $\dfrac{120}{x-8}+\dfrac{120}{x+8}=8$
c. 32 miles per hour

53. -336 **54.** $2a^{5n}-2a^{2n}b^n+2a^{3n}b^{2n}-2b^{3n}$ **55.** $y^3+2y^2+3y+4+\dfrac{5y-5}{y^2-2y+1}$

56. $\dfrac{x^2+x}{x-1}$ **57.** $x^n(1+2x^2+x^n)$ **58.** $(2x^2+5y^2)(2x^2-7y^2)$ **59.** $(x-1)(x^2+4x+7)$

60. $x^{n+12}y^{6-4n-n^2}$ **61.** 0.4 **62.** $\{x|x\leqslant1\}$ **63.** $\dfrac{a}{a+2}\sqrt[3]{a(a^2-3)(a+2)}$

64. $\dfrac{x-2\sqrt{x-1}}{2-x}$ **65.** $\dfrac{1}{i^5}=-i$ **66.** $x=\dfrac{y\pm\sqrt{8-y^2}}{2}$ **67.** $v=\pm\sqrt{\dfrac{2(E-P)}{m}}$

68. $\{-1, 1+\sqrt[3]{4}\}$ **69.** $(-\infty,-11]\cup(-3,1)$

70. a. Length of equal sides: x
Length of base: $36-2x$
b. $(18-x)^2+12^2=x^2$
c. 10 inches

Exercise 8.1 [page 275] **1.** **a.** $(0, 7)$ **b.** $(2, 9)$ **c.** $(-2, 5)$
3. **a.** $(0, {}^{-3}/_2)$ **b.** $(2, 0)$ **c.** $(-5, {}^{-21}/_4)$ **5.** $(-3, -7), (0, -4), (3, -1)$
7. $(1, 1), (2, 3/4), (3, 3/5)$ **9.** $(1, 0), (2, \sqrt{3}), (3, 2\sqrt{2})$ **11.** $y = 6 - 2x; \{2, -2\}$
13. $y = \dfrac{x + 2}{x}; \{0, 2\}$ **15.** $y = \dfrac{4}{x - 1}; \{4/3, 4/7\}$ **17.** $y = \dfrac{2}{x^2 - x - 4}; \{-1, -1/2\}$
19. $y = \dfrac{\pm\sqrt{x + 8}}{2}; \left\{\pm\dfrac{\sqrt{7}}{2}, \pm\dfrac{\sqrt{11}}{2}\right\}$ **21.** $y = \pm\dfrac{\sqrt{3x^2 - 4}}{2}; \left\{\pm\sqrt{2}, \pm\dfrac{\sqrt{23}}{2}\right\}$ **23.** $(-4, 1)$
25. $(-1, 3)$ **27.** $(3, 8)$ **29.** $(3, 1/8)$ **31.** $k = 7/2$

Exercise 8.2 [page 281]

1.

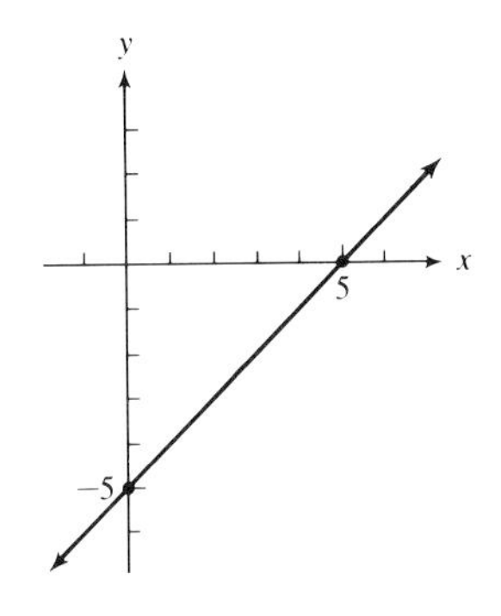

3.

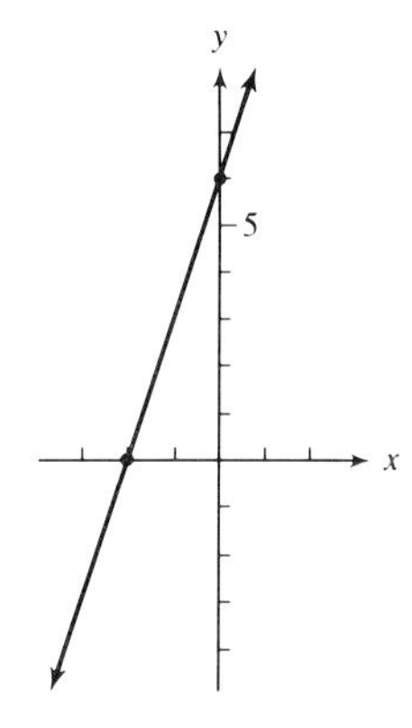

5.

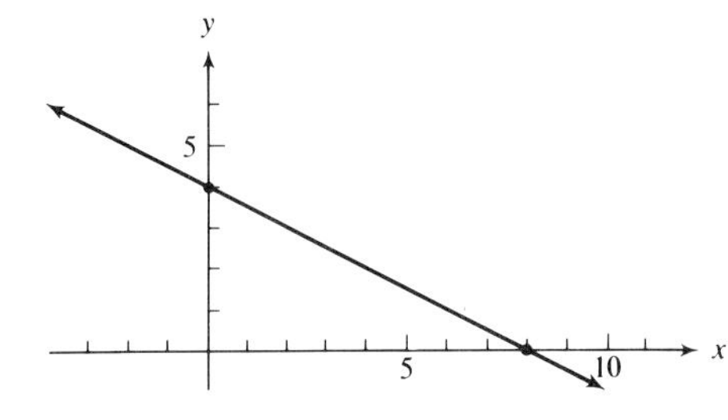

7.

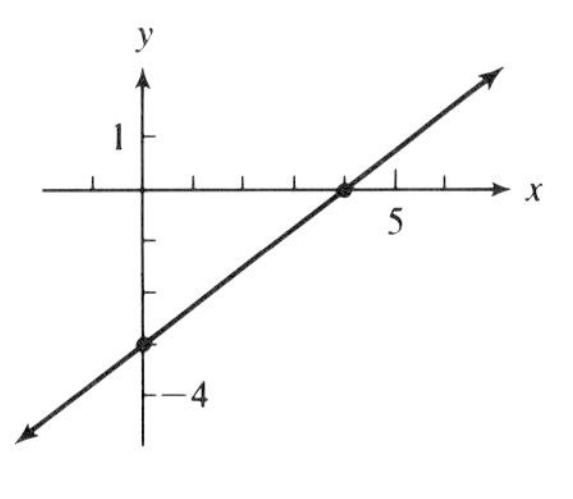

9.

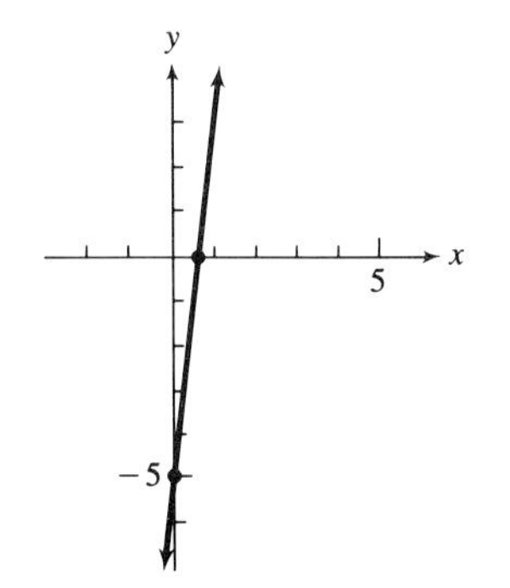

11.

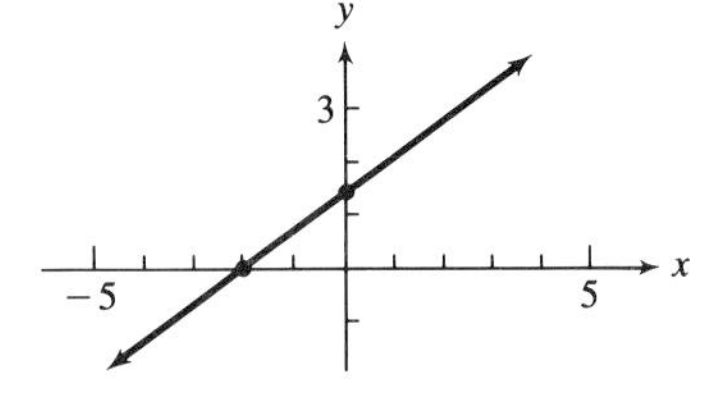

13.

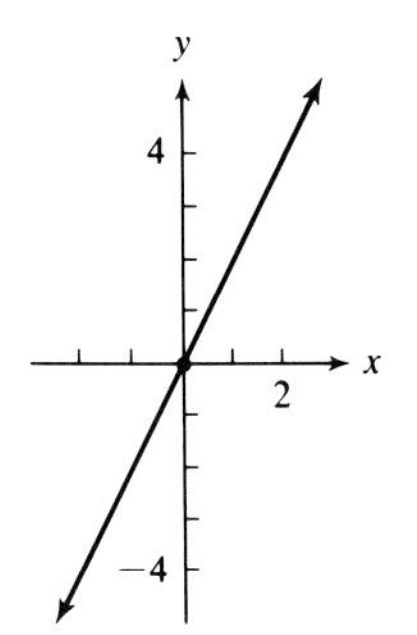

15.

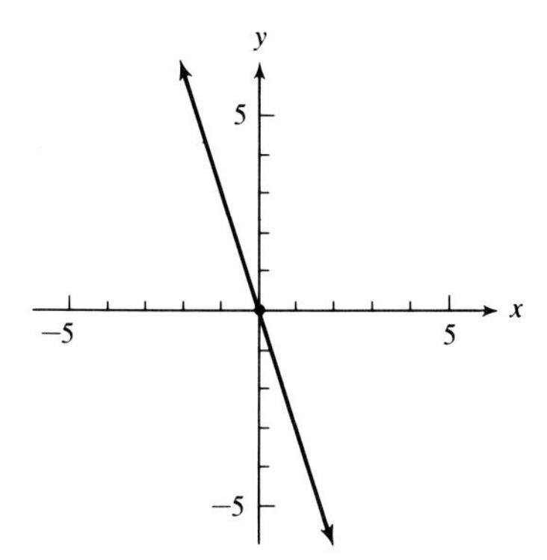

17.

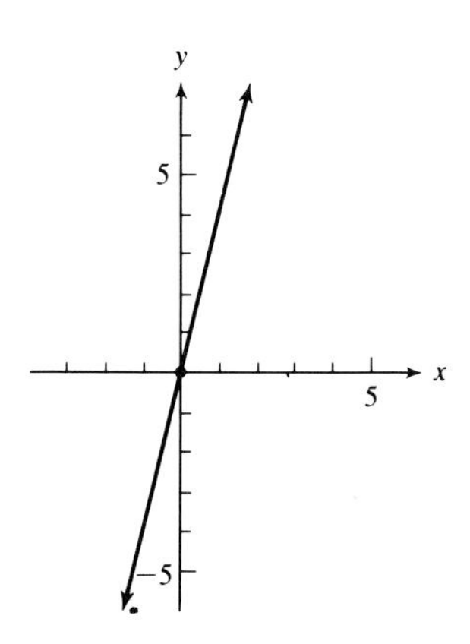

19.

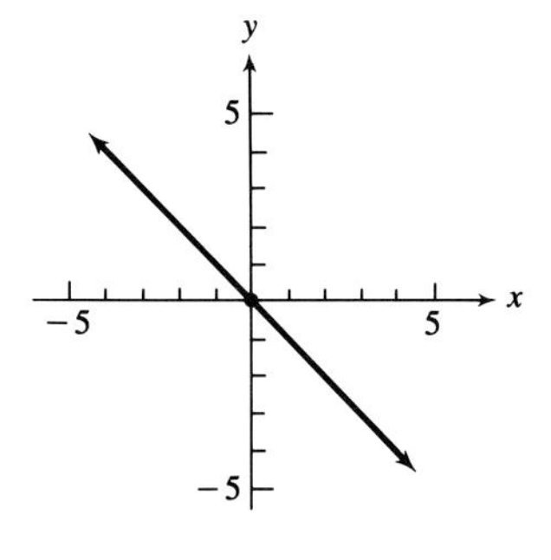

21.

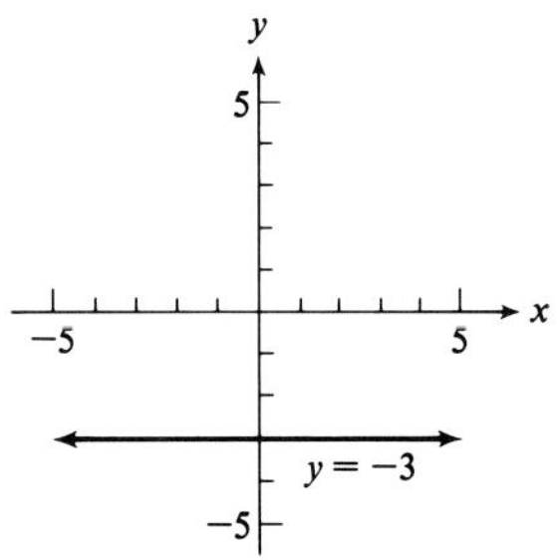

23.

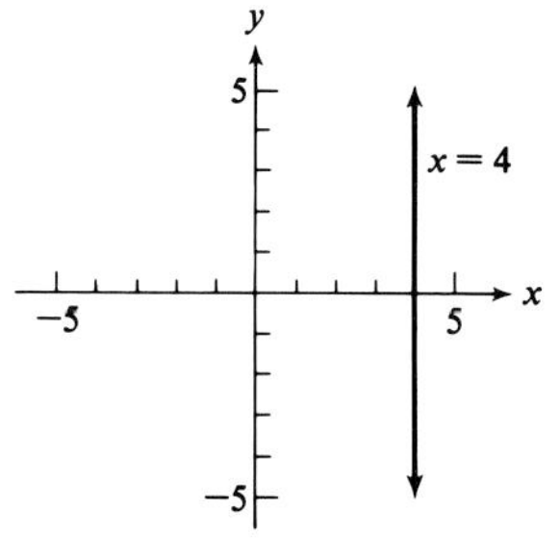

25.

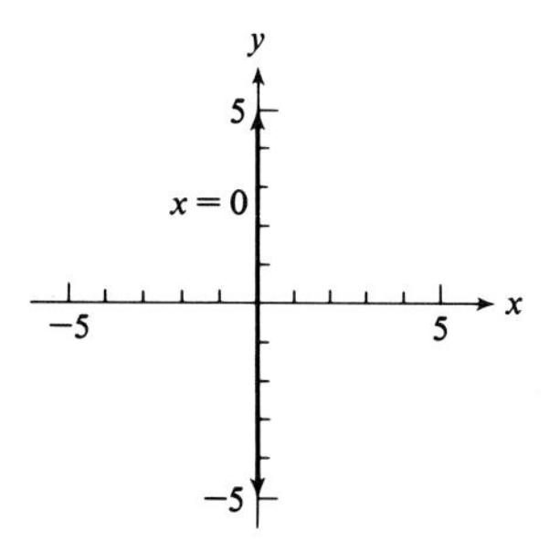

27.

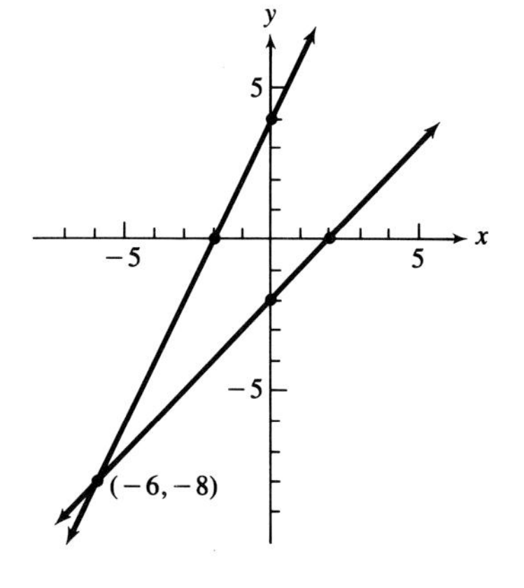

29.

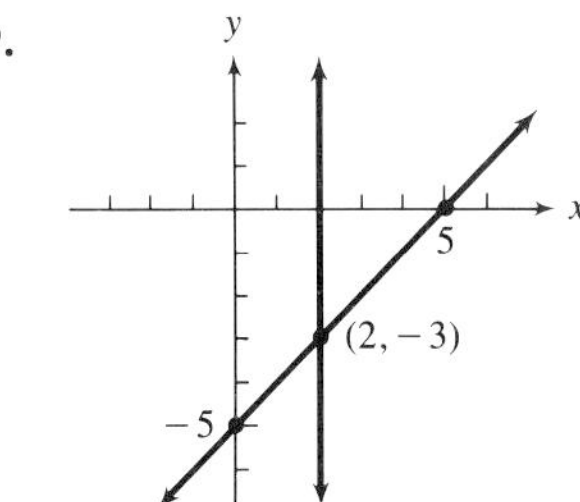

Exercise 8.3 [page 287]

1. distance: 5; slope: $\frac{4}{3}$

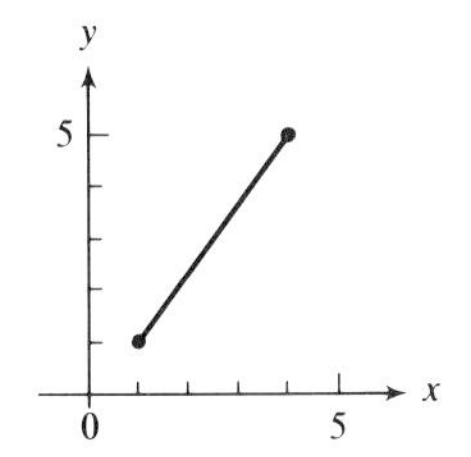

3. distance: 13; slope: $\frac{12}{5}$

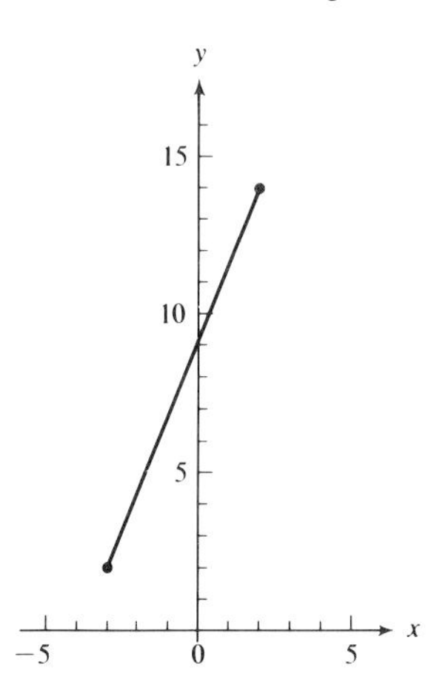

5. distance: $\sqrt{5}$; slope: $\frac{-1}{2}$

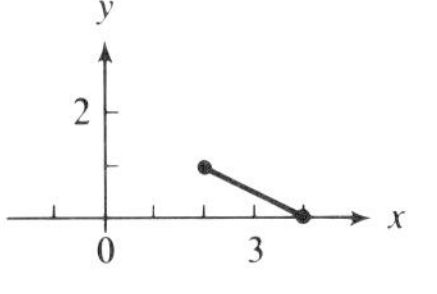

7. distance: $\sqrt{61}$; slope: $\frac{-5}{6}$

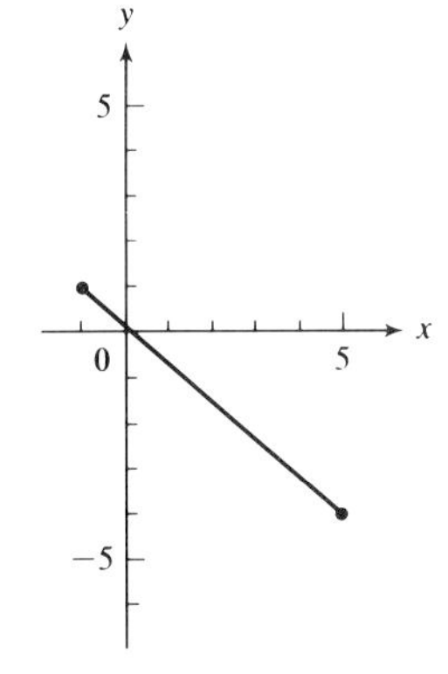

9. distance: 5; slope: 0

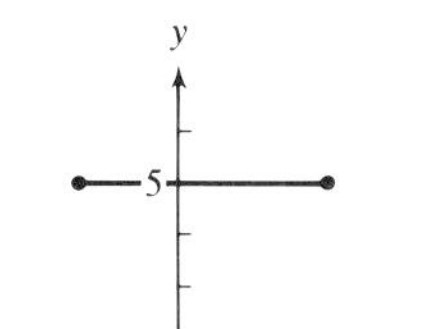
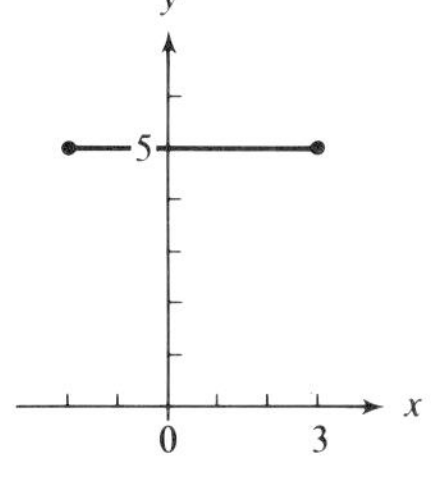

11. distance: 10; slope; not defined

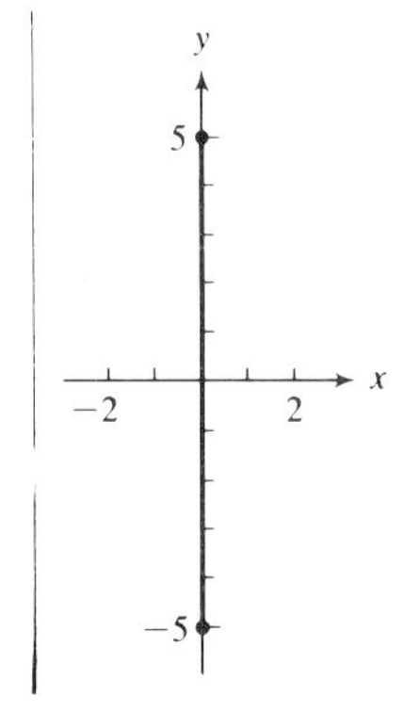

13. $15 + 9\sqrt{5}$

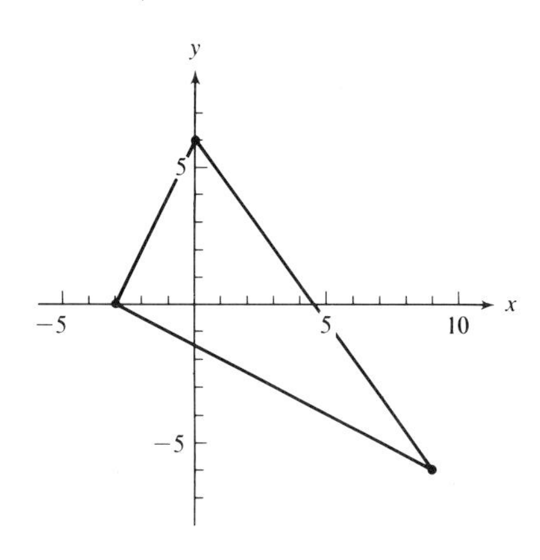

15. 48

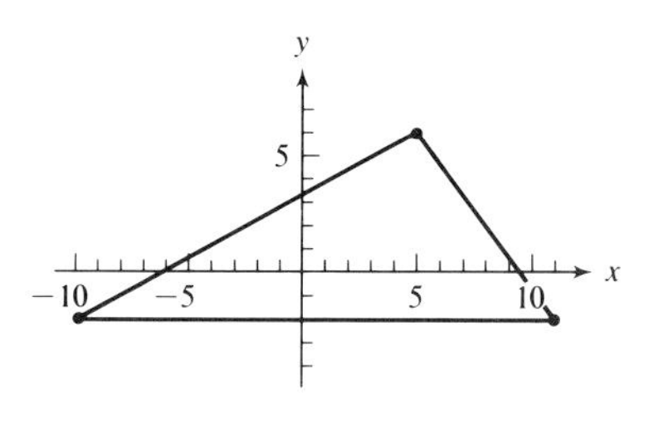

17. The slope of the segment with endpoints (5, 4) and (3, 0) is 2. The slope of the segment with endpoints (−1, 8) and (−4, 2) is 2. Therefore, the line segments are parallel.

19. Identify the given points as $A(0, -7)$, $B(8, -5)$, $C(5, 7)$, and $D(8, -5)$. Then, the slopes of AB and CD are

$$\text{slope } AB = \frac{-7-(-5)}{0-(8)} = \frac{1}{4} \quad \text{and} \quad \text{slope } CD = \frac{7-(-5)}{5-8} = -4;$$

and

$$\left(\frac{1}{4}\right)(-4) = -1.$$

21. **a.** 1 **b.** ¼ 23. **a.** −6 **b.** 0

25. Identify the vertices as $A(0, 6)$, $B(9, -6)$, and $C(\ 3, 0)$. Then, by the distance formula, $(AB)^2 = 225$ and $(BC)^2 + (AC)^2 = 180 + 45 = 225$; the triangle is a right triangle.

27. Let the given points be $A(2, 4)$, $B(3, 8)$, $C(5, 1)$, and $D(4, -3)$. Since AB and CD have equal slopes $(m = 4)$ and BC and AD have equal slopes $(m = -7/2)$, there are two pairs of parallel sides. Hence, $ABCD$ is a parallelogram.

29. $k = -28$

Exercise 8.4 [page 294] 1. $2x - y + 5 = 0$ 3. $x + y - 4 = 0$ 5. $x - 2y + 6 = 0$
7. $3x + 2y - 1 = 0$ 9. $y + 5 = 0$ 11. $x + 3 = 0$ 13. $x - 7y = -18$
15. $x - y = 2$ 17. $2x - 5y = 14$ 19. $4x - 3y = 12$

21. $y = -x + 3$; slope: -1; y-intercept: 3 23. $y = \frac{-3}{2}x + \frac{1}{2}$; slope: $\frac{-3}{2}$; y-intercept: $\frac{1}{2}$

25. $y = \frac{1}{3}x - \frac{2}{3}$; slope: $\frac{1}{3}$; y-intercept: $\frac{-2}{3}$ 27. $y = \frac{8}{3}x$; slope: $\frac{8}{3}$; y-intercept: 0

29. $y = 0x - 2$; slope: 0; y-intercept: -2 **31.** $2x - 3y = 6$ **33.** $2x - y = -4$

35. $2x + y = -6$ **37.** $6x - 4y = -3$

39. $x - 2y = 0$

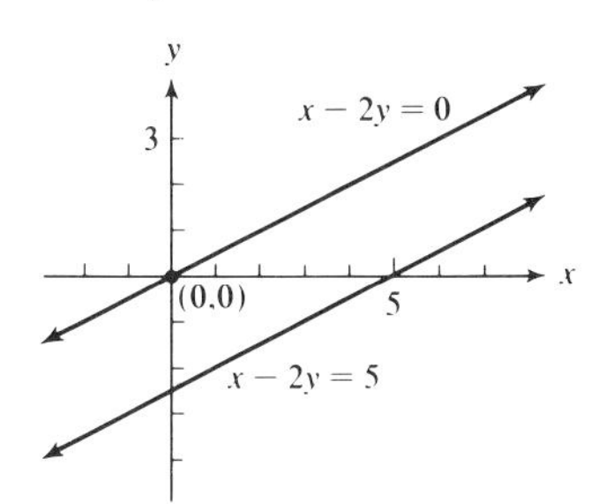

41. $2x + y = 0$

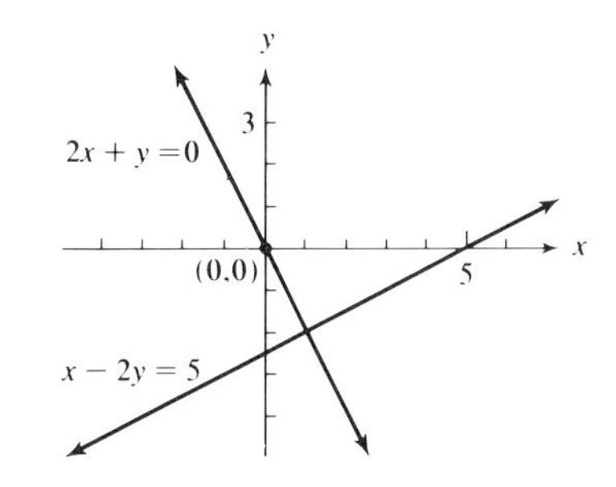

Exercise 8.5 [page 298]

1.

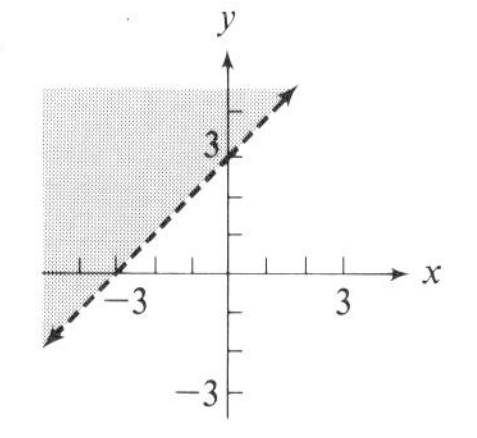

3.

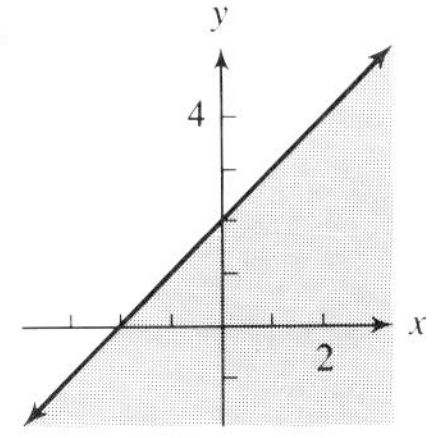

5.

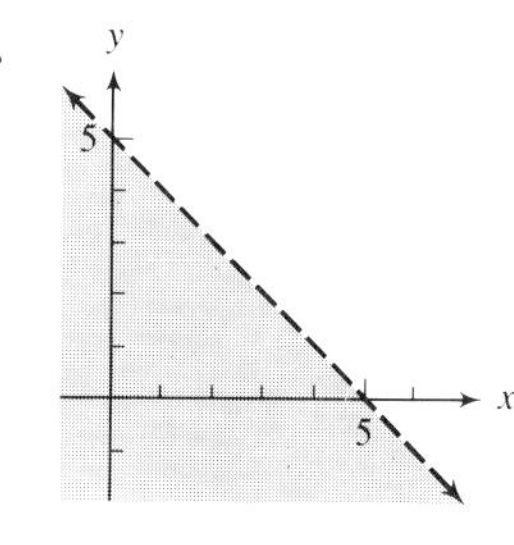

7.

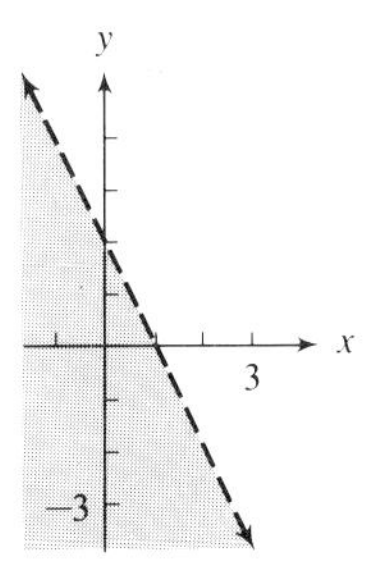

9.

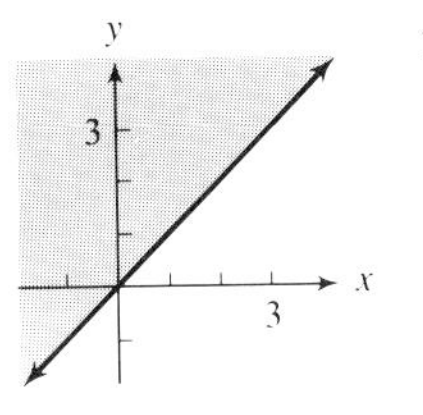

11.

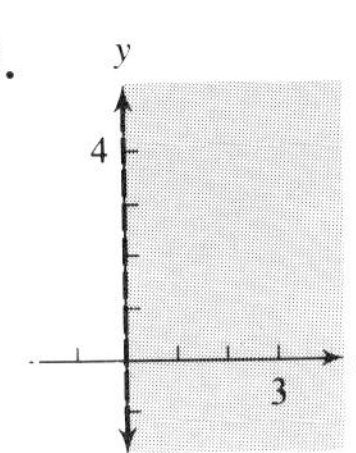

13.

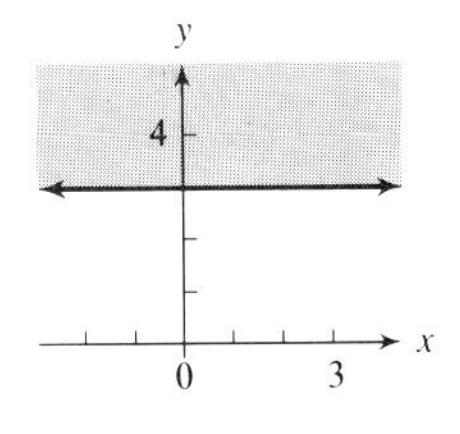

15.

17.

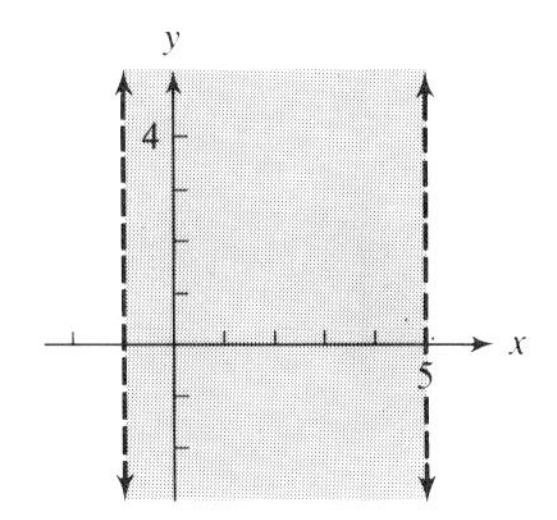

19.

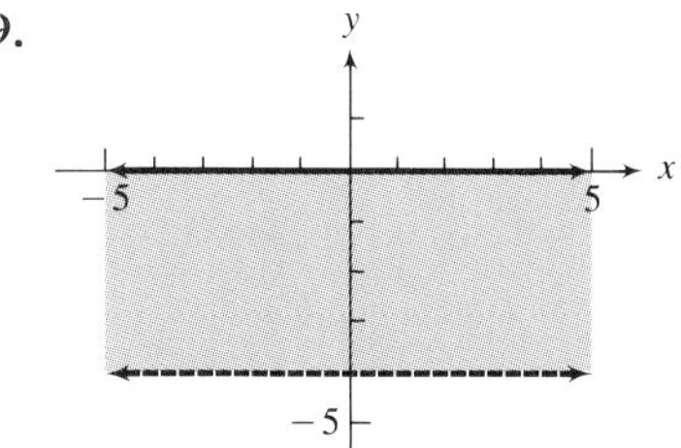

21.

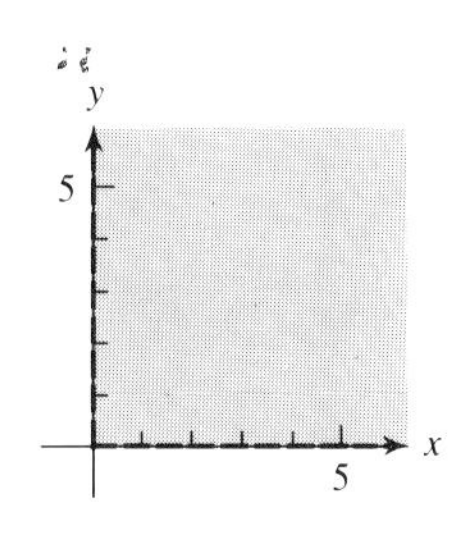

23.

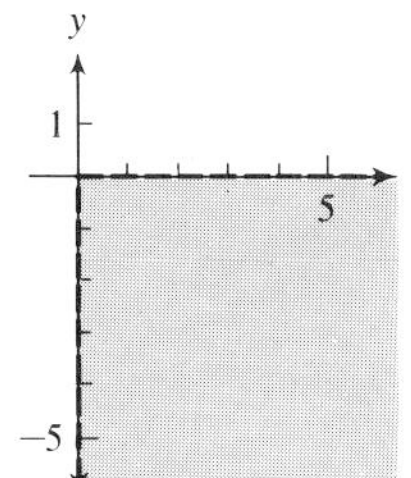

25.

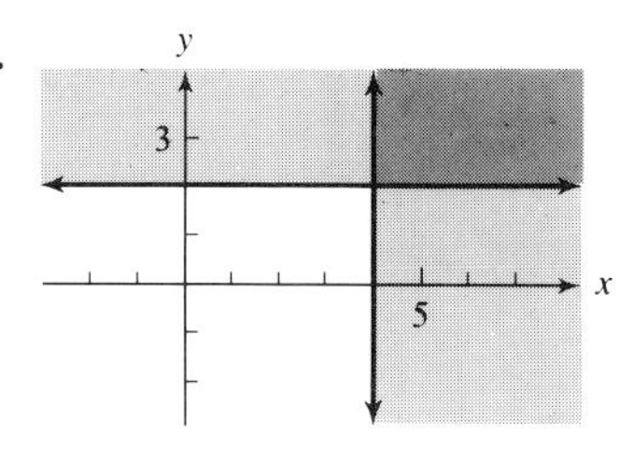

27.

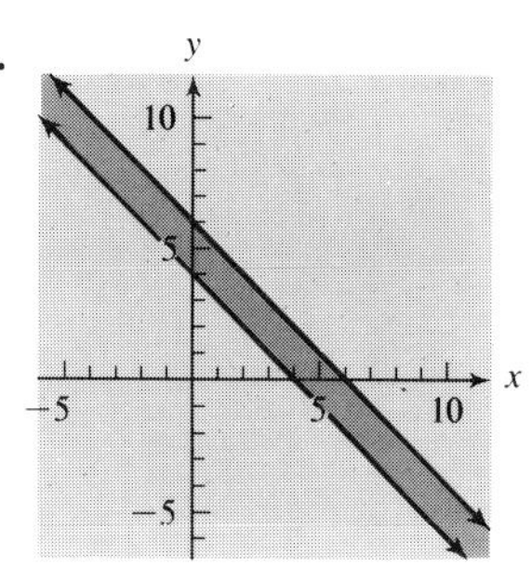

29.

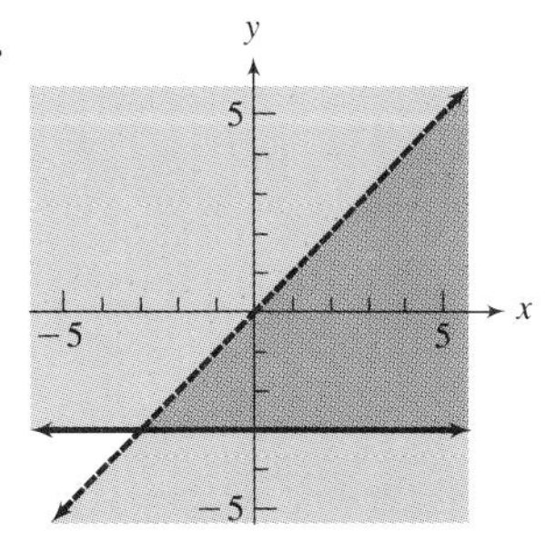

Review Exercises [page 300] **1. a.** (0, −2) **b.** (6, 0) **c.** (3, −1)

2. (2, 1), (4, 5), and (6, 9) **3.** $y = \dfrac{3}{x - 2x^2}$ $(x \neq 0, x \neq ½)$ **4.** $y = \pm\dfrac{1}{3}\sqrt{x^2 - 4}$

5.

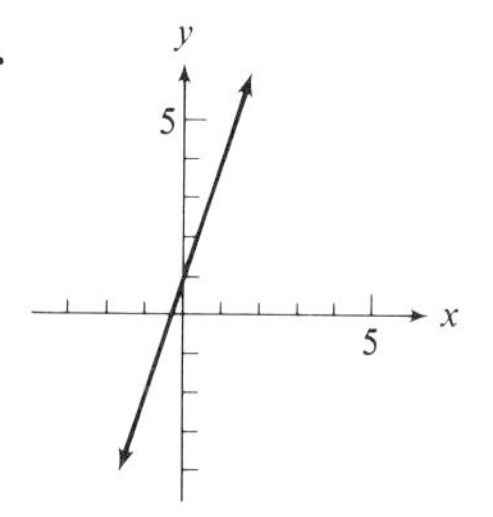

6.

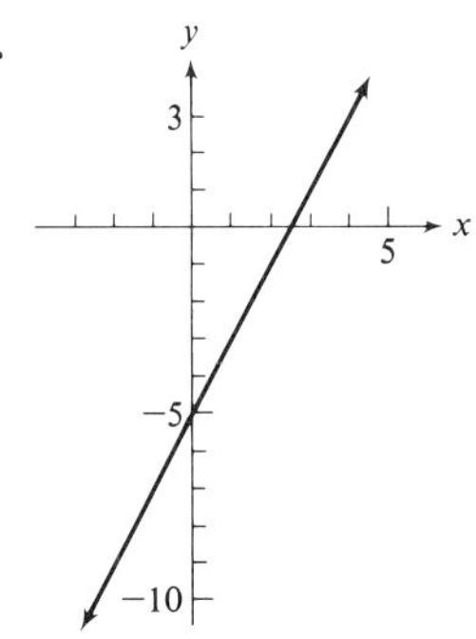

7.

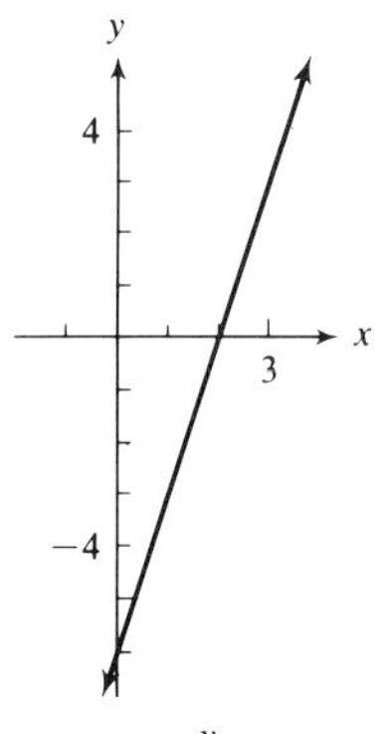

8.

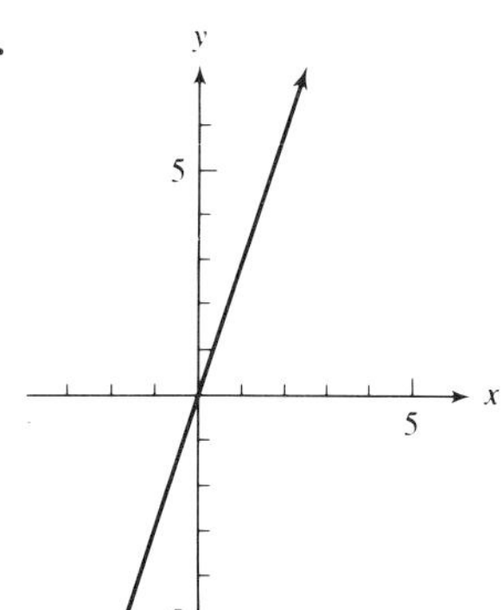

9.

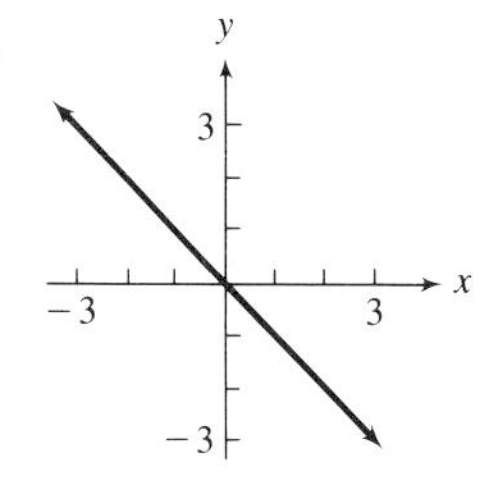

10.

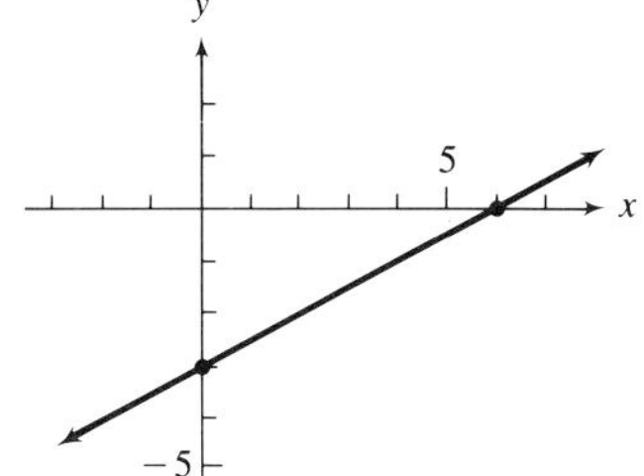

11.

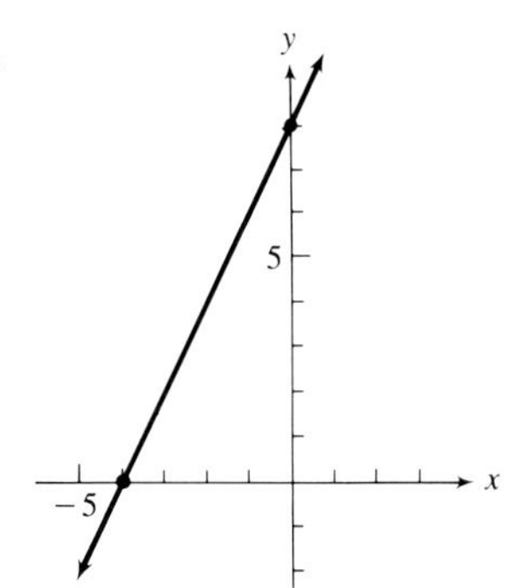

12.

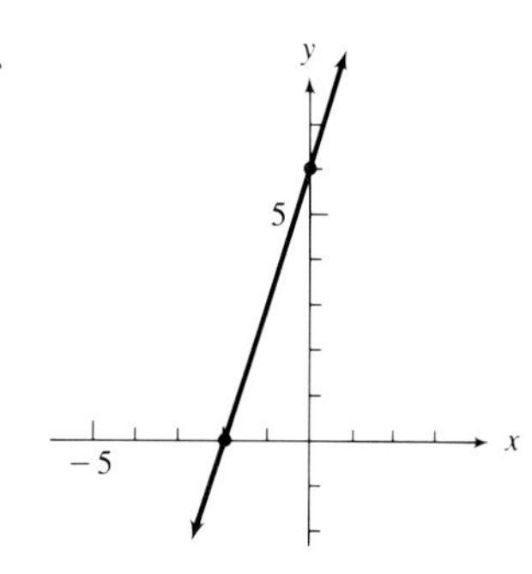

13.

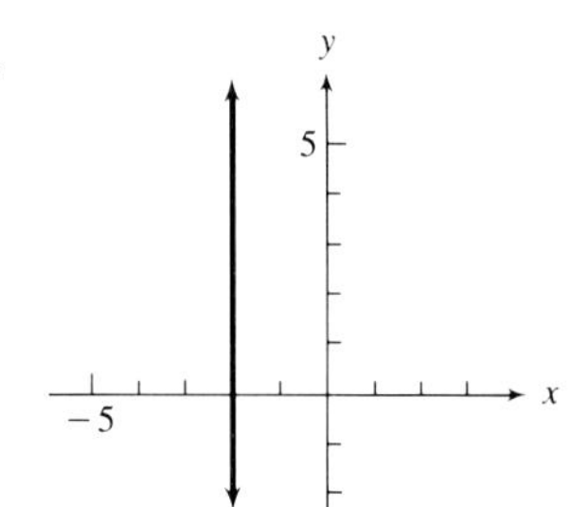

14.

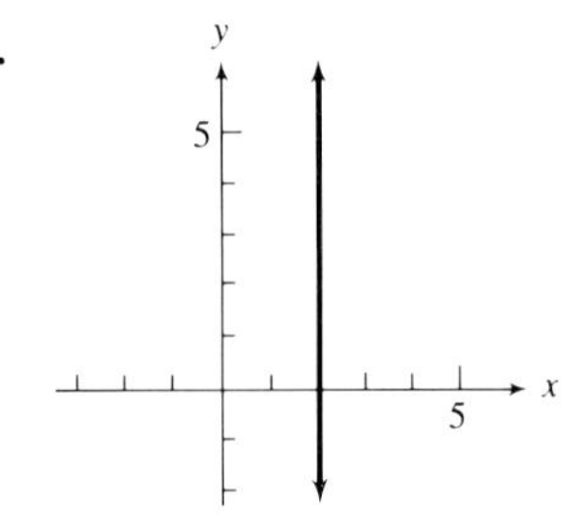

15. **a.** $\sqrt{178}$ **b.** $\frac{13}{3}$ **16.** **a.** $3\sqrt{5}$ **b.** -2

17. **a.** Slope of $P_1P_2 = 1$ and slope of $P_3P_4 = 1$; therefore, P_1P_2 and P_3P_4 are parallel.
b. Slope of $P_1P_2 = 1$ and slope of $P_1P_3 = -1$; since $1(-1) = -1$, the lines are perpendicular.

18. $2x - y - 1 = 0$ **19.** $x + 2y = 0$ **20.** **a.** $y = 3x - 4$ **b.** slope: 3; y-intercept: -4

21. **a.** $y = -\frac{2}{3}x + 2$ **b.** slope: $-\frac{2}{3}$; y-intercept: 2 **22.** $6x - y = 19$ **23.** $x - 4y = -12$

24. $2x - 5y = -10$

25.

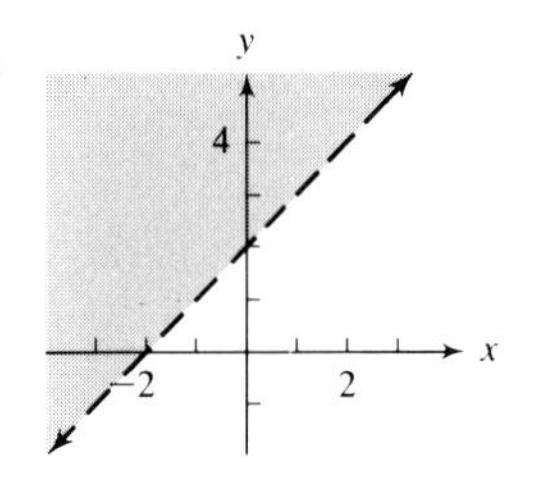

26.

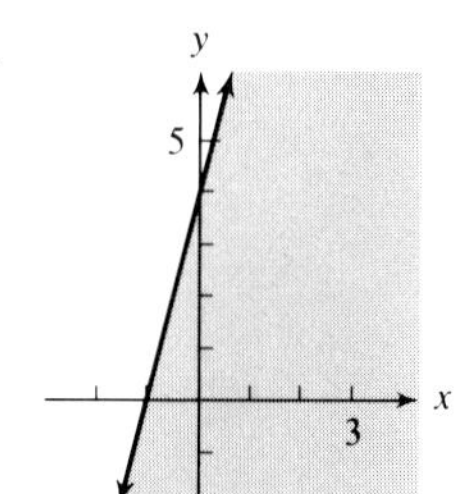

27.

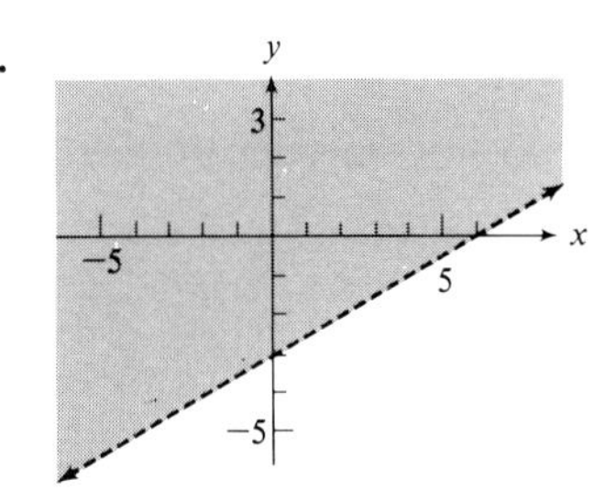

28.

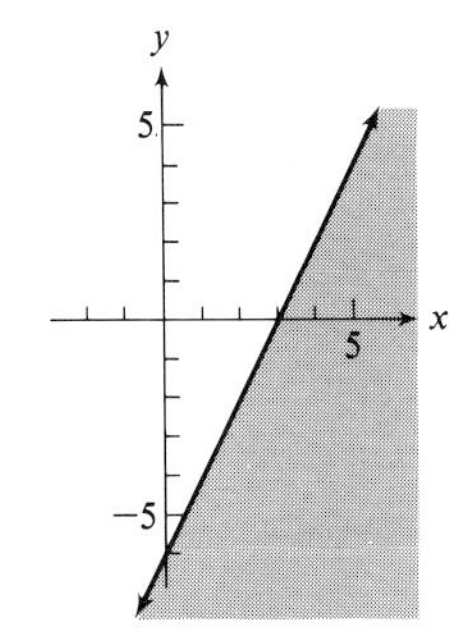

29.

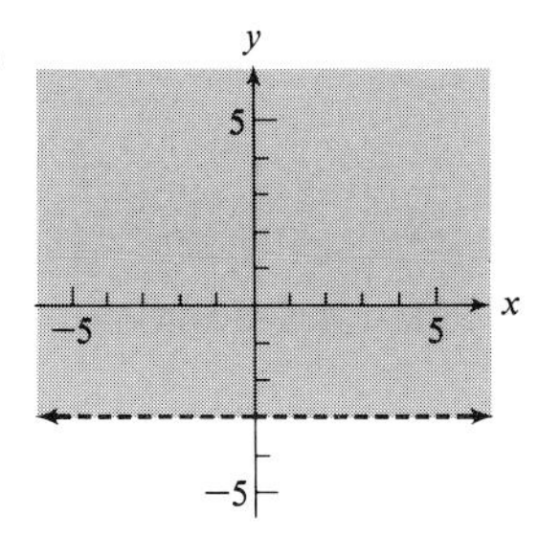

30.

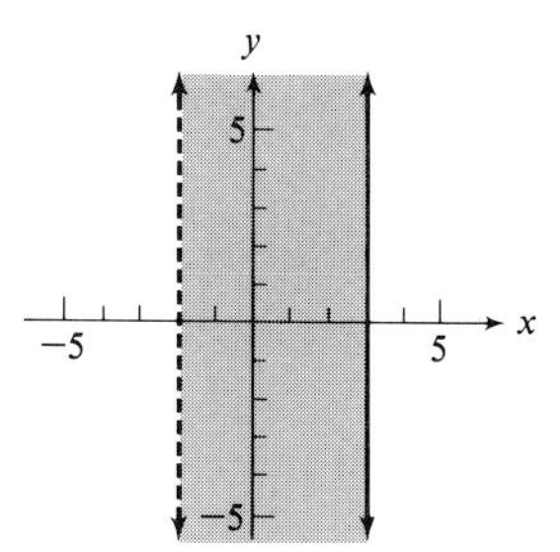

31. $k = \dfrac{-9}{7}$

32.

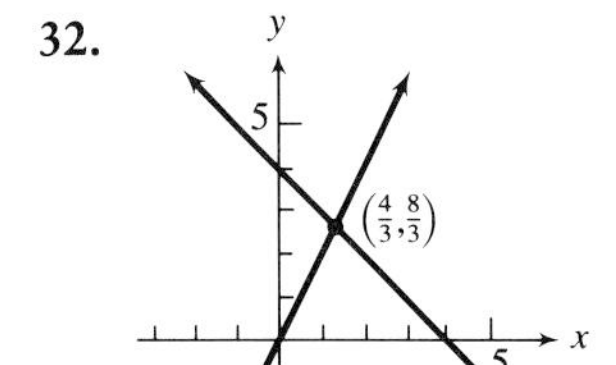

33. $2x - y - 7 = 0$

34.

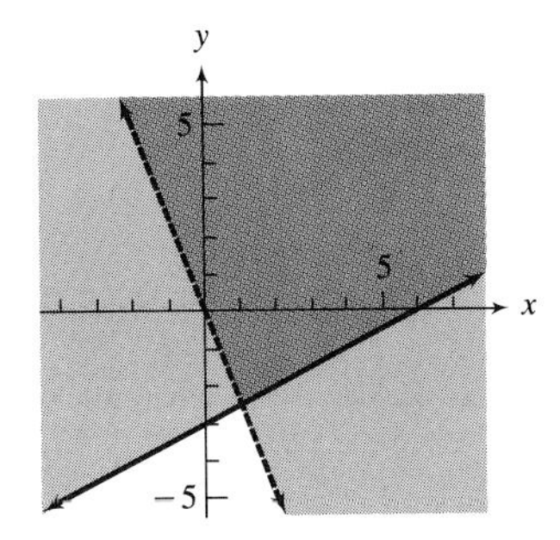

Exercise 9.1 [page 311]

1. $\{(3, 2)\}$

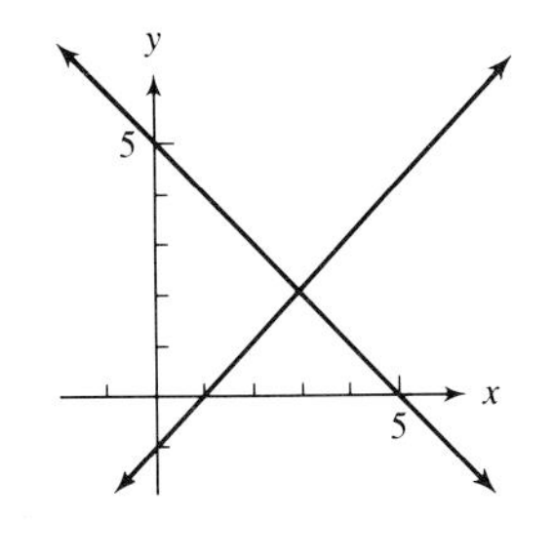

3. $\{(2, 1)\}$

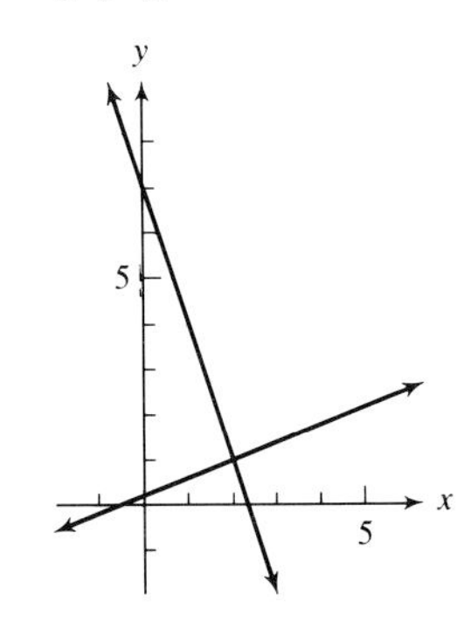

5. $\{(-5, 4)\}$

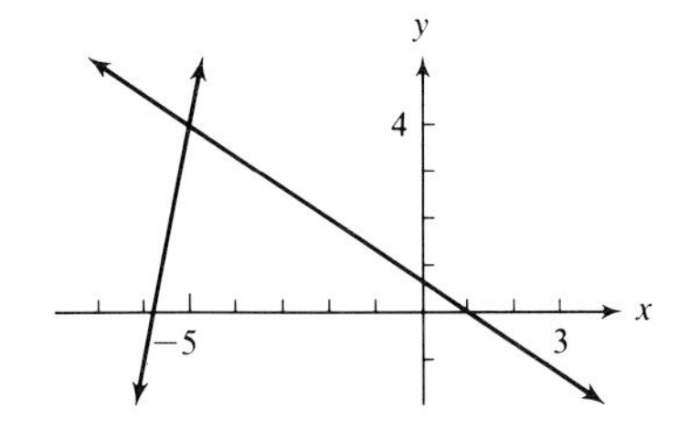

7. $\{(1, 2)\}$

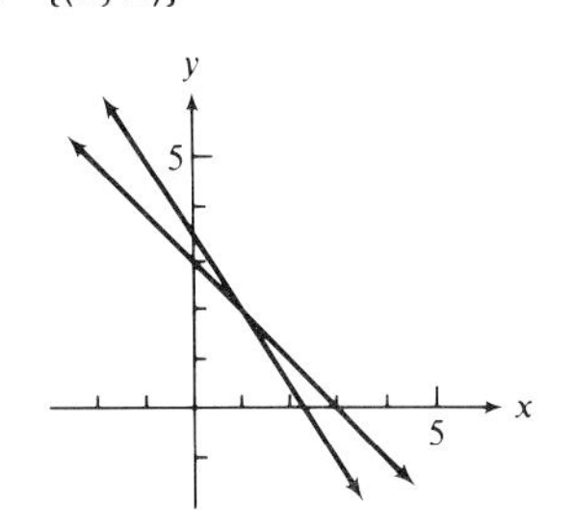

9. $\{(1, 0)\}$

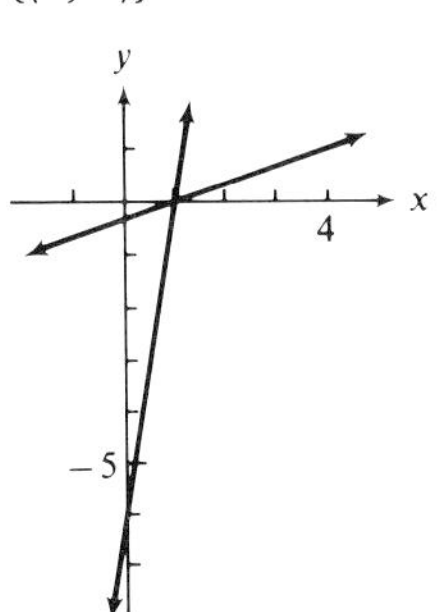

11. $\{(1, 2)\}$

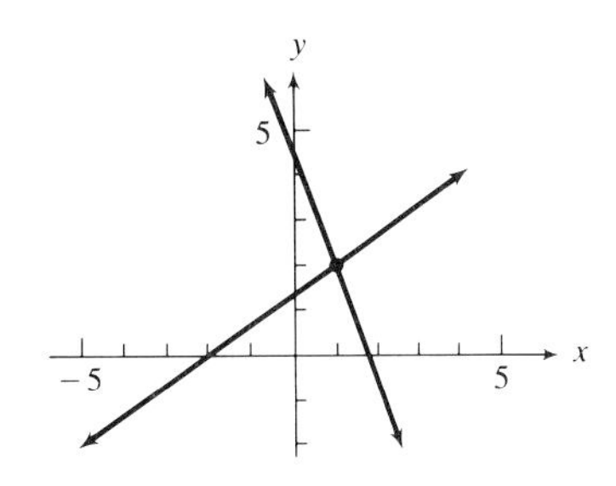

13. $\{(0, 3/2)\}$

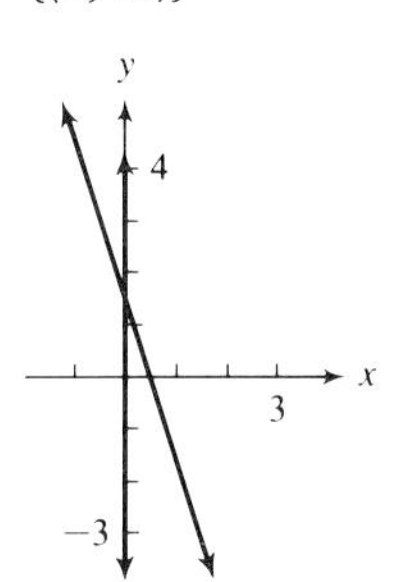

15. $\{(2/3, -1)\}$

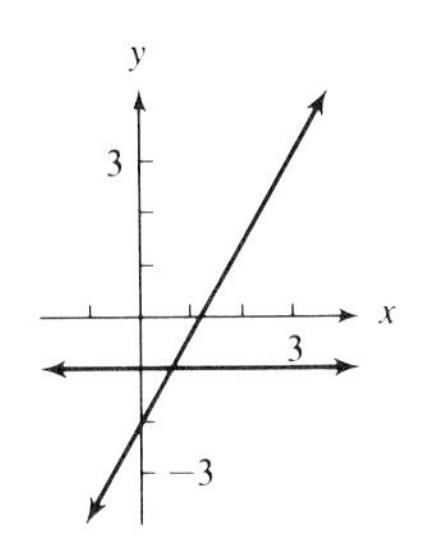

17. $\{(1, 2)\}$

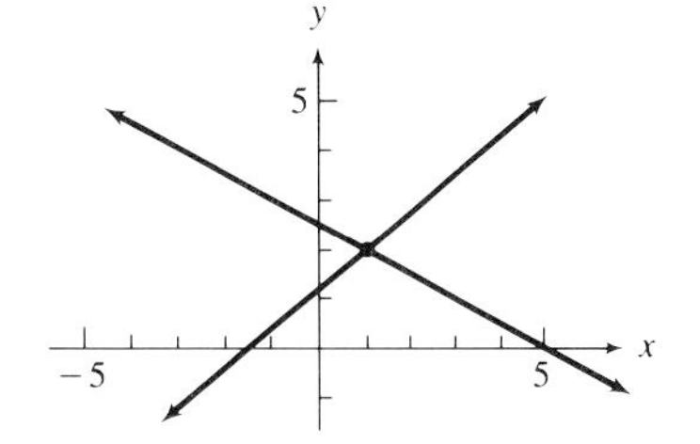

19. $\{(-19/5, -18/5)\}$

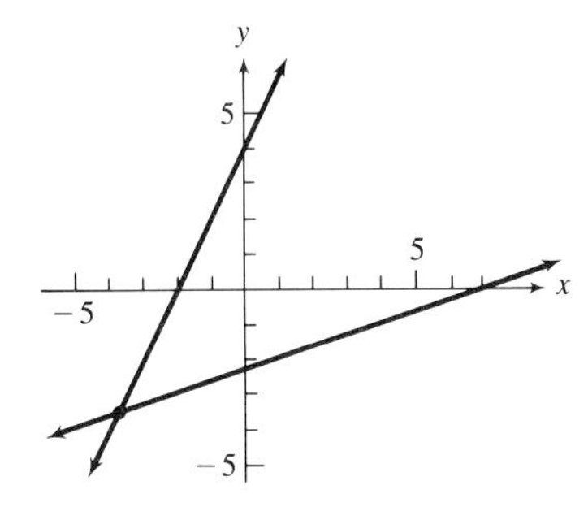

21. infinitely many solutions **23.** $\varnothing$; no solutions **25.** one solution **27.** $\varnothing$; no solutions
29. one solution **31.** one solution

33.

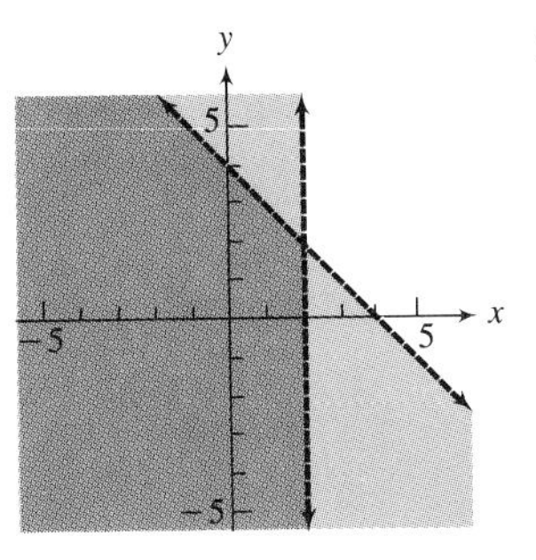

35.

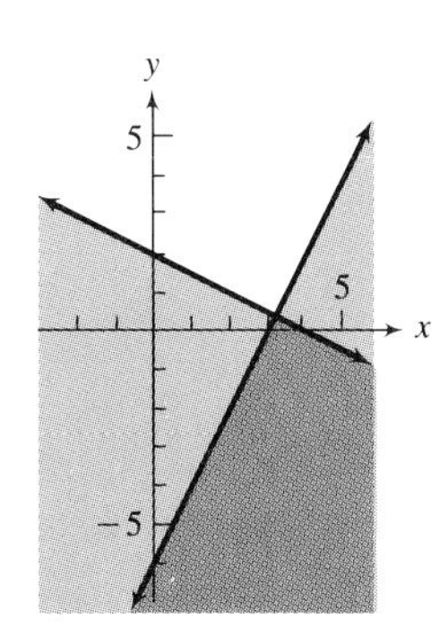

37.

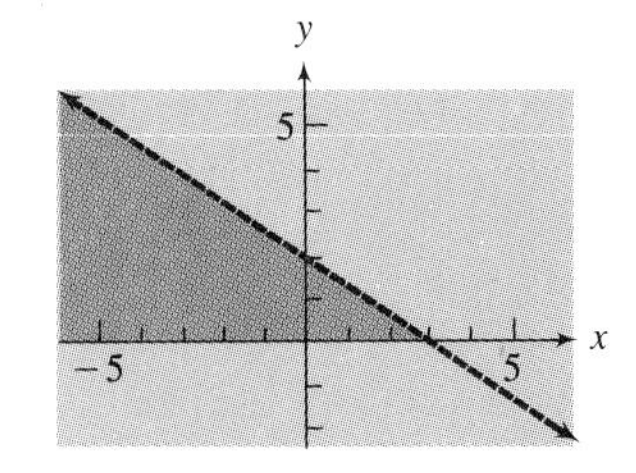

39. 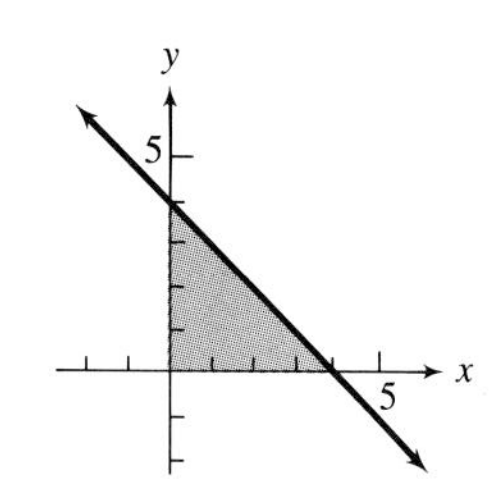

41. 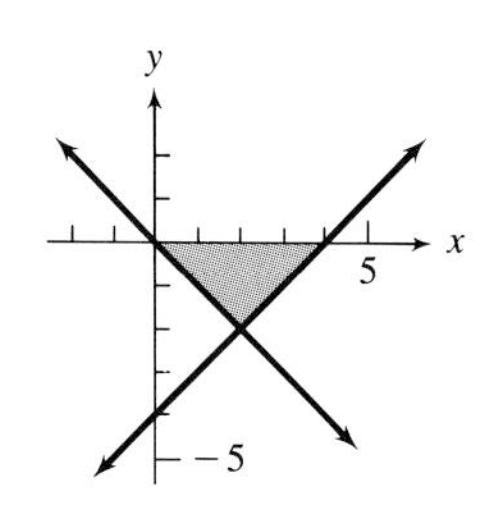

43. 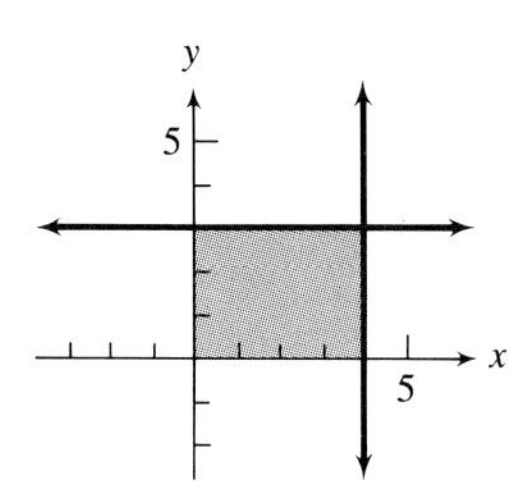

45. $\{(1/5, 1/2)\}$ 47. $\{(-5, 3)\}$ 49. $\{(1/6, 1/2)\}$

Exercise 9.2 [page 317] 1. $\{(1, 2, -1)\}$ 3. $\{(2, -2, 0)\}$ 5. $\{(2, 2, 1)\}$ 7. $\{(3, -1, -1)\}$ 9. $\{(1, -3, 4)\}$ 11. $\{(0, -2, 3)\}$ 13. $\{(4, -2, 2)\}$ 15. $\{(1, 1, 0)\}$ 17. $\{(1/2, -1/2, 1/3)\}$ 19. no unique solution 21. $\{(1/2, 0, 3)\}$ 23. no unique solution 25. $\{(-1, 3, 0)\}$ 27. no unique solution 29. $\{(1/2, 1/2, 3)\}$ 31. $\{(1, 1, 1)\}$ 33. $\{(1, 2, -2)\}$

Exercise 9.3 [page 323] 1. 1 3. −12 5. 2 7. 5 9. $\{(1, 1)\}$ 11. $\{(2, 2)\}$ 13. $\{(6, 4)\}$ 15. inconsistent, $\varnothing$ 17. $\{(4, 1)\}$ 19. $\left\{\left(\frac{1}{a+b}, \frac{1}{a+b}\right)\right\}$

21. $\begin{vmatrix} a & a \\ b & b \end{vmatrix} = ab - ab = 0$

23. $\begin{vmatrix} a_1 & b_1 \\ a_2 & b_2 \end{vmatrix} = a_1b_2 - a_2b_1;\quad -\begin{vmatrix} b_1 & a_1 \\ b_2 & a_2 \end{vmatrix} = -(b_1a_2 - b_2a_1) = -b_1a_2 + b_2a_1 = a_1b_2 - a_2b_1$

25. $\begin{vmatrix} ka & a \\ kb & b \end{vmatrix} = kab - kba = 0$

27. Since $D_x = \begin{vmatrix} c_1 & b_1 \\ c_2 & b_2 \end{vmatrix} = c_1b_2 - c_2b_1 = 0$, then $b_1c_2 = b_2c_1$.

Since $D_y = \begin{vmatrix} a_1 & c_1 \\ a_2 & c_2 \end{vmatrix} = a_1c_2 - a_2c_1 = 0$, then $a_1c_2 = a_2c_1$.

Forming the proportion $\frac{b_1c_2}{a_1c_2} = \frac{b_2c_1}{a_2c_1}$, if c_1 and c_2 are not both 0, then $\frac{b_1}{a_1} = \frac{b_2}{a_2}$,

from which $a_1b_2 = a_2b_1$ and $a_1b_2 - a_2b_1 = 0$. Since $D = \begin{vmatrix} a_1 & b_1 \\ a_2 & b_2 \end{vmatrix} = a_1b_2 - a_2b_1$,

it follows that $D = 0$.

Exercise 9.4 [page 328] 1. 3 3. 9 5. 0 7. −1 9. −5 11. 0 13. x^3 15. 0 17. $-2ab^2$ 19. $\{(1, 1, 1)\}$ 21. $\{(1, 1, 0)\}$ 23. $\{(1, -2, 3)\}$ 25. $\{(3, -1, -2)\}$ 27. no unique solution 29. $\{(1, -1/3, 1/2)\}$ 31. $\{(-5, 3, 2)\}$

33. Expanding about the elements of the third column of the given determinant produces

$$a\begin{vmatrix} y & y \\ z & z \end{vmatrix} - b\begin{vmatrix} x & x \\ z & z \end{vmatrix} + c\begin{vmatrix} x & x \\ y & y \end{vmatrix} = 0 + 0 + 0 = 0.$$

35. For the left side: $\begin{vmatrix} 1 & 2 & 3 \\ 4 & 5 & 6 \\ 0 & 0 & 1 \end{vmatrix} = 1\begin{vmatrix} 1 & 2 \\ 4 & 5 \end{vmatrix} = 5 - 8 = -3.$

For the right side: $-\begin{vmatrix} 4 & 5 & 6 \\ 1 & 2 & 3 \\ 0 & 0 & 1 \end{vmatrix} = -1\begin{vmatrix} 4 & 5 \\ 1 & 2 \end{vmatrix} = -1(8 - 5) = -1(3) = -3.$

37. Expanding about the elements of the first row produces

$$x\begin{vmatrix} -1 & 1 \\ 3 & 1 \end{vmatrix} - y\begin{vmatrix} 4 & 1 \\ 2 & 1 \end{vmatrix} + 1\begin{vmatrix} 4 & -1 \\ 2 & 3 \end{vmatrix} = 0,$$

$$-4x - 2y + 14 = 0,$$

which is equivalent to

$$2x + y = 7,$$

and is satisfied by $(4, -1)$ and $(2, 3)$.

Exercise 9.5 [page 336]

1. $\begin{bmatrix} 1 & -3 \\ 0 & 7 \end{bmatrix}$ **3.** $\begin{bmatrix} 2 & 6 \\ 4 & 0 \end{bmatrix}$ **5.** $\begin{bmatrix} 1 & -2 & 2 \\ 0 & 7 & -5 \\ 0 & 9 & -11 \end{bmatrix}$ **7.** $\begin{bmatrix} -1 & 4 & 3 \\ 3/2 & 0 & -5/2 \\ 3/2 & 0 & 3/2 \end{bmatrix}$

9. $\begin{bmatrix} -2 & 1 & -3 \\ 0 & 4 & -6 \\ 0 & 0 & 5 \end{bmatrix}$ **11.** $\{(2, 3)\}$ **13.** $\{(-2, 1)\}$ **15.** $\{(3, -1)\}$ **17.** $\{(-7/3, 17/3)\}$

19. $\{(1, 2, 2)\}$ **21.** $\{(2, -1/2, 1/2)\}$ **23.** $\{(-3, 1, -3)\}$ **25.** $\{(5/4, 5/2, -1/2)\}$

Exercise 9.6 [page 340] One possible system of equations is shown for each problem. Other systems that yield correct solutions are possible.

1. a. Larger number: x
Smaller number: y

b. $x - y = 6$
$x + y = 24$

c. Numbers are 15 and 9

3. a. One integer: x
Next consecutive integer: y

b. $x + 1 = y$
$\frac{1}{3}x + \frac{1}{2}y = 33$

c. Integers are 39 and 40

5. a. Number of adults: x
Number of children: y

b. $x + y = 82$
$1.50x + 0.85y = 93.10$

7. a. Amount invested at 10%: x
Amount invested at 8%: y

b. $x + y = 2000$
$0.10x + 0.08y = 184$

c. 36 adults and 46 children

c. \$1200 at 10% and \$800 at 8%

9. a. Number of first-class passengers: x
Number of tourist passengers: y

b. $x + y = 42$
$80x + 64y = 2880$

c. 12 first-class passengers and 30 tourist passengers

11. a. Rate of automobile: x
Rate of airplane: y

b. $x + 120 = y$
$\frac{420}{x} = \frac{1260}{y}$

c. Rate of automobile: 60 miles per hour and rate of airplane: 180 miles per hour

13. a. One number: x
Second number: y
Third number: z

b. $x + y + z = 15$
$y = 2x$
$z = y$

c. The numbers are 3, 6, and 6

15. a. Number of nickels: x
Number of dimes: y
Number of quarters: z

b. $x + y + z = 85$
$0.5x + 0.10y + 0.25z = 6.25$
$x = 3y$

c. 60 nickels, 20 dimes, and 5 quarters

17. a. One side: x
Second side: y
Third side: z

b. $x + y + z = 155$
$x + 20 = y$
$z + 5 = y$

c. Side x: 40 inches; side y: 60 inches; and side z: 55 inches

19. a. Age of red wine: x

b. Age of white wine: y
1976: $y = 2x$
1986: $y + 10 = (x + 10) + 4$

c. $x = 4$ and $y = 8$;
White wine: $1976 - 8 = 1968$;
Red wine: $1976 - 4 = 1972$

21. a. $C = 20 + 0.40x$

b. $R - 1.20x$

c. 25 records

23. $a(-1) + b(2) + 3 = 0$
$a(-3) + b(0) + 3 = 0$;
$a = 1$ and $b = -1$

25. $a(-2) + b(0) + c(4) = 1$
$a(6) + b(-1) + c(0) = 1$
$a(0) + b(3) + c(0) = 1$;
$a = 2/9$, $b = 1/3$, and $c = 13/36$

Review Exercises [page 344] **1.** $\{(1/2, 7/2)\}$ **2.** $\{(1, 2)\}$ **3.** $\{(12, 0)\}$ **4.** $\{(1/2, 3/2)\}$
5. a. unique solution b. dependent **6.** a. inconsistent b. unique solution

7. a.

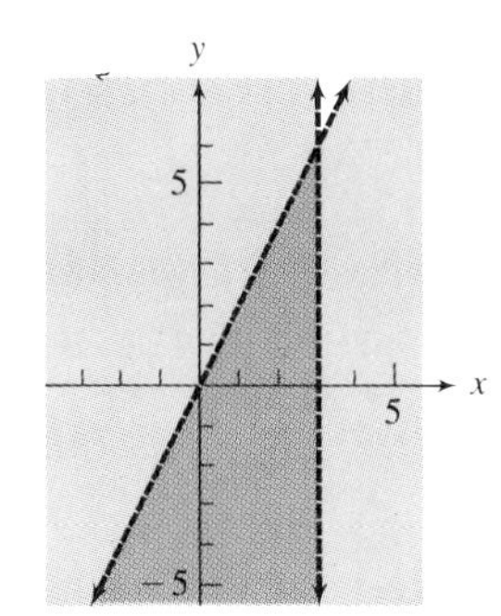

b.

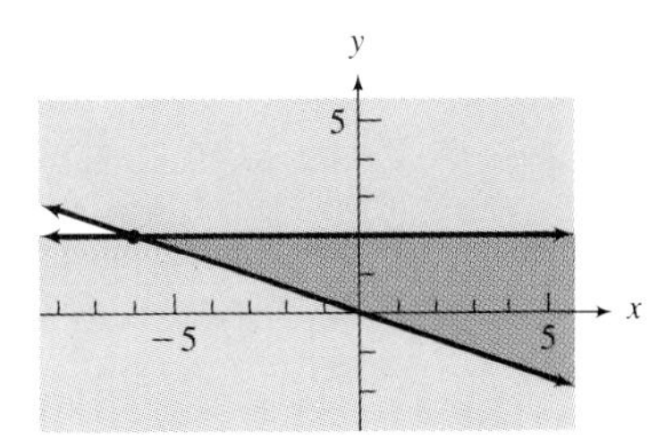

8. a.

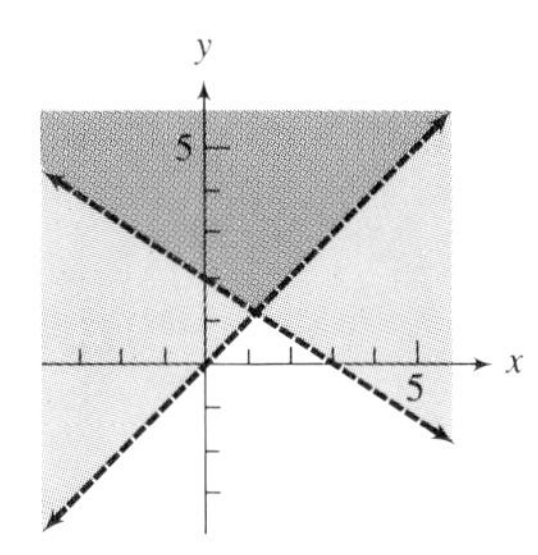

b.

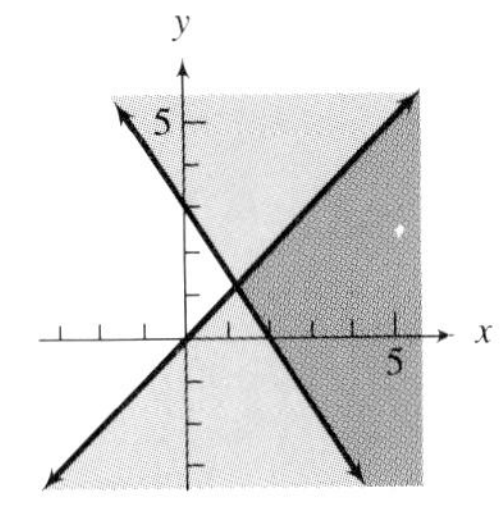

9. $\{(2, 0, -1)\}$ 10. $\{(2, 1, -1)\}$ 11. $\{(2, -5, 3)\}$ 12. $\{(4/17, -2/17, -84/17)\}$
13. $\{(-2, 1, 3)\}$ 14. $\{(2, -1, 0)\}$ 15. -13 16. 24 17. $\{(8, 1)\}$ 18. $\{(-7, 4)\}$
19. $\{(1, 2)\}$ 20. $\{(6, 0)\}$ 21. -2 22. -19 23. $\{(1, 2, 3)\}$ 24. $\{(2, -1, 3)\}$
25. $\{(1, 2, -1)\}$ 26. $\{(1, 2, -1)\}$ 27. $\{(3, -1)\}$ 28. $\{(4, 0)\}$ 29. $\{(4, 1)\}$
30. $\{(0, 3, -1)\}$ 31. $\{(-1, 0, 2)\}$ 32. $\{(1, 0, 0)\}$

33. a. Number of dimes: x
Number of quarters: y
b. $x - y = 25$
$0.10x + 0.25y = 4.95$
c. 32 dimes; 7 quarters

34. a. Number of first-class tickets: x
Number of tourist tickets: y
b. $x + y = 64$
$280x + 160y = 12{,}160$
c. 16 first-class tickets;
48 tourist tickets

35. a. Amount invested at 10%: x
Amount invested at 12%: y
b. $x + y = 8000$
$0.10x + 0.12y = 844$
c. \$5800 invested at 10%;
\$2200 invested at 12%

36. a. Amount invested at 14%: x
Amount invested at 11%: y
b. $x + y = 2400$
$0.14x - 0.11y = 111$
c. \$1500 invested at 14%;
\$900 invested at 11%

37. a. Rate of automobile: x
Rate of airplane: y

38. a. Speed of one woman (slower): x
Speed of second woman: y

b. $y - x = 180$

$$\frac{840}{y} = \frac{210}{x}$$

c. Rate of automobile: 60 miles per hour; rate of airplane: 240 miles per hour

b. $y - x = 5$

$$\frac{200}{y} = \frac{180}{x}$$

c. Speed of first woman: 45 miles per hour; speed of second woman: 50 miles per hour

39. a. Length of shortest side: x
Length of second side: y
Length of longest side: z

b. $x + y + z = 30$
$y - x = 7$
$z - y = 1$

c. Lengths are 5 cm, 12 cm, and 13 cm

40. a. Measure of smallest angle: x
Measure of second angle: y
Measure of third angle: z

b. $x + y + z = 180$
$y - x = 20$
$z = 3x$

c. Measures are 32°, 52°, and 96°

41.

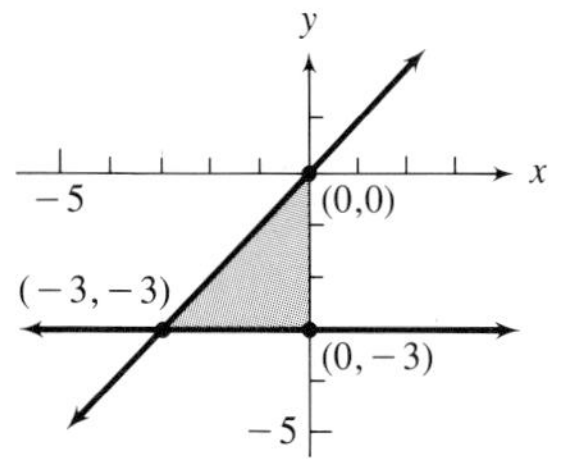

42. $\{(\frac{22}{35}, \frac{11}{14})\}$ **43.** $a = \frac{1}{2}, b = 3$

44. $a = \frac{5}{2}, b = \frac{1}{2}$ **45.** $a = \frac{1}{4}, b = \frac{1}{2}, c = 0$ **46.** $a = \frac{1}{3}, b = \frac{1}{3}, c = \frac{1}{6}$

Exercise 10.1 [page 349]

1.

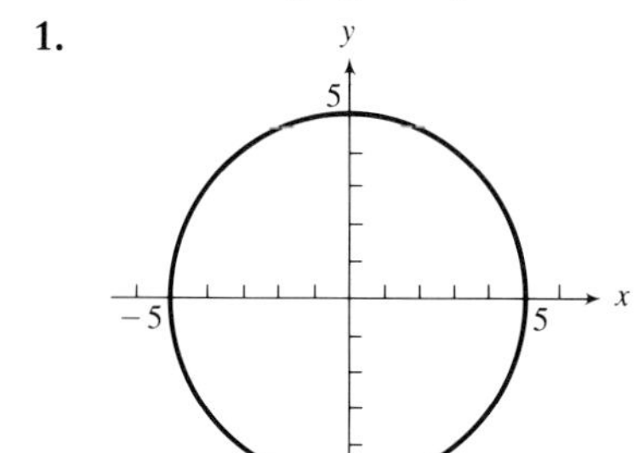

3.

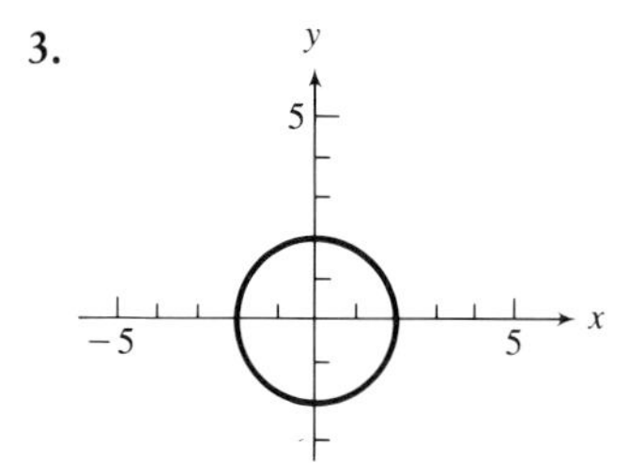

5.

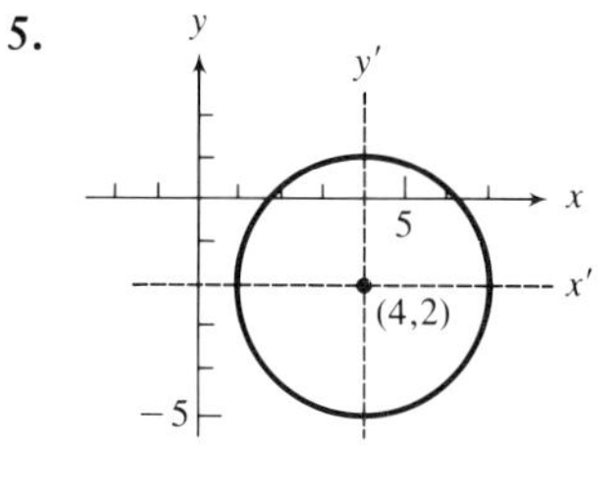

7.

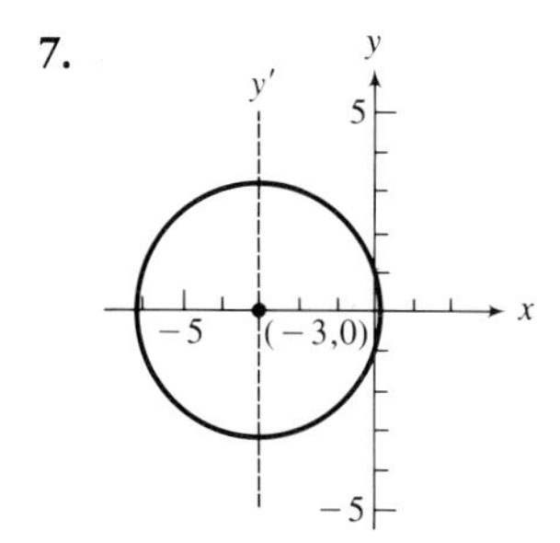

9.

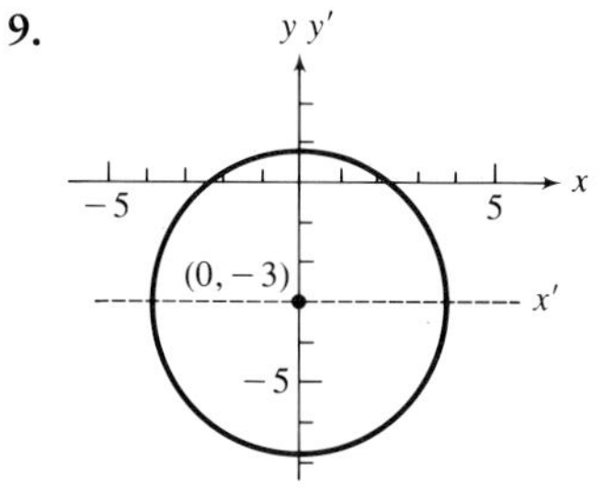

11.

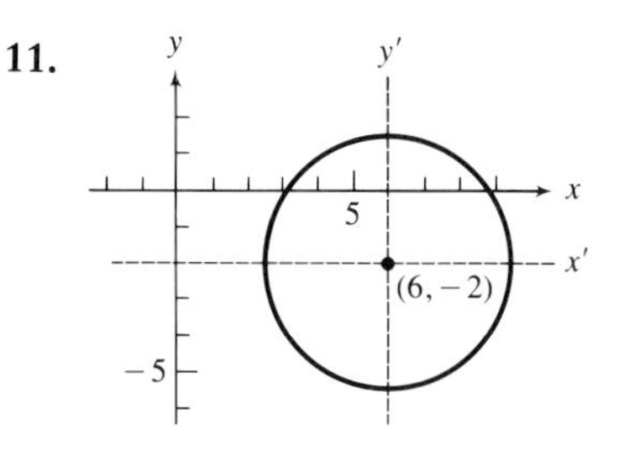

13.

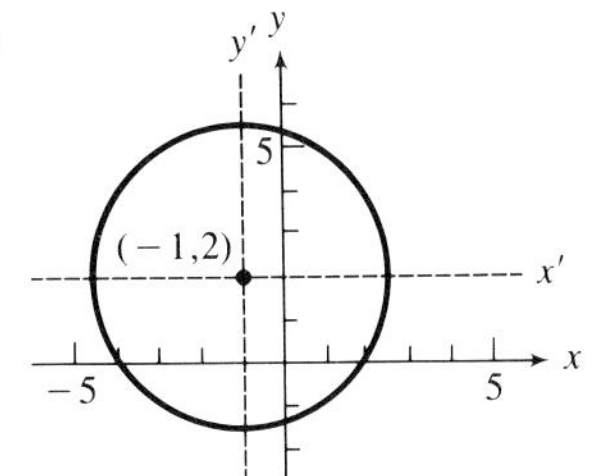

15.

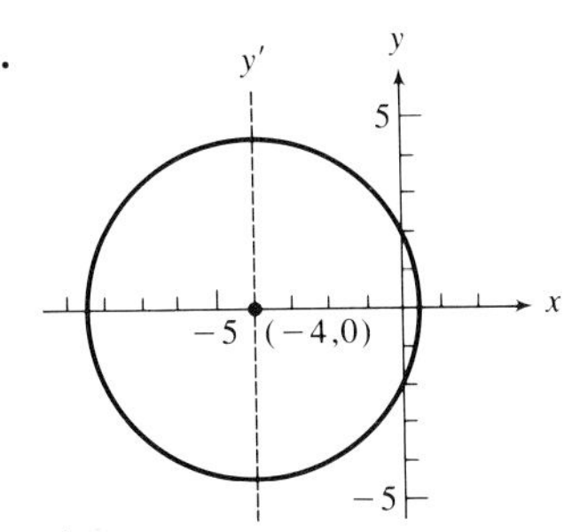

17.

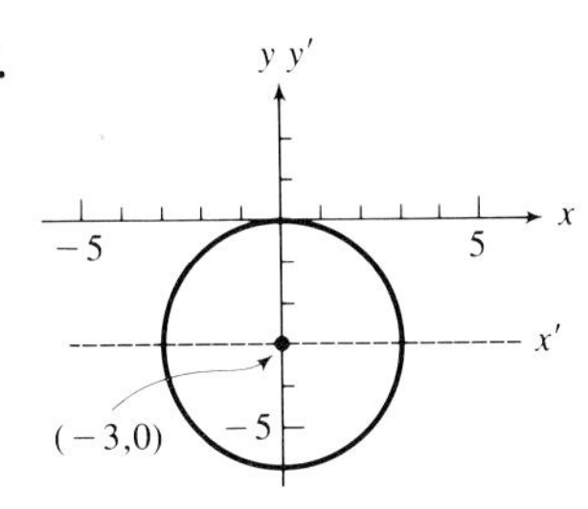

19.

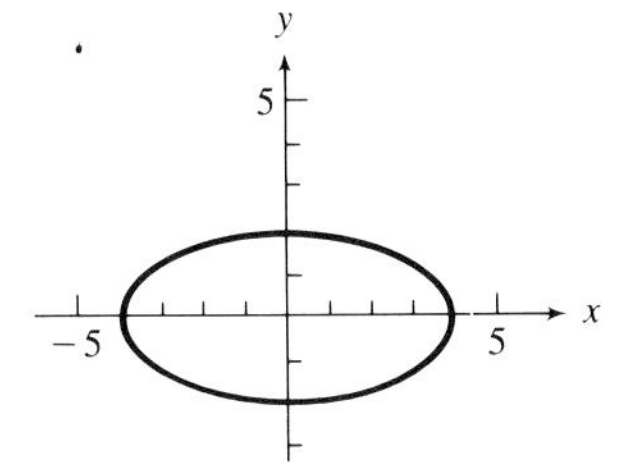

21.

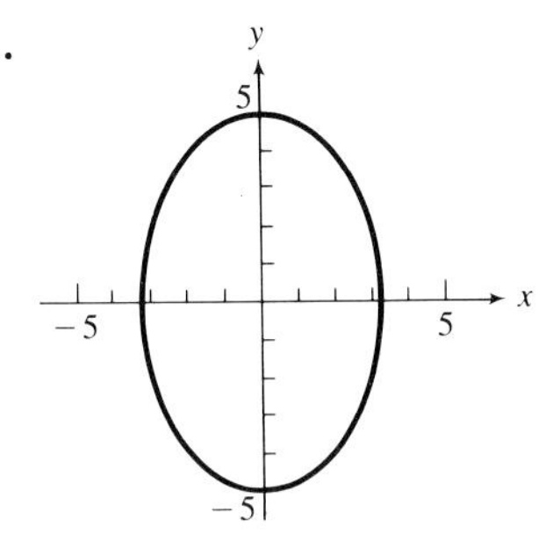

23.

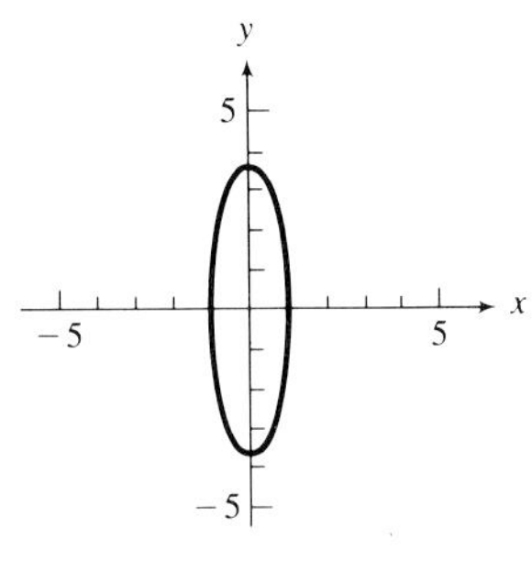

25.

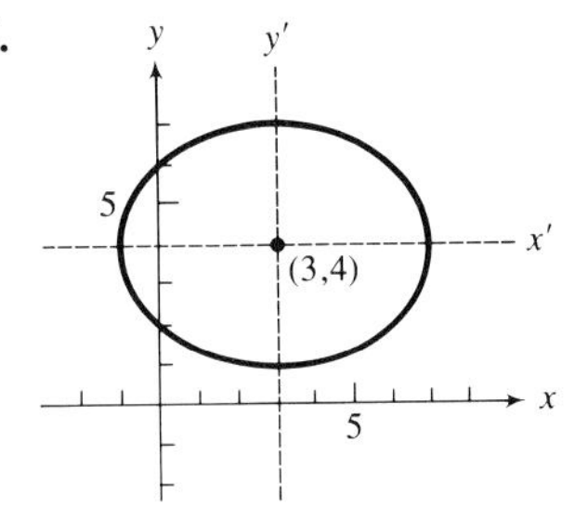

27.

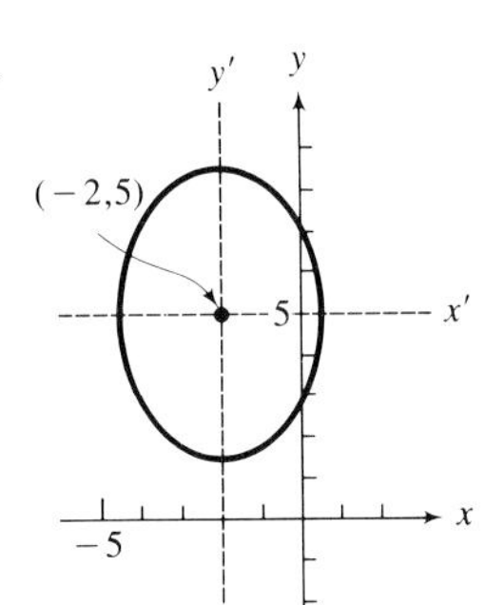

29.

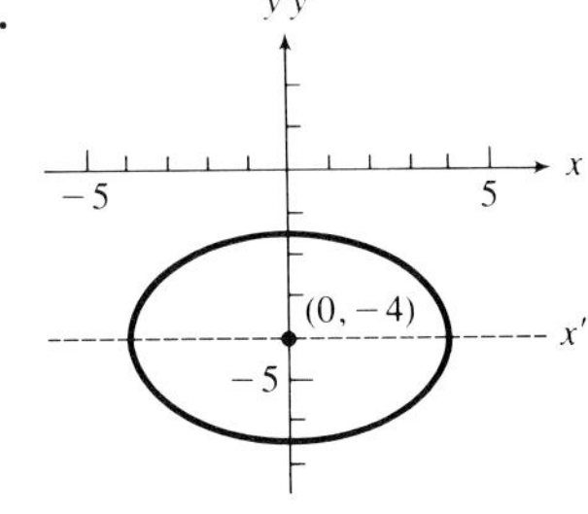

31.

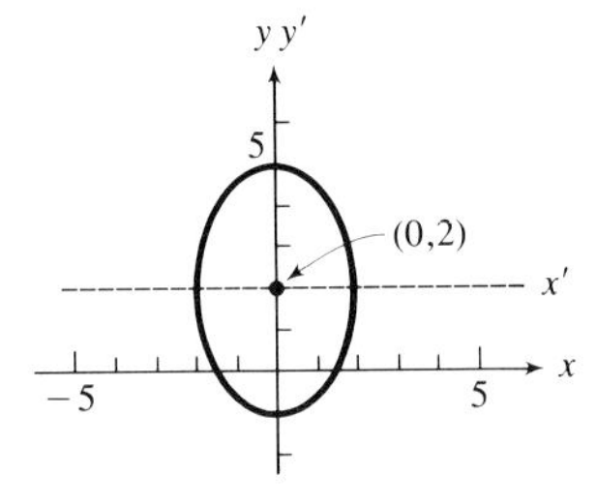

33.

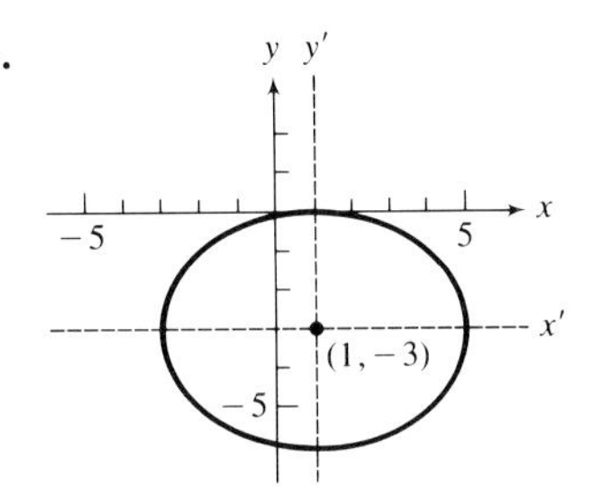

35.

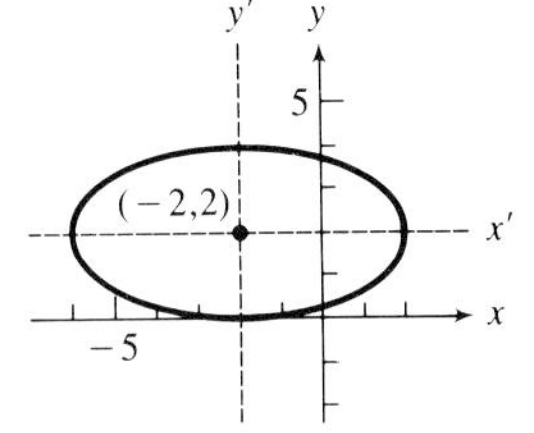

37.

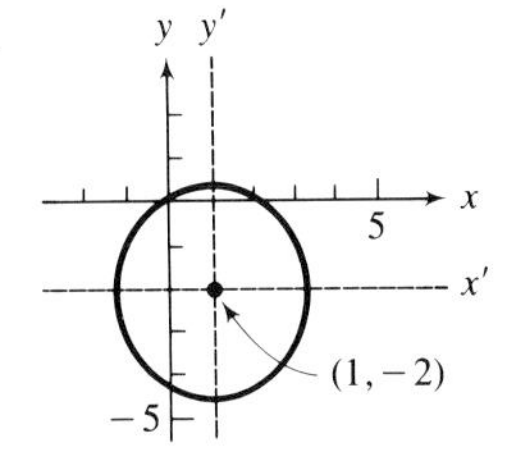

39.

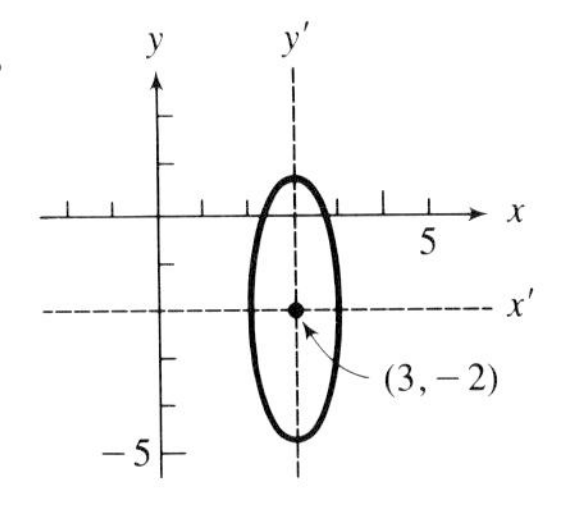

Exercise 10.2 [page 357] **1.** **a.** upward **b.** x-intercepts, 1 and 4; y-intercept, 4
3. **a.** downward **b.** x-intercepts, -2 and 3; y-intercept, 6
5. **a.** upward **b.** x-intercepts, -3 and $\frac{1}{2}$; y-intercept, -3
7. **a.** downward **b.** x-intercepts, $\frac{1}{3}$ and -2; y-intercept, 2
9. **a.** upward **b.** x-intercepts, 0 and 4; y-intercept, 0
11. **a.** downward **b.** x-intercepts, none; y-intercept, -8 **13.** $(2, -5)$ **15.** $(0, 8)$
17. $(-3, 0)$ **19.** $(-3, -4)$ **21.** $(-3, -6)$ **23.** $(\frac{3}{2}, +\frac{19}{4})$

25.
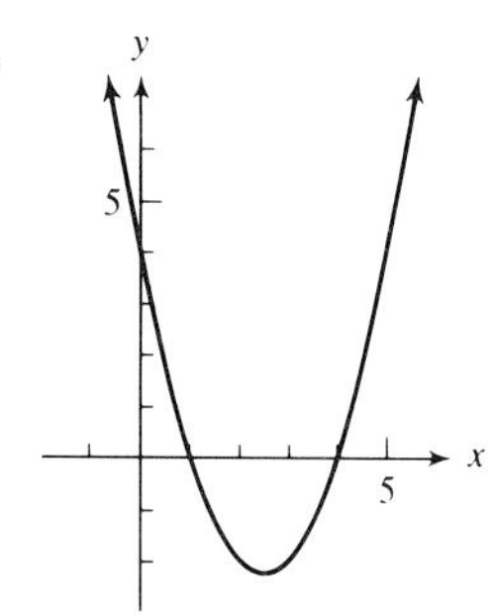

27.
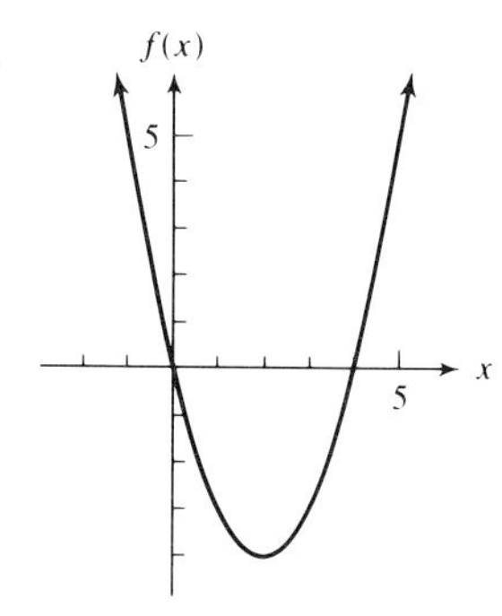

29.
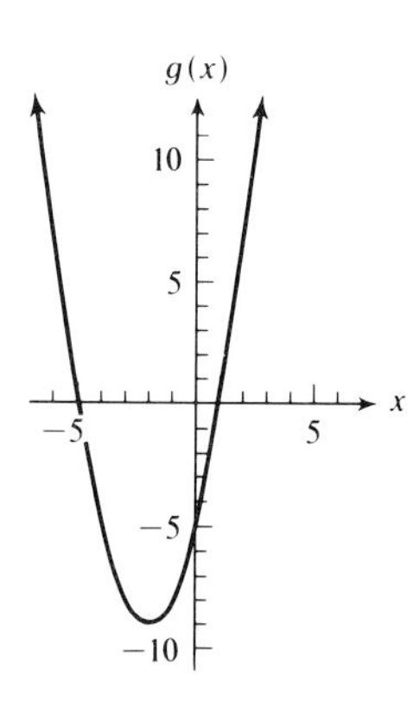

31.
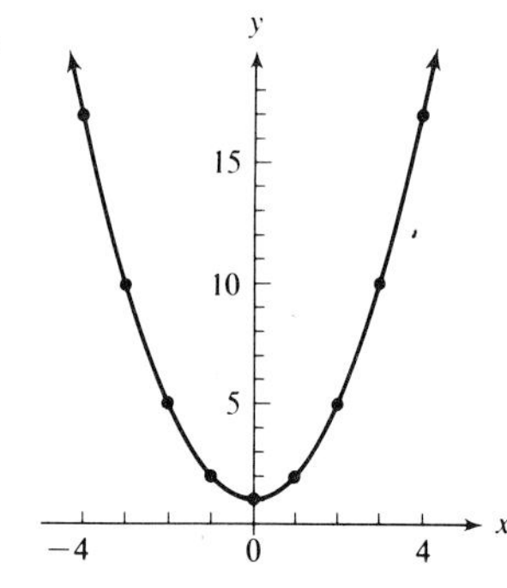

33.
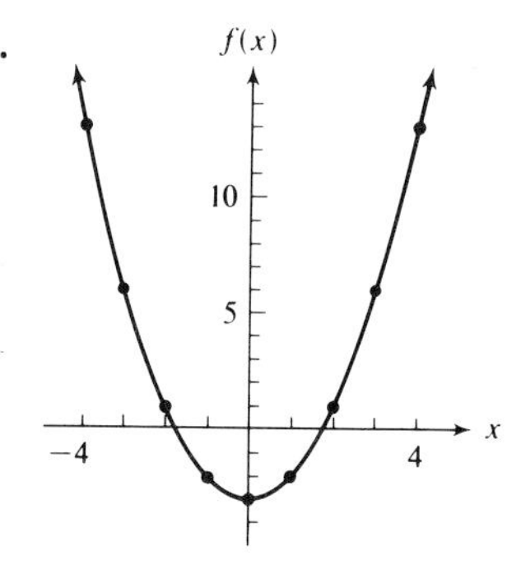

35.
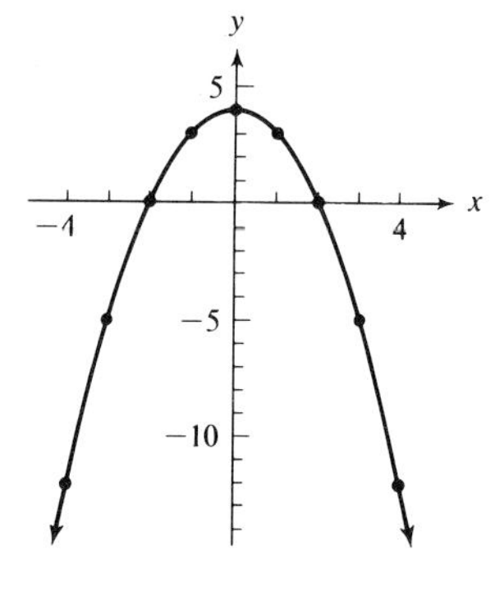

37.
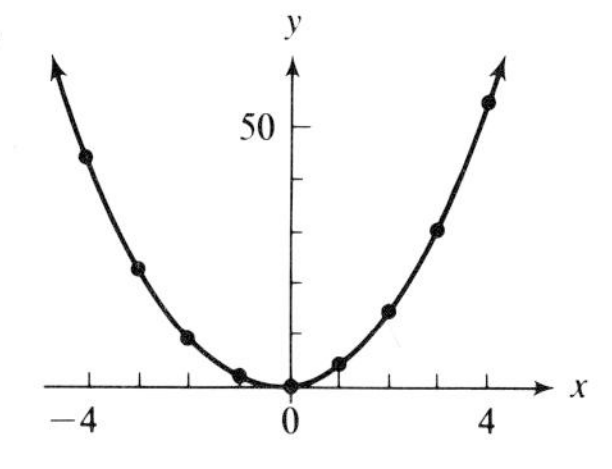

39.
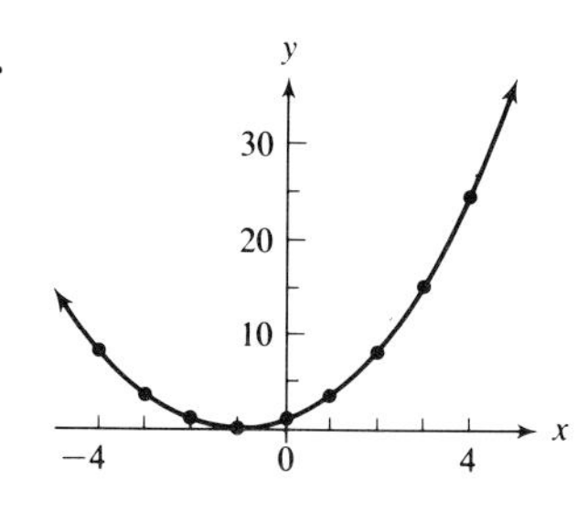

41.
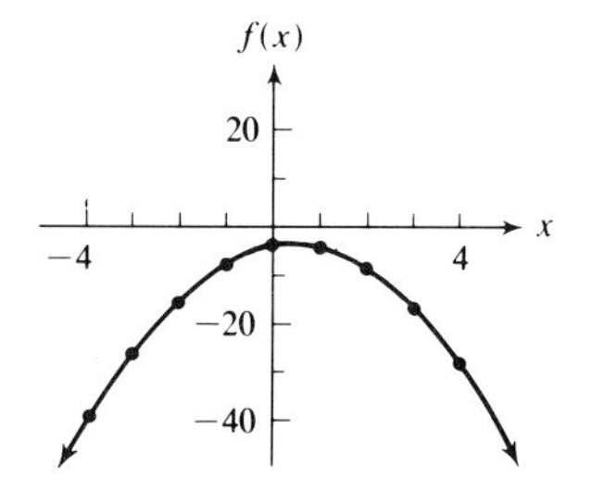

43.

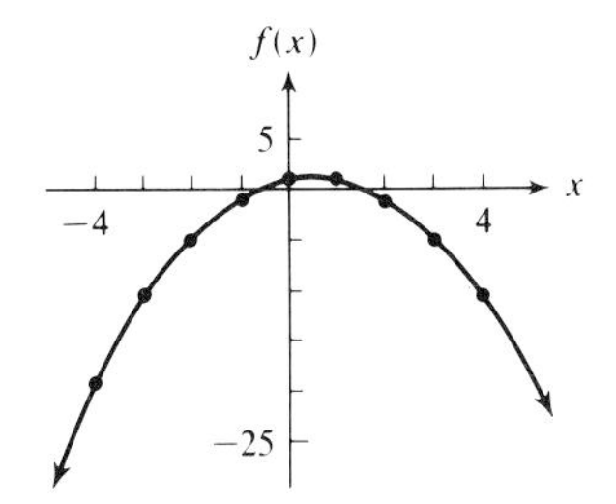

45. 6 and 6

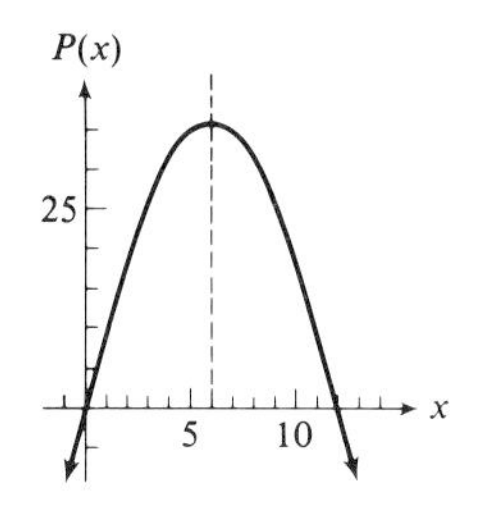

47. 2 seconds

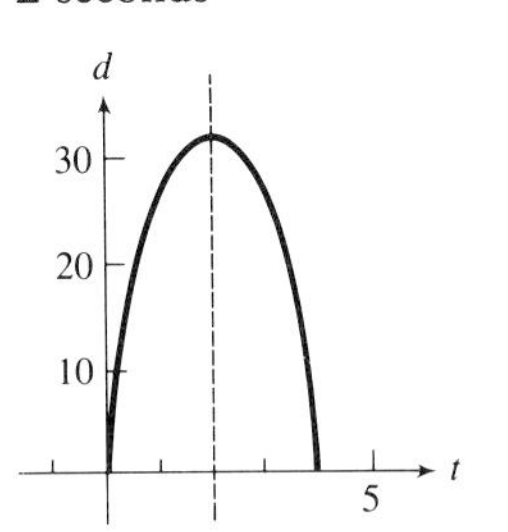

49.

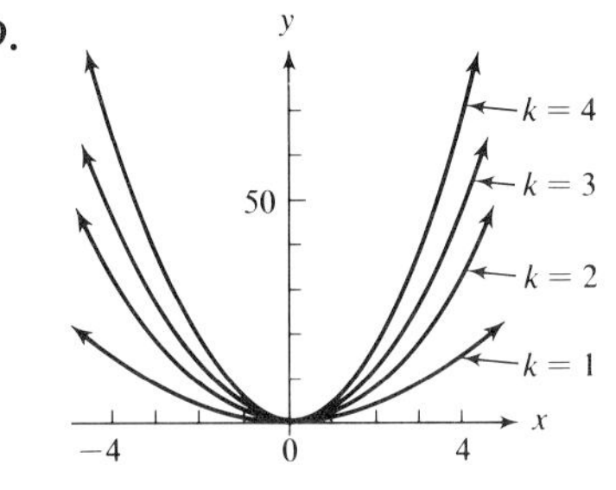

51.

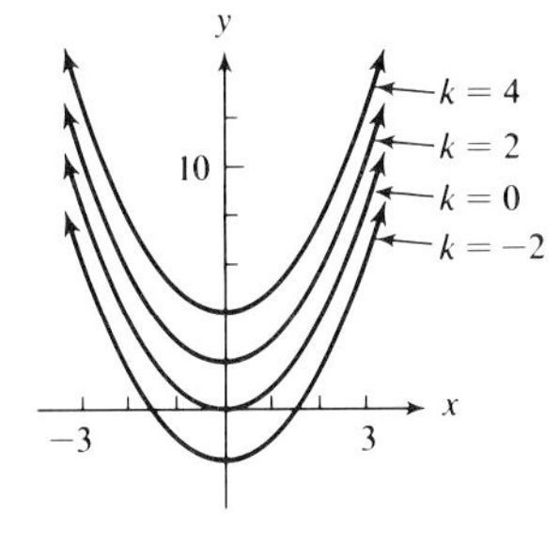

53. $a = 1, b = 2, c = 3$

Exercise 10.3 [page 368]

1.

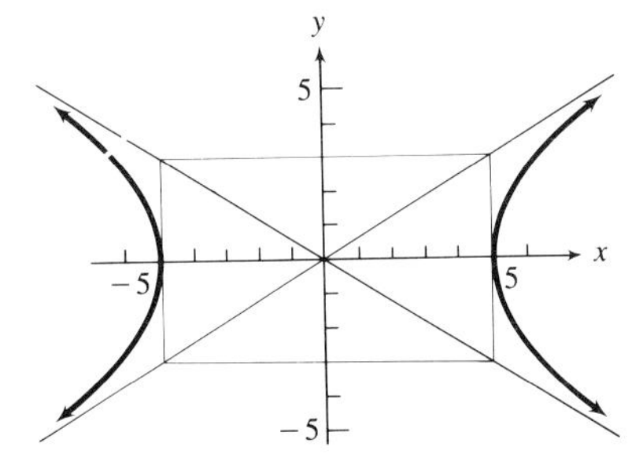

3.

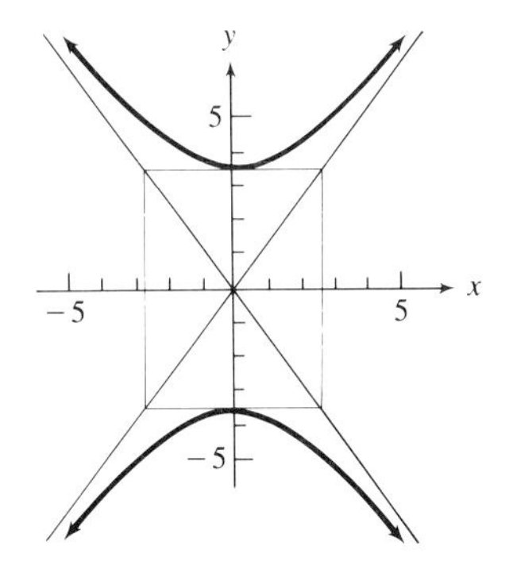

5.

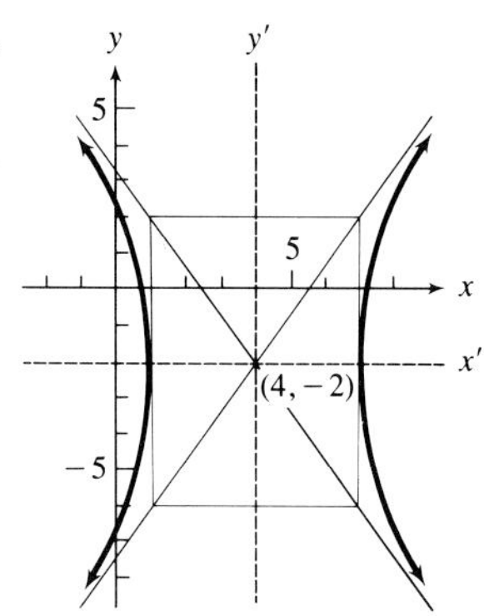

7.

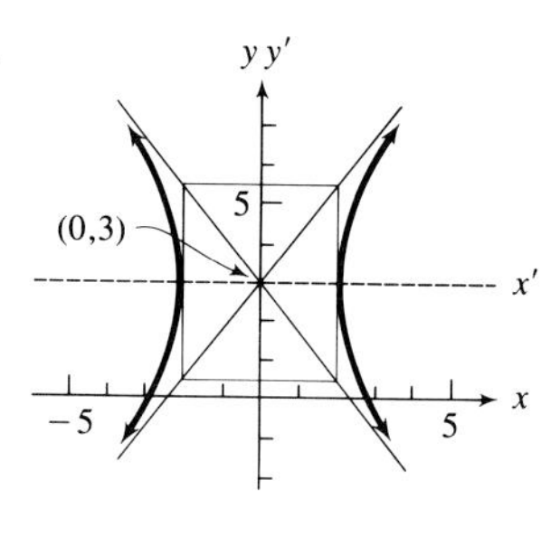

9.

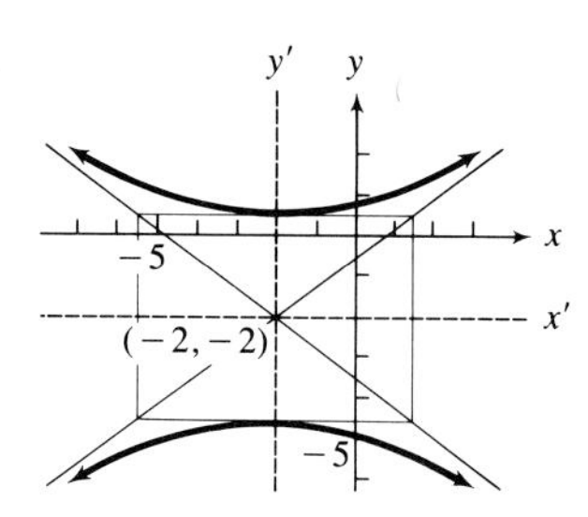

11.

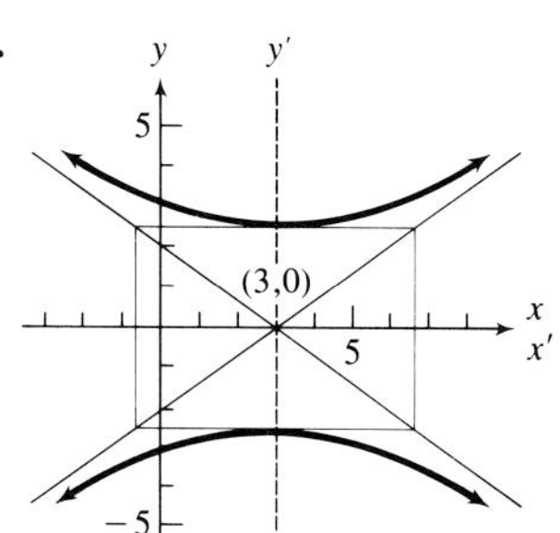

13.

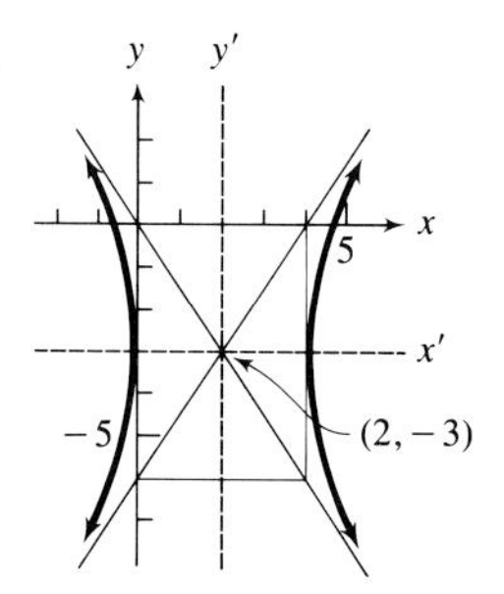

15.

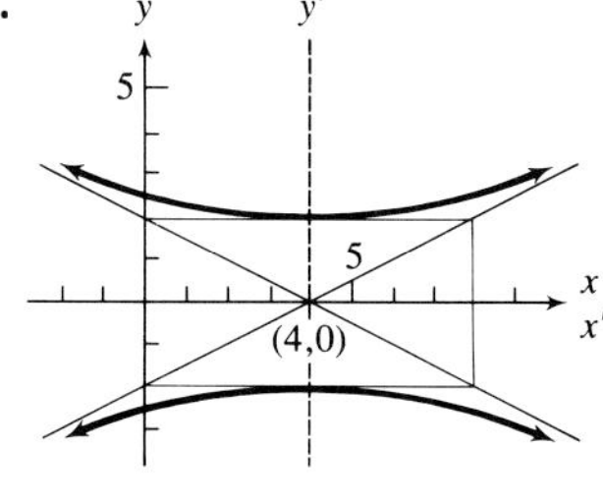

17.

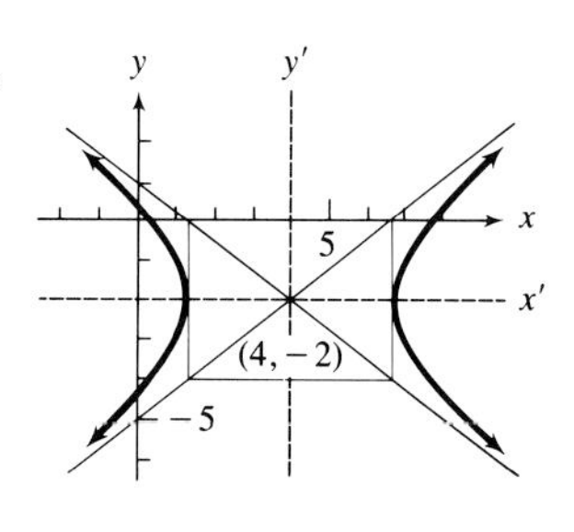

19.

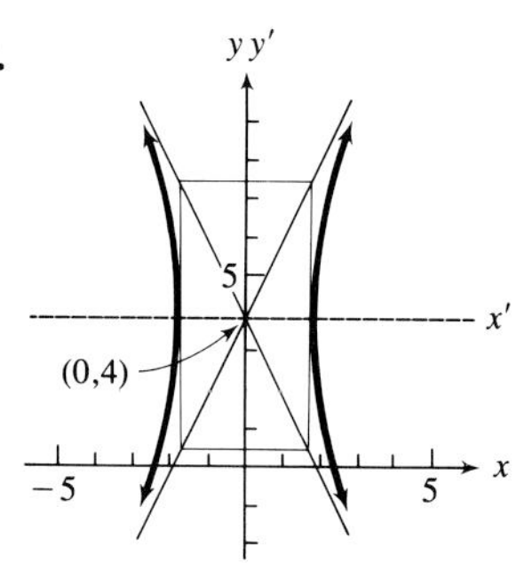

21. circle **23.** hyperbola **25.** ellipse **27.** parabola **29.** hyperbola **31.** parabola
33. parabola **35.** circle **37.** ellipse **39.** circle

41.

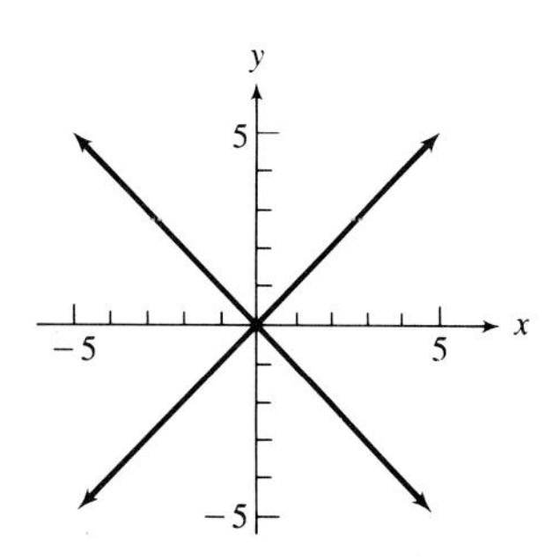

43.

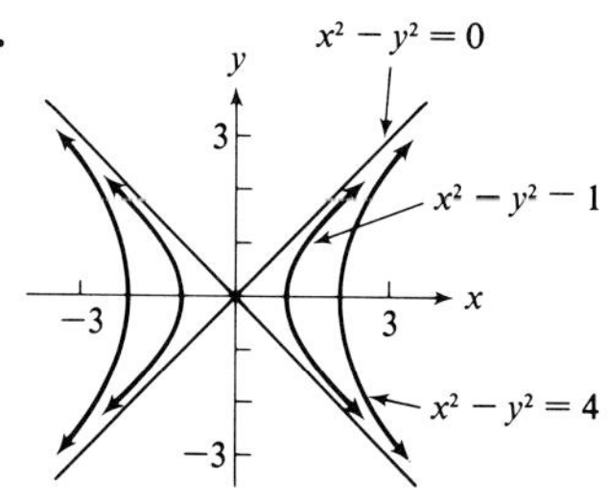

45. $a = -6, b = 8$, and $c = 9$

Exercise 10.4 [page 376]

1. $\{(-1, -4), (5, 20)\}$

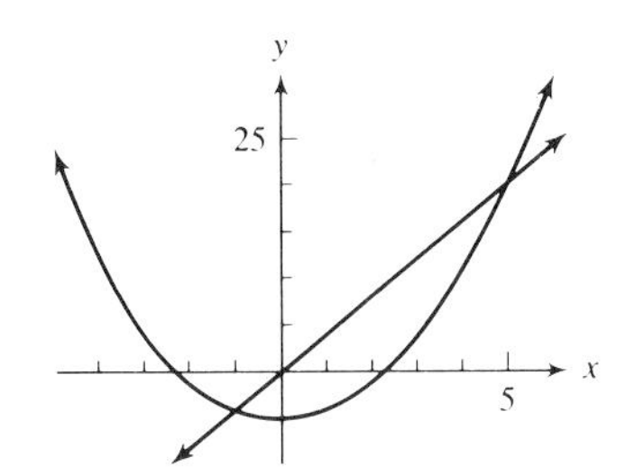

3. $\{(2, 3), (3, 2)\}$

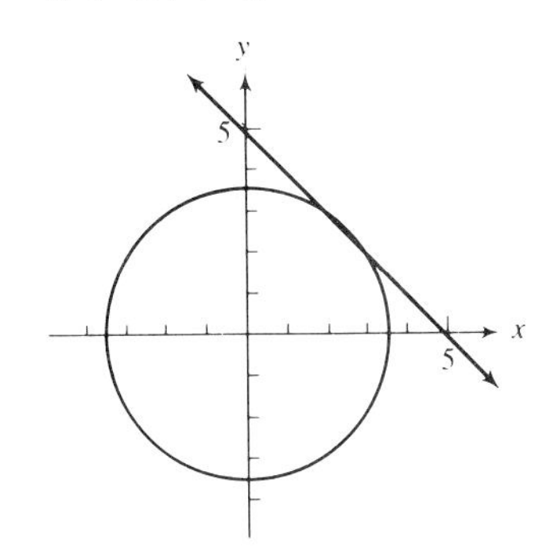

5. $\{(-3, 4), (4, -3)\}$

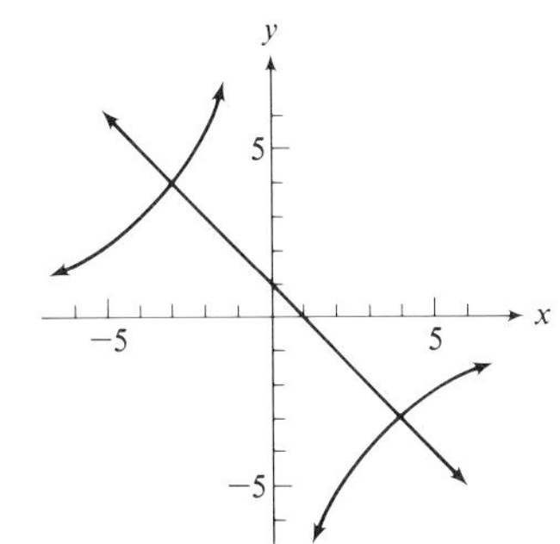

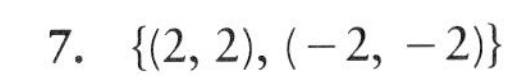

7. $\{(2, 2), (-2, -2)\}$

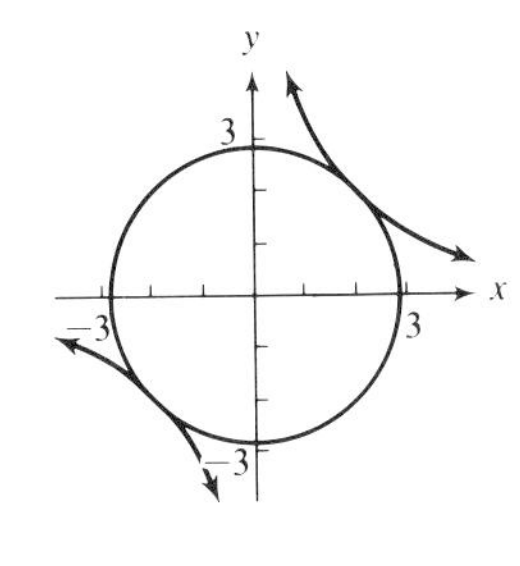

9. $\{(i\sqrt{7}, 4), (-i\sqrt{7}, 4)\}$

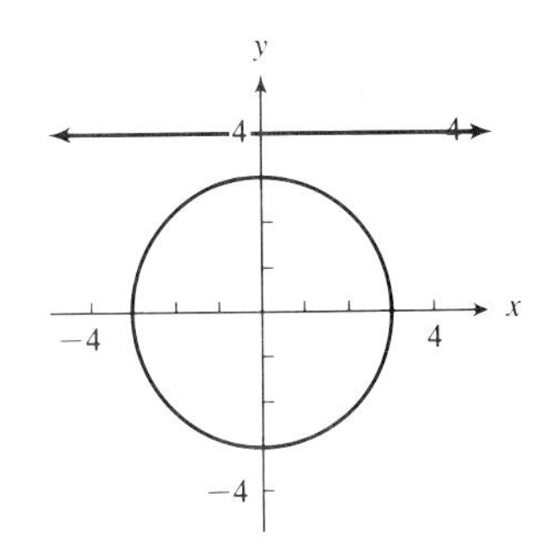

11. $\left\{\left(\frac{-1+\sqrt{13}}{2}, \frac{1+\sqrt{13}}{2}\right), \left(\frac{-1-\sqrt{13}}{2}, \frac{1-\sqrt{13}}{2}\right)\right\}$

13. $\{(3, 1), (2, 0)\}$ 15. $\{(-1, -3)\}$ 17. 2; 3 19. length, 12 inches; width, 1 inch

21. a. Amount of investment: A

Rate of investment: r

b. $rA = 32$

$(r - 0.005)(A + 200) = 35$

c. \$800 invested at 4%

Exercise 10.5 [page 380]

1. $\{(1, 3), (-1, 3), (1, -3), (-1, -3)\}$
3. $\{(1, 2), (-1, 2), (1, -2), (-1, -2)\}$ 5. $\{(3, \sqrt{2}), (-3, \sqrt{2}), (3, -\sqrt{2}), (-3, -\sqrt{2})\}$
7. $\{(2, 1), (-2, 1), (2, -1), (-2, -1)\}$ 9. $\{(\sqrt{3}, 4), (-\sqrt{3}, 4), (\sqrt{3}, -4), (-\sqrt{3}, -4)\}$
11. $\{(2, -2), (-2, 2), (2i\sqrt{2}, i\sqrt{2}), (-2i\sqrt{2}, -i\sqrt{2})\}$ 13. $\{(1, -1), (-1, 1), (i, i), (-i, -i)\}$
15. $\{(1, -2), (-1, 2), (2, -1), (-2, 1)\}$ 17. $\{(3, 1), (-3, -1), (-2\sqrt{7}, \sqrt{7}), (2\sqrt{7}, -\sqrt{7})\}$
19. $\left\{\left(\frac{2\sqrt{3}}{3}, \frac{\sqrt{3}}{3}\right), \left(\frac{-2\sqrt{3}}{3}, -\frac{\sqrt{3}}{3}\right), (-6i, 4i), (6i, -4i)\right\}$ 21. a. 1 b. 2 c. 4

Exercise 10.6 [page 384]

1.

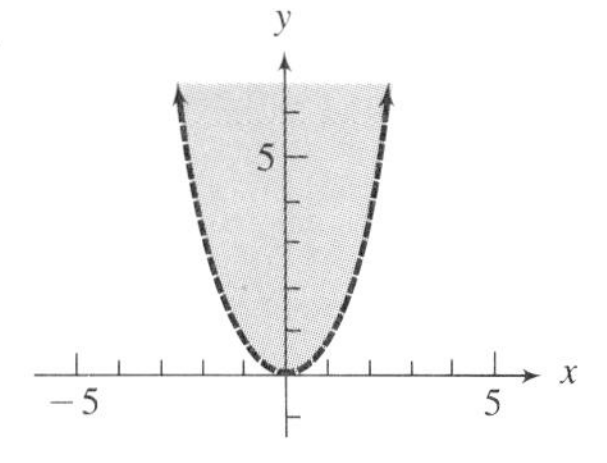

3.

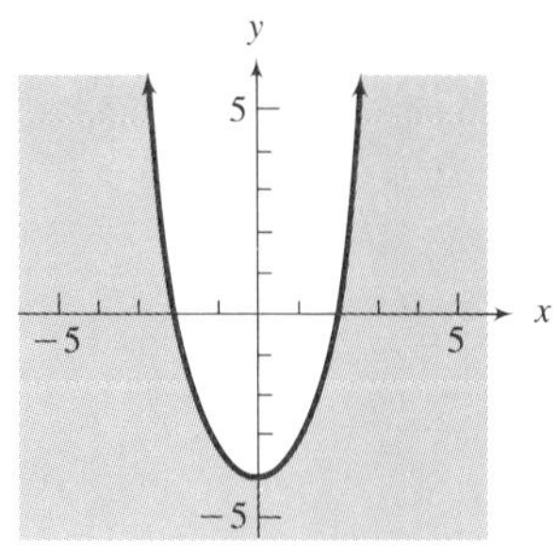

5.

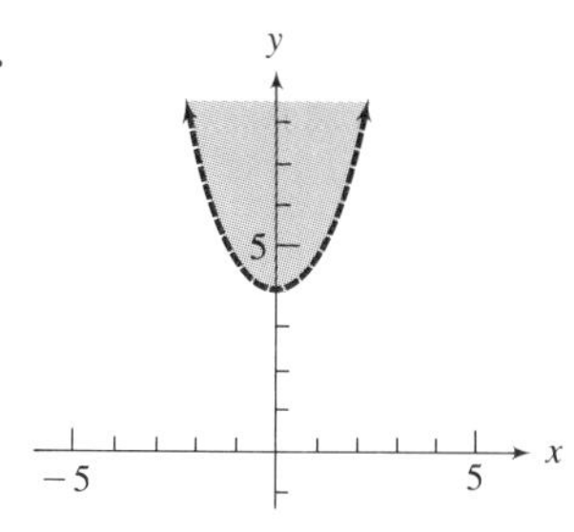

7.

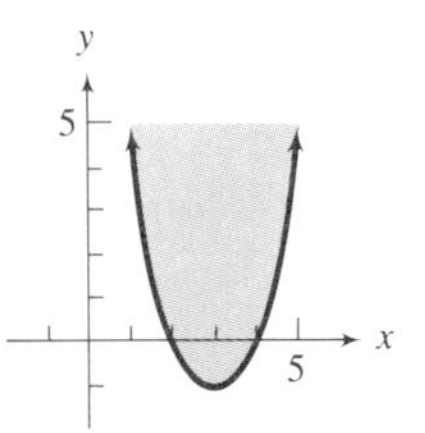

9.

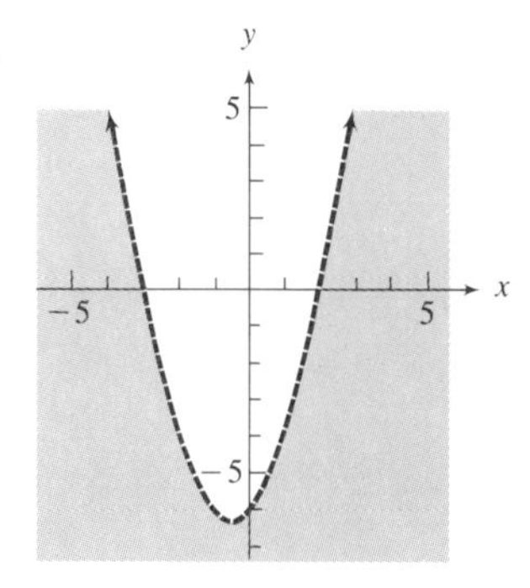

11.

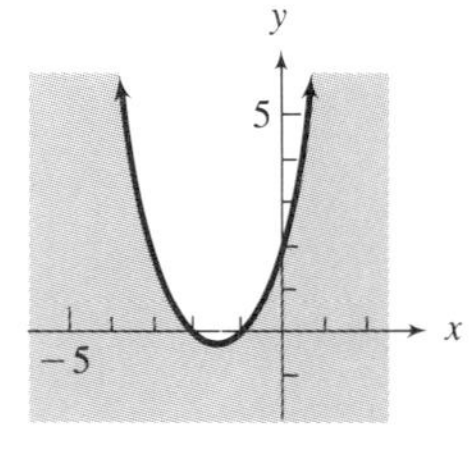

13.

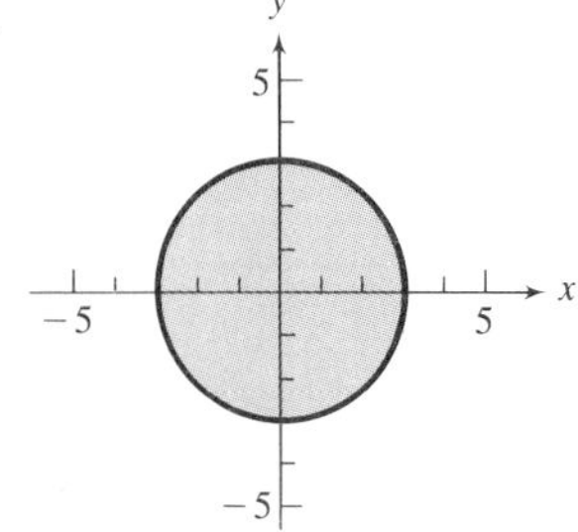

15.

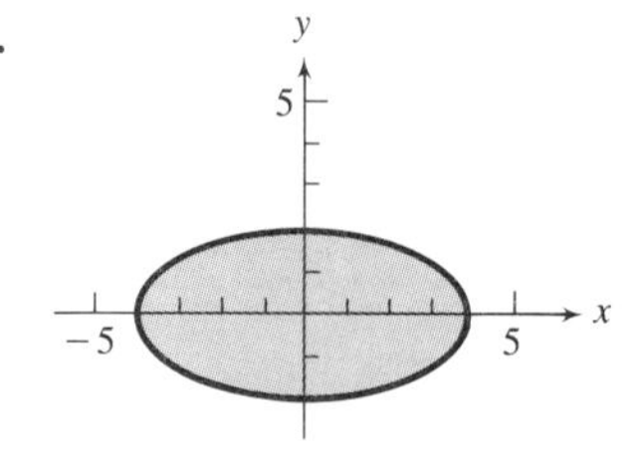

17.

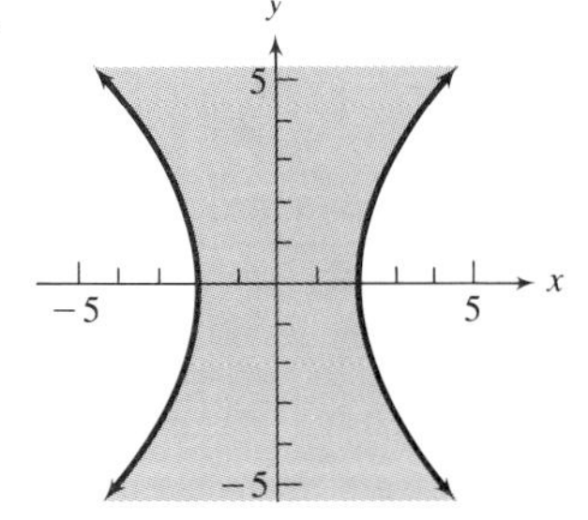

19.

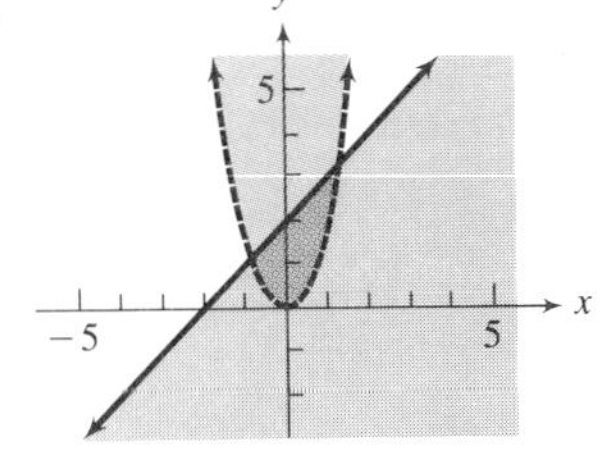

21.

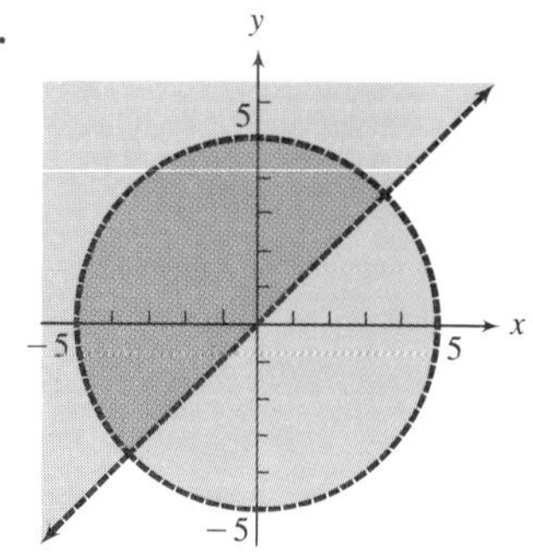

23.

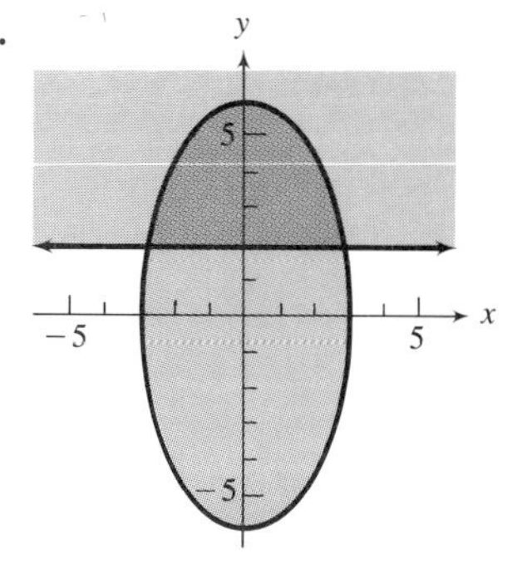

25.

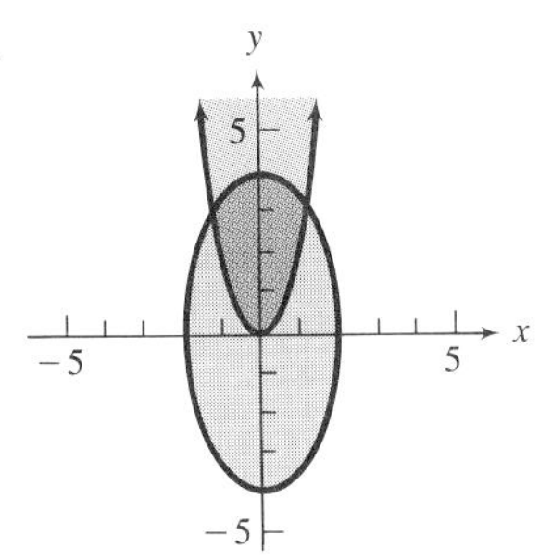

27.

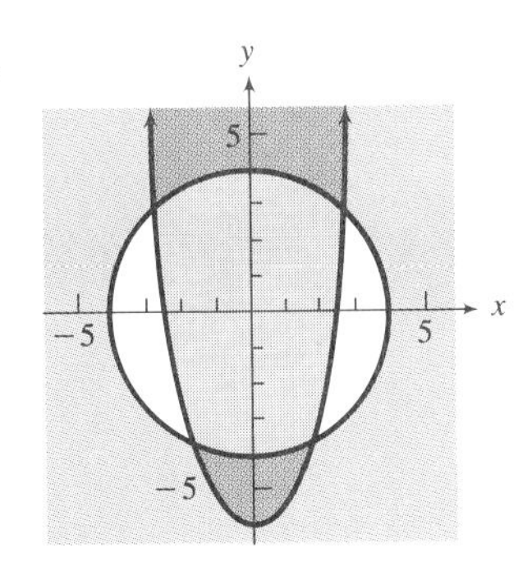

29.

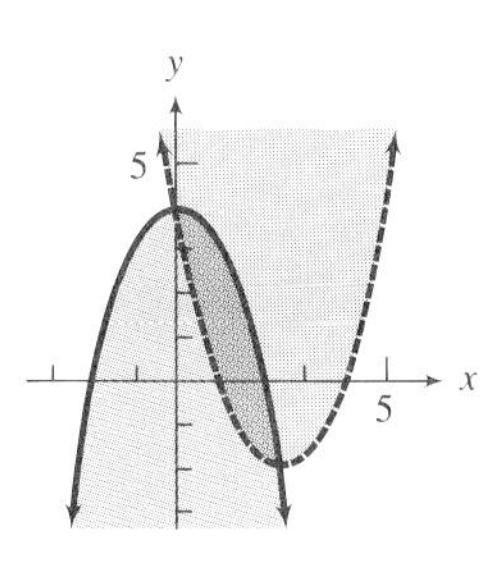

Review Exercises [page 388]

1.

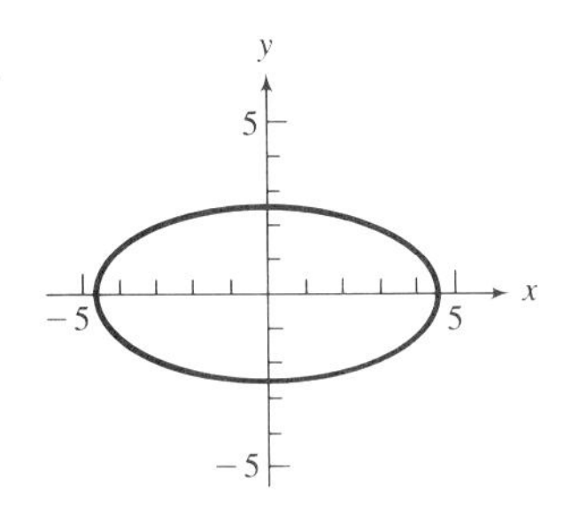

2.

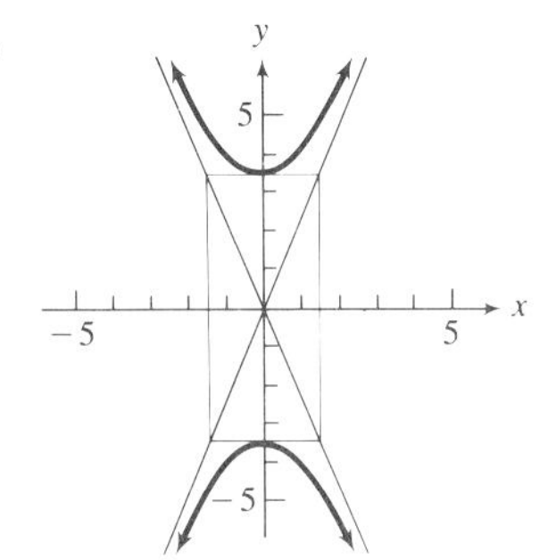

3.

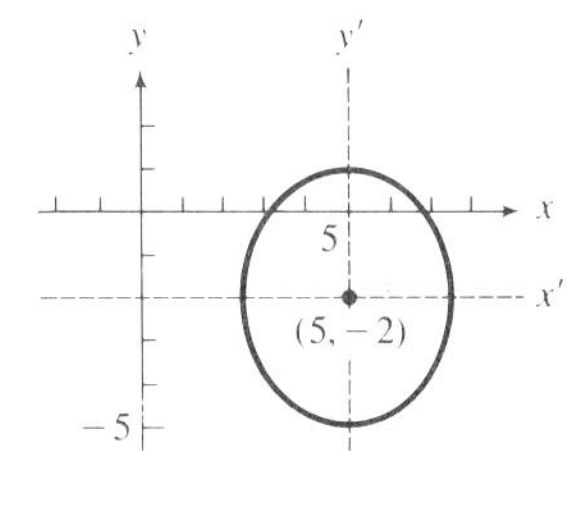

4.

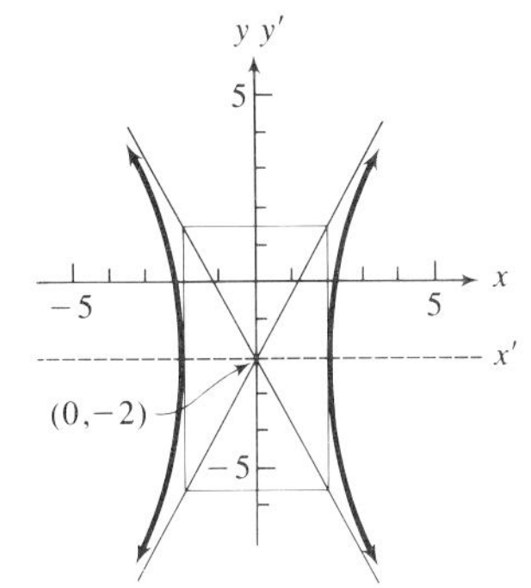

5.

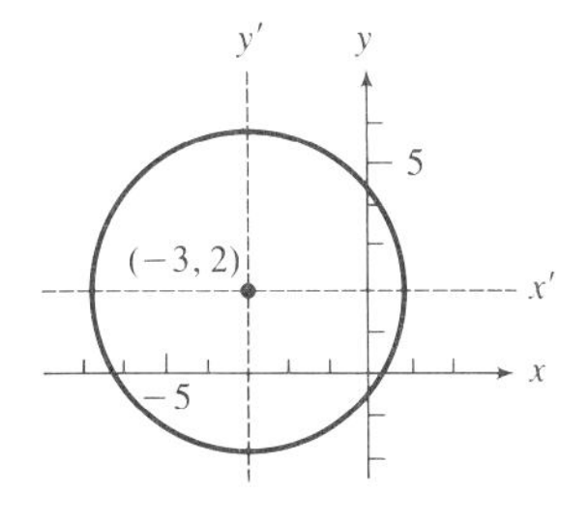

6.

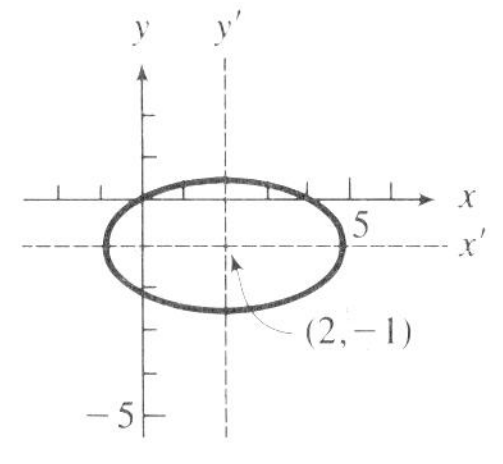

7.

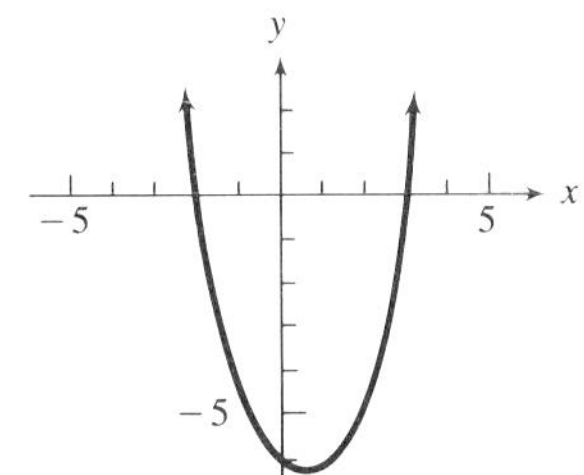

8.

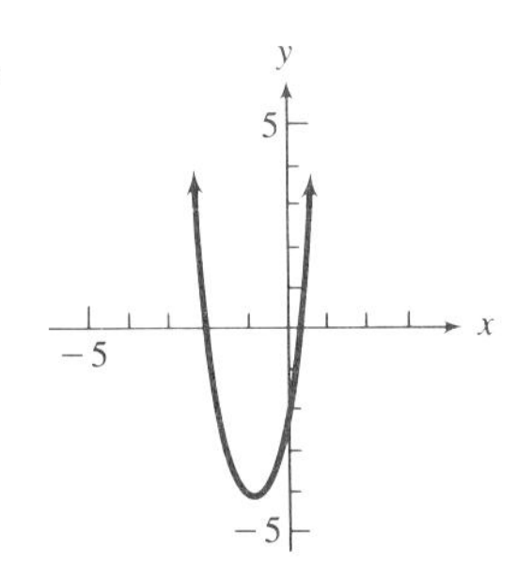

9.

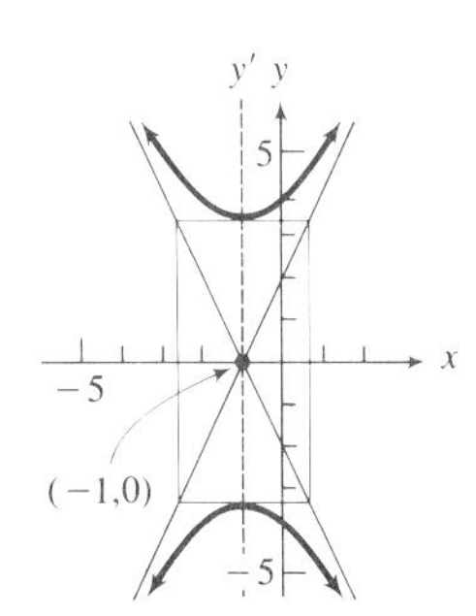

10.

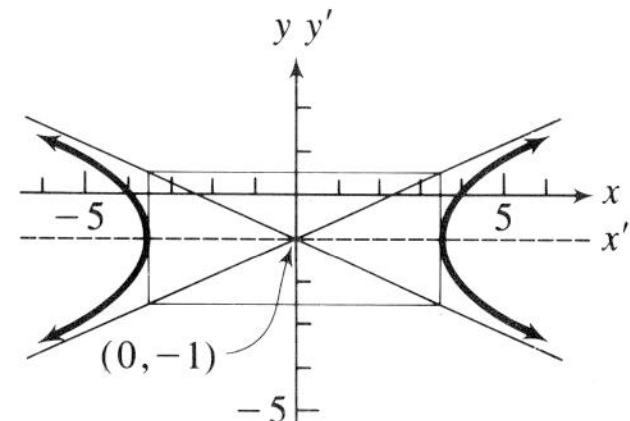

11. $\left\{\left(3, \frac{\sqrt{3}}{3}\right), \left(3, -\frac{\sqrt{3}}{3}\right)\right\}$ 12. $\{(4i, -2), (-4i, -2)\}$ 13. $\{(4, -13), (1, 2)\}$

14. $\{(6/7, -37/7), (2, -3)\}$ 15. $\{(1, \sqrt{5}), (1, -\sqrt{5}), (-1, \sqrt{5}), (-1, -\sqrt{5})\}$

16. $\{(2, 3), (2, -3), (-2, 3), (-2, -3)\}$ 17. $\left\{(1, -2), (-1, 2), \left(2\sqrt{3}, \frac{-\sqrt{3}}{3}\right), \left(-2\sqrt{3}, \frac{1}{\sqrt{3}}\right)\right\}$

18. $\{(2, 1), (-2, -1), (i, -2i), (-i, 2i)\}$ 19. width: 7 centimeters; length: 10 centimeters

20. width: 2 feet; length: 7 feet

21.

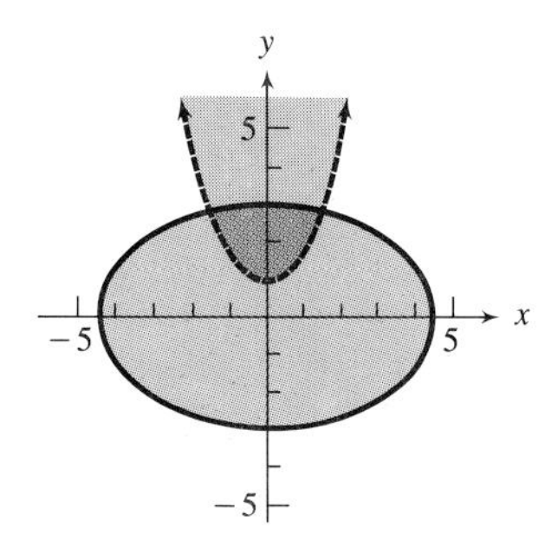

22.

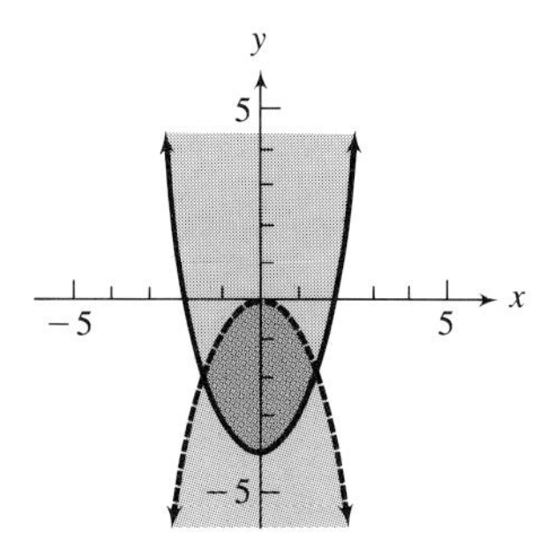

23. $y = x^2 - x - 6$ 24. $x^2 + y^2 - 4x - 2y - 5 = 0$ 25. $b = \frac{-a^2 - 16}{4}$

Cumulative Review Exercises, Chapters 1–10 [page 391]

1. -32 2. $(3x)^{1/4} + xy^{2/3}$ 3. $3\sqrt[5]{\frac{x}{(2x - y)^2}}$ 4. a^3 5. -16 6. $6x^2 + 2x$

7. $\frac{-b^{13}c^8t^7}{a^5s^4}$ 8. $\frac{2y^2z^5}{3}\sqrt[3]{15xy^2z}$ 9. $(a - 3)(a + 2b)$ 10. $a^2b(a^2 + b^2)(a + b)(a - b)$

11. $(3a^2 + b)(9a^4 - 3a^2b + b^2)$ 12. $(x - 3)(x - 2)(2x - 1)$ 13. $\frac{-x - 6}{(x - 3)(x - 1)(x + 2)}$

14. $\frac{3x}{2(x + 2)(x - 4)}$ 15. $\frac{a^3}{-2(a - 2)(a - 1)}$ 16. $3t^2 + 6t + 5 + \frac{13}{t - 2}$ 17. $\frac{x^2y^2}{(x^2 - y^2)^2}$

18. $x^{-1/2} - 2x^{1/4} + x$ 19. $1 - 2y$ 20. $\frac{x + \sqrt{2x} + \sqrt{3x} + \sqrt{6}}{x - 3}$ 21. $5 + 12i$

22. $\{3, 7\}$ 23. $\left\{\frac{16}{81}, 256\right\}$ 24. $\left\{\frac{3 \pm i\sqrt{15}}{4}\right\}$ 25. $l = \frac{3V - Qh}{h(a + b)}$ 26. $y = \frac{-2}{2x^2 - x + 5}$

27. $(-2, -1)$ **28.** $(-\infty, 3) \cup (4, \infty)$ **29.** $(-\infty, -\frac{1}{2}] \cup [\frac{7}{2}, \infty)$

30. $16x^2 - 8x - 3 = 0$ **31.** $\{(12, -4)\}$ **32.** $\{(8, -6)\}$

33. $\{(4, \sqrt{3}), (4, -\sqrt{3}), (-4, \sqrt{3}), (-4, -\sqrt{3})\}$ **34.** $\varnothing$; no solutions **35.** $\{(-1, 2, 1)\}$

36. infinitely many solutions **37.** $\{(2, 0, 4)\}$ **38.** $\{(-2, 3, -4)\}$ **39.** $\{(5, 1, -1)\}$

40. $\{(3, 1), (-3, -1)\}$ **41.** $\{(1, 1), (-\frac{1}{8}, -\frac{5}{4})\}$

42.

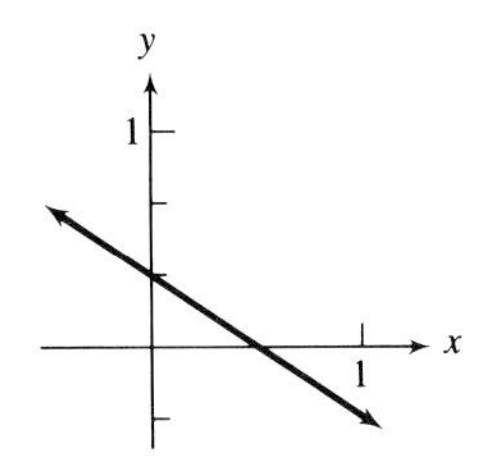

43.

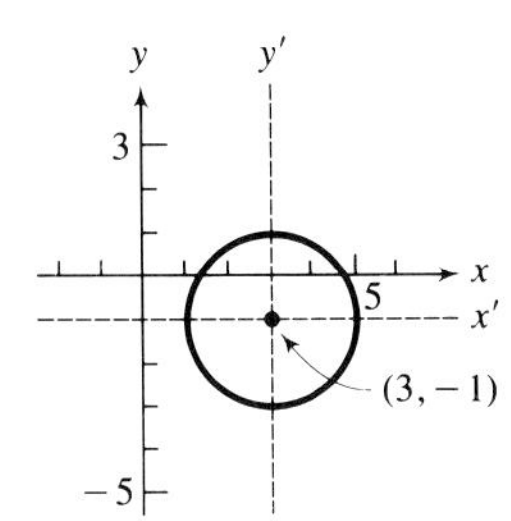

44.

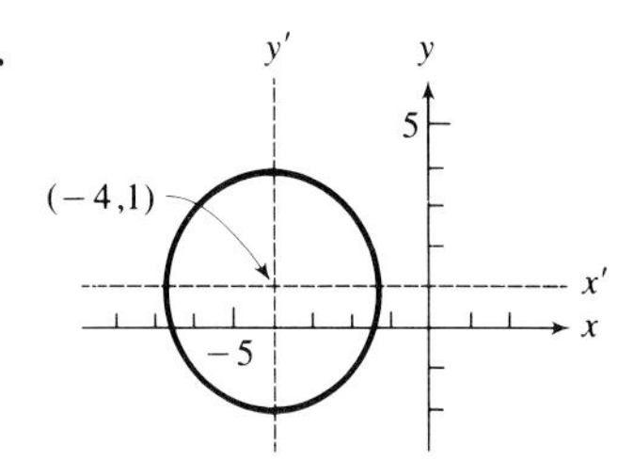

45.

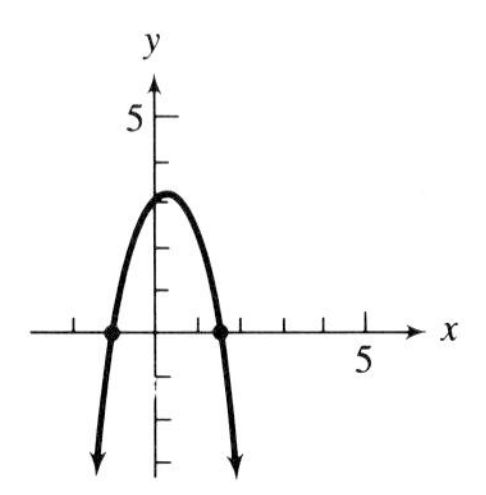

46.

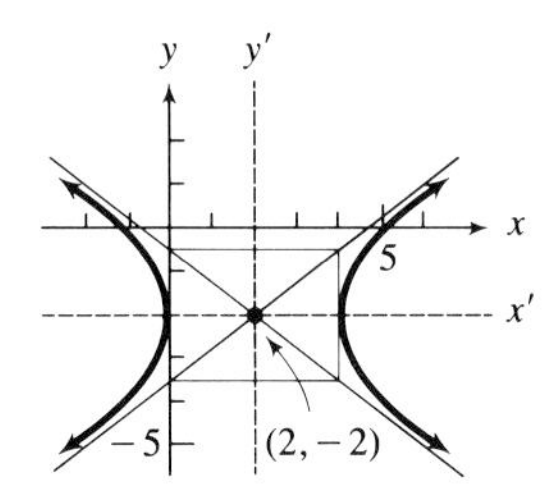

47.

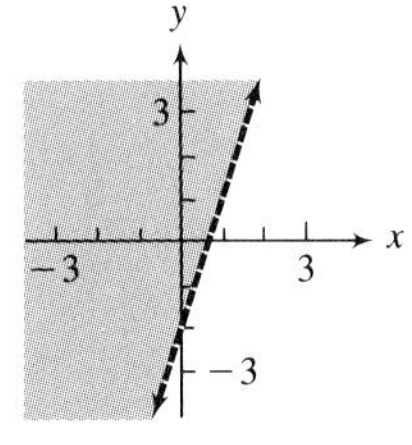

48.

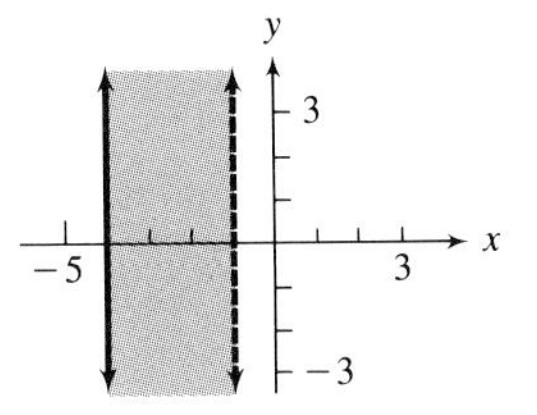

49.

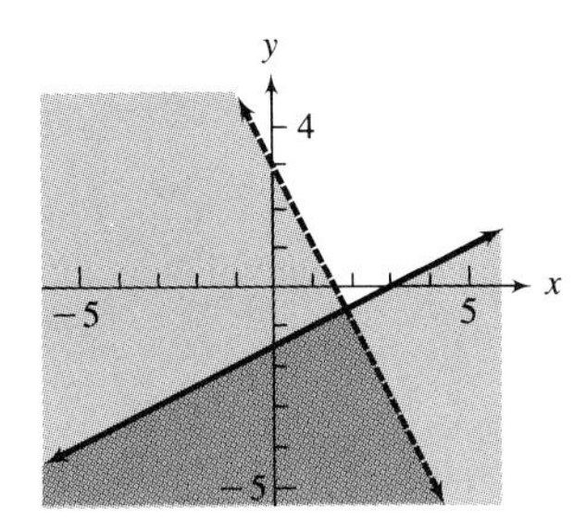

50.

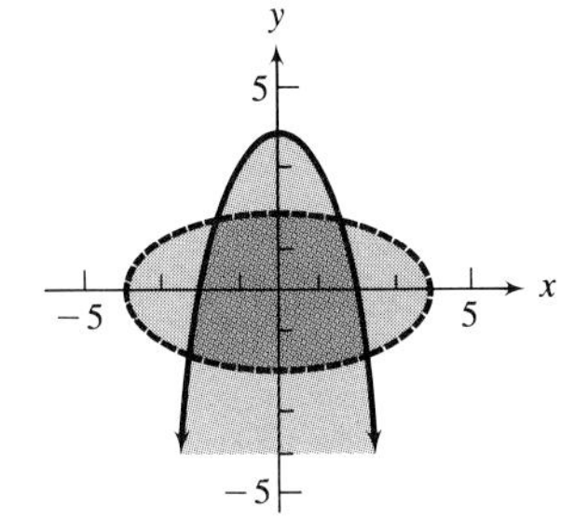

51. $2x + 3y = 11$ **52.** $9x - 2y = 3$

53. The slope of the line segment with endpoints $(-3, 2)$ and $(4, -1)$ is $-\frac{3}{7}$.
The slope of the line segment with endpoints $(-2, -5)$ and $(4, 9)$ is $\frac{7}{3}$.
Therefore, the line segments are perpendicular.

54. $1 + \sqrt{10} + \sqrt{17}$

55. **a.** Width of rectangle: x
Length of rectangle: $x + 5$

56. **a.** Width of margin at sides: x
Width of margin at top: $2x$

b. $x(x + 5) + 60 = (x + 5)^2$

c. 7 feet by 12 feet

57. a. Length of rectangle: x
Width of rectangle: y

b. $2x + 2y = 58$
$xy = 208$

c. 13 meters by 16 meters

b. $(11 - 4x)(8 - 2x) = 52$

c. ¾ inch at sides; 3⁄2 inch at top and bottom

58. 400 square inches

$A(x) = 40x - x^2$

A(x)
500
$A(x) = 40x - x^2$
50
x

59. a. Number of student tickets: x
Number of general tickets: y

b. $x + y = 350$
$5x + 16y = 4940$

c. 60 student tickets and 290 general admission.

60. a. First angle: x
Second angle: y
Third angle: z

b. $x = 2y$
$z = x + 10$
$x + y + z = 180$

c. 68°, 34°, and 78°

61. $-14x + 7y - 2$ 62. $x^{3n} + 2x^n - x^{3n+1} - 2x^{n+1}$ 63. $a^{n-3}b^{2m-1}$

64. $(3x^n + 4)(x^n - 5)$ 65. $k = -2$

66. $\begin{cases} \dfrac{2}{2x - y} & \text{if } 2x > y \\ \dfrac{2}{y - 2x} & \text{if } 2x < y \end{cases}$ 67. 0 68. $y = \pm\frac{2}{5}\sqrt{25 - x^2}$

69. $c = \dfrac{abd}{ab + bd - ad}$ 70. $(-\infty, -11/2) \cup (-1/2, 2)$ 71. $(-2, 8)$

72. The line segments joining $(-1, -1)$ and $(9, 7)$ and $(5, 12)$ and $(-5, 4)$ have slope $4/5$. The line segments joining $(9, 7)$ and $(5, 12)$, and $(-5, 4)$ and $(-1, -1)$ have slope $-5/4$. Therefore, the adjacent sides are perpendicular, and the figure is a rectangle.

73. $2x - 3y + 20 = 0$

74. $0 = \begin{vmatrix} x & y & 1 \\ x_1 & y_1 & 1 \\ x_2 & y_2 & 1 \end{vmatrix} = x(y_1 - y_2) - y(x_1 - x_2) + (x_1y_2 - x_2y_1)$,

so $y(x_1 - x_2) + y_1x_2 = x(y_1 - y_2) + x_1y_2$

or $y(x_1 - x_2) - y_1x_1 + y_1x_2 = x(y_1 - y_2) - x_1y_1 + x_1y_2$

or $(y - y_1)(x_1 - x_2) = (x - x_1)(y_1 - y_2)$

or $y - y_1 = \dfrac{y_1 - y_2}{x_1 - x_2}(x - x_1)$, which is an equation for the line.

75. $\{(-1/3, 1/2)\}$ 76. $a = -2, b = 5, c = -3$ 77. $(x - 1)^2 + (y + 2)^2 = 25$

78.

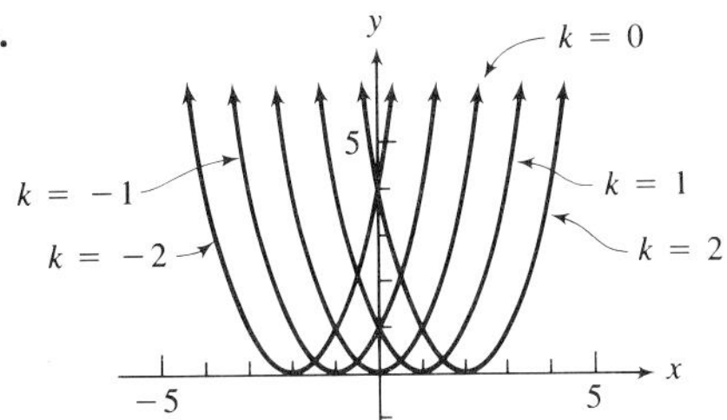

79.

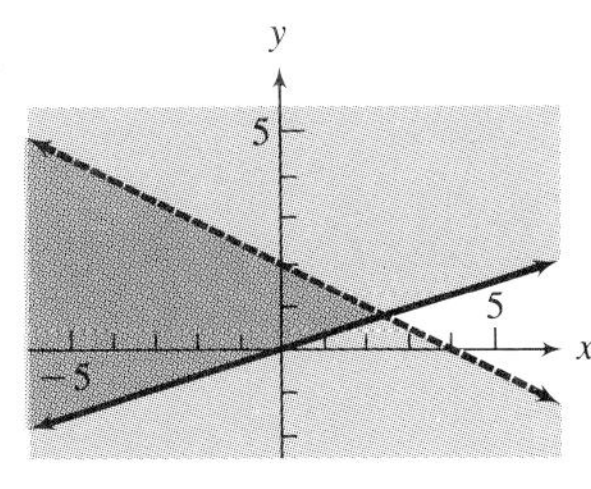

80.

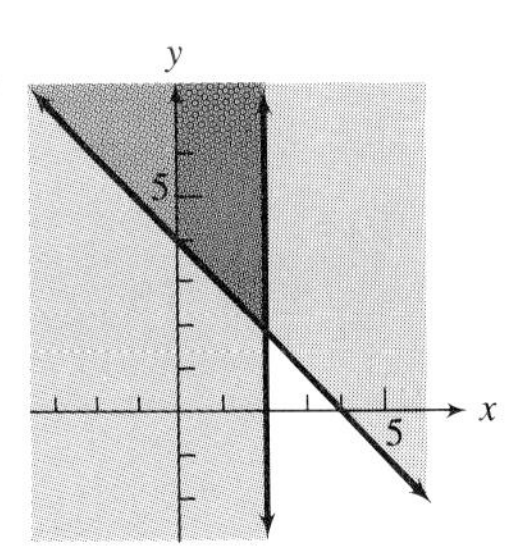

Exercise 11.1 [page 397] **1.** **a.** domain: $\{-2, -1, 0, 1\}$; range: $\{3, 4, 5, 6\}$ **b.** relation is a function **3.** **a.** domain: $\{5, 6, 7, 8\}$; range $\{1\}$ **b.** relation is a function

5. **a.** domain: $\{2, 3\}$; range: $\{3, 4\}$ **b.** relation is not a function

7. **a.** domain: $\{0, 2, 3\}$; range: $\{0, 1, 2, 3\}$ **b.** relation is not a function

9. **a.** $y = 6 - 2x$ **b.** function **c.** $x \in R$

11. **a.** $y = \dfrac{x-8}{2}$ **b.** function **c.** $x \in R$

13. **a.** $y = \dfrac{4}{x-2}$ **b.** function **c.** $x \in r, x \neq 2$

15. **a.** $y = \dfrac{6}{x-4}$ **b.** function **c.** $x \in R, x \neq 4$

17. **a.** $y = \pm\sqrt{8 - 2x^2}$ **b.** not a function **c.** $-2 \leqslant x \leqslant 2$

19. **a.** $y = \pm 2\sqrt{x^2 - 4}$ **b.** not a function **c.** $x \leqslant -2$ or $x \geqslant 2$

21. **a.** $y > x^2 + 3$ **b.** not a function **c.** $x \in R$ **23.** 9 **25.** 3 **27.** -6

29. $5a - 3$; $5b - 3$; $5a + 5b - 3$ **31.** $a^2 + 3a$; $b^2 + 3b$; $a^2 + 2ab + b^2 + 3a + 3b$

33. **a.** $3x + 3h - 4$ **b.** $3h$ **c.** 3

35. **a.** $x^2 + 2hx + h^2 - 3x - 3h + 5$ **b.** $2hx + h^2 - 3h$ **c.** $2x + h - 3$

37. **a.** $x^3 + 3x^2h + 3xh^2 + h^3 + 2x + 2h - 1$ **b.** $3x^2h + 3xh^2 + h^3 + 2h$ **c.** $3x^2 + 3xh + h^2 + 2$ **39.** $f(-2) \approx -3$; $f(3) \approx 2$; $f(5) \approx 4$; zero: 1 **41.** -2

43. 1, 4 **45.** $-2, 3$

Exercise 11.2 [page 404] **1.** **a.** $y = kx$ **b.** $100/3$ **3.** **a.** $y = \dfrac{kx^2}{z}$ **b.** $96/5$

5. **a.** $s = 8t$ **b.** 56 feet **7.** **a.** $L = \dfrac{6000}{l}$ **b.** 2000 pounds

9. **a.** $C = 5000 + 10n$ **b.** 11,400 books **11.** **a.** $C = 7.50(4s)$ **b.** \$9000

13. **a.** $C = 52.50s$ **b.** \$3150 **15.** **a.** $V = \dfrac{1}{27}\pi h^3$ **b.** 64π cu cm

17. **a.** $A = \dfrac{1}{2}d^2$ **b.** 50 sq in. **19.** **a.** $A = 4r^2 - \pi r^2$ **b.** $144 - 36\pi$ sq in.

21. **a.** $A = x^2 - 4x + 4$ **b.** 784 sq in. **23.** $A = \frac{1}{32}l^2$ **25.** $A = \frac{2x^2 - 100x + 2500}{4\pi}$

27. $C = \frac{15w^2 + 1500}{w}$ **29.** $A = 50\sqrt{x^2 - 36x + 468} + 30x$

Exercise 11.3 [page 410] **1.** 1, 3, 9 **3.** $\frac{1}{16}$, 1, 16 **5.** 16, 1, $\frac{1}{16}$ **7.** $\frac{1}{100}$, $\frac{1}{10}$, 1

9. 0.14, 19.68 **11.** 0.83, 0.12

13.

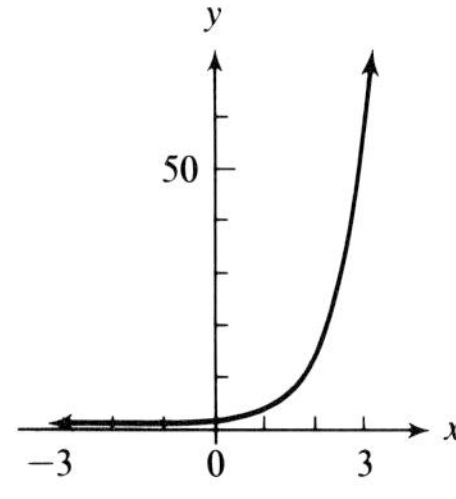

15.

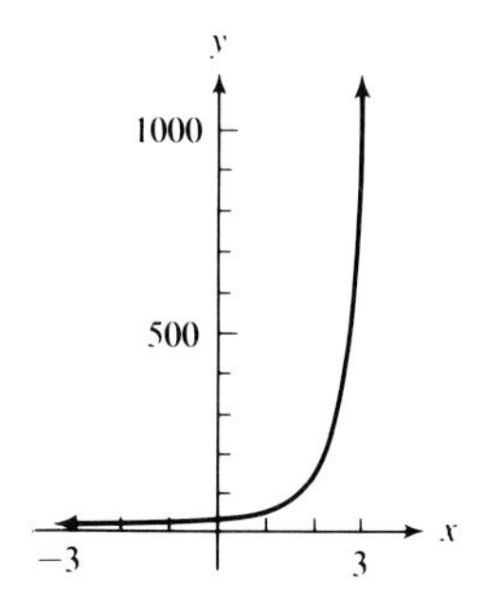

17.

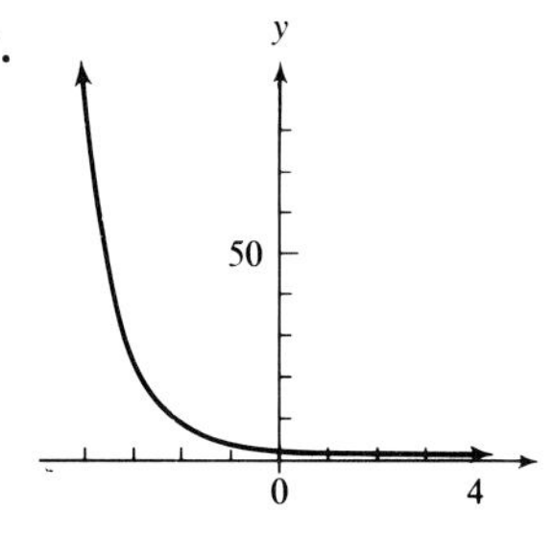

19.

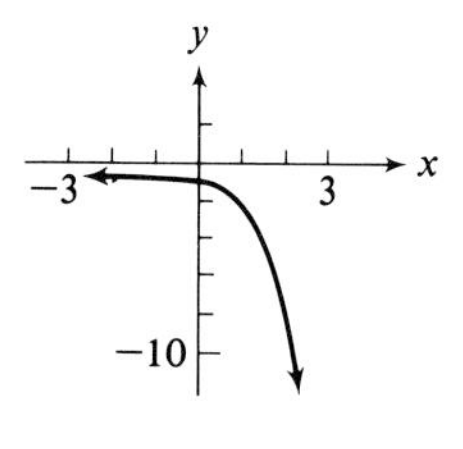

21.

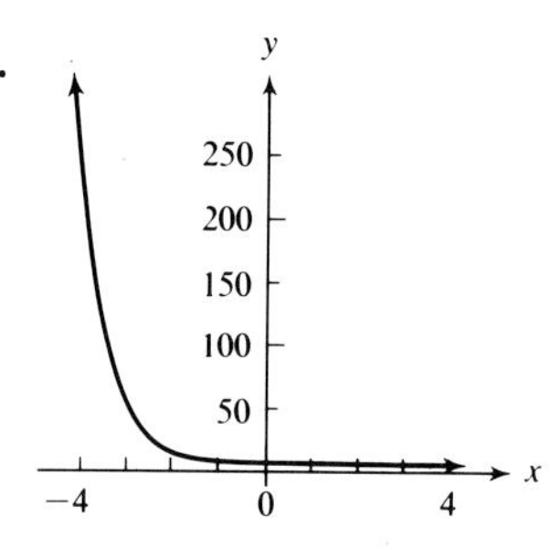

23.

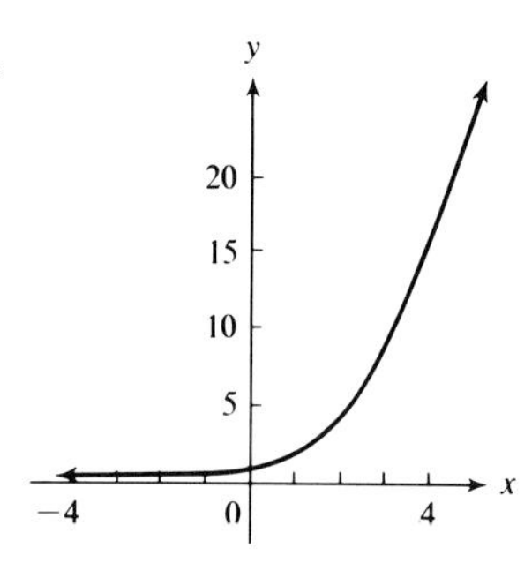

25.

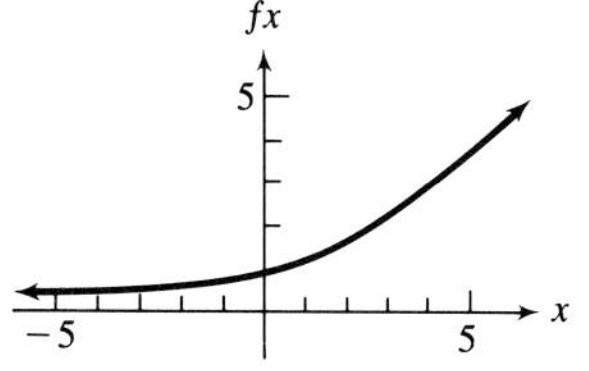

27.

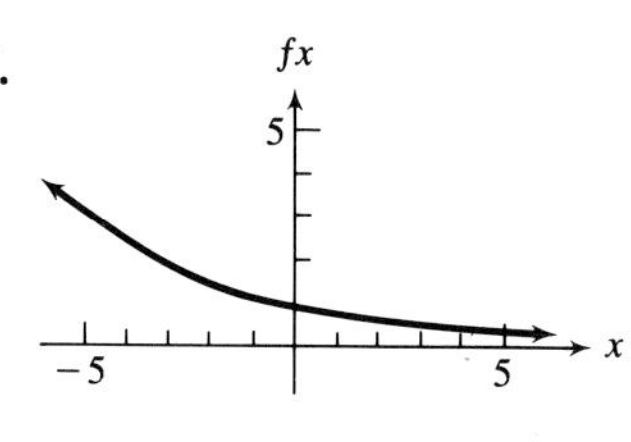

29.

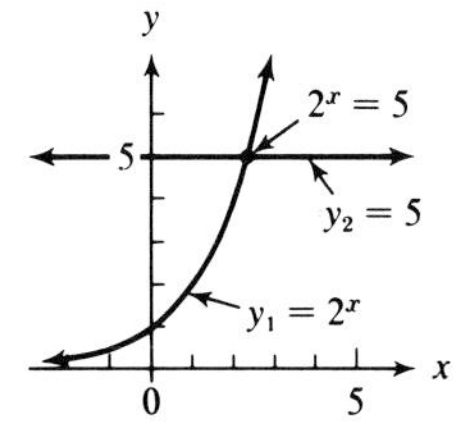

31. An increasing function for all $a > 1$ and a decreasing function for all $0 < a < 1$.

33.

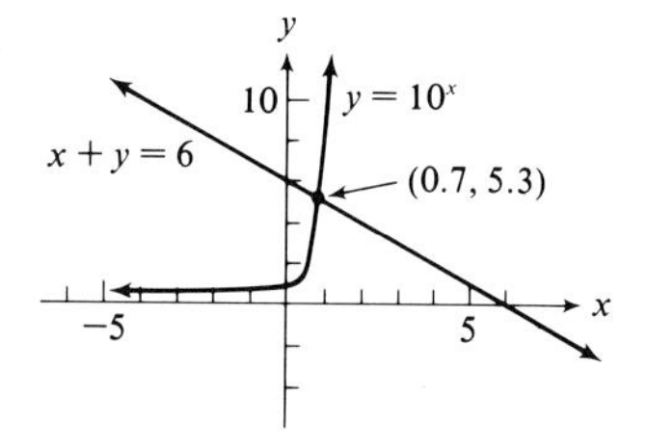

Exercise 11.4 [page 416] 1. $\{(-2, -2), (2, 2)\}$; a function

3. $\{(3, 1), (3, 2), (4, 3)\}$; not a function 5. $\{(1, 1), (2, 2), (3, 3)\}$; a function

7. a. $2y + 4x = 7$

b.

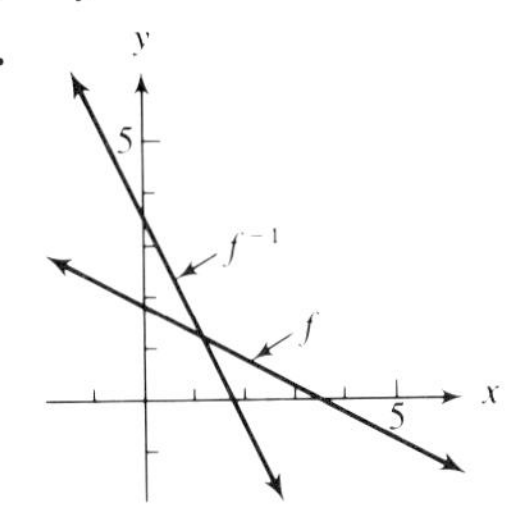

c. a function

9. a. $y - 3x = 6$

b.

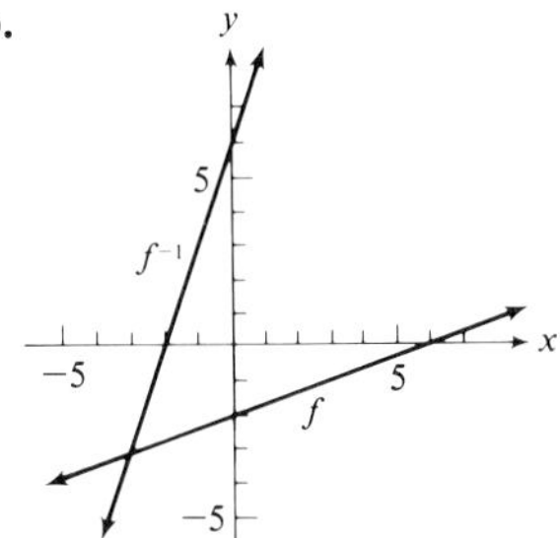

c. a function

11. a. $x = y^2 - 4y$

b.

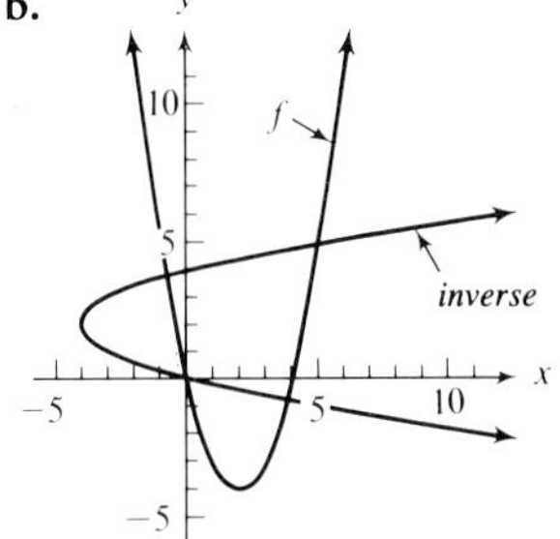

c. not a function

13. a. $x = \sqrt{4 + y^2}$

b.

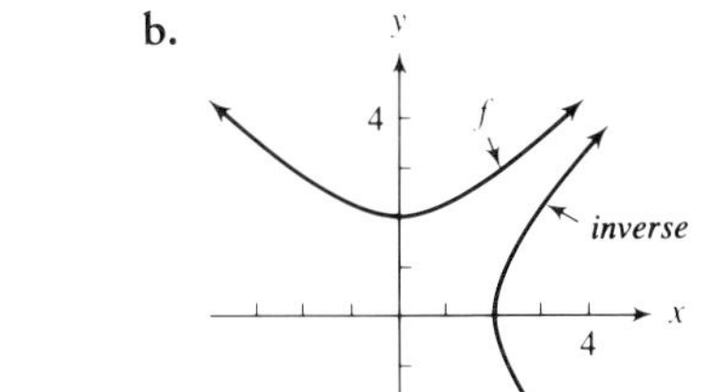

c. not a function

15. a. $x = |y|$

b.

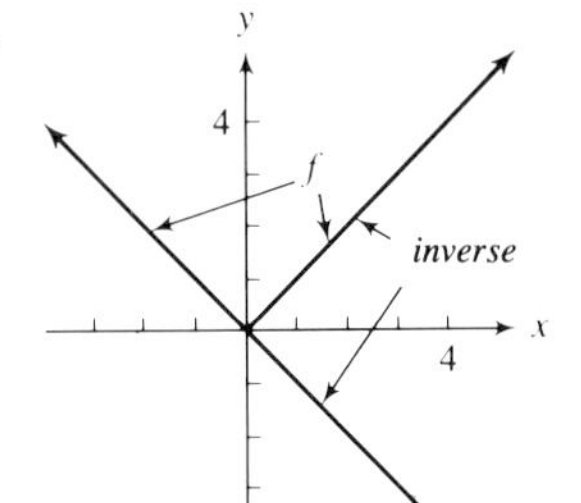

c. not a function

17.

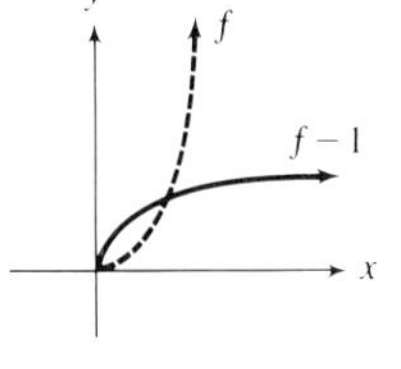

19.

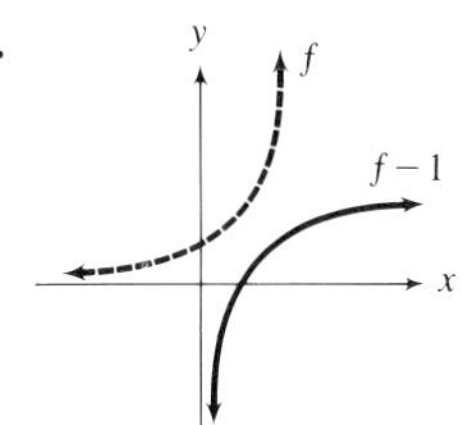

21.

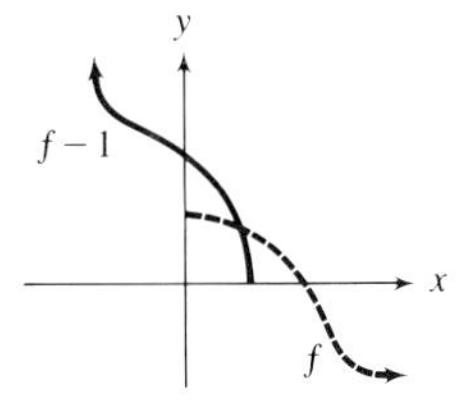

23. $f^{-1}: y = x$; $f[f^{-1}(x)] = f[x] = x$;
$f^{-1}[f(x] = f^{-1}[x] = x$

25. $f^{-1}: y = \dfrac{4 - x}{2}$; $f[f^{-1}(x)] = f\left[\dfrac{4 - x}{2}\right] = -2\left[\dfrac{4 - x}{2}\right] + 4 = -4 + x + 4 = x$;

$f^{-1}[f(x)] = f^{-1}[-2x + 4] = 2 - \dfrac{1}{2}[-2x + 4] = 2 + x - 2 = x$

27. $f^{-1}: y = \dfrac{4x + 12}{3}$; $f[f^{-1}(x)] = f\left[\dfrac{4x + 12}{3}\right] = \dfrac{3}{4}\left[\dfrac{4x + 12}{3}\right] - 3 = x + 3 - 3 = x$;

$f^{-1}[f(x)] = f^{-1}\left[\dfrac{3x - 12}{4}\right] = \dfrac{4}{3}\left[\dfrac{3x - 12}{4}\right] + 4 = x - 4 + 4 = x$

Review Exercises [page 422] **1. a.** domain: {3, 4}; range: {5, 6}; not a function
b. domain: {2, 3, 4, 5}; range: {3, 4, 5}; function

2. a. domain: $x \in R, x \neq 1$; function **b.** domain: $\{-2 \leq x \leq 2\}$; not a function

3. a. 2 **b.** h **4. a.** 10 **b.** $4x + 2h - 3$

5. a. $y = \dfrac{144}{t^2}$ **b.** $9/4$ **6. a.** $y = \dfrac{5s^2}{t}$ **b.** $45/4$ **7. a.** $s = \dfrac{7t^2}{4}$ **b.** 63 cm

8. a. $V = \dfrac{4T}{P}$ **b.** 32 cu units **9. a.** $A = \dfrac{\sqrt{3}}{4}s^2$ **b.** $4\sqrt{3}$ sq cm

10. a. $A = \dfrac{1}{2}x\sqrt{144 - x^2}$ **b.** $16\sqrt{2}$ sq cm **11. a.** $\dfrac{1}{1000}$ **b.** 8

12. a. **b.**

13. a. {(7, 3), (8, 4), (4, 8)} is a function
b. {(6, 2), (8, 3), (8, 5)} is not a function

14. $x = y^2 + 9y$ does not define a function

15.

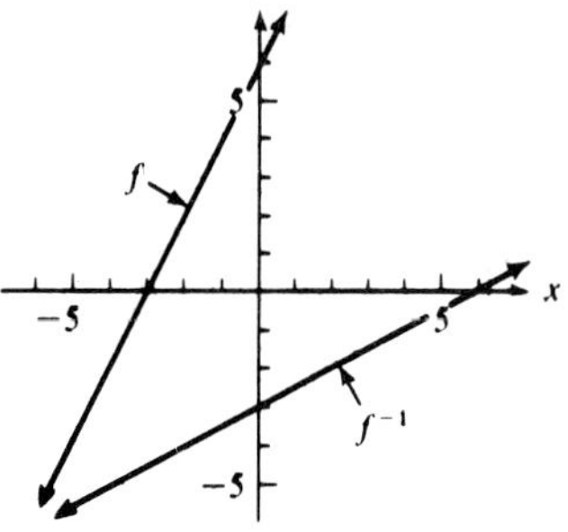

16.

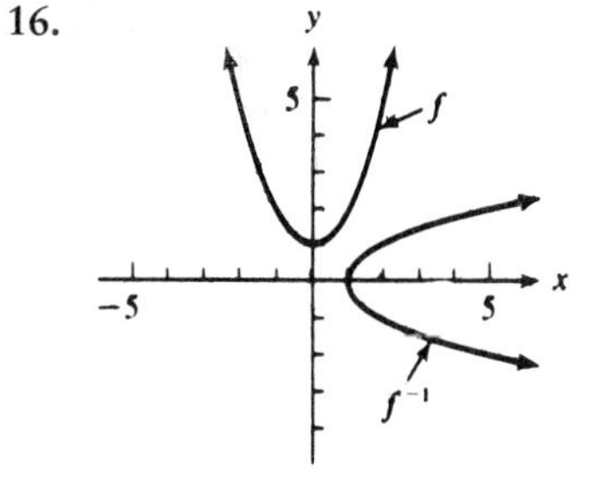

17. **18.**

19. The inverse defined by $x = y + 3$ can be written as $f^{-1}(x) = x - 3$. Hence,

20. The inverse defined by $x - 2y = 4$ can be written as $f^{-1}(x) = \dfrac{x - 4}{2}$. Hence,

$f[f^{-1}(x)] = (x - 3) + 3 = x;$
$f^{-1}[f(x)] = (x + 3) - 3 = x$

$f[f^{-1}(x)] = 4 + 2\left(\frac{x - 4}{2}\right) = x;$

$f^{-1}[f(x)] = \frac{(4 + 2x) - 4}{2} = x.$

Exercise 12.1 [page 425] 1. $\log_4 16 = 2$ 3. $\log_3 27 = 3$ 5. $\log_{1/2} \frac{1}{4} = 2$

7. $\log_8 \frac{1}{2} = -\frac{1}{3}$ 9. $\log_{10} 100 = 2$ 11. $\log_{10} 0.1 = -1$ 13. $2^6 = 64$ 15. $3^2 = 9$

17. $\left(\frac{1}{3}\right)^{-2} = 9$ 19. $10^3 = 1000$ 21. $10^{-2} = 0.01$ 23. 2 25. 3 27. ½

29. -1 31. 1 33. 2 35. -1 37. 2 39. 2 41. 64 43. -3 45. 100

47. 4 49. 1 51. 0 53. 1 55. 0 57. $x > 9$

Exercise 12.2 [page 430] 1. $\log_b 2 + \log_b x$ 3. $\log_b 3 + \log_b x + \log_b y$ 5. $\log_b x - \log_b y$

7. $\log_b x + \log_b y - \log_b z$ 9. $3 \log_b x$ 11. $\frac{1}{2}\log_b x$ 13. $\frac{2}{3}\log_b x$

15. $2 \log_b x + 3 \log_b y$ 17. $\frac{1}{2}\log_b x + \log_b y - 2 \log_b z$ 19. $\frac{1}{3}\log_{10} x + \frac{2}{3}\log_{10} y - \frac{1}{3}\log_{10} z$

21. $\frac{3}{2}\log_{10} x - \frac{1}{2}\log_{10} y$ 23. $\log_{10} 2 + \log_{10} \pi + \frac{1}{2}\log_{10} l - \frac{1}{2}\log_{10} g$

25. $\frac{1}{2}\log_{10}(s - a) + \frac{1}{2}\log_{10}(s - b)$ 27. 0.7781 29. -0.3980 31. 0.9542 33. 0.8751

35. -8.0970 37. 1.8751 39. $\log_b 4$ 41. $\log_b 2$ 43. $\log_b \frac{x^2}{y^3}$ 45. $\log_b \frac{x^3 y}{z^2}$

47. $\log_{10} \frac{\sqrt{xy}}{z}$ 49. $\log_b x^{-2}$ or $\log_b \frac{1}{x^2}$ 51. $\{500\}$ 53. $\{4\}$ 55. $\{3\}$

57. Substitute arbitrary numbers, for example, 1 and 10, for x and y in the given equation.

$$\log_{10}(1 + 10) \stackrel{?}{=} \log_{10} 1 + \log_{10} 10$$

$$\log_{10} 11 \stackrel{?}{=} 0 + 1$$

$$\log_{10} 11 \neq 1$$

Since $\log_{10} 11 \neq 1$, $\log_{10}(x + y) \neq \log_{10} x + \log_{10} y$ for all $x, y > 0$.

59. Show that the left and right members of the equality reduce to the same quantity.
For the left side:

$$\log_b 4 + \log_b 8 = \log_b 2^2 + \log_b 2^3$$
$$= 2 \log_b 2 + 3 \log_b 2 = 5 \log_b 2.$$

For the right side:

$$\log_b 64 - \log_b 2 = \log_b 2^6 - \log_b 2$$
$$= 6 \log_b 2 - \log_b 2 = 5 \log_b 2.$$

61. For the left side:

$$2 \log_b 6 - \log_b 9 = \log_b 6^2 - \log_b 9$$
$$= \log_b 36 - \log_b 9 = \log_b \frac{36}{9} = \log_b 4.$$

For the right side:

$$2 \log_b 2 = \log_b 2^2 = \log_b 4.$$

63. For the left side:

$$\frac{1}{2}\log_b 12 - \frac{1}{2}\log_b 3 = \frac{1}{2}(\log_b 12 - \log_b 3)$$
$$= \frac{1}{2}\left(\log_b \frac{12}{3}\right) = \frac{1}{2}(\log_b 2^2) = \log_b 2.$$

For the right side:

$$\frac{1}{3}\log_b 8 = \log_b 8^{1/3} = \log_b 2.$$

Exercise 12.3 [page 435] **1.** 2 **3.** −4 **5.** 0 **7.** 0.8280 **9.** 1.9227 **11.** 2.5011 **13.** 0.9101 − 1 **15.** 0.9031 − 2 **17.** 2.3945 **19.** 4.10 **21.** 36.7 **23.** 0.0642 **25.** 0.00718 **27.** 0.297 **29.** 0.0503 **31.** 0.00205 **33.** 7.52 **35.** 784 **37.** 0.0357 **39.** 1.5373 **41.** 0.5655 **43.** 4.4817 **45.** 0.0907 **47.** 1.3610 **49.** 2.7726 **51.** 0.0837 − 1 or −0.9163 **53.** 1.1735 **55.** 6.0496 **57.** 90.017

Exercise 12.4 [page 444] **1.** 2 **3.** −4 **5.** 1 **7.** 4 **9.** 1.7348 **11.** 3.3700 **13.** −1.1367 **15.** −0.15851 **17.** 41.687 **19.** 0.13490 **21.** 129.72 **23.** 0.58157 **25.** 25.704 **27.** 3.3113 **29.** 0.050119 **31.** 1.3610 **33.** 2.7726 **35.** −0.91629 **37.** 1.4918 **39.** 10.381 **41.** 0.30119 **43.** 0.67032 **45.** 4.1371 **47.** 1.8776 **49.** 0.074274

Exercise 12.5 [page 450] **1.** 0.693 **3.** 1.37 **5.** 3.83 **7.** 0.642 **9.** 3.81 **11.** −1.20 **13.** 0.977 **15.** −3.72 **17.** 0.768 **19.** −0.684 **21.** 0.828 **23.** 7.52 **25.** −2.97 **27.** 1.65 **29.** −3.07 **31.** 2.63 **33.** 2.89 **35.** 2.06 **37.** 2.81 **39.** 0.89 **41.** ±1.40 **43.** −2.10 **45.** $t = \frac{1}{k}\ln y$ **47.** $t = 2(\ln T - \ln R)$

49. $k = e^{T/T_0} - 10$ **51.** $\ln N = \frac{\log_{10} N}{\log_{10} e} \approx \frac{\log_{10} N}{0.43429} \approx 2.303 \log_{10} N$

Exercise 12.6 [page 456] **1.** 9.60 inches of mercury **3.** 1.91 miles **5.** 3.34 miles **7. a.** 19,960,000 **b.** 25,349,000; 32,193,000; 40,884,000 **9.** 1.56% **11.** 18.7 years **13.** 12.1 grams **15.** 99.3 volts **17.** 2.77 seconds **19.** 34.7 hours **21.** 17.5% **23.** 13.5% **25.** $15,529.25; $16,035.68 **27.** 16 years

Review Exercises [page 462] 1. a. $\log_9 27 = \frac{3}{2}$ b. $\log_{4/9} \frac{2}{3} = \frac{1}{2}$ 2. a. $5^4 = 625$ b. $10^{-4} = 0.0001$ 3. a. 2 b. 8 4. a. 0.01 b. −1 5. a. 2 b. −1 6. a. 3 b. −3 7. a. $\log_b 3 + 2\log_b x + \log_b y$ b. $\log_b y + \frac{1}{2}\log_b x - 2\log_b z$ 8. a. 1.5562 b. 0.6276 9. a. $\log_b \frac{x^5}{y^2}$ b. $\log_b \sqrt[3]{\frac{xz^2}{y^4}}$ 10. a. {1} b. {5} 11. a. −0.147 b. 3.26 12. a. 431 b. 0.0379 13. a. 17.2 b. 0.285 14. a. 2.29 b. 0.273 15. a. 1.95 b. −0.916 16. a. 2.08 b. 14.9 17. 0.114 18. 4.13 19. 0.453 20. 7.05 21. 4.61 22. 8.05 23. 4.53 24. 1.78 25. 2.46 26. 7.32 27. 0 mg; 3.9 mg 28. 461.6 lumens 29. 13.5 years 30. 1.7 seconds 31. 0

32. Left side: $2\log_b 8 - \log_b 4 = \log_b \frac{8^2}{4} = \log_b 16.$

Right side: $4\log_b 2 = \log_b 2^4 = \log_b 16.$

33. $t = -\frac{1}{k}\ln\left(\frac{N}{N_0}\right)$ 34. $t = ke^{(N - C)/N_0}$ 35. 6.2 36. 2.5×10^{-6}

Exercise 13.1 [page 467] 1. −4, −3, −2, −1 3. −½, 1, 7/2, 7 5. 2, 3/2, 4/3, 5/4 7. 0, 1, 3, 6 9. −1, 1, −1, 1 11. 1, 0, −⅓, ½ 13. 1 + 4 + 9 + 16 15. 3 + 4 + 5 17. 2 + 6 + 12 + 20 19. −½ + ¼ − ⅛ + 1/16 21. 1 + 3 + 5 + ⋯ 23. 1 + ½ + ¼ + ⋯ 25. $\sum_{i=1}^{4} i$ 27. $\sum_{i=1}^{4} x^{2i-1}$ 29. $\sum_{i=1}^{5} i^2$ 31. $\sum_{i=1}^{\infty} \frac{i}{i+1}$ 33. $\sum_{i=1}^{\infty} \frac{i}{2i-1}$ 35. $\sum_{i=1}^{\infty} \frac{2^{i-1}}{i}$

Exercise 13.2 [page 473] 1. 11, 15, 19; $s_n = 4n - 1$ 3. −9, −13, −17; $s_n = 3 - 4n$ 5. $x + 2, x + 3, x + 4$; $s_n = x + n - 1$ 7. $x + 5a, x + 7a, x + 9a$; $s_n = x + 2an - a$ 9. $2x + 7, 2x + 10, 2x + 13$; $s_n = 2x + 3n - 2$ 11. $3x, 4x, 5x$; $s_n = nx$ 13. 31 15. 15/2 17. −92 19. 2; 3; 41 21. twenty-eighth term 23. 1; 8 25. 16 27. 9; 14; 19 29. 63 31. 806 33. −6 35. 1938 37. 196 bricks 39. 2, 7, 12 41. 14/15

43. The first n odd natural numbers form an arithmetic sequence with $a = 1$ and $d = 2$. Hence, the sum of the first n odd natural numbers is

$$S_n = \frac{n}{2}[2a + (n-1)d] = \frac{n}{2}[2 + (n-1)2]$$

$$= \frac{n}{2}[2 + 2n - 2] = \frac{n}{2}[2n] = n^2.$$

Exercise 13.3 [page 480] **1.** 128, 512, 2048; $s_n = 2(4)^{n-1}$ **3.** $\frac{16}{3}, \frac{32}{3}, \frac{64}{3}$; $s_n = \frac{2}{3}(2)^{n-1}$ **5.** $-\frac{1}{2}, \frac{1}{4}, -\frac{1}{8}$; $s_n = 4(-\frac{1}{2})^{n-1}$

7. $-\dfrac{x^2}{a^2}, \dfrac{x^3}{a^3}, -\dfrac{x^4}{a^4}$; $s_n = \dfrac{a}{x}\left(-\dfrac{x}{a}\right)^{n-1}$ **9.** 1536 **11.** $-243a^{20}$ **13.** 3 **15.** 3, 9 **17.** 18 or -18

19. 16, 8, 4 or $-16, 8, -4$ **21.** 1092 **23.** $\frac{31}{32}$ **25.** $\frac{364}{729}$

27.

29. 1 hour: 20 $(10 \cdot 2)$;
2 hours: 40 $(10 \cdot 2^2)$;
3 hours: 80 $(10 \cdot 2^3)$;
4 hours: 160 $(10 \cdot 2^4)$;
n hours: $(10 \cdot 2^n)$.

Exercise 13.4 [page 486] **1.** 24 **3.** does not exist **5.** $\frac{9}{20}$ **7.** $\frac{8}{49}$ **9.** 2 **11.** $\frac{1}{3}$
13. $\frac{31}{99}$ **15.** $\frac{2408}{999}$ **17.** $\frac{29}{225}$ **19.** \$220.80; \$222.55 **21.** 20 centimeters
23. 30 feet
25. After 2 years: $P(1 + r)(1 + r) = P(1 + r)^2$;
after 3 years: $P(1 + r)^2(1 + r) = P(1 + r)^3$;
after n years: $P(1 + r)^n$.

Exercise 13.5 [page 493] **1.** $8 \cdot 7 \cdot 6 \cdot 5 \cdot 4 \cdot 3 \cdot 2 \cdot 1$ **3.** $2 \cdot 4 \cdot 3 \cdot 2 \cdot 1$
5. $6 \cdot 5 \cdot 4 \cdot 3 \cdot 2 \cdot 1$ **7.** 120 **9.** 72 **11.** 15 **13.** 28 **15.** 3! **17.** $\dfrac{6!}{2!}$ **19.** $\dfrac{8!}{5!}$
21. $n(n - 1)(n - 2) \cdot \cdots \cdot 3 \cdot 2 \cdot 1$ **23.** $3n(3n - 1)(3n - 2) \cdot \cdots \cdot 3 \cdot 2 \cdot 1$
25. $(n - 2)(n - 3)(n - 4) \cdot \cdots \cdot 3 \cdot 2 \cdot 1$ **27.** $x^5 + 15x^4 + 90x^3 + 270x^2 + 405x + 243$
29. $x^4 - 12x^3 + 54x^2 - 108x + 81$ **31.** $8x^2 - 6x^2y + \dfrac{3}{2}xy^2 - \dfrac{1}{8}y^3$
33. $\dfrac{1}{64}x^6 + \dfrac{3}{8}x^5 + \dfrac{15}{4}x^4 + 20x^3 + 60x^2 + 96x + 64$
35. $x^{20} + 20x^{19}y + \dfrac{20 \cdot 19}{2!}x^{18}y^2 + \dfrac{20 \cdot 19 \cdot 18}{3!}x^{17}y^3$
37. $a^{12} + 12a^{11}(-2b) + \dfrac{12 \cdot 11}{2!}a^{10}(-2b)^2 + \dfrac{12 \cdot 11 \cdot 10}{3!}a^9(-2b)^3$
39. $x^{10} + 10x^9(-\sqrt{2}) + \dfrac{10 \cdot 9}{2!}x^8(-\sqrt{2})^2 + \dfrac{10 \cdot 9 \cdot 8}{3!}x^7(-\sqrt{2})^3$ **41.** $-3003a^{10}b^5$
43. $3360x^6y^4$ **45.** $21x^{10}y^4$ **47.** $63a^5b^4$

49. a. $1^{-1} + (-1)(1^{-2})x + 1(1^{-3})x^2 + (-1)(1^{-4})x^3$ or $1 - x + x^2 - x^3$
b. $1 - x + x^2 - x^3$ c. results are equal

Exercise 13.6 [page 498] 1. 24 3. 362,880 5. 120 7. 30,240 9. 2058
11. 720 13. 648 15. 448 17. 17,576 19. 676,000 21. 60 23. 120
25. 45,360 27. 6720 29. 30

Exercise 13.7 [page 505] 1. 21 3. 220 5. 220 7. 1365 9. 7 11. 2,598,960
13. 15 15. 84 17. $\binom{13}{6}\cdot\binom{13}{3}\cdot\binom{13}{4} = 350{,}904{,}840$ 19. 7425

21. $\binom{n}{n-r} = \dfrac{n!}{(n-r)![n-(n-r)]!} = \dfrac{n!}{(n-r)!r!} = \dfrac{n!}{r!(n-r)!} = \binom{n}{r}$ 23. 7

25. From Section 13.5, we see that the coefficients of the binomial expansion of $(a + b)^4$ are 1, 4, 6, 4, and 1. Since

$$\binom{4}{0} = \frac{4!}{0!4!} = 1, \quad \binom{4}{1} = \frac{4!}{1!3!} = 4, \quad \binom{4}{2} = \frac{4!}{2!2!} = 6,$$
$$\binom{4}{3} = \frac{4!}{3!1!} = 4, \quad \text{and} \quad \binom{4}{4} = \frac{4!}{4!0!} = 1,$$

the combinations are equal to the respective coefficients of the binomial expansion $(a + b)^4$.

Review Exercises [page 511] 1. $1, -\frac{1}{2}, \frac{1}{3}, -\frac{1}{4}$ 2. $2 + 6 + 12 + 20$ 3. $4n + 1$
4. a. 94 b. 138 5. 68 6. $5\left(\frac{9}{5}\right)^{n-1}$ 7. a. $-81/8$ b. $-26/27$ 8. $121/243$
9. $\frac{1}{2}$ 10. $\frac{4}{9}$ 11. 56 12. $x^{10} - 20x^9y + 180x^8y^2 - 960x^7y^3$ 13. $-15{,}360x^3y^7$
14. a. 125 b. 60 15. 360 16. 1260 17. 15 18. 126 19. 16 20. 24
21. 1539 22. 2, 5, 8 23. $2^{(n/4)+1}$ 24. 56 feet 25. $\dfrac{2}{(n+2)(n+1)}$
26. $(2n - 1)(2n - 2)(n + 1)$

Cumulative Review Exercises, Chapters 1–13 [page 513]

1. -6 2. $y = \dfrac{6}{x^2 - 4}$; domain: $x \in R, |x| \neq 2$; relation is a function 3. $-2, 7$
4. $y = \pm\sqrt{x^2 - 1}$; the inverse is not a function 5. 825 6. $3x^5y^6 - 3x^6y^{10}$ 7. 8×10^3
8. $2yz^3\sqrt[4]{4x^3y^3}$ 9. $2x\sqrt[3]{2y^2} + 4\sqrt[3]{x}$ 10. $\pm(8x^3 - 2x)$ 11. $\dfrac{x - 4}{(x + 4)(x^2 - 4)}$
12. $\dfrac{x^3 - x^2 - 7x - 2}{x^2 - 9}$ 13. $\dfrac{-(16x^2 + 4x + 1)}{16x^2}$ 14. $x^4 - 2x^3 + 2x^2 - 3x + 6 - \dfrac{15}{x + 2}$
15. $-x^2(x - 2)(x + 3)$ 16. $(x - 1)(x + 1)(2x - 1)$ 17. $7x^2y(3x - 5y)(x - y)$
18. $6a^4 + 5a^3 - 14a^2 + a + 2$ 19. $-6 + 4\sqrt{6}$ 20. $\dfrac{2^{1/5}}{x^{1/5}y^{3/5}}$ 21. $\dfrac{a^{7/2}b^{27}}{c^{21}}$

22. $\frac{7}{13} + \frac{9}{13}i$ **23.** $\frac{-\sqrt{15}}{6} - \frac{\sqrt{3}}{3}i$ **24.** 1.8277 **25.** 0.0002 **26.** 0.2725

27. $\log_b \frac{\sqrt{x}}{y^2z^3}$ **28.** 2, −1, 8/9, −1 **29.** $3 - 5/2 + 7/3 - 9/4 + 11/5 - 13/6 + 15/7$ **30.** 14,924

31. 61 **32.** $-5280x^4y^{21}$ **33.** $\frac{1609}{990}$ **34.** $y = \pm\sqrt{\frac{3x-2}{6}}$ **36.** $\left\{\frac{-3 \pm i\sqrt{15}}{4}\right\}$

37. $\left\{\frac{3 \pm \sqrt{3}}{4}\right\}$ **38.** $\left\{\frac{3 \pm \sqrt{3}}{4}\right\}$ **39.** $\left\{\frac{20}{3}\right\}$ **40.** $\{\pm\sqrt{38}\}$ **41.** {5, 8} **42.** {1/2}

43. {4} **44.** {0.7711} **45.** {0.3996} **46.** {−0.4583} **47.** {−0.7909} **48.** {5}

49. (9/5, ∞) **50.** (−∞, 7/6) ∪ (11/6, ∞) **51.** (−∞, −3] ∪ (−1, ∞)

52. {(−1, 5)} **53.** {(−2, −1), (−10/7, −10/7)} **54.** {(−2, −4), (−4, −2), (2, 4), (4, 2)}

55. {(−3, 2, 0)} **56.** {(12, −8, −15)} **57.** {(5, 3, −2)}

58. **59.**

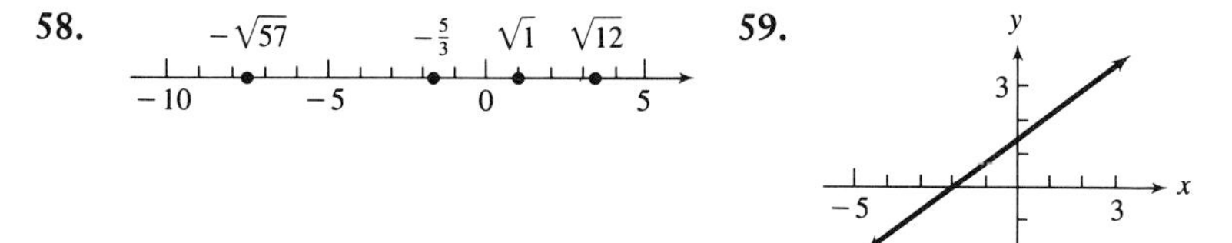

60.

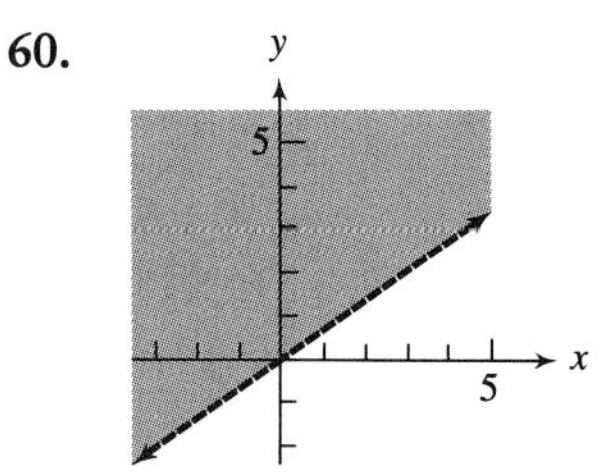

61. **62.**

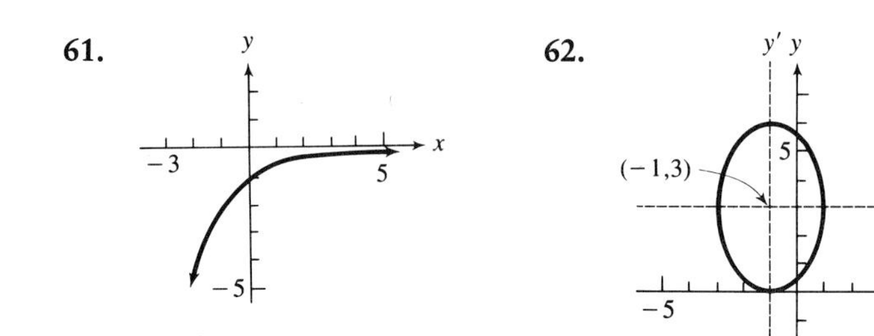

63.

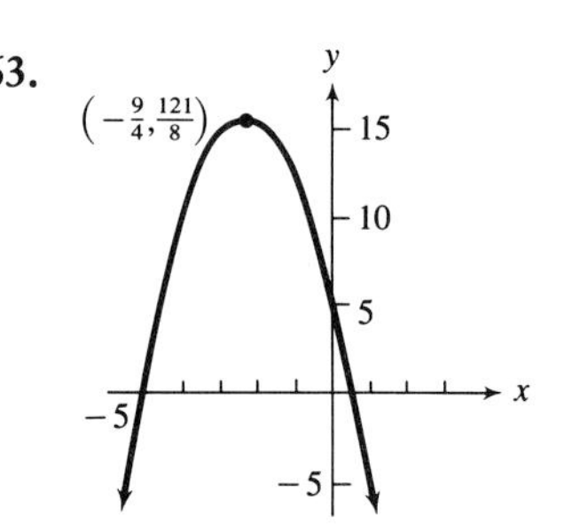

64.

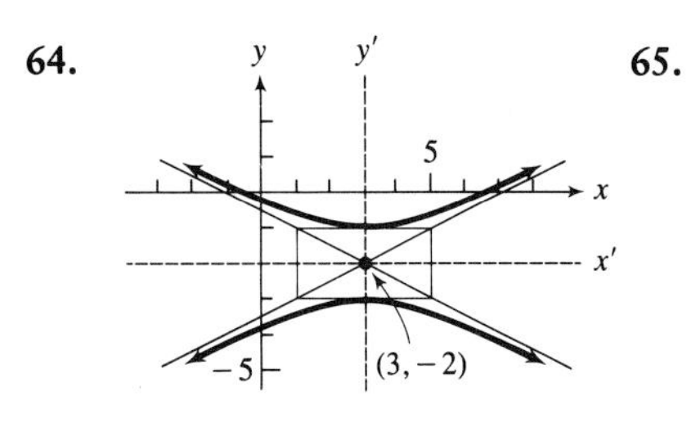

65.

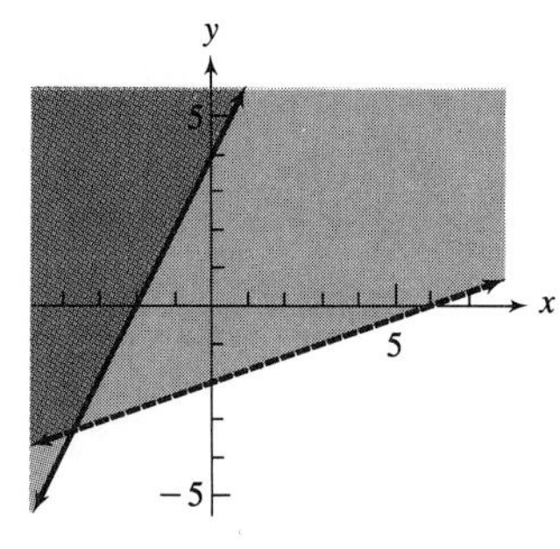

66. $11x + 2y + 23 = 0$ **67.** 15 **68.** $S(x) = 4x^2$; $S(20) = 1600$ **69.** $-3/4$ **70.** \$599.39

71. 2.8% **72.** 1151 years **73.** $23 \cdot 24^2 \cdot 10^4 = 132{,}480{,}000$

74. $\binom{20}{2}\binom{30}{3} = 771{,}400$

75. a. Crates shipped to Boston: x
Crates shipped to Chicago: y
Crates shipped to Los Angeles: z

b. $x + y + z = 55$
$10x + 5y + 12z = 445$
$x - 2z = 0$

c. 20 crates to Boston; 25 to Chicago and 10 to Los Angeles

76. -323 **77.** $(-1, 3]$ **78.** x^{n^2-n-2} **79.** $x^{2n-6}y^{-2n}$ **80.** $\dfrac{-x^2 + 3x + 10}{x^2 - 5x - 6}$

81. $\dfrac{y}{x}\sqrt[4]{x^2(x^2y^2 - 1)}$ **82.** $y^{n-1}(1 - 3y + 6y^{n+1})$ **83.** $6(3a^{3n} + 2)(2a^{3n} - 1)$

84. $f^{-1}(x) = 4 - 2x$; $f^{-1}[f(x)] = 4 - 2\left(\dfrac{8 - 2x}{4}\right) = x$;

$$f[f^{-1}(x)] = \frac{1}{4}[8 - 2(4 - 2x)] = x$$

85. $\{x \mid -3 < x < 3\}$ **86.** $\sum_{k=1}^{\infty} k^2x^{k-1}$ **87.** 2.02

88. $\begin{vmatrix} a_1 & b_1 \\ a_1 + a_2 & b_1 + b_2 \end{vmatrix} = a_1(b_1 + b_2) - b_1(a_1 + a_2)$

$$= a_1b_1 + a_1b_2 - b_1a_1 - b_1a_2 = a_1b_2 - b_1a_2 = \begin{vmatrix} a_1 & b_1 \\ a_2 & b_2 \end{vmatrix}$$

89. $x = \dfrac{-k \pm \sqrt{k^2 - 4k + 16}}{2}$ **90.** $\{-1/3, 1/5\}$ **91.** $k = \dfrac{-1}{t}\ln\left(\dfrac{P - C}{P_0}\right)$

92. $\{6\}$ **93.** $\{(-1, 1/3, 2/3)\}$

94.

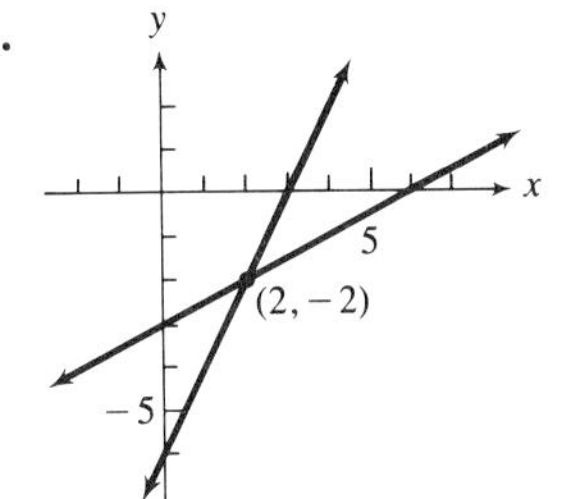

95.

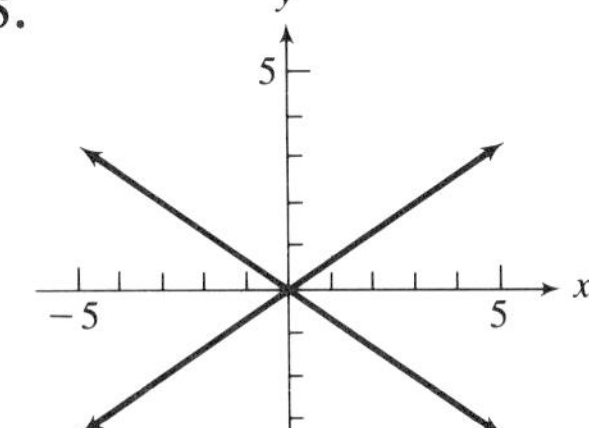

96.

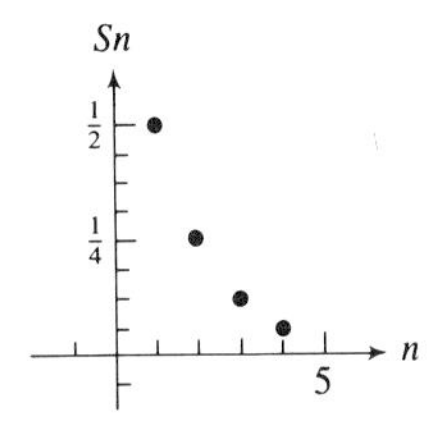

97. The length of the side joining $(-6, 2)$ and $(-2, 5)$ is 5 units, and the length of the side joining $(-2, 5)$ and $(1, 1)$ is 5 units. Therefore the triangle is isosceles.

98. $x - 3y = 10$ **99.** 8.7% **100.** 8806

Exercise A.1 [page 522] 1. 74; −46 3. 5; 44

5.

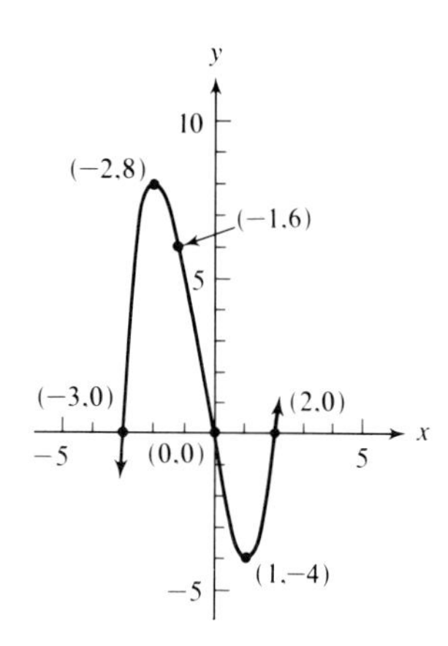

7.

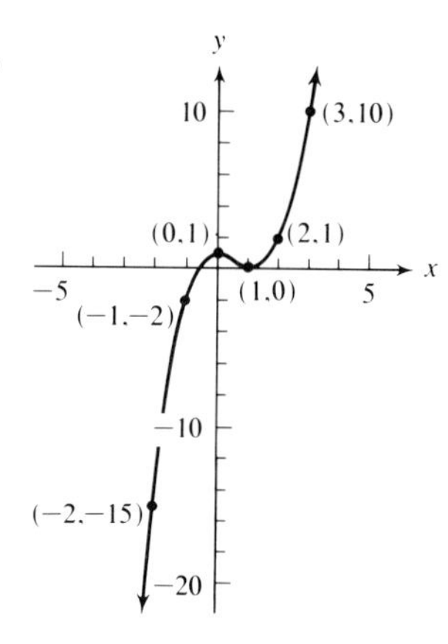

9.

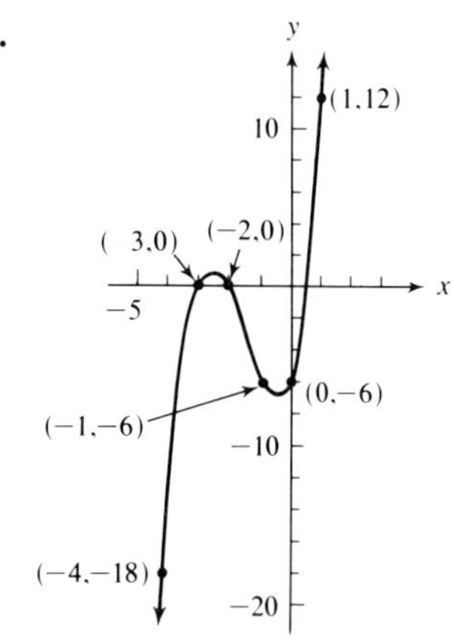

11.

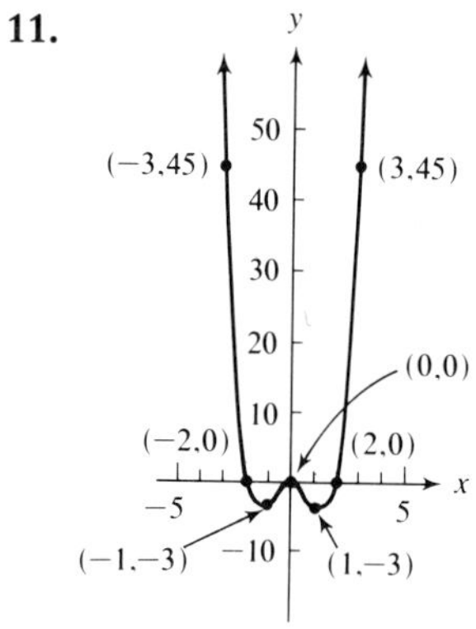

13. no 15. yes 17. −1 and −2 19. 0, 2, and 4

21.

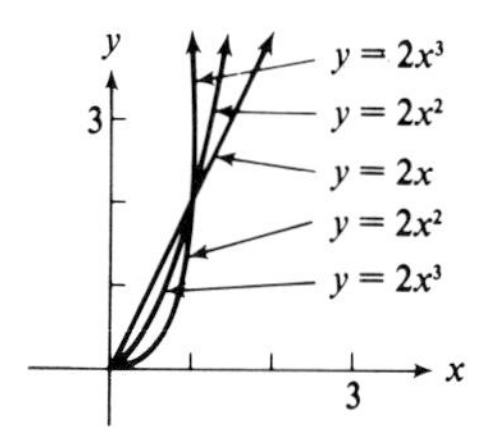

Exercise A.2 [page 526] 1. $x = -3$ 3. $x = 2, x = -3$ 5. $x = -2, x = 3$

7. vertical asymptotes: $x = 3, x = -3$
horizontal asymptote: $y = 0$

9. vertical asymptote: $x = \frac{1}{2}$
horizontal asymptote: $y = \frac{1}{2}$

11. vertical asymptote: $x = -1, x = 4$
horizontal asymptote: $y = 2$

13. horizontal asymptote: $y = 0$
vertical asymptote: $x = -3$
y-intercept: $1/3$

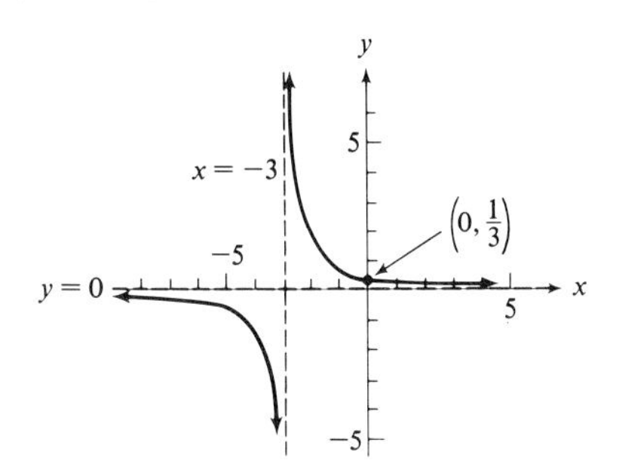

15. horizontal asymptote: $y = 0$
vertical asymptotes: $x = 4, x = -1$
y-intercept: $-1/2$

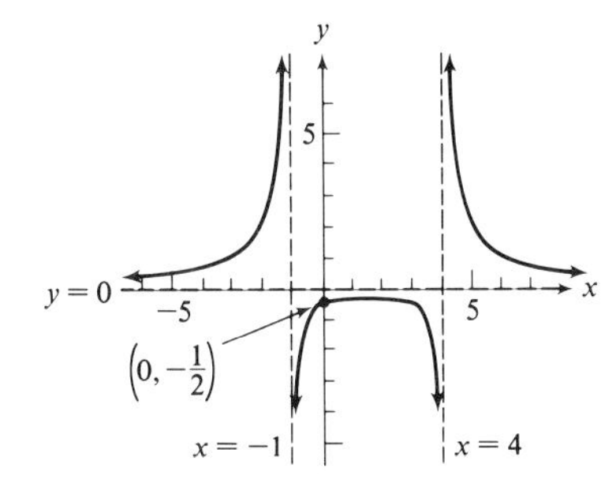

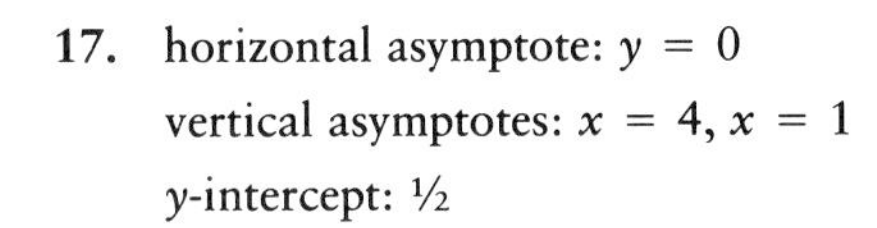

17. horizontal asymptote: $y = 0$
vertical asymptotes: $x = 4, x = 1$
y-intercept: $1/2$

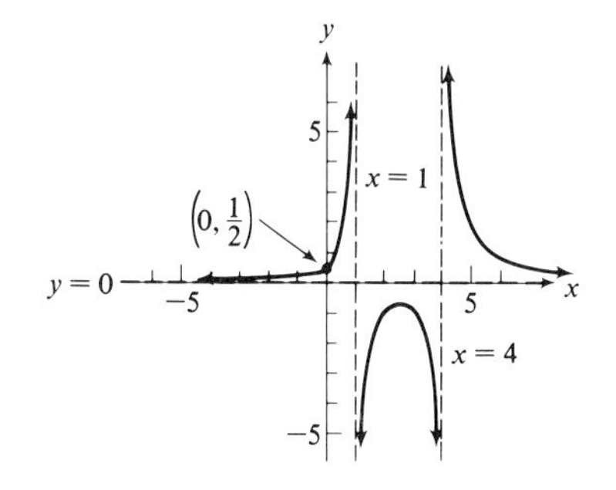

19. horizontal asymptote: $y = 1$
vertical asymptote: $x = -3$
intercepts: $x = 0, y = 0$

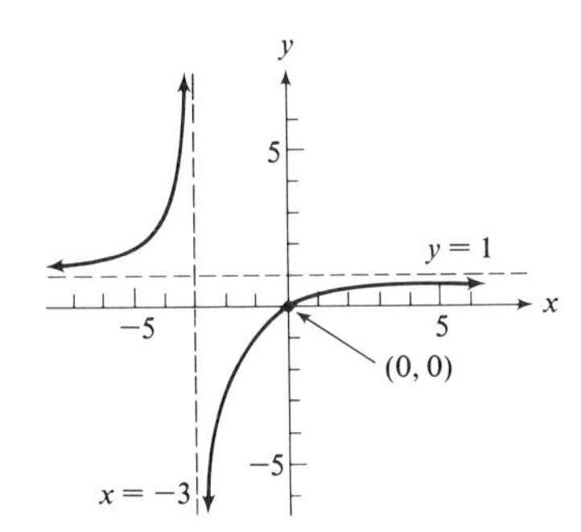

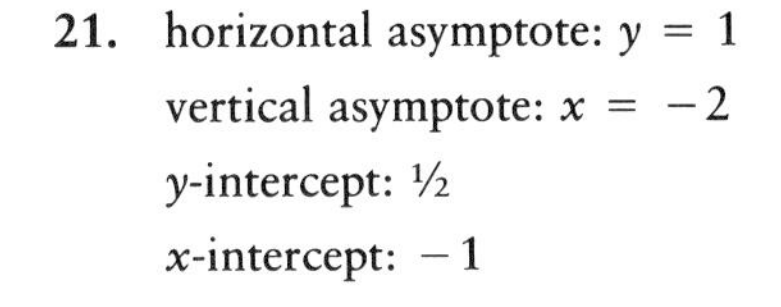

21. horizontal asymptote: $y = 1$
vertical asymptote: $x = -2$
y-intercept: $1/2$
x-intercept: -1

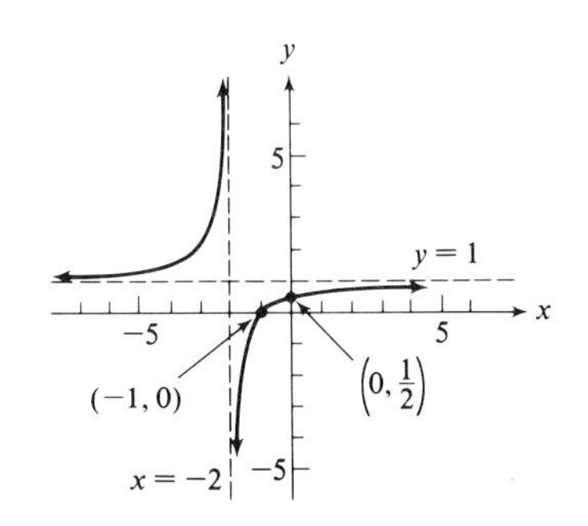

23. horizontal asymptote: $y = 0$
vertical asymptotes: $x = 2, x = -2$
intercepts: (0, 0)

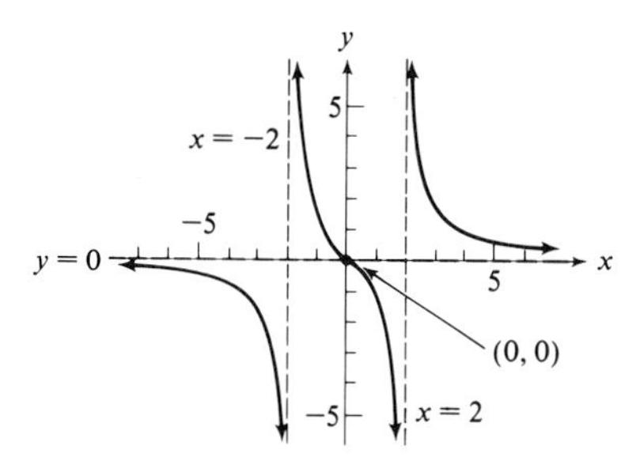

25. horizontal asymptote: $y = 0$
vertical asymptotes: $x = -1, x = -4$
y-intercept: $-\frac{1}{2}$
x-intercept: 2

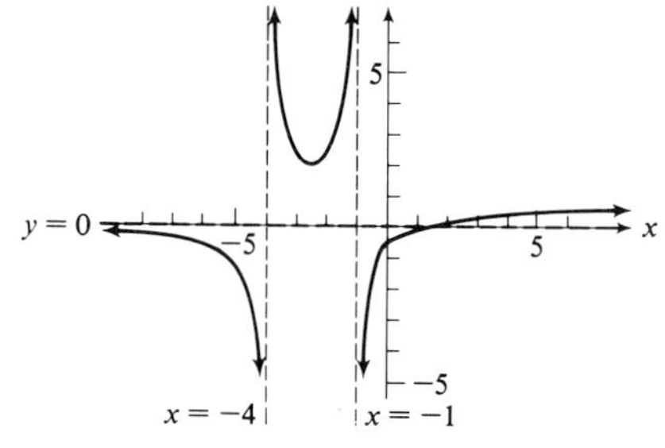

27.

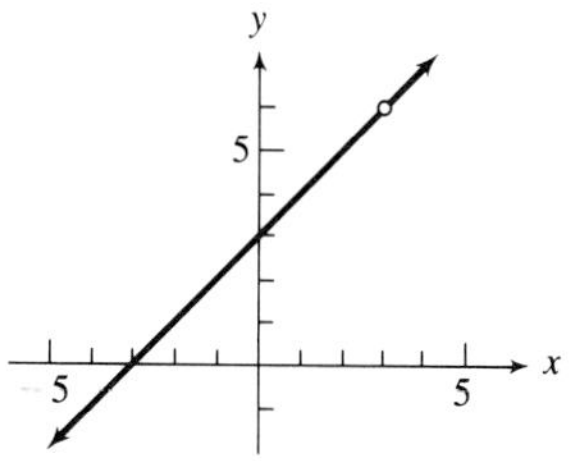

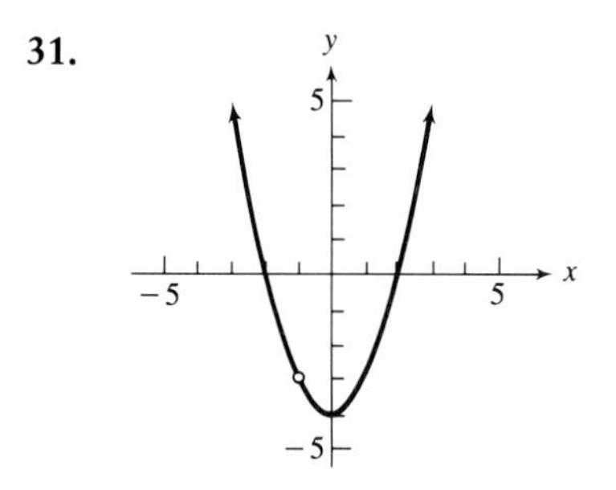

29.

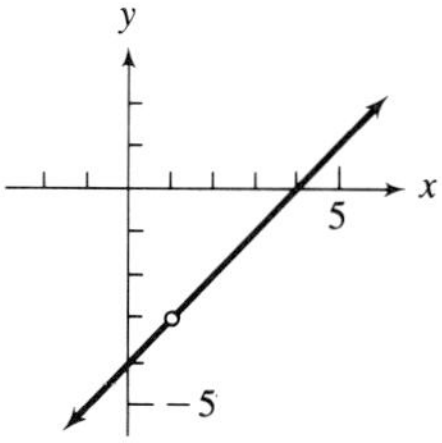

31.

33.

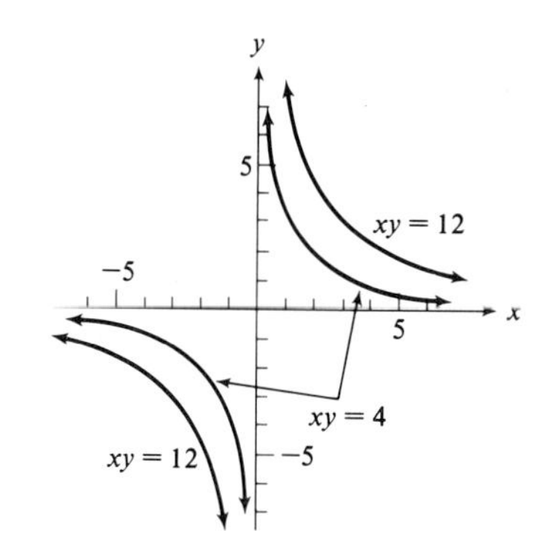

Exercise A.3 [page 531] 1. 1 3. 8

5.

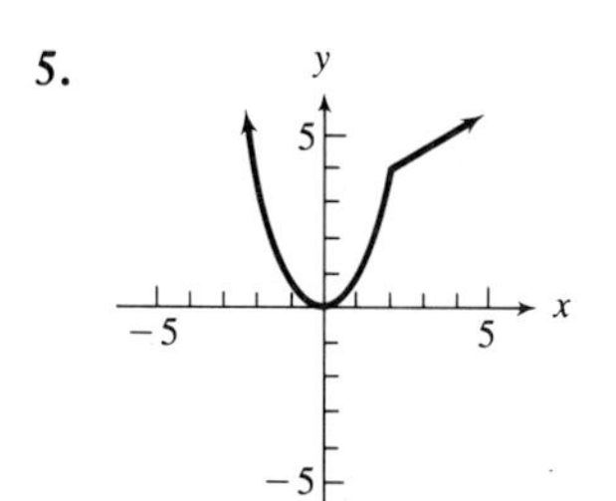

7.

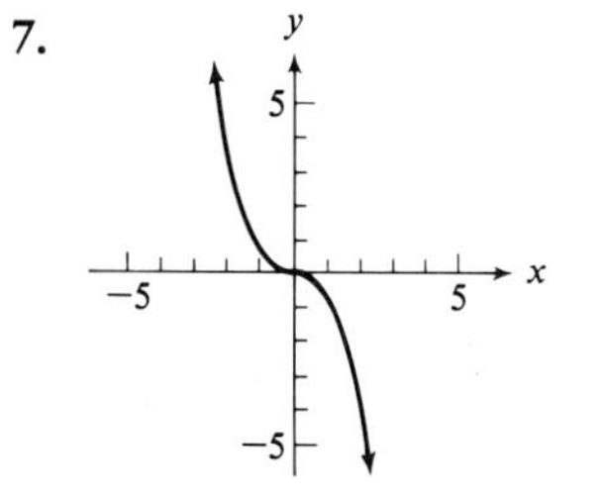

9.

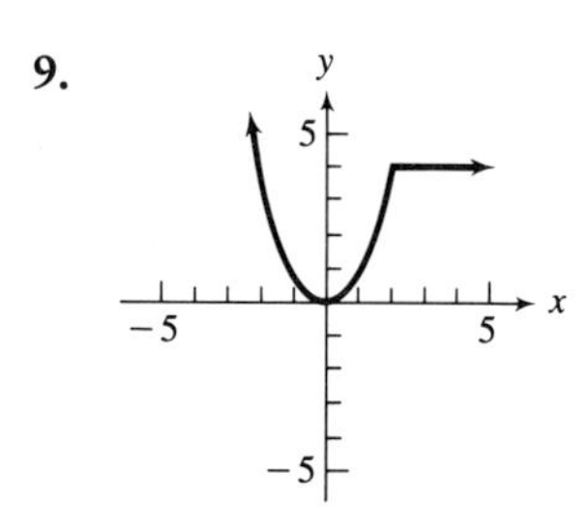

11.

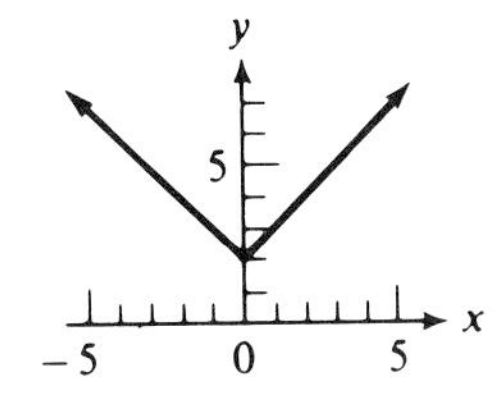

13.

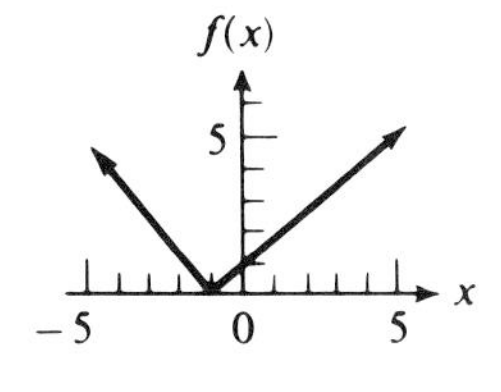

15.

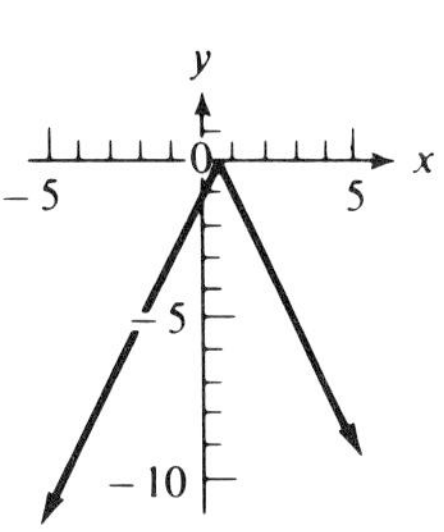

17.

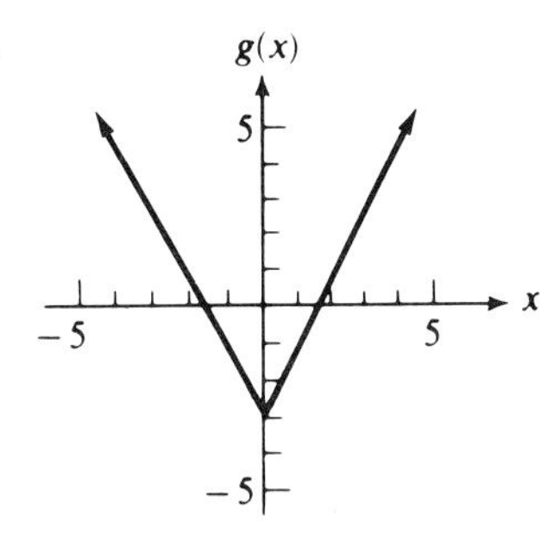

19.

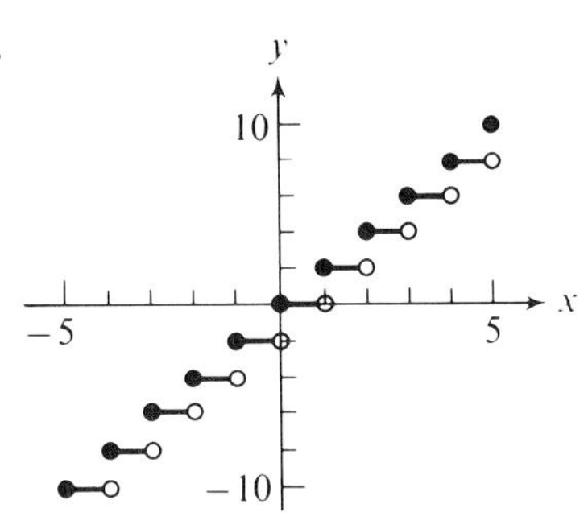

21.

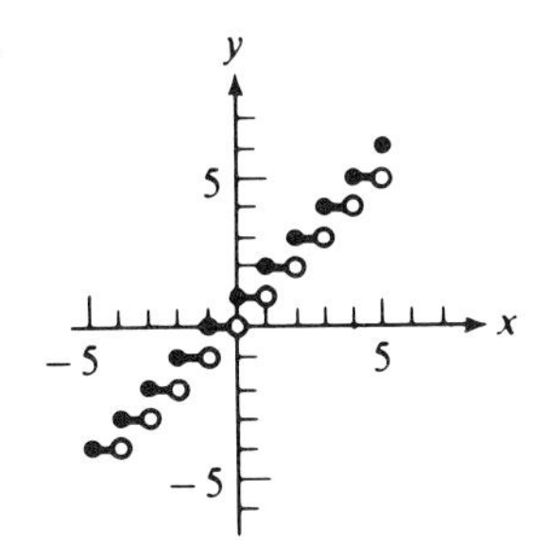

23.

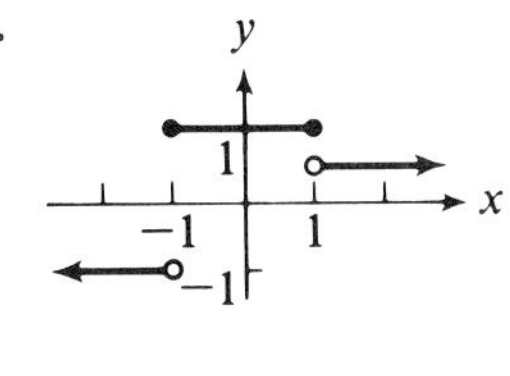

25.

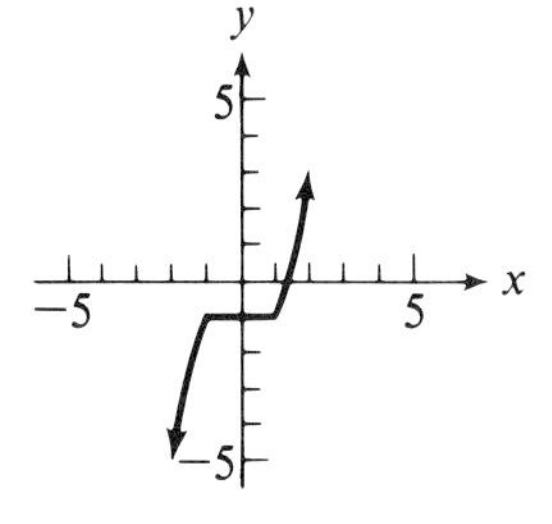

27.

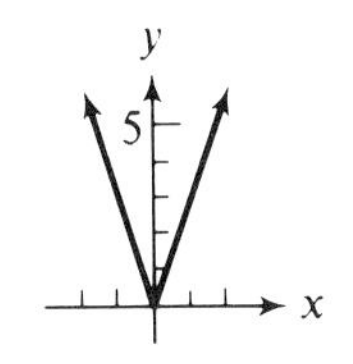

29.

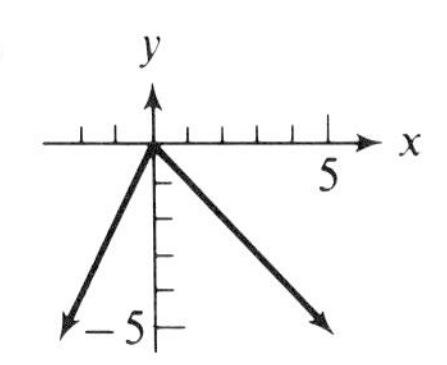

31.

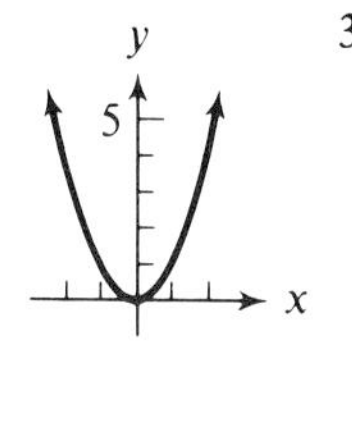

33.

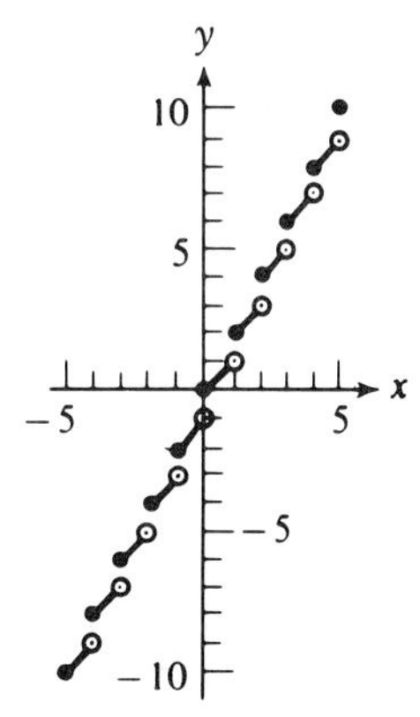

Review Exercises [page 533] **1.** 19; -25 **2.** 4; 8

3.

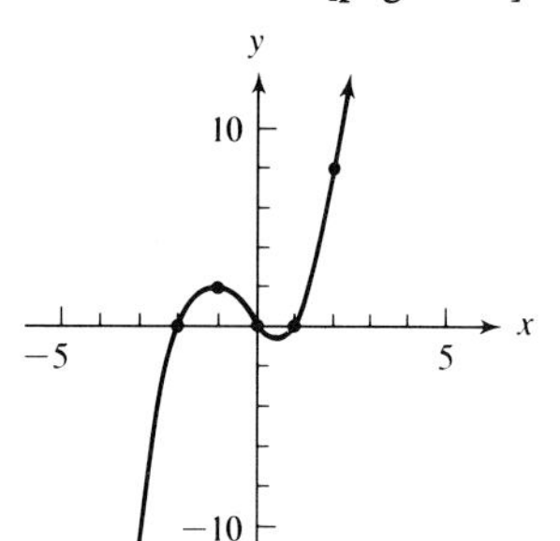

4.

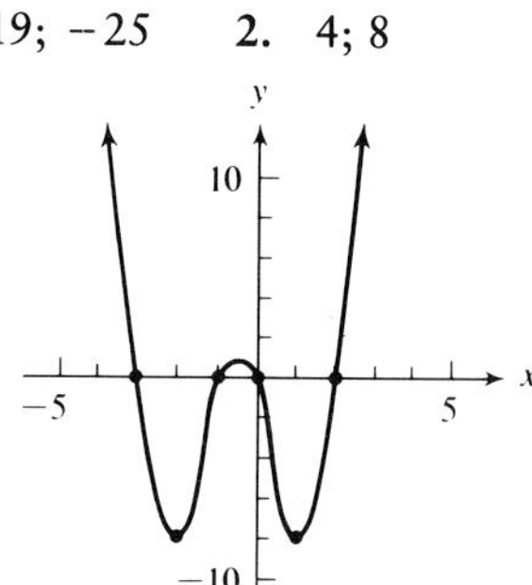

5. no **6.** $(x + 1);(x - 3)$

7.

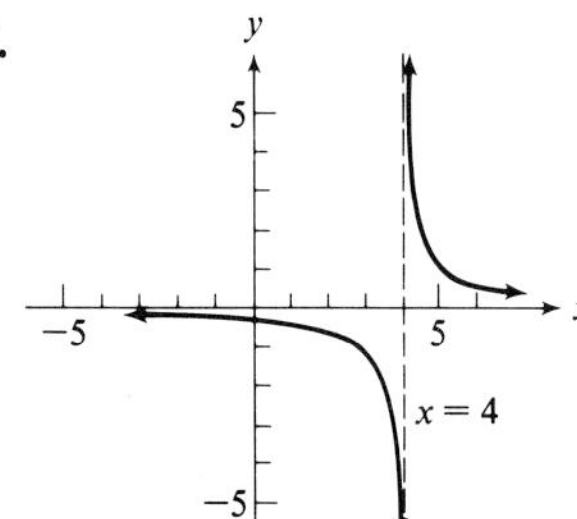

8.

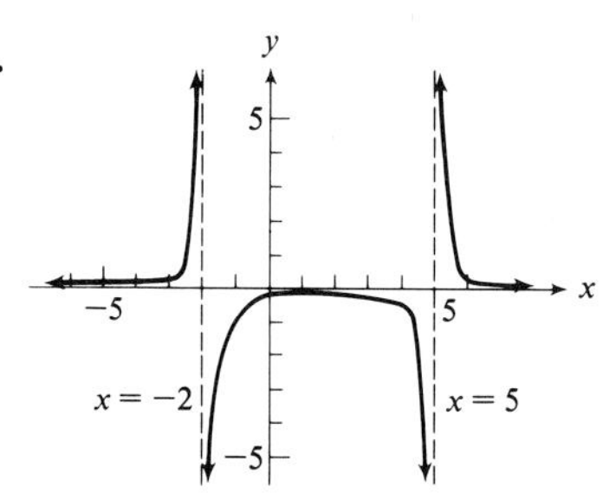

9.

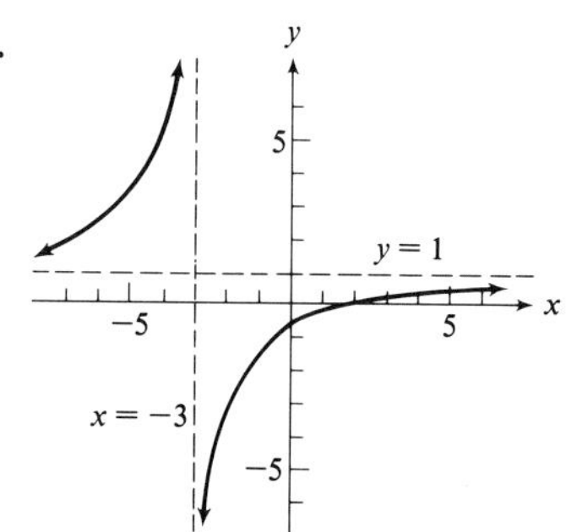

10.

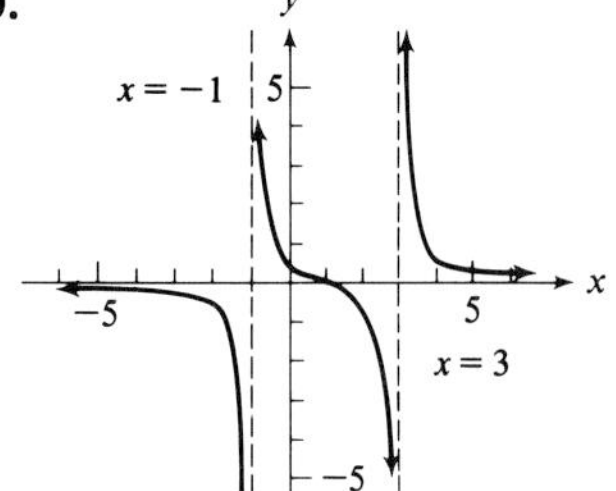

11.

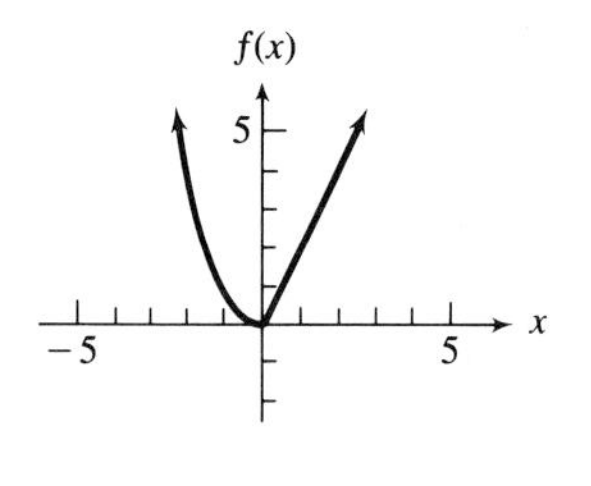

12.

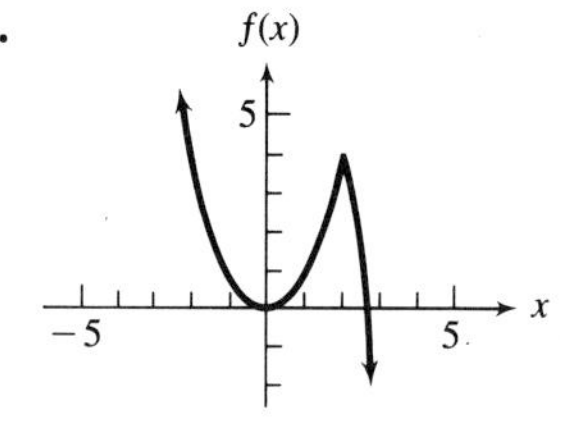

13.

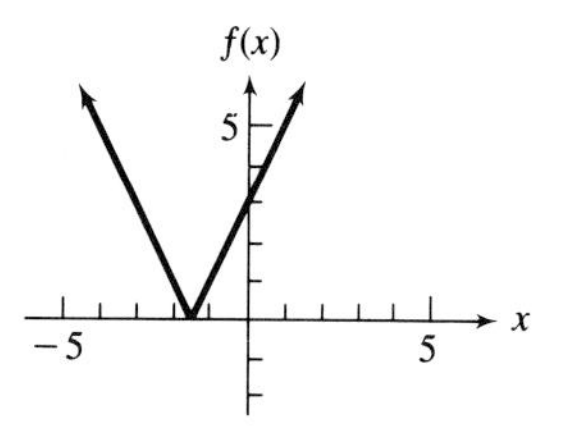

14.

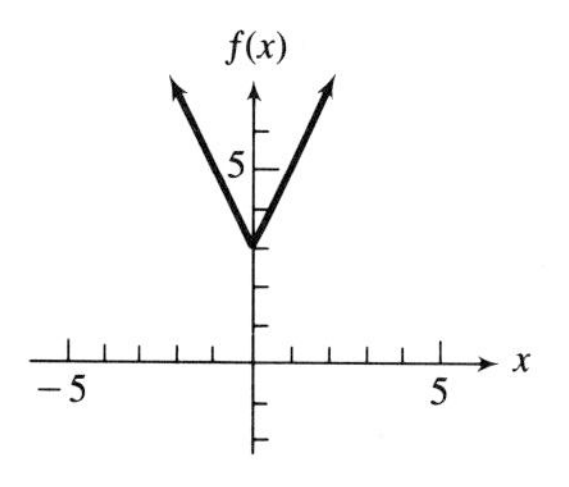

15.

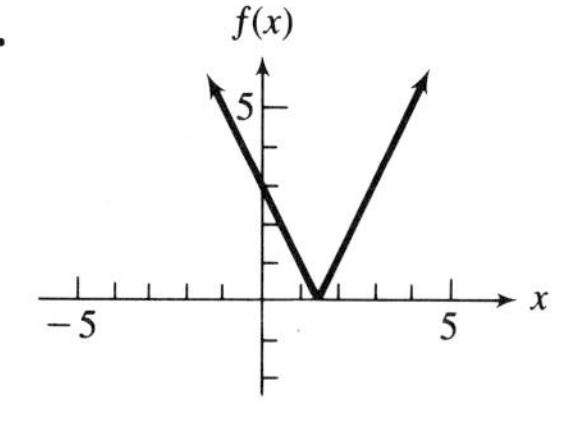

16.

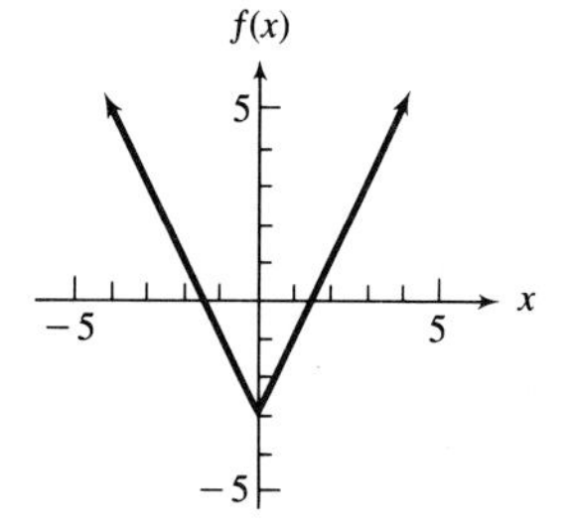

17.

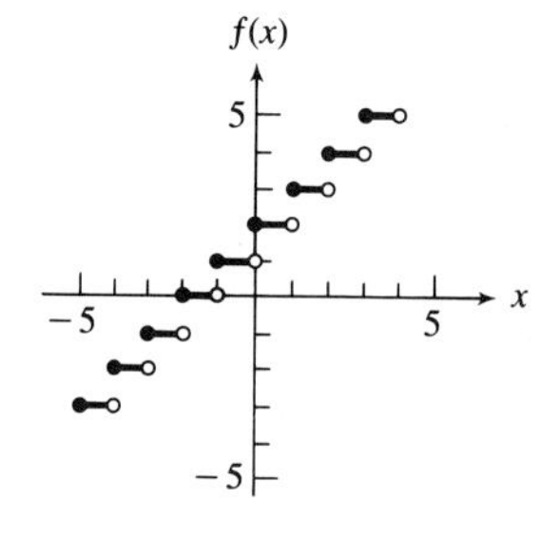

18.

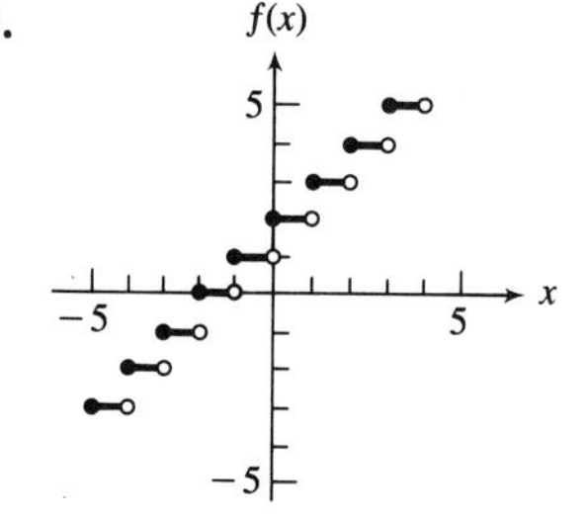

19. $\{-2, 1, 2\}$ **20.** $\{0, -3, -2, 2, 3\}$

INDEX

(continued from front endpapers)

Systems of Linear Equations

A system of the form

$$a_1 x + b_1 y = c_1$$
$$a_2 x + b_2 y = c_2$$

has one and only one solution if

$$\frac{a_1}{a_2} \neq \frac{b_1}{b_2}$$

has no solution if

$$\frac{a_1}{a_2} = \frac{b_1}{b_2} \neq \frac{c_1}{c_2}$$

and has an infinite number of solutions if

$$\frac{a_1}{a_2} = \frac{b_1}{b_2} = \frac{c_1}{c_2}$$

Determinants

$$\begin{vmatrix} a_1 & b_1 \\ a_2 & b_2 \end{vmatrix} = a_1 b_2 - a_2 b_1$$

Cramer's rule for $a_1 x + b_1 y = c_1$, $a_2 x + b_2 y = c_2$;

$$x = \frac{D_x}{D} = \frac{\begin{vmatrix} c_1 & b_1 \\ c_2 & b_2 \end{vmatrix}}{\begin{vmatrix} a_1 & b_1 \\ a_2 & b_2 \end{vmatrix}}; \quad y = \frac{D_y}{D} = \frac{\begin{vmatrix} a_1 & c_1 \\ a_2 & c_2 \end{vmatrix}}{\begin{vmatrix} a_1 & b_1 \\ a_2 & b_2 \end{vmatrix}}$$

$$\begin{vmatrix} a_1 & b_1 & c_1 \\ a_2 & b_2 & c_2 \\ a_3 & b_3 & c_3 \end{vmatrix} = a_1 \begin{vmatrix} b_2 & c_2 \\ b_3 & c_3 \end{vmatrix} - b_1 \begin{vmatrix} a_2 & c_2 \\ a_3 & c_3 \end{vmatrix} + c_1 \begin{vmatrix} a_2 & b_2 \\ a_3 & b_3 \end{vmatrix}$$

Cramer's rule for a linear system of three equations in three variables:

$$x = \frac{D_x}{D}; \quad y = \frac{D_y}{D}; \quad z = \frac{D_z}{D}.$$

Matrices

A matrix can be written as a row-equivalent matrix by:

1. Multiplying the entries of any row by a nonzero real number
2. Interchanging two rows
3. Multiplying the entries of any row by a real number and adding the results to the corresponding elements of another row

Logarithmic Functions

The logarithmic function

$$x = b^y \quad \text{or} \quad y = \log_b x \quad (x > 0)$$

is the inverse of the exponential function

$$y = b^x \quad (y > 0)$$

Laws of Logarithms

I $\log_b(x_1 x_2) = \log_b x_1 + \log_b x_2$

II $\log_b \dfrac{x_2}{x_1} = \log_b x_2 - \log_b x_1$

III $\log_b x^p = p \log_b x$

$$\ln x = \log_e x = \frac{\log_{10} x}{\log_{10} e} = 2.303 \log_{10} x$$

Sequences and Series

First term a, number of terms r, nth term s_n, sum of n terms S_n

Arithmetic progression with common difference d:

$$s_n = a + (n - 1)d$$

$$S_n = \frac{n}{2}(a + s_n) = \frac{n}{2}[2a + (n - 1)d]$$

Geometric progression with common ratio r $(r \neq 1)$:

$$s_n = ar^{n-1} \qquad S^n = \frac{a - ar^n}{1 - r} = \frac{a - rs_n}{1 - r}$$

Infinite geometric progression with $|r| < 1$:

$$S_\infty = \frac{a}{1 - r}$$

Binomial expansion:

$$(a + b)^n = a^n + \frac{n}{1!} a^{n-1} b + \frac{n(n-1)}{2!} a^{n-2} b^2 + \frac{n(n-1)(n-2)}{3!} a^{n-3} b^3 + \cdots$$

where $1! = 1$, $2! = 2 \cdot 1$, $3! = 3 \cdot 2 \cdot 1$, etc.